Symmetries in Physics

Alejandro Frank
Kurt Bernardo Wolf
Editors

Symmetries in Physics

Proceedings of the International Symposium
Held in Honor of Professor Marcos Moshinsky
at Cocoyoc, Morelos, México, June 3–7, 1991

With 51 Figures

Springer-Verlag
Berlin Heidelberg New York
London Paris Tokyo
Hong Kong Barcelona
Budapest

Editors

Dr. Alejandro Frank
Instituto de Ciencias Nucleares
Universidad Nacional Autónoma de México
Apdo. Postal 20-543, 01000 México DF

Dr. Kurt Bernardo Wolf
Instituto de Investigaciones en Matemáticas
Aplicadas y en Sistemas, Universidad Nacional
Autónoma de México, Apdo. Postal 139-B
62191 Cuernavaca, Morelos, México DF

Technical Editor

Arturo Sánchez y Gandara
Ediciones Técnico-Científicas
Sociedad Mexicana de Física
Facultad de Ciencias, UNAM, Apdo. Postal 20-364
01000 México DF

ISBN-13: 978-3-642-77286-3 e-ISBN-13: 978-3-642-77284-9
DOI: 10.1007/978-3-642-77284-9

Softcover reprint of the hardcover 1st editiont 1992

Typesetting: Camera ready by author

55/3140-5 4 3 2 1 0 Printed on acid-free paper

Preface

Marcos Moshinsky was born on 20 April 1921, in Kiev, Ukraine, and emigrated to Mexico at the age of four. He began work at the Universidad Nacional Autónoma de México on 1 January 1942, as a laboratory assistant working on the measurement of cosmic rays. He pursued his graduate studies at Princeton University, and wrote his thesis under the supervision of Professor Eugene Wigner. Since 1949, and in spite of many visits and temporary posts held abroad, Moshinsky has been based in Mexico.

Through example and encouragement, Moshinsky may be credited to a large extent with the shaping of Mexican scientific research. He has directed 40 B.Sc., M.Sc., and Ph.D. theses, and published over 200 scientific articles and four books; he holds all the Mexican science prizes, and several international ones, being a member of 11 academies of learning. Talent and circumstance have placed Marcos Moshinsky at the origin of several of the enterprises of the Mexican and Latin American scientific communities: he was founding editor of the *Revista Mexicana de Física* from 1952 to 1967; the Escuela Latinoamericana de Física was initiated and five times organized by him in Mexico; he was founding member and later president of the Academia de la Investigación Científica (1962–1963), the Sociedad Mexicana de Física (1967–1969), and the Centro Internacional de Física y Matemáticas Aplicadas (1986–), in Cuernavaca.

Marcos Moshinsky has worked to understand the symmetries of physics, shaping the tools of the theory of groups. His favorite example is the harmonic oscillator, because it underlies atoms, nuclei, and elementary particles; indeed, it is inextricably woven into the phase space formulation of quantum mechanics. Over the years, Moshinsky developed an impressive network of colleagues and friends, many meeting at the International Colloquia on Group Theoretic Methods in Physics and the Mexican Schools on Nuclear Physics, which have influenced many fields and propagated his work throughout them. The International Symposium on Symmetries in Physics gathered together 45 of his close collaborators, friends and students, 17 of them working in Mexico. This proceedings volume, and its companion volume (II), to appear shortly from the American Institute of Physics, collects together the refereed contributions from the participants. It is current work and forms a presentation of various areas of physics that bear genetically encoded material: the pursuits, methods, and style of Marcos Moshinsky. The academic commentary written by Flores, Frank, and Seligman tracks his path among symmetries and systems.

In contrast to many other renowned scientists, Moshinsky has been able to avoid administration. He has been most effective, therefore, in voicing his ideas and concerns to the scientific community at large. At this symposium held in his honor, he was awarded, through Drs. Vladimir and Margarita Man'ko, the Andrei Sakharov Medal for his active support of human rights in the 1970s and 1980s in the Soviet Union.

The symposium was organized through the Centro Internacional de Física y Matemáticas Aplicadas, with the support of Mexican federal institutions and the International Centre for Theoretical Physics. Several of Moshinsky's collaborators are now executives in science administration in Latin America. We thus had the opportunity to convene a round table on science and technology in this region; we reproduce the contributions in the Appendix to this volume. We see diverging science policies between Colombia, Brazil, and Mexico, and therefore invite readers to examine their positions carefully, because the consequences will soon become apparent.

The logo of the symposium was MM120, because 70 + 50 years add to 120, the age of fulfillment in the Jewish tradition, as Marcos Moshinsky once remarked. Professor John Archibald Wheeler, who could not attend the symposium, sent a letter from which we extracted here a few lines for this preface that tell of the character of the man.

The International Symposium on Symmetry in Physics was supported by the Dirección Adjunta de Investigación Científica, Consejo Nacional de Ciencia y Tecnología (CONACYT), by the Dirección General de Investigación Científica y Superación Académica, Secretaría de Educación Pública, and by the Coordinación de la Investigación Científica, Universidad Nacional Autónoma de México, as well as with the patronage of the International Centre for Theoretical Physics (Trieste). We hereby express the recognition of the Organizing Committee, Drs. Jorge Flores, Alejandro Frank, Miguel José Yacamán, Thomas H. Seligman, and Kurt Bernardo Wolf.

The book layout was produced by EDICIONES TÉCNICO CIENTÍFICAS with the Springer LaTeX macro package. Its manager, Fís. Arturo Sánchez y Gándara, and K.B. Wolf wish to acknowledge CONACYT project P107 CCOT89/4517, *Tipografía Científica Automatizada*, the Sociedad Mexicana de Física, and Professor W. Beiglböck, Springer-Verlag (Heidelberg) for funds and help in the acquisition of expertise in the art of scientific publication. We also acknowledge the support of CONACYT for printing and distribution, and the Instituto de Física, de Ciencias Nucleares, and de Investigaciones en Matemáticas Aplicadas y en Sistemas, UNAM, for bearing the composition costs.

Cuernavaca, México

Alejandro Frank and Kurt Bernardo Wolf

November 1991

PROFESSOR MARCOS MOSHINSKY

Dear Marcos,

You seventy? I can't believe it! Or, as Mrs. Niels Bohr exclaimed when she heard that Robert Frisch was retiring, "I can't understand it. All the young men that I know are retiring."

You, I'm sure, are not retiring and never will retire. You have too much creative drive, too much sense of responsibility, too much love for our wonderful subject.

I want to thank you especially for three things. You have brought group theory to bear on almost every branch of physics. You have trained three generations of scientisits in a great country for which I have a life-long affection. Last but not least, you have made a very warm spot for yourself in the hearts of all your old Princeton colleagues, including not least Eugene Wigner and me.

Warmest good wishes from all of us in this little Einstein town and from Janette and me.

Yours

John Archibald Wheeler

PRINCETON UNIVESITY

Academic Honors

PRIZES:	*Premio de la Academia de la Investigación Científica*	*Mexico, 1961*
	Premio Elías Sourasky	*Mexico, 1966*
	Premio Nacional de Ciencias	*Mexico, 1968*
	Premio Luis Elizondo	*Mexico, 1971*
	Premio UNAM, Investigación en Ciencias Exactas	*Mexico, 1985*
	Premio Príncipe de Asturias	*Spain, 1988*
	Premio Bernardo A. Houssay	*Organization of American States, 1991*
	Andrei Sakharov Medal for Human Rights	*USSR, 1991*

MEMBER OF: *Academia Brasileira de Ciencia*
Academia de Ciencias de América Latina
Academia de la Investigación Científica
American Academy of Arts and Science
American Physical Society
El Colegio Nacional
European Academy of Sciences, Arts, and Letters
Pontificial Academy of Sciences
Sociedad Matematica Mexicana
Sociedad Mexicana de Física
Third World Academy of Sciences

President,	*Academia de la Investigación Científica*	*1962–1963*
	Sociedad Mexicana de Física	*1967–1969*
	Centro Internacional de Física y Matemáticas Aplicadas	*1986–*
Editor,	*Revista Mexicana de Física*	*1952–1967*
Asociate Editor,	*Nuclear Physics*	*1958–1965*
	Journal of Mathematical Physics	*1971–1973*

VISITING PROFESSOR: semesters at the Institut Henri Poincaré (1954), Princeton University (1959–1960), State University of New York at Stony Brook (1967–1968), Université de Montréal (1973 and 1977). Lecturer in more than 100 universities and research centers.

Participants, Lectures,* Addresses

VALDIR AGUILERA-NAVARRO
Generalizing the BCS universal constants to high-temperature superconductivity [with M. de Llano][1]

Instituto de Física Teórica
Caixa Postal 5956
01405 São Paulo, **Brazil**
aguilera@ift.vesp.ansp.br

CARLOS BUNGE
Atomic physics calculations without group theory

Instituto de Física, UNAM
Apdo. Postal 20–364
01000 México DF
Mexico

OCTAVIO CASTAÑOS
The symplectic model and potential-energy surfaces [with P. Hess][1]
Nœther's theorem and dynamical groups [with A. Frank and R. López Peña][2]

Instituto de Ciencias Nucleares, UNAM
Apdo. Postal 20–543
01000 México DF
Mexico ocasta@unamvm1

ELPIDIO CHACÓN
Bases for irreps of unitary and pseudo-unitary groups obtained by Raccah's tensor-coupling method

Instituto de Física, UNAM
Apdo. Postal 20–364
01000 México DF
Mexico

JEREMY DRAAYER
Pseudospin in perspective[2]

Physics and Astronomy
Lousiana State University
Baton Rouge, LA 70803
USA phdryr@lsuvm

JORGE FLORES
Group theory and the harmonic oscillator: The work of Marcos Moshinsky [with A. Frank and T.H. Seligman][1]
Doorway states in elastic media [with T.H. Seligman, J.M. Alvarez-Tostado, O. Novaro, and J.L. Mateos][1]

Centro Universitario de Comunicación de la Ciencia,
and
Instituto de Física, UNAM
Apdo. Postal 20–364
01000 México DF
Mexico

---

* Authors marked with [1] have their contributions published in this volume, those marked [2] appear in the volume II of the MM120 Proceedings.

Alejandro Frank
[see, with J. Flores and T.H. Seligman][1]
[see, with O. Castaños and R. López Peña][2]

Instituto de Ciencias
Nucleares, UNAM
Apdo. Postal 20–543
01000 México DF
Mexico

Gastón García-Calderón
Aspects of nonexponential contributions to quantum decay[1]
[see also, with M. Moshinsky and L. Loyola][1]

Instituto de Física, UNAM
Apdo. Postal 20–364
01000 México DF
Mexico garcal@unamvm1

Walter Greiner
Nuclear shock waves as the key mechanism for exploring the nuclear equation of state

Inst. Theoretische Physik
Universität Frankfurt
Postfach 111932
D-6000 Frankfurt-am-Main
Germany

K.T. Hecht
Generalizations of the vector coherent state method[2]

Department of Physics
University of Michigan
Ann Arbor, MI 48109
USA hecht@umiphys

Peter Hess
[see, with O. Castaños][1]

Instituto de Ciencias
Nucleares, UNAM
Apdo. Postal 20–543
01000 México DF
Mexico

Francesco Iachello
Spectrum generating algebras and dynamic symmetries in hadron physics[2]

Department of Physics
Yale University
P.O. Box 6666-8167
New Haven, CT 06511
USA ami%nst@yalevm

Gerhard Jacob
Science and Technology Development in Brazil (Round Table)[1]

Conselho Nacional de
Desenvolvimento Científico
e Tecnológico
Caixa Postal 6186
Brasília DF, **Brazil**

Miguel José Yacamán
The new science policy in Mexico (Round Table)[1]

Dirección Adjunta
de Desarrollo Científico
Consejo Nacional
de Ciencia y Tecnología
Mexico

PETER KRAMER
Concepts of symmetry in quasicrystals[2]

Inst. Theoretische Physik
Universität Tübingen
D-7400 Tübingen
Germany ptsl001@convex.zdv.uni.tuebingen.de

DECIO LEVI
Inhomogeneous Toda lattices and their symmetries[2]

Dipartimento di Fisica
Università di Roma
Piazzale A. Moro 2,
00185 Roma, Italy
levi%roma1.infn.it@iboinfn

FRANÇOIS LEYVRAZ
A new decomposition for canonical transformations and its implications for their unitary representations [with T.H. Seligman][2]

Instituto de Física, UNAM
Apartado Postal 139-B
Cuernavaca, Morelos, 62191
Mexico lfranc@unamvm1

MANUEL DE LLANO
Fermion clustering in an exactly-soluble N-body model for hadronic, nuclear, and superconductivity physics [with J. Dukelsky and C. Essebag][1]
[see also, with V. Aguilera][1]

Physics Department
North Dakota S. University
Fargo, ND 58105
USA nu105535@ndsuvm1

JAMES D. LOUCK
The SU(3) *generalization of Raccah's* $SU(2j+1) \supset SU(2)$ *group-subgroup embedding [with L.C. Biedenharn]*[1]

Theoretical Division
Los Alamos National Lab.
P.O Box 1663
Los Alamos, NM 87545
USA

VLADIMIR I. MAN'KO
Nonstationary oscillator in quantum mechanics[1]

P.N. Lebedev Institute
USSR Academy of Sciences
117924 Moscow
USSR

F. ALBERT MATSEN
Isomorphism in atomic and nuclear physics[2]

Department of Physics
University of Texas
Austin, TX 78712
USA

PIER A. MELLO
The scattering approach to quantum electronic transport[1]

Instituto de Física, UNAM
Apartado Postal 139-B
Cuernavaca, Morelos, 62191
Mexico mello@unamvm1

ROSA MARÍA MÉNDEZ-MORENO
Relativistic equations in external fields
[with M. Moreno][1]

Departamento de Física,
Fac. de Ciencias, UNAM
Apdo. Postal 20–364
Mexico mendezv@unamvm1

ALFONSO MONDRAGÓN
Scaling in the strong shock compression of ordinary material and nuclear matter[1]

Instituto de Física, UNAM
Apdo. Postal 20–364
01000 México DF
Mexico

MATÍAS MORENO
[see, with R.M. Méndez-Moreno][1]

Instituto de Física, UNAM
Apdo. Postal 20–364
01000 México DF
Mexico

MARCOS MOSHINSKY
The decay process: an exactly soluble example and its implications
[with G. García-Calderón and G. Loyola][1]

Instituto de Física, UNAM
Apdo. Postal 20–364
01000 México DF
Mexico

YUVAL NE'EMAN
A paralelism between quantized gravity and the IR limit in QCD[1]

Physics and Astronomy
University of Tel Aviv
69978 Tel Aviv
Israel b21a@taunos

OCTAVIO NOVARO
Symmetry-avoided crossings and their rôle in the catalic activity of transition metals[1]
[see also, with J. Flores et al.][1]

Instituto de Física, UNAM
Apdo. Postal 20–364
01000 México DF
Mexico

HERCH M. NUSSENZVEIG
Moshinsky functions, resonances, and tunneling[1]

Departamento de Fisica
U. Pontificia Católica
Caixa Postal 38071
22452 Rio de Janeiro
Brazil nuss@lncc

JOSEF PALDUS
Valence bond theory of molecular electronic structure[2]

Applied Mathematics
University of Waterloo
Waterloo, ON N2L 3G1
Canada paldus@watdcs

Jiří Patera
Graded contractions of representations of Lie algebras[2]

Centre de Recherches en
Mathématiques Appliquées
Université de Montréal
C.P. 6128
Montréal, PQ H3C 3J7
Canada patera@umtlvr

Stuart Pittel
Deriving nuclei from quarks [with J. Engel, J. Dukelsky, and P. Ring][1]

Bartol Research Foundation
University of Delaware
Newark, DE 19711
USA

Jerzy F. Plebański
The Wheeler-Reinich-Mischner conditions in nonlinear electrodynamics [with M. Przanowski][1]

Departamento de Física
CINVESTAV
Apartado Postal 14–740
07000 México DF
Mexico

Christianne Quesne
Symmetry and dynamical Lie algebras in classical and quantum mechanics[1]

Physique Nucléaire
Théorique et Mathématique
Université Libre
B-1050 Bruxelles
Belgium

G. Rosensteel
Dynamical symmetry of rotating systems[2]

Department of Physics
Tulane University
New Orleans, LA 70118
USA gtr.phy.tulane.edu

Thomas Seligman
[see, with J. Flores and A. Frank][1]
[see, with J. Flores et al.][1]
[see, with F. Leyvraz][2]

Instituto de Física, UNAM
Apartado Postal 139-B
Cuernavaca, Morelos 62191
Mexico selig@unamvm1

Robert T. Sharp
Generating functions for the $SO(7) \supset SU(2)^3$ *branching rules*

Physics Department
Mc Gill University
3600 University Avenue
Montreal, PQ H3A 2T8
Canada
brow@physics.mcgill.ca

ADAM SZCZEPANIAK
Hamiltonian formulation of a simple covariant harmonic oscillator for bosons and fermions[1]

Supercomputer Computations Research Institute
Florida State University
Tallahassee
FL 32306 – 4052, USA
adam@scri1.scri.fsu.edu

IGAL TALMI
Binding energies of nuclei and atoms[1]

Nuclear Physics
Weizmann Institute
76100 Rehovoth
Israel fntalmi@weizmann

GALILEO VIOLINI
Some reflections on Science in Latin America (Round Table)[1]

Centro Internacional
de Física
Apartado Aéreo 49490
Bogotá
Colombia cif@andescol

PAVEL WINTERNITZ
Symmetries and nonlinear phenomena in physics[2]

Centre de Recherches en
Mathématiques Appliquées
Université de Montréal
C.P. 6128
Montreal, PQ H3C 3J7
Canada 4855@umtlvr.bitnet

KURT BERNARDO WOLF
Canonical transformations in optics vis-à-vis those in mechanics[1]

IIMAS–UNAM
Apartado Postal 139-B
Cuernavaca, Morelos 62191
Mexico bwolf@unamvm1

CARL E. WULFMAN
Continuous symmetries of characteristic equations[2]

Department of Physics
University of the Pacific
Stockton, CA 95211
USA
cwulfman@madvax.uop.edu

BRIAN G. WYBOURNE
q-deformation of symmetric functions and Hecke algebras of type A_n[2]

Department of Physics
University of Canterbury
Christchurch 1
New Zealand
userjtus@ualtamts.bitnet

Other Authors in this Volume

J.M. ALVAREZ-TOSTADO
J.L. MATEOS
[see, with J. Flores et al.]

Instituto de Física, UNAM
Apdo. Postal 20–364
01000 México DF
Mexico

L.C. BIEDENHARN
[see, with J.D. Louck]

Department of Physics
Duke University
Durham, NC 27706
USA

A. BOHM
The relativistic oscillator and mass formulas
[with P. Kielanowski]

Center for Particle Theory
University of Texas
Austin, Texas 78712
USA bohm@utaphy

J. DUKELSKY
[see, with M. de Llano]
[see, with S. Pittel]

Física Teórica
U. Autónoma de Madrid
28049 Madrid
Spain

J. ENGEL
[see, with S. Pittel]

Bartol Research Foundation
University of Delaware
Newark, DE 19711
USA

C. ESSEBAG
[see, with M. de Llano]

Física Teórica
U. Autónoma de Madrid
28049 Madrid
Spain

P. KIELANOWSKI
[see, with A. Bohm]

Departamento de Física
CINVESTAV
Apartado Postal 14–740
07000 México DF
Mexico

G. LOYOLA
[see, with G. García-Calderón and M. Moshinsky]

Instituto de Física, UNAM
Apdo. Postal 20–364
01000 México DF
Mexico

M. Przanowski
[see, with J.F. Plebanski]

Institute of Physics
Łódź Technical University
93–005 Łódź
Poland

P. Ring
[see, with S. Pittel]

Physikdepartment
Technische U. München
D–8046 Garching
Germany

A. Salam
Marcos Moshinsky and the future of science in Mexico (Round Table)

International Centre for
Theoretical Physics
P.O. Box 586, Miramare
34100 Trieste
Italy

Contents

1

Group Theory and the Harmonic Oscillator: The Work of Marcos Moshinsky

Jorge Flores[1], Alejandro Frank[2], and Thomas H. Seligman[3]

1.1 Introduction

Physics in Mexico is very young. Only a handful of papers had been published in the international literature by Mexican physicists before the end of World War II. It is therefore not difficult to imagine the scientific desert Marcos Moshinsky found after returning from Princeton in 1949. Nevertheless he started publishing work of good quality already in 1950 and has done so ever since for the following 41 years, including the present 1991 when we are celebrating his seventieth anniversary.

The quality and relevance of his work have made him earn worldwide recognition and the respect of his peers. It would be long to list the numerous national and international honors he has received. The scientific contributions of Moshinsky to mathematical physics have been published along 40 years, conforming today 201 publications, all of which are characterized by Marcos' personal style and elegance. It is interesting to note that his publication rate has actually increased. In his first decade as a professional physicist (1950–1959) he published 32 papers; in the second one (1960–1969) the figure grew to 42 articles; from 1970 to 1979, 49 of his papers appeared in print, and in the period from 1980 to 1989 the impressive amount of 61 articles were published. Extrapolating, when we celebrate the MM140 symposium, Marcos' list of publications will consist of 270 research papers, which will have produced around 4000 citations in the physics literature!

Although in what follows we shall concentrate on the scientific contributions of Marcos Moshinsky from a technical point of view, it should be stressed that his role in many other areas related to the development of

[1]Instituto de Física and CUCC, UNAM.
[2]Instituto de Ciencias Nucleares, UNAM.
[3]Instituto de Física, Laboratorio de Cuernavaca, UNAM.

Mexico and Latin America has been equally significant. He guided and strongly influenced the academic careers of dozens of Mexican and foreign physicists; he founded and directed the Revista Mexicana de Física and the Latin American School of Physics; he has published hundreds of newspaper articles devoted to science policy and to the popularization of science; he has given countless conferences and seminars, and has lectured in the classroom for nearly half a century. Any of these activities would have made him a well-known scholar, but put together, they make Professor Moshinsky one of the most distinguished scientists in Latin America.

1.2 Schematic theory of nuclear reactions

When Marcos Moshinsky returned from Princeton to the University of Mexico after obtaining his Ph.D. degree in physics under the advise of Professor Eugene P. Wigner, he worked by himself. Even so, in the period 1951–52 he produced eight papers, four of them in *Physical Review* [7, 8, 9, 10].[4] We recently consulted the last of them, whose attractive title is "Diffraction in Time"; the volume in the library of the Instituto de Física in which this article appears shows the effects of having been read many times, and pages 625 to 631 are still there but unbound. This means that many of Marcos' pupils have glanced through it for almost forty years. In fact, papers he published with some of his students in 1970 [84, 95], 1971 [79], 1976 [117] and 1981 [134] still use the techniques he developed at that time and, as a matter of fact, his contribution to this volume is based on those papers.

The idea behind these articles was to use boundary conditions to describe the time evolution of elastic scattering and two-particle nuclear reactions. A time-dependent description for the process of disintegration of a compound particle was so obtained.

He was then guided by the analogy between vibrational problems in which bodies of different dimensionalities are involved and the quantum mechanical problem in which the number of particles is not conserved. Three-particle disintegrations are then shown to remind us of what Beta decay theory leads to. Marcos went afterwards deep into the problem of transient effects and discussed what happens quantum-mechanically when a shutter is opened. The transient current resembles Fresnel diffraction by a straight edge, so he named the effect *diffraction in time*. He used it later in the study of the time-dependent behavior of isobaric analogue states using a classical model [79, 84, 95], to understand the time-energy uncertainty relations [117], and to discuss the decay of a compound particle and verify

[4] Reference numbers in brackets correspond to Moshinsky's list of publications, at the end of this article.

the Einstein-Podolsky-Rosen argument [134].

In most of the papers mentioned, Marcos used a time, coordinate, and momentum-dependent function that he called Basic Interaction Green (BIG) function. H.M. Nussenzveig gave them the name of Moshinsky functions and he discusses them extensively in his article in these proceedings.

1.3 The Moshinsky brackets

At the beginning of the second half of the century, the understanding of nuclear structure was an important problem, at the forefront of physics. Calculating nuclear wave functions and spectra was a difficult endeavor and nuclear models, such as the shell model, the liquid drop model and others, were invented. With the methods available before the publication of the "Table of Transformation Brackets" [39], the evaluation of the matrix elements of the effective interaction involved a computational labor which quickly became prohibitive.

The transformation bracket $\langle nl, NL, \lambda \mid n_1, l_1, n_2 l_2, \lambda\rangle$ is the coefficient by which a two-particle wave function $|n_1 l_1, n_2 l_2, \lambda\rangle$ is expressed in terms of another two particle wave function, but now refered to relative and center-of-mass coordinates. The single-particle wave functions correspond to a common potential which is a harmonic oscillator.

What Moshinsky did was to evaluate the transformation bracket for angular momenta l_1 and l_2 coupled to a total value λ. He first discussed the bracket for the particular case $n_1 = n_2 = 0$, obtaining the following closed formula [35]:

$$\begin{aligned}\langle nl, NL, \lambda | 0 l_1, 0 l_2, \lambda\rangle &= \sqrt{\frac{l_1\,! l_2\,!}{(2l_1)\,!(2l_2)\,!}\frac{(2l+1)(2L+1)}{2^{l+L}}} \\ &\times \sqrt{\frac{(n+l)\,!}{n\,!(2n+2l+1)\,!}\frac{(N+L\,!}{N\,!(2N+2L+1)\,!}} \\ &\times (-1)^{n+l+L-\lambda}\sum_z (2z+1)A(l_1 l, l_2 L, z) \\ &\times W(lLl_1l_2;\lambda z),\end{aligned}$$

where W is a Racah coefficient and an explicit formula is given for the coefficient A. He then obtained the recursion relation linking $\langle nl, NL, \lambda | n_1 + 1, l_1, n_2 l_2, \lambda\rangle$ with the transformation bracket corresponding to n_1. Since a similar formula was obtained also for increasing n_2, the general Moshinsky bracket was then available.

The transformation bracket allows us to write the matrix elements of a rather general interaction in the following fashion, which Moshinsky obtained in 1960 [36]:

$$\langle n_1l_1, n_2l_2, \lambda|V(r)|n_1'l_1', n_2'l_2', \lambda\rangle = \sum_p C_c(n_1l_1, n_2l_2; n_1'l_1', n_2'l_2', \lambda; p)\, I_p,$$

where C_c is given in terms of the Moshinsky brackets by

$$C_c = \sum_{nlNLn'} \langle nl, NL, \lambda|n_1l_1, n_2l_2\lambda\rangle\langle n'l, NL, \lambda|n_1'l_1'n_2'l_2', \lambda\rangle B(nl, n'l, p).$$

Here B is a coefficient for which a closed formula was obtained [36] and I_p is the Talmi integral. The above result is valid for any central potential, but of course the case of non central interactions can also be dealt with. As a matter of fact, the first realistic application of the transformation brackets was to analyze the effect of a tensor force in the spectrum of ^{210}Bi (Flores and Mello, 1963).

The Moshinsky brackets were tabulated by Thomas Brody in 1959 using a rather primitive electronic machine, an IBM 650, the first scientific computer in Latin America. The job must have been rather ardous, as is clear from what they say in their book [39]:

> *The authors dedicate this book to each other, with the fervent hope that they will never have to perform another such task.*

After their publication, the tables were used immediately by many nuclear physicists. They received a large amount of citations: from 1962 to 1980 we have detected 605 references. The tables became so well known that although they were used and mentioned explicitly in the text of many articles, they did not appear in their list of references, thereby escaping from the Science Citation Index.

1.4 Marcos' harmonic oscillator

With the transformation brackets, Moshinsky established himself as a first-class user of the harmonic oscillator and its symmetry properties. The association of Marcos with the oscillator would be a large and fruitful one.

The long series of papers entitled "Group Theory of Harmonic Oscillators" [38, 40, 58, 63, 66, 78] and "Harmonic Oscillator States and the Alpha particle" [75, 76], as well as the book "The Harmonic Oscillator in Physics: From Atoms to Quarks" [77], form the backbone of his work. As is quite characteristic for most of Marcos' work, the entire development is carried out algebraically and in great detail. Their clarity and basic simplicity make them valuable background for many developments in both representation theory and nuclear physics.

Marcos started this work in collaboration with Bargmann [38, 40] where n-particle states are constructed taking advantage not only of the $U(3)$

symmetry of the oscillator but of the $U(n)$ group of transformations in particle space.

A set of commuting integrals of motion for n particles in an oscillator are explicitly constructed and a complete set of states defined by their highest weight state is given in terms of these integrals of motion. These integrals of motion are shown to remain valid when a quadrupole-quadrupole interaction is added to the oscillator. The chain $U(3) \supset O(3) \supset O(2)$ is introduced and a quantum number necessary to resolve the multiplicity is given. Thus, a complete theory was developed for the collective motion of nuclei in the sense of the nuclear $SU(3)$ model (Elliott, 1956).

In the ensuing papers of the series, the construction of states is carried to much greater detail through the chains of groups

$$\begin{array}{lllllll} U(3n) & \supset & U(3) & \times & U(n), & & \\ U(3) & \supset & O(3) & \supset & O(2), & & \\ U(n) & \supset & U(n-1) & \supset & U(n-2) & \supset & U(1), \end{array}$$

or

$$U(n) \supset O(N) \supset S_n \supset S_{n-2} \times S_2.$$

Here S_n is the group of permutations in particle space. It allows one to construct explicitly the symmetry-adapted orbital states that are needed to construct antisymmetric states when the spin-isospin part of the wave function is included [57]. The subgroup $S_{n-2} \times S_2$ of S_n indicates the road to the classical shell-model treatment in terms of fractional parentage coefficients for oscillator functions that are fully developed all the way to the multi-shell case.

On the other hand, the problem of translational invariance is also addressed and discussed in the shell-model framework. This lead to a long series of papers on the three- and four-particle problems [55, 65, 66, 74, 75, 76, 88, 89, 90, 93, 106]. Here the techniques suggested in previous papers are carried out explicitly. Various choices of coordinate systems and subgroups of S_4 associated to these, make this problem a ground for Marcos' great sense of simplification.

Another interesting aspect of the harmonic oscillator that Marcos has implemented successfully several times, is to use it as test ground for approximations. In a set of papers called "How good is the... approximation?" he dealt with the Hartree-Fock method [70, 80], the Born-Oppenheimer approximation [73], and other methods commonly used in atomic physics [99]. The idea behind these papers is simple: use a solvable many-body problem and then deal with it assuming that a certain approximate method is valid. Compare then the approximate answer with the exact one, thereby testing the method. The soluble many-body problem used by Moshinsky was that of n spin-$\frac{1}{2}$ particles moving in a common harmonic oscillator potencial and interacting through harmonic-oscillator forces. The exact eigenstates and eigenvalues were obtained by Marcos for n particles corresponding to closed sub-shells. The main result is that the Hartree-Fock energy remains reasonably close to the exact value, but that the overlap between the approximate and the exact wave functions decreases dramatically with n.

1.5 Group theory and nuclear structure

The papers with Bargmann [38, 40] mark the starting point in Marcos' marriage with group theory and its applications to different physical problems. He continued his use of the harmonic oscillator by introducing what he called then the "caricature" of a nuclear problem [44]: many particles in a single l-shell interacting through a two-body potential are replaced by a system of particles moving in a $(2l+1)$-dimensional harmonic oscillator. The interaction is expressed in terms of the generators of $\mathrm{U}(2l+1)$ in such a way that the caricature problem has the same eigenvalues as the original problem. The method has the advantage that permutational symmetry can be taken easily into account. The technique was applied to study the spectrum of $2s$–$1d$ shell in nuclei [46, 52, 59].

Nuclear structure has always been of interest to Marcos Moshinsky. He started dealing with a model interaction, a linear combination of pairing force plus quadrupole-quadrupole interaction [43], and applied his techniques first to ^{20}Ne [46] and then to several nuclei in the $2s$–$1d$ shell [52, 59], including negative-parity particle-hole states [60]. All this work was summarized in a book [61]: "Group Theory and The Many Body Problem". In this series of papers, the main objective was to find a set of exact solutions to the many-body problem defined by a model interaction. A group-theoretical classification of the wave functions allows the exact solution to be obtained. In the case of the pairing force, the chain

$$U(n) \supset O(n) \supset O(3)$$

solves the problem if we consider only configurations space, and for a quadrupole-quadrupole interaction, the adequate chain of groups is

$$U(n) \supset SU(3) \supset O(3).$$

The transformation matrix from one basis of irreducible representations to the others were obtained by Moshinsky and his collaborators. The matrix for the complete model interaction could be then computed and the model tested in different cases. These group-theoretical techniques can be also used for more realistic interactions. Marcos reported [72] on some results for nuclei in the $2s$–$1d$ shell for an effective interaction derived from a hard-core potential.

Besides the harmonic oscillator, Moshinsky's work on group theory during the sixties moved in several directions. He computed the Wigner coefficients for the $SU(3)$ group [47], obtained bases for irreducible representations for $U(n)$ and the symmetric group, and discussed lowering and raising operators for the unitary group, among other results.

In the paper published in the issue of Reviews of Modern Physics to celebrate Prof. Eugene Wigner's sixtieth anniversary, a closed formula for the Wigner coefficients of $SU(3)$ was obtained for the first time [47]. This was an important result and the paper has received around 100 citations in the literature, some of them very recently. The result was very important for calculations related to the octet model in elementary particle physics; therefore, it should have received more attention. Since the formula obtained by Marcos was written in the language of Elliott's $SU(3)$ model for nuclei and not in the notation used by elementary particle physicists, the latter ignored Moshinsky's result.

Another interesting result obtained by Marcos at the beginning of the sixties concerned operators that lower or raise the irreducible vector spaces of $U(n-1)$ in an irreducible vector space of $U(n)$. These lowering operators of $U(n)$ are the exact analog of L_- for the $O(3) \supset O(2)$ chain of groups. He was led into this problem because he was interested in the complete classification of many-particle states. Moshinsky first obtained the solution for $SU(4)$, that is, for supermultiplet theory [49].

In references [53] and [54] he used Gel'fand states and obtained the normalized lowering and raising operators for the canonical subgroup $U(n-1)$ of $U(n)$. These operators are polynomial functions of the generators $C_\mu^{\mu'}$ of $U(n)$. As an example, L_n^m such that

$$L_n^m \left| \begin{matrix} h_1 & h_2 & \dots & \dots & h_n \\ q_1 & \dots & q_m & \dots & q_{n-1} \end{matrix} \right\rangle \simeq \left| \begin{matrix} h_1 & h_2 & \dots & \dots & h_n \\ q_1 & \dots & q_m - 1 & \dots & q_{n-1} \end{matrix} \right\rangle ,$$

is given by

$$L_n^m = \left[\sum_{p=1}^{n-m-1} \sum_{\mu_i=1}^{n-1} C_{\mu_1}^m C_{\mu_2}^{\mu_1} \dots C_{\mu_p}^{\mu_{p-1}} C_n^{\mu_p} \right] \left[\prod_{i=1}^{p} \mathcal{E}_{m\mu_i}^{-1} \right] \left[\prod_{\mu=m+1}^{n-1} \mathcal{E}_{m\mu} \right] ,$$

$$\mathcal{E}_{\mu\mu'} = C_{\mu'}^{\mu'} - C_\mu^\mu + \mu' - \mu .$$

He then uses these lowering operators to obtain special Gel'fand states that are bases for irreducible representations of the symmetric group [57]. States with a given spin and isospin and definite permutational symmetry can be thus constructed.

The lowering operators are also used to generate complete bases for non-orthogonal orbitals [86], a situation that exists, for example, when a many-center molecule is considered. Marcos introduces a dual basis in the second quantized picture that plays with respect to the original non-orthogonal basis the same role as the reciprocal lattice in crystallography; the dual basis is also non-orthogonal. Since then, this method has been used frequently in atomic and molecular calculations (Paldus, 83).

In references [80] and [82] the generalization of the concept of quasispin to arbitrary groups is achieved. Moshinsky treats in detail the case of identical fermions in a single j-shell and introduces the concept of complementary group. This concept is defined as follows. Let the direct product of two groups G_1 and G_2 be contained in a larger group G. Then G_1 and G_2 are complementary to each other within a given irreducible representation of G, if there exists a one-to-one correspondence between all irreducible representations of G_1 and G_2 contained in G. The treatment of many-body forces, the state-labeling problem and the quasiparticle picture are then discussed using this concept.

1.6 Classical canonical transformations and their unitary representation

After many years of intensive work on the harmonic oscillator and its symmetry group $U(n)$, Marcos was rather naturally pushed toward the investigation of the meaning of this symmetry from the point of view of the group rather than the Lie algebra. This in turn leads to the classical problem and its symmetry properties. Indeed, it can be shown that the subgroup $U(n)$, embedded in the group $Sp(2n, R)$ of homogeneous linear canonical transformations

$$\frac{1}{2}\begin{pmatrix} U+U^* & -iU-U^* \\ iU-U^* & U+U \end{pmatrix}, \qquad U \in U(n),$$

is the symmetry group of the n-dimensional harmonic oscillator. This fact allows us to understand the sudden interest of Marcos in the quantum pictures of classical canonical transformations. He immediately proceeded to derive the representations of the linear transformations group. His result, both by the direct route of derivation and by the useful presentation of the result, is a classical example of the work of Marcos; the fact that we learned later of previous derivations by Itzykson (1967) and Bargmann (1968) does not reduce the importance of Moshinsky's work, as both are quite mathematical and had little impact.

Marcos grasped the full importance of the subject as outlined in Dirac's book (1947). He proceeded to explore many problems in the light of these representations. This led him directly to the study of complex extensions [109], of non-linear [111] and non-bijective canonical transformations [120]. Progress in this direction appears in reference [111], where implicit forms of Dirac's equations for these representations are given that allow solving a wide class of problems. The basic trick is to find an implicit form of the classical equations for the canonical transformation that is amenable to a reasonable quantization. The adequate generalization of Dirac's work became known as the MM equations, which have the following implicit form:

$$H\left(\frac{\partial}{\partial p'},p'\right)\langle p'|U|p''\rangle = \left[\overline{H}^{\dagger}\left(\frac{\partial}{\partial p''},p''\right)\right]^{*}\langle p'|U|p''\rangle,$$
$$G\left(\frac{\partial}{\partial p'},p'\right)\langle p'|U|p''\rangle = \left[\overline{G}^{\dagger}\left(\frac{\partial}{\partial p''},p''\right)\right]^{*}\langle p'|U|p''\rangle,$$

where the functions H and G, taken in the classical sense, must fullfil the relation

$$\{H,G\}_{q,p} = \{\overline{H},\overline{G}\}_{\overline{q},\overline{p}}, \qquad H(q,p) = \overline{H}(\overline{q},\overline{p}), \qquad G(q,p) = \overline{G}(\overline{q},\overline{p}).$$

Whenever H and G can be chosen appropriately, the MM equations make sense and one can proceed to write down differential equations for the representation integral kernel.

To grasp the general situation it seemed reasonable to discuss the transformation to action and angle variables for bound systems, and the transformation to energy and time for free systems. Here again Dirac's formulation was used. This question was first understood, in Marcos way, for several examples [131] starting, needless to say so, with the harmonic oscillator [123]. The Coulomb system [127] proved quite tedious, but it is an important step on the way to the general treatment in terms of the WKB approximation [132].

On this path another important step became necessary: If one wishes to mantain phase space as extending from minus infinity to plus infinity, clearly the transformation to action and angle variables is not bijective. Thus an entire line of research into the structure of phase space started; this produced a theory of folliation by involving groups of ambiguity and the "ambiguity spin" related to the irreducible representations of these groups. Indeed, if we know the group that connects either all points that are images of a given point in the original space or all original points mapped into the same image, then the quantum problem for both spaces can be decomposed according to the irreducible representations of these groups. Specifying in addition to phase space coordinates the irreducible representation of these so-called ambiguity groups of transformations, all ambiguities are avoided in the source and image spaces if the groups are abelian; otherwise, row or column indices of the irreducible representations must also be used. The additional labels stemming from these representations and their rows are termed ambiguity spins. They neatly solve the problem of non-bijectiveness by embedding.

A further line that resulted from this work was extensive work on the symplectic algebra of $Sp(2n,R)$ as a spectrum-generating algebra for the harmonic oscillator and other related problems. These are algebras that can be expressed in terms of raising and lowering operators. Their properties can be put to extensive use both to calculate matrix elements [94] and to understand basic features of the problem at hand. Again, techniques similar to those used earlier for $U(n)$ involving chains of subgroups are used and yield new results [104]. Both the MM equations and the concept of ambiguity spin play a major role in the discussions of accidental degeneracies described in the next section.

1.7 Rendering accidental degenerancy non-accidental

The symmetry and dynamical algebras of the oscillator lead, as we mentioned earlier, to extensive studies of the representations of canonical transformations. It is therefore not surprising that Marcos inverted this line of thought to develop and fully understand the symmetry groups and the dynamical algebras of other systems that display accidental degeneracy.

Doing it his way, Marcos first attacked explicitly the simplest of these problems, namely the two-dimensional oscillator with rational frequency ratio [102]. The classical and quantum mechanical Lie algebras for the symmetry group as well as the finite transformations and their unitary representations are constructed explicitly. A new, richer representation of $SU(2)$ is given in these terms and the splitting of spectra is understood in terms of the ambiguity group.

These ideas were then extensively applied to more complicated cases, and in a later paper [108] a unified approach for two-dimensional problems was presented, that in some sense was programatic for the general problem. General principles according to which invariance algebras may be constructed were given and their extension to dynamical algebras was discussed. Then the construction of the group of canonical transformations responsible for the accidental degeneracy was given, thus permitting a full understanding of the symmetry underlying this degeneracy, that now probably should no longer be called accidental. Marcos pursued this line of research intermittently, investigating more complicated problems that defied this general technique.

In a recent paper [194], he discusses simple examples corresponding to hamiltonians of the form $H = N + M$ and $N \pm |M|$, where N is the number operator of a two-dimensional oscillator and M the corresponding angular momentum operator. This example is typical for some of the problems encountered in previous papers. It can happen that ladder operators that connect states of the same energy also connect other states, or all states, or it may happen that they do not close. With some complicated renormalizations of ladder operators, satisfactory algebras are obtained. In any case, the procedures are, as Marcos points out, more art than science, this gives what we believe to be a realistic note to his magic touch for oscillator algebras.

1.8 Collective models

In 1975 Marcos Moshinsky became interested in the nuclear collective model of Bohr and Mottelson (1969), which pictures collective motions as vibrations of a quantized liquid drop. The natural basis for the diagonalization of the Hamiltonian is that of a five-dimensional harmonic oscillator with well-defined angular momentum, for which only particular states had been determined. Marcos was able to solve the problem completely [115, 119] making use of an elegant method to introduce the appropriate $U(5) \supset O(5) \supset O(3)$ classification and, in addition, resolving the missing label problem involved. These two articles marked the beginning of Marcos' fruitful incursion into the field of nuclear collective models which would last for more than a decade.

Having solved the mathematical problems associated to quadrupole vibrations in nuclei, he went on to analyze a new nuclear model, the so-called interacting boson model, IBM (Iachello and Arima, 1987), whose natural framework is group theory and where his previous results were of immediate relevance. In this model the building blocks are again quadrupole bosons (d-bosons) but supplemented by monopole bosons (s-bosons), which together give rise to an $U(6)$ dynamical algebra. Besides the $U(5)$ chain discussed above, this algebra contains two additional chains of physical interest, the $U(6) \supset O(6) \supset O(5) \supset O(3)$ and the $U(6) \supset SU(3)$ chains. He proceeded to show that the most general IBM Hamiltonian can be expressed in terms of invariant operators in the three chains and was able to compute analytic expressions for all matrix elements of interest [124]. These results were very significant for subsequent applications of the model.

One of the outstanding questions regarding the IBM model was its relationship with the collective Bohr-Mottelson model. Marcos was the first to establish such a connection, for which he used a Holstein-Primakoff realization of the $U(6)$ algebra, expressing the Hamiltonian and other operators in terms of quadrupole degrees of freedom [133, 145].

Following these studies, he turned his attention to the problem of defining the collective subspace associated to a many-body system. Building on earlier results by Vanagas (1980) and Dzublik *et al.* (1972), he studied the case where the particles interact through harmonic oscillator forces. The Hamiltonian may be written in terms of collective coordinates associated to the quadrupole matrix of the system plus other variables associated to an $O(n)$ subgroup of the original $U(3n)$ symmetry group of the n particle system. The $O(n)$ subgroup labels the particle indices and may be used thus to characterize the appropriate permutational symmetry. The large n result coincides with the IBM oscillator Hamiltonian for the scalar representation of $O(n)$ [149, 150].

In subsequent papers he showed that general Hamiltonians for systems of n-particles can be formulated in the enveloping algebra of the symplectic group $Sp(6n)$ which admits, among others, the $Sp(6) \times O(n)$ subgroup, while collective Hamiltonians can be formulated in the $Sp(6)$ enveloping algebra. The collective states are thus characterized by definite irreducible representations [157]. He went on to develop algorithms to determine the matrix elements of the $Sp(6)$ generators in bases characterized by irreducible representations of this group and its subgroups [159, 161], thus opening the way to the microscopic determination of collective excitations for both closed and open shell nuclei. These results were obtained working in configuration space and his next step was to consider the boson realization of the algebras, generalizing the results of Holstein and Primakoff to the $Sp(2n)$ algebras [166, 167]. An essential ingredient for the transformation from a Dyson to a Holstein-Primakoff realization involves the explicit determination of the K operator. Although K^2 had been derived in terms of recursion relations, no explicit and closed form had been found. Marcos expressed this operator as an overlap of coherent states of the symplectic group, leading to its matrix elements when expanded in terms of appropriate eigenstates [169]. He first applied this method to the $Sp(4)$ algebra [170, 176] and then to the physically relevant $Sp(6)$ algebra. He obtained closed forms for the matrix elements of the generators for arbitrary irreducible representations [177].

1.9 Structure of matter in strong magnetic fields

In the eighties Marcos Moshinsky and some of his students in that period became interested in this subject matter. It started with the study of accidental degeneracy in the Zeeman effect [162], pseudo atoms and atoms in strong magnetic fields [163], and continued with the analysis of the stability of nuclei in strong magnetic fields [180, 181], coherent states and, accidental degeneracy for charged particles in strong magnetic fields [188, 189].

1.10 Relativistic oscillators

As we have pointed out, the harmonic oscillator system has always played a central role in Marcos' approach to physics. His third monography, "The Harmonic Oscillator in Modern Physics", published in 1969, summarized his view at the time on its applications to nuclear, atomic and molecular physics and included an analysis of the form factor of the proton as a three-quark system [74]. An extensive application in hadron physics would of course require an appropriate relativistic treatment.

His first publication addressing this question was connected with a relativistic generalization of his ideas regarding collective coordinates in a many body system, defining a Lorentz invariant transformation in space-time analogous to the one of Dzublik *et al.* (1972) for three-dimensional space [153]. He later proposed a relativistic symplectic model based on a covariant harmonic oscillator equation (Kim and Noz, 1981) and applied it to the mass spectra of a system of three spinless quarks [187]. The introduction of the spin degrees of freedom in this context turns out to be quite awkward, however. Very recently, he went back to the one-particle Dirac equation and was able to define a non-minimal coupling, linear in both moment and coordinates, that reduces to the ordinary oscillator plus a strong spin-orbit coupling term for the large component of the wave function [190]. Although similar equations were proposed before (Ito *et al.*, 1967; Swamy, 1969), Marcos realized its importance and went on to analyse the two-, three- and many-body versions with the aim of applying them to the structure of mesons, baryons and nuclei, respectively [191, 192, 195]. Among other aspects of the problem, he also studied the relativistic invariance of the system and the characterization of the states by irreducible representations of the Poincaré group [195, 196], a time-dependent relativistic two-body theory [198], and a new version of the equation which includes a spin-orbit coupling from the outset [201].

The linearly coupled Dirac equation was appropriately named "Dirac oscillator" and constitutes an elegant, exactly solvable relativistic generalization of the harmonic oscillator Schrödinger equation that had already aroused so much interest along Marcos' academic life.

We cannot conclude this short summary of Marcos Moshinsky's scientific work without adding some personal remarks. We believe that Marcos' dimension as a scientist is endorsed by his published work and the testimonies of his scientific peers and colleagues. There is, however, another aspect which is perhaps less well known: his human dimension. His intellectual honesty, commitment to his work, and loyalty to his friends and students have been an inspiration for all of us.

1.11 References

[1] Problemas de condiciones a la frontera de características discontinuas. *Bol. Soc. Mat. Mex.* **4**, 1 (1947).

[2] *Relativistic interactions by means of boundary conditions.* Ph.D Dissertation, Princeton University (1949).

[3] Vibración de una membrana circular de dos densidades. *Rev. Ing. Hid.* **3**, 5 (1949).

[4] Interference phenomena for particles obeying Bose or Fermi statistics. *Proc. A. Phil. Soc.* **94**, 53 (1950).

[5] Interactions of Birkhoff's gravitational field with the electromagnetic and pair fields. *Phys. Rev. 80,* **514,** (1950).

[6] Auto-adjunticidad de problemas vectoriales de condiciones a la frontera. *Bol. Soc. Mat. Mex.* **7**, 1 (1950).

[7] Boundary conditions for the description of nuclear reactions. *Phys. Rev.* **81**, 347 (1951).

[8] Boundary conditions and time dependent states. *Phys. Rev.* **84,** 525 (1951).

[9] Quantum mechanics in Fock space. *Phys. Rev.* **84,** 533 (1951).

[10] Efectos transitorios en la dispersión por una esfera rígida. *Rev. Mex. Fís.* **1,** 29 (1952).

[11] Sobre una clase de transformadas de interés en la dispersión nuclear. *Rev. Mex. Fís.* **1,** 114 (1952).

[12] Diffraction in time. *Phys. Rev.* **88,** 625 (1952).

[13] On a dynamical theory of scattering. In: *Novas Técnicas de Pesquisa en Física.* Academia Brasileira de Ciencias (1954), pp. 279-285.

[14] Difracción en el tiempo y la ecuación de Dirac. *Rev. Mex. Fís.* **1,** 151 (1952).

[15] On matrix boundary value problems [with J. Adem]. *Quart. App. Math.* **9,** 424 (1952).

[16] Dispersión de neutrones por cristales en vibración. *Rev. Mex. Fís.* **2,** 137 (1953).

[17] Poles of the S matrix for resonance reactions. *Phys. Rev.* **91,** 984 (1953).

[18] Reacciones nucleares entre partículas con carga. *Rev. Mex. Fís.* **2,** 244 (1953).

[19] Vibraciones térmicas de los cristales y la difracción de neutrones. *Rev. Mex. Fís.* **3,** 1 (1954).

[20] Polos de la función S para dispersión de momento angular arbitrario. *Ann. Acad. Brasileira de Ciencias,* **25,** 343 (1953).

[21] Description de la diffraction-hachage par une distribution de sources. *C.R. Acad. Sci. Paris,* **283,** 2395 (1954).

[22] Definition et propriétés analytiques des matrices R et S associées aux forces tensorielles. I. Cas de la matrices R [with W. Laskar]. *C.R. Acad. Sci. Paris,* **283,** 2496 (1954).

[23] II. Cas de la matrice S [with W. Laskar]. *C.R. Acad. Sci. Paris,* **239,** 29 (1954).

[24] Forces tensoriélles dépendant de la vitesse. *J. de Phys. et Radium.* **15,** 264 (1954).

[25] Difracción en el tiempo asociada a una distribución de fuentes. *Rev. Mex. Fís.* **3,** 236 (1954).

[26] Dispersión de ondas sonoras por una gota de líquido. *Rev. Mex. Fís.* **4,** 95 (1955).

[27] Movimientos colectivos y las reacciones nucleares. *Rev. Mex. Fís.* **5,** 1 (1956).

[28] Collective motions in nuclei [with A.M. Mallman]. Notes on the Summer Course in Theoretical Nuclear Physics (1956).

[29] Velocity dependent forces and nuclear structure. I. Central forces. *Phys. Rev.* **106**, 117 (1957).

[30] Repulsive core and nuclear shell theory [with M. Bauer]. *Nuclear Physics* **4,** 615 (1957).

[31] Fuerzas nucleares con centro repulsivo y el modelo de capas del núcleo. Efectos de segundo orden. *Rev. Mex. Fís.* **6,** 185 (1957).

[32] Velocity dependent forces and nuclear structure. II. Spin dependent forces. *Phys. Rev.* **109,** 933 (1958).

[33] Short range forces and nuclear shell theory. *Nucl. Phys.* **8,** 19 (1958).

[34] La interacción spin-órbita entre nucleones y el acoplamiento spin-órbita en el modelo de capas del núcleo. *Rev. Mex. Fís.* **7,** 95 (1958).

[35] Transformation brackets for harmonic oscillator functions. *Nucl. Phys.* **13,** 104 (1959).

[36] Matrix elements in nuclear shell theory. *Nucl. Phys.* **17,** 16 (1960).

[37] Invariance principles and beta decay [with C.A. Mallmann and J. de Oyarzábal]. Notes of the Latin American school of Physics (1959).

[38] Group theory of harmonic oscillators. I. The collective modes [with V. Bargmann]. *Nuclear Phys.* **18,** 697 (1960).

[39] *Tables of transformation brackets* [with T.A. Brody] Monografías del Instituto de Física (1960). Reprinted by Gordon & Breach (1969).

[40] Group theory of harmonic oscillators. II. The integrals of motion of the quadrupole-quadrupole interaction [with V. Bargmann]. *Nucl. Phys.* **23,** 177 (1961).

[41] Dispersion relations and the causality concept. [with J.M. Lozano]. *N. Cimento* **20,** 59 (1961).

[42] Simetrías y reglas de suma de los paréntesis de transformación [with T.A. Brody]. *Rev. Mex. Fís.* **9,** 181 (1960).

[43] On exact solutions for the pairing plus quadrupole-quadrupole interactions. Proceedings of the Rutherford Jubilee Conference (1961).

[44] The harmonic oscillator and supermultiplet theory. I. The single shell picture. *Nucl. Phys.* **31,** 384 (1962).

[45] A soluble many body problem for particles in a Coulomb field. *Phys. Rev.* **126,** 1880 (1962).

[46] Estudio teórico de los niveles de Ne^{20} [with E. Chacón]. 4° Simposio de Energía Nuclear (1962). Translation: Theoretical analysis of the energy levels in Ne^{20}. *Phys. Lett.* **1,** 330 (1962).

[47] Wigner coefficients for the SU_3 group and some applications. *Rev. Mod. Phys.* **34,** 813 (1962).

[48] Bases for the irreducible representations of the unitary groups and some applications. *J. Math. Phys.* **4,** 1128 (1963).

[49] Complete classification of states of supermultiplets theory [with J. Nagel]. *Phys. Lett.* **5,** 173 (1963).

[50] Group theory and many body problems with finite number of states. *Phys. Lett.* **6,** 305 (1963).

[51] *Group theory and collective motions.* Notes of the Latin American School of Physics (1962).

[52] Group theoretical methods and a model interaction in nuclear shell theory. Proceedings of the International Conference on Nuclear Physics (1964).

[53] Operators that lower or raise the irreducible vector space of U_n [with J. Nagel]. *J. Math. Phys.* **6,** 682 (1965).

[54] Normalization coefficients for lowering and raising operators and the matrix elements of the generators of U_n [with J. Nagel]. *Rev. Mex. Fís.* **14,** 29 (1965).

[55] The three body problems and the SU_4 group [with E. Chacón]. *Rev. Mex. Fís.* **14,** 119 (1965).

[56] Recurrence relations for the Wigner coefficients of unitary groups [with T.A. Brody and I. Renero]. *J. Math. Phys.* **6,** 1540 (1965).

[57] Gel'fand states and the irreducible representations of the symmetric group. *J. Math. Phys.* **7,** 691 (1966).

[58] Group theory of harmonic oscillators. III. States with permutational symmetry [with P. Kramer]. *Nuclear Physics* **82,** 241 (1966).

[59] Group theory and nuclear structure in the $2s$–$1d$–shell [with M. Berrondo and J. Pineda]. Proceedings of the Second Conference on Structure of Light and Medium Nuclei (U. of Kansas, 1967).

[60] Studies on nuclear structure in the $2s$–$1d$–shell. III. Group theory and particle hole states [with J. Flores]. *Nucl. Phys.* **A93,** 81 (1967).

[61] Group theory and the many body problems. In: *Physics of Many Particle Systems*, (Gordon & Breach, 1966); reproduced in expanded book form with the same title by Gordon & Breach (1967).

[62] Some matrix elements and normalization coefficients in SU_n [with T.A. Brody and I. Renero]. *Rev. Mex. Fís.* **15,** 145 (1966).

[63] Group theory of harmonic oscillators and nuclear structure [with P. Kramer]. In: *Group Theory and Applications.* Ed. by E.M. Loebl (Academic Press, 1968).

[64] Representation of finite U_3 transformations [with E. Chacón]. *Phys. Lett.* **23,** 567 (1966).

[65] Energy levels of the 4 nucleon system [with P. Kramer]. *Phys. Lett.* **23,** 574 (1966).

[66] Group theory of harmonic oscillators. IV. Dynamics of the two, three and four particle problem [with P. Kramer]. *Nucl. Phys.* **A107,** 481 (1968).

[67] Racah coefficients and states with permutational symmetry [with E. Chacón]. In: *Racah Memorial Volume* (North Holland Publishing Co., 1968).

[68] The harmonic oscillator in atomic and molecular physics. I. General approach and application to atoms and molecules of up to 4 electrons [with O. Novaro]. *J. Chem. Phys.* **48,** 4162 (1968).

[69] Group theory and nuclear structure. Lectures Notes (ICTP, 1967). In: *Fundamentals of Nuclear Theory* (IAEA, 1967).

[70] How good is the Hartree-Fock approximation? *Am. J. Phys.* **36,** 52 (1967).

[71] A general approach to fractional parentage coefficients [with V. Syamala Devi]. *J. Math. Phys.* **10,** 455 (1969).

[72] Group theory and nuclear structure in the $2s$–$1d$-shell. Proceedings of the Tokyo Conference on Nuclear Structure (1967).

[73] How good is the Born-Oppenheimer approximation? [with C. Kittel] *Proc. Natl. Acad. Sci.* **60,** 1110 (1968).

[74] Harmonic oscillator states for three particle systems. Applications to the form factor of the proton as a system of three quarks [with V.C. Aguilera-Navarro and W. Yeh]. *Rev. Mex. Fís.* **4,** 241 (1968).

[75] Harmonic oscillator states and the α particle. I. Form factor for symmetric states in configuration space. *Ann. Phys.* **51,** 312 (1969).

[76] Harmonic oscillator states and the α particle. II. Configuration space states of arbitrary symmetry [with P. Kramer and V.C. Aguilera-Navarro]. *Ann. Phys.* **54,** 379 (1969).

[77] *The Harmonic Oscillator in Modern Physics: From Atoms to Quarks.* Gordon & Breach, New York (1969). Russian Translation (MIR, 1972).

[78] Group theory of harmonic oscillators. V. The harmonic oscillator shell model [with P. Kramer]. *Nucl. Phys.* **A125,** 321 (1969).

[79] Mechanical models for reactions involving isobaric analogue and doorway states [with P.A. Mello]. *Am. J. Phys.* **39,** 54 (1971).

[80] Generalization to arbitrary groups of the relations between seniority and quasispin [with C. Quesne]. *Phys. Lett.* **29B,** 482 (1971).

[81] How good is the Hartree-Fock approximation? II. The case of closed shells [with A. Calles]. *Am. J. Phys.* **38,** 456 (1970).

[82] Non-invariance groups in the second quantization picture and their applications [with C. Quesne]. *J. Math. Phys.* **11,** 5 (1970).

[83] The pseudo-atom, a soluble many body problem [with O. Novaro and A. Calles]. *J. de Physique* **31,** 125 (1970).

[84] Time dependent behavior of a classical model for reactions involving isobaric analogue states [with P.A. Mello]. *Rev. Mex. Fís.* **19,** 157 (1970).

[85] Oscillator systems [with C. Quesne]. Proceedings of the XV Solvay Conference, (Brussels, 1971).

[86] Group theory and second quantization for non-orthogonal orbitals [with T.H. Seligman]. *Ann. Phys.* **66,** 311 (1971).

[87] $O(4)$ and $U(3)$ symmetry breaking in the $2s$–$2p$–shell [with E. Chacón, O. Novaro and C. Wulfman]. *Phys. Rev.* **3,** 166 (1971).

[88] Correlation effects in the α particle [with V.C. Aguilera-Navarro]. *Phys. Lett.* **32B,** 336 (1970).

[89] Transformation brackets for three and four nucleon systems. Proceedings of the Conference on Clustering Phenomena. Bochum (IAEA, 1969).

[90] Group theory and the few nucleon problem. In: *1968 Cargese Lectures* (Gordon & Breach, 1969).

[91] Applications of group theory to problems of atomic physics. In: *New Directions of Atomic Physics* (Yale University Press 1971).

[92] Linear canonical transformations and their unitary representation [with C. Quesne]. *J. Math. Phys.* **12,** 1772 (1971).

[93] The charge form factor of the mass-3 nuclei [with C. Quesne and A.D. Jackson]. *Rev. Mex. Fís.* **20,** 43 (1971).

[94] Canonical transformations and matrix elements [with C. Quesne]. *J. Math. Phys.* **12,** 1780 (1971).

[95] Schematic theory for isobaric analogue states [with P.A. Mello]. *Rev. Mex. Fís.* **19,** 339 (1970).

[96] Implications of classical ergodicity in quantum mechanics [with P.A. Mello]. *Ann. Phys.* **71**, 1 (1972).

[97] Canonical transformations and quantum mechanics. Notes of the Latin American School of Physics (México, 1971).

[98] Canonical transformations and the radial oscillator and Coulomb problems [with T.H. Seligman and K.B. Wolf]. *J. Math. Phys.* **13,** 901 (1972).

[99] Test of the many electron theory of atoms in an exactly solvable model [with A. Calles and O. Sinanoğlu]. *Phys. Lett.* **40A,** 79 (1972).

[100] Canonical transformations and quantum mechanics. *SIAM J. Appl. Math.* **25,** 193 (1973).

[101] Comments on charge distribution in multipion production [with E. Chacón]. *Phys. Rev.* **D7,** 2783 (1973).

[102] Canonical transformations and accidental degeneracy. I. The anisotropic oscillator [with J.D. Louck and K.B. Wolf]. *J. Math. Phys.* **14,** 692 (1973).

[103] Canonical transformations and accidental degeneracy. II. The isotropic oscillator in a sector [with J.D. Louck and K.B. Wolf]. *J. Math. Phys.* **14,** 696 (1973).

[104] Relation between hyperspherical and harmonic oscillator many-body matrix elements [with D. Levi]. *N. Cimento* **204,** 107 (1974).

[105] An application of linear canonical transformations: Coherent states [with P.A. Mello]. *Rev. Mex. Fís.* **22,** 257 (1973).

[106] Relation between the hyperspherical harmonic and the harmonic oscillator method for the three body problem [with E. Chacón and D. Levi]. *Rev. Mex. Fís.* **22,** 291 (1973).

[107] Isotopic spin conservation and charge distribution in multipion production [with J. Patera, R.T. Sharp, and P. Winternitz]. *Phys. Rev.* **D10,** 1587 (1974).

[108] Canonical transformations and accidental degeneracy. III. A unified approach to the problem [with J. Patera and P. Winternitz]. *J. Math. Phys.* **16,** 82 (1975).

[109] Complex extensions of canonical transformations and quantum mechanics [with P. Kramer and T.H. Seligman]. *Group Theory its Applications* Ed. by E.M. Loebl. Academic Press, (1975).

[110] Lie Algebras in the Heisenberg and Schrödinger pictures and two body matrix elements. Proceedings of the 3rd Conference of Group Theoretical Methods in Physics (Marseille, 1973).

[111] Nonlinear canonical transformations and their representation in quantum mechanics [with P.A. Mello]. *J. Math. Phys.* **16,** 2017 (1975).

[112] Canonical transformation and accidental degeneracy. IV. Problems with continuous spectra [with J. Patera]. *J. Math. Phys.* **16,** 1866 (1975).

[113] Equivalence of a class of Wigner coefficients of $SU(1,1)$ with those of $SU(2)$ [with E. Chacón and D. Levi]. *J. Math. Phys.***16,** 1876 (1975).

[114] Everything you ever wanted to know about $SU(3) \supset O(3)$ [with J. Patera, R.T. Sharp, and P. Winternitz]. *Ann. Phys.* **95,** 139 (1975).

[115] $U(5) \supset O(5) \supset O(3)$ and the exact solution of the problem of quadruple vibrations of the nucleus [with E. Chacón and R.T. Sharp]. *J. Math. Phys.* **17,** 668 (1976).

[116] Canonical transformations and their representations in quantum mechanics. Proceedings of the Bonn Conference on Geometrical Quantization (1976).

[117] Diffraction in time and the time-energy uncertainty relation. *Am. J. Phys.* **44,** 1037 (1976).

[118] Lie Algebras in the Schrodinger picture and radial integrals [with E. Chacón and D. Levi]. *J. Math. Phys.* **17,** 1919 (1976).

[119] Group theory of the collective model of the nucleus [with E. Chacón]. *J. Math. Phys.* **18,** 870 (1977).

[120] Non-bijective canonical transformations and their representation in quantum mechanics [with P. Kramer and T.H. Seligman]. *J. Math. Phys.* **19,** 683 (1978).

[121] Ambiguities in canonical transformations of classical systems and the spectra of quantum observables [with T. H. Seligman]. Proceedings of the 1977 Bonn Conference on Differential Geometric Methods in Mathematical Physics (Springer Verlag, 1978).

[122] The gradient formula for the $O(5) \supset O(3)$ chain of groups [with O. Castaños and A. Frank]. *J. Math. Phys.* **19,** 1781 (1978).

[123] Canonical trasnformations to action and angle variables and their representation in quantum mechanics [with T.H. Seligman]. *Ann. Phys.* **114,** 243 (1978).

[124] Group theory of the interacting boson model of the nucleus [with O. Castaños, E. Chacón, and A. Frank]. *J. Math. Phys.* **20,** 35 (1979).

[125] Group theory of the collective model of the nucleus. Proceedings of the Nuclear Physics Conference (Oaxtepec, 1978).

[126] Nuclear reactions with collective degree of freedom [with E. Aguilera, O. Castaños, and A. Frank]. Proceedings of the Nuclear Physics Conference (Oaxtepec, 1978).

[127] Canonical transformations to action and angle variables and their representation in quantum mechanics. II. The Coulomb problem [with T.H. Seligman]. *Ann. Phys.* **120,** 430 (1979).

[128] The ambiguity group for canonical transformations in classical mechanics [with T. H. Seligman]. Proceedings of the Conference on Group Theory and Applications, Austin, Texas. Springer-Verlag, Lecture Notes in Physics, Vol. 97 (1979).

[129] Quadratic Hamiltonians in phase space and their eigenstates [with P. Winternitz]. *J. Math. Phys.* **21,** 1667 (1980).

[130] Irreducible decomposition of the direct products of q arbitrary representations of the symplectic group $Sp(2n)$ and its relation with the orthogonal group $O(q)$ [with E. Chacón and P. Winternitz]. *Kinam* **1,** 259 (1979).

[131] Canonical transformations to action and angle variables and their representation [with T.H. Seligman]. *J. Phys. A.: Math. Gen* **12,** L135 (1979).

[132] Canonical transformations to action and angle variables. III. The general problem [with T.H. Seligman and J. Deenen]. *Ann. Phys. (N.Y.)* **127,** 458 (1980).

[133] Confrontation of nuclear collective models. *Nucl. Phys.* **A338,** 156 (1980).

[134] Decay of a compound particle and the Einstein, Podolsky, and Rosen argument [with J. Flores, E. Henestroza, and P.A. Mello]. *Am. J. Phys.* **49,** 59 (1981).

[135] Wigner distribution functions and the representation of canonical transformations in quantum mechanics [with G. García-Calderón]. *J. Phys. A: Math. Gen.* **13,** L185 (1980).

[136] Review of the group theory behind the interacting boson model of the nucleus. In: *Symmetries in Science*, Ed. by B. Gruber and R.S. Millman (Plenum, 1980).

[137] Group theory and nuclear models. *Kinam* **2,** 79 (1980).

[138] Group theory and the confrontation of nuclear collective models. In: *Group Theoretical Methods in Physics.* Cocoyoc, 1980. Ed. by K.B. Wolf (Springer-Verlag, Lecture Notes in Physics Vol. 135, 1980) pp. 382–398.

[139] On the intrinsic shape of nuclei [with E. Chacón and S. Rubinstein]. In: *Group Theoretical Methods in Physics.* Cocoyoc, 1980. Ed. by K.B. Wolf, Springer-Verlag, Lecture Notes in Physics Vol. 135 (1980) pp. 382–398.

[140] Confrontation of macroscopic and microscopic nuclear collective models [with E. Chacón and V. Vanagas]. *J. Math. Phys.* **22,** 605 (1981).

[141] On the structure of phase space [with T.H. Seligman]. *J. Math. Phys.* **22,** 1338 (1981).

[142] Is it possible to define an intrinsic quadrupole moment? Can it be measured? Does it imply a characteristic shape? Proceedings of the 4th Winter Meeting in Nuclear Physics (Oaxtepec, México 1981).

[143] Canonical transformations relating the oscillator and Coulomb problems and their relevance for collective motions [with T.H. Seligman]. *J. Math. Phys.* **22,** 1526 (1981).

[144] Dynamical group, eigenstates and matrix elements for the collective Hamiltonian projected from a many body system [with E. Chacón]. *Kinam* **3,** 3 (1981).

[145] Confrontation of the confrontations of nuclear collective models [with O. Castaños, A. Frank, and P. Hess]. *Phys. Rev.* **C24,** 1367 (1981).

[146] Group theory and collective degrees of freedom in many body systems. In: *Proceedings of the II International Conference on Recent Progress in Many Body Theories* (Springer-Verlag, 1981).

[147] Canonical transformations and their representation in quantum mechanics. In: *Group theory and its applications in physics.* Proceedings of the 1980 Latin American School of Physics (American Institute of Physics Conference Proceedings, Vol. 71, 1981) pp.312-349.

[148] Macroscopic and microscopic nuclear collective Hamiltonians. Their symmetry groups and the canonical transformation relating them. Proceedings of the 1981 Colloquium on Group Theoretical Methods in Physics. *Physica* **114A**, 322 (1982).

[149] Microscopic derivation of nuclear collective variables [with O. Castaños, E. Chacón, P. Hess, and A. Frank]. *Phys. Rev.* **C25**, 1611 (1982).

[150] Complete set of states for microscopic nuclear collective models. [with O. Castaños, E. Chacón, A. Frank, and P. Hess]. *J. Math. Phys.* **23,** 2537 (1982).

[151] Unified approach to nuclear collective models Lecture Notes of the Granada School on the Interacting Boson Model (Springer-Verlag, 1982).

[152] Does accidental degeneracy imply a symmetry group [with C. Quesne]. *Ann. Phys.* **148,** 462 (1983).

[153] Relativistic collective variables for many body systems [with P.O. Hess, W. Greiner, and G. Schmidt]. *J. Phys. G: Nucl. Phys.* **8,** L179 (1982).

[154] Accidental degeneracies and symmetry groups. *Found. Phys.* **13,** 73 (1983).

[155] $SU(3)$ and $SU(5)$ dynamical symmetries in the extended interacting boson model [with Sun Hong-Zhou, A. Frank, and P. van Isacker]. *Kinam,* **5,** 135 (1983).

[156] A hidden symmetry in collective excitations of many body systems [with O. Castaños and A. Frank]. Group Theoretical Methods in Physics Proceedings, Istanbul, Turkey, 1983. Springer-Verlag, Lecture Notes in Physics Vol. 180, (1983).

[157] Collectivity and Geometry. I. General Approach. *J. Math. Phys.* **25,** 1555 (1984).

[158] Collectivity and Geometry. II. The two dimensional case [with E. Chacón and P. Hess]. *J. Math. Phys.* **25,** 1565 (1984).

[159] Collectivity and Geometry. III. the three dimensional case in the $Sp(6) \supset Sp(2) \times O(3)$ chain for closed shells [with O. Castaños and E. Chacón]. *J. Math. Phys.* **25,** 2815 (1984).

[160] Geometry of Nuclear Collective Motions. Proceedings of Group Theoretical Methods in Physics, Trieste, 1983. Springer-Verlag, Lecture Notes in Physics Vol. 201 (1984).

[161] Analytic expressions for the matrix elements of the generators of $Sp(6)$ in the $Sp(6) \supset U(3)$ basis [with O. Castaños and E. Chacón]. *J. Math. Phys.* **25,** 1211 (1984).

[162] Accidental degeneracies in the Zeeman effect and the symmetry groups [with N. Méndez, E. Murow, and J.W.B. Hughes]. *Ann. Phys.* **155,** 231 (1984).

[163] Pseudoatoms and atoms in strong magnetic fields [with N. Méndez and E. Murow]. *Ann. Phys.* **163,** 1 (1985).

[164] Symmetry constrained bosons and collectivity. *Nucl. Phys.* **A421,** 81 (1984).

[165] Constrained bosons for collective states in open shell nuclei [with E. Chacón and O. Castaños]. Proceedings of the XIII Colloquium on Group Theoretical Methods in Physics, University of Maryland (World Publishing Co., Singapore, 1984).

[166] Boson realization of $Sp(4)$. I. The matrix formulation [with O. Castaños, E. Chacón, and C. Quesne]. *J. Math. Phys.* **26,** 2107 (1985).

[167] Boson realization of symplectic algebras *J. Phys. A: Math. Gen.* **18,** 21 (1985).

[168] Are there boson degrees of freedom in collective shell model states? Proceedings of the Drexel Conference on Nuclear Shell Model Theory (World Publishing Co., Singapore, 1984).

[169] Generating kernel for boson realization of symplectic algebras [with O. Castaños and P. Kramer]. *J. Phys. A: Math. Gen.* **18,** L493 (1985).

[170] Boson realization of $sp(4, R)$. II. The generating kernel formulation [with O. Castaños and P. Kramer]. *J. Math. Phys.* **27,** 924 (1986).

[171] Accidental degeneracy and symmetry Lie algebras [with R. Dirl]. *J. Phys. A: Math. Gen.* **18,** 2423 (1985).

[172] Collectivity and Geometry. IV. $Sp(6) \supset Sp(2) \times O(3)$ basis states for open shells [with R.T. Sharp and M. Nicolescu]. *J. Math. Phys.* **26,** 2995 (1985).

[173] Canonical transformations to action and angle variables and their representation in quantum mechanics. IV. Periodic potentials [with J. Flores, G. López, and G. Monsiváis]. *Ann. Phys.* **172,** 156 (1986).

[174] Critical analysis of algebraic collective models. Proceedings of the Dubrovnik Conference on Nuclear Structure, Reactions and Symmetries (World Scientific Vol.2, 1986) pp. 981–986.

[175] Structure of phase space and quantum mechanics. Proceedings of the Phase Space Conference, Maryland (World Scientific, 1986).

[176] Matrix representation of the generators of symplectic algebras. I. The case of $Sp(4, R)$ [with O. Castaños]. *J. Phys. A: Math. Gen.* **20,** 513 (1987).

[177] Matrix representation of the generators of symplectic algebras. II. The general case with explicit results for $Sp(6, R)$ [with E. Chacón]. *J. Phys. A: Math. Gen.* **20** (1987).

[178] Collectivity and Geometry. V. Spectra and shapes in the two dimensional case [with E. Chacón and P. Hess]. *J. Math. Phys.* **28,** 2223 (1987).

[179] Group theory of the symplectic nuclear model. Proceedings of the Nuclear Physics Conference (Oaxtepec, 1987).

[180] Stability of deuterons in strong magnetic fields [with G. Loyola]. *Mod. Phys. Lett. A (Singapore)* **3**, 3435 (1988).

[181] Stability of nuclei in strong magnetic fields Proceedings of the Nuclear Physics Conference (Oaxtepec, 1988) p. 233.

[182] Collectivity and geometry VI. Spectra and shapes in the three dimensional case [with E. Chacón and P. O. Hess]. *J. Math. Phys.* **30**, 970 (1989).

[183] Wigner distribution functions and the representation of non–bijective canonical transformations in quantum mechanics [with R. Dirl and P. Kasperkovitz]. *J. Phys.* **A21**, 1835 (1988).

[184] Collectivity and geometry. In: *Modern Developments in Nuclear Physics*, Novosibirsk Conference on Contemporary Problems in Nuclear Structure Physics, 1987. Ed. by O.P. Sushkov (World Scientific, Singapore, 1988) p. 353.

[185] The symplectic nuclear model. Proceedings of the Varna Conference on Group Theoretical Methods in Physics. Springer-Verlag, Lecture Notes in Physics Vol. 313 (1988) p. 414.

[186] Las tres caras de la espectroscopía: atómica, nuclear y sub-nuclear [with A. Sánchez]. *Rev. Mex. Fís.* **34**, 511 (1988).

[187] Relativistic symplectic model for scalar-quark systems. Proceedings of the Conference on Space-Time Symmetries. *Nucl. Phys. B* (*Proc. Suppl.*) **6**, 76 (1989).

[188] Coherent states and accidental degeneracy for a charged particle in a magnetic field [with G. Loyola and A. Szczepaniak]. *Am. J. Phys.* **57**, 811 (1989).

[189] Accidental degeneracy and the structure of matter in strong magnetic fields [with G. Loyola and A. Szczepaniak]. In: *Symmetry in Science III* (Plenum Press, 1989) pp. 311–322.

[190] The Dirac oscillator [with A. Szczepaniak]. *J. Phys. A: Math Gen.* **22**, L817 (1989).

[191] The two body Dirac oscillator [with G. Loyola and A. Szczepaniak]. Anniversary Volume in Honor of J.J. Giambiaggi (World Scientific, Singapore, 1990).

[192] The Dirac oscillator and its contribution to the baryon mass formula [with G. Loyola, A. Szczepaniak, C. Villegas, and N. Aquino]. Proceedings of the Rio de Janeiro International Workshop on Relativistic Aspects of Nuclear Physics (World Scientific, Singapore, 1990) pp. 271–307.

[193] Symmetry Lie algebra of the Dirac oscillator [with C. Quesne]. *J. Phys. A: Math. Gen.* **23**, 2263 (1990).

[194] Art or Science: The symmetry Lie algebra for a Hamiltonian with accidental degeneracy [with C. Quesne and G. Loyola]. *Ann. Phys.* **198**, 103 (1990).

[195] Relativistic mass formula for baryons [with G. Loyola and C. Villegas]. Proceedings of the XIII Oaxtepec Conference on Nuclear Physics (1990) pp. 187–196.

[196] Anomalous basis for representations of the Poincaré Group [with G. Loyola and C. Villegas]. *J. Math Phys.* (1990).

[197] Comparison between perturbative and exact transitions induced by an interaction [with G. Loyola, and C. Villegas]. *Am. J. Phys.* (1990).

[198] Exact time dependent transitions to negative energy states, due to an interaction [with G. Loyola and C. Villegas]. Festschrift In Honor of the 70th birthday of Jayme Tiomno (World Scientific, 1990).

[199] Dynamical model for heavy ion collisions with a single resonance [with G. Loyola and C. Villegas]. *Phys. Rev. C.* (1990).

[200] Relativistic invariance of a many body system with a Dirac oscillator interaction [with G. Loyola and C. Villegas]. Proceedings of Moscow Colloquium on Group Theory and its Applications (Springer-Verlag, Lecture Notes in Physics Vol. 382, 1991.) pp. 339–345.

[201] The decay process: an exactly soluble example and its implications [with G. García Calderón and G. Loyola]. *These Proceedings.*

2

Generalizing the BCS Universal Constants to High-Temperature Superconductivity[1]

Valdir C. Aguilera-Navarro[2] and Manuel de Llano[3]

ABSTRACT Weak-coupling BCS superconductivity theory is shown to admit a more general T_c formula wherein T_c approaches zero somewhat faster than with the BCS T_c-formula. This alows recent empricial values for both organic and ceramic superconductors of the gap-to-T_c ratio larger than the universal BCS value of 3.53 not to be inconsistent with weak electron-boson coupling. Similarly, other universal constants related to specific heat and critical magnetic field will be modified.

2.1 Introduction

A variety of experimental techniques have been devised since the appearance of the BCS [1] theory of superconductivity to measure the theory's mos significant quantity, the zero-temperature gap parameter $\Delta(0)$, which plays the same role as temperature in smoothing out an otherwise sharp fermionic distribution. Weak-coupling BCS theory predicts the dimensionless energy-gap-to-transition-temperature-T_c ratio, $2\Delta(0)/k_BT_c$, where k_B is Boltzmann's constant, to be the universal dimensionless number $2\pi^{-\gamma} \simeq 3.53$ where $\gamma \simeq 0.5772$ is the Euler constant. Early tunneling experiments with elemental superconductors (maximum $T_c = 9.3K$) yielded values [2] for $2\Delta(0)/k_BT_c$ very close to the BCS value for all superconducting elements, except the presumably strongly-coupled ones like Hg and Pb which yielded ratio values somewhat larger than 3.53. Uncertainties related to the correct formalism and/or dynamical mechanism to describe the new ceramic-

---

[1]Work partially supported by FAPESP Brazil, and by an NSF short-term visit grant.

[2]Instituto de Física Teórica, UNESP, São Paulo, Brazil.

[3]On leave from Physics Department, North Dakota State University, USA.

oxide superconductors have spurred renewed vigorous interest in measuring $2\Delta(0)/k_BT_c$ by tunneling [3], far-infrared reflectivity [4], NMR [5], Raman scattering [6] and high-resolution photoemission [7]. Though there remains a wide variation in the reported values [6], [8] for the ratio —from 0 to about 13— photoemission studies [9]-[12] on the compound BiSrCaCuO ($T_c = 80$ to 110 K) suggest a value of around 8, a number also obtained for YBaCuO ($T_c \simeq 90$ K) from reflectivity work [4], with results suggesting a $\Delta(T)$ isotropic in the CuO basal plane. Enhancements found in the newer organic [13] and ceramic superconductors of the ratio above the 3.53 BCS value are commonly cited as avidence for *strong* coupling.

In this paper we show that within standard weak-coupling BCS theory, values of $2\Delta(0)/k_BT_c$ in excess of 3.53 need not be indicative of strong coupling, since this familiar value only holds when $T_c \ll \Theta_D$, the Debye temperature characteristic of the material, a condition well satisfied only in the low-temperature elemental superconductors.

2.2 Generalized BCS T_c-formula

For a system of electrons interacting pairwise via the BCS model attractive interaction of strength V, with single-spin electron density-of-states at the Fermi surface $N(0)$ [which for an ideal gas reduces to $(m/\hbar^2)^{3/2}E_F^{1/2}/\sqrt{2}\pi^2$, with E_F the Fermi energy], the usual pairing parameter is defined as $\lambda \equiv N(0)V$. The temperature- dependent gap parameter $\Delta(T)$ is then specified by the BCS gap equation [1], [14], [15]

$$1 = \lambda \int_0^{\hbar\omega_D} \frac{d\xi}{\sqrt{\xi^2 + \Delta^2(T)}} \tanh \frac{\sqrt{\xi^2 + \Delta^2(T)}}{2k_BT}, \tag{2.1}$$

which is limited to $\lambda \longrightarrow 0$. Here $\xi \equiv \varepsilon_k$, with ε_k the single-electron spectrum, μ is the chemical potential, and $\hbar\omega_D$ is the maximum possible energy of a phonon. For $T = 0$, (2.1) reduces to

$$1 = \lambda \int_0^{\hbar\omega_D} \frac{d\xi}{\sqrt{\xi^2 + \Delta^2(0)}} = \lambda \sinh^{-1}\left[\frac{\hbar\omega_D}{\Delta(0)}\right], \tag{2.2}$$

so that inverting,

$$\Delta(0) \underset{\lambda\to 0}{\longrightarrow} 2\hbar\omega_D e^{-1/\lambda}. \tag{2.3}$$

Since $\Delta(T_c) \equiv 0$, (2.1) additionally provides an implicit equation for T_c given by

$$\frac{1}{\lambda} = \int_0^{\hbar\omega_D} \frac{d\xi}{\xi} \tanh \frac{\xi}{2k_BT_c} = \int_0^{Z} dx \frac{\tanh x}{x}, \tag{2.4}$$

where $Z \equiv \hbar\omega_D/2k_BT_c \equiv \Theta_D/2T_c$ is a parameter whose magnitude is not associated *a priori* with that of the coupling parameter λ. Integrating by parts gives

Superconductor	T_C[K]	Θ_D[K]	Z
Al	1.17	385	164.5
Pb	7.2	85	5.9
YBaCuO	90	$\geq$ 288 [16]	1.6
TlCaBaCuO	125	$\geq$ 363 [17]	1.45
BiSrCaCuO	100	$\geq$ 230 [18]	1.15
RTSC	300	300	0.5

TABLE I. Critical temperatures T_c Debye temperatures Θ_D, both in kelvins [K], and the ratio $Z \equiv \Theta_D/2T_c$, for several superconductors. RTSC refers to an imaginary substance. For the cuprate compunds, the smallest reported Θ_D value is listed

$$\frac{1}{\lambda} = [\tanh x \ln x]_0^Z - \int_0^Z dx \ln x \,\mathrm{sech}^2 x, \tag{2.5}$$

a result usually [14] expressible in terms of the digamma function. If $Z \gg 1$ is assumed, then

$$\frac{1}{\lambda} \underset{Z \gg 1}{\longrightarrow} \ln Z - \int_0^\infty dx \ln x \,\mathrm{sech}^2 x \equiv \ln \frac{\Theta_D}{2T_c} - (\ln \frac{\pi}{4} - \gamma). \tag{2.6}$$

Exponentiating and solving for T_c gives the famous BCS T_c- formula [1]

$$T_c = \frac{2e^\gamma}{\pi} \Theta_D e^{-1/\lambda} \simeq 1.13 \Theta_D e^{-1/\lambda}. \tag{2.7}$$

This formula is valid for weak-coupling and vanishingly small T_c/Θ_D. Our main purpose here is to show that this latter restriction is very easily removed, leading to remarkable consequences in the physical interpretation of recent empirical findings in both organic and ceramic superconductors.

Exponentiating (2.5) without approximation, and rearranging, leads to a new (generalized) BCS T_c-formula

$$T_c = C(\Theta_D/2t_c)\Theta_D \exp\left[-\frac{1}{\lambda}\coth(\Theta_D/2T_c)\right], \tag{2.8}$$

which is really a transcendental equation in T_c. Here, the dimensionless coefficient $C(\Theta_D/2T_c)$ is defined as

$$C(Z) \equiv \frac{1}{2}\exp\left[-\coth \int_0^Z dx \ln x \,\mathrm{sech}^2 x\right] \tag{2.9}$$

$$\underset{Z \to \infty}{\longrightarrow} \frac{1}{2}\exp\left[-\int_0^\infty dx \ln x \,\mathrm{sech}^2 x\right] \equiv \frac{2e^\gamma}{\pi} \simeq 1.13. \tag{2.10}$$

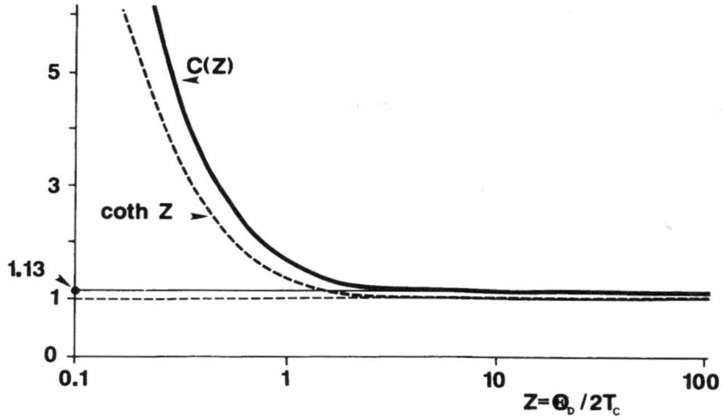

FIGURE 2.1. Illustration of how rapidly the two functions $C(Z)$, Eqs. (2.9) and (2.10), and coth Z, approach their asymptotic values of 1.13 and 1, respectively, as $Z \equiv \Theta_D/2T_C$ increases. Note that the abcscissa is on a log scale.

Thus (2.10) reduces to (2.7) when $\Theta_D/2T_c \to \infty$. This limit, however, is *not* well satisfied by the new cuprate materials, as we illustrate in Table 2.1, where the lowest reported [16]-[18] Debye temperature Θ_D was used to get the smallest possible Z for these materials. The compound "RTSC" refers to an imaginary material with $T_c = \Theta_D = 300$ K. Figure 2.1 shows a plot of $C(Z)$ *vs.* Z (full curve); it quickly reaches the asymptotic value of 1.13 for $Z \gtrsim 5$. Also displayed in the figure is the familiar function coth Z (dashed curve).

Finally, using (2.3) and (2.8), the gap-to-T_c ratio becomes the non-universal number

$$\frac{2\Delta(0)}{k_B T_c} = \frac{4}{C(\Theta_D/2T_C)} \exp\left(\frac{1}{\lambda}[\coth(\Theta_D/2T_c) - 1]\right) \qquad (2.11)$$

$$\underset{\Theta_D/2T_c \to \infty}{\longrightarrow} \frac{4}{C(\infty)} \simeq \frac{4}{1.13} \simeq 3.53. \qquad (2.12)$$

Hence, the dependence upon coupling λ cancels out in the gap-to-T_c ratio —making it a universal number— only in the precise limit stated before as being well satisfied in elemental superconductors. A non-universal gap-to-T_c ratio also emerges from strong-coupling Eliashberg theory [8], where the coupling constant is eliminated in favor of a smallness parameter similar to our Z^{-1}. The generalizationn (2.11), however, is still valid for weak-coupling. Formula (2.11) *vs.* $1/\lambda$ is graphed in Figure 2.2 for the six materials listed in Table 2.1. The value $\lambda = 1/2$ is marked as a vertical dashed line; for $\lambda > 1/2$ ($1/\lambda < 2$), simple arguments [19] would give electronn-phonon- induced lattice instability, and so are conceivably too large to even consider seriously. Open circles mark intersections with the value of 8 for the gap-to- T_c ratio reported by various experimental groups

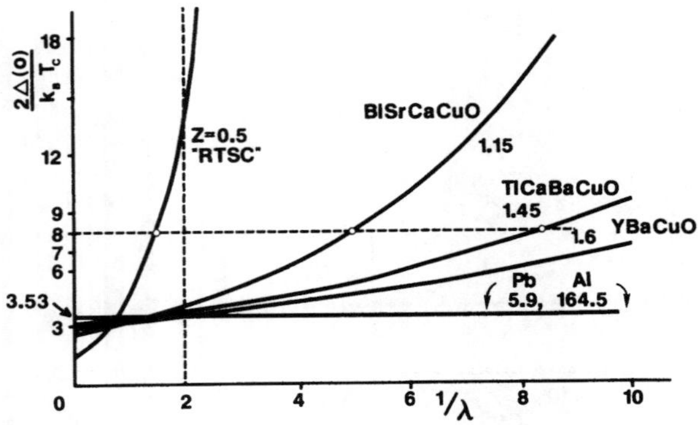

FIGURE 2.2. Variation with inverse coupling $1/\lambda$ of the gap-to-T_c ratio $2\Delta(0)/k_BT_c$ as given by Eq. (2.11), for several typical-superconductor $Z \equiv \Theta_D/2T_c$ values taken from Table 2.1. The case RTSC refers to an imaginary substance. Open circles mark intersections at a ratio value of 8 [7].

using different techniques to measure $2\Delta(0)$ [7]. Note that values in excess of the BCS value of 3.53 marked on the ordinate axis can always be obtained for sufficiently small coupling λ, so long as $Z < \infty$.

Other BCS universal constants [8], [15] (related to the critical magnetic field and to the specific heat anomaly at T_c) can also be expected to suffer generalizations similar to (2.11) for finite $\Theta_D/2T_c$. These are, with their respective weak-coupling (and low-T_c) BCS values given on the right-hand-side:

$$\frac{T_c C_n(T_c)}{H_c^2(0)} = \frac{e^{2\gamma}}{6\pi} \simeq 0.168, \tag{2.13}$$

$$\left[\frac{C_s - C_n}{C_n}\right]_{T_c} = \frac{12}{7\zeta(3)} \simeq 1.43 \tag{2.14}$$

$$\left[\frac{T}{C_n(T)}\frac{d}{dT}\{C_s(T) - C_n(T)\}\right]_{T_c} \simeq 3.77, \tag{2.15}$$

$$\min_{T/T_c}\left\{\frac{H_c(T)}{H_c(0)} - \left[1 - \left(\frac{T}{T_c}\right)^2\right]\right\} \simeq -0.037, \tag{2.16}$$

where C_n and C_s are the normal and superconducting specific heats, respectively, $H_c(T)$ the critical magnetic field, and $\zeta(3) \simeq 1.202$ the Riemannn zeta function. As (2.12), equations (2.13) through (2.16) are strictly valid only if $T_c = 0$ and their generalization to high T_c values is in progress.

2.3 Conclusion

We therefore conclude that actual values of $2\Delta(0)/k_BT_c$ in excess of the weak-coupling (and low-T_c/Θ_D) BCS value of 3.53 need not imply strong-coupling of a single (or joint) boson-exchange interaction in metallic, ceramic or organic superconductors. Our conclusion agrees with various recent experiments [20]-[22] suggesting the absence of strong-coupling in cuprate superconductors.

2.4 References

[1] J. Bardeen, L.N. Cooper, and J.R. Schrieffer, *Phys. Rev.* **108**, 1175 (1957).

[2] R. Meservey and B.B. Schwartz, *Superconductivity*, Vol. I, R.D. Parks ed. (Marcel Dekker, N.Y., 1969).

[3] R. Escudero *et al.*, *Physica* **C166**, 15 (1990).

[4] Z. Schlesinger *et al.*, *Phys. Rev. Lett.* **65**, 801 (1990); **59**, 1958 (1987).

[5] W.W. Warren, Jr. *et al.*, *Phys. Rev. Lett.* **59**, 1860 (1987); D. Brinkmann, *Physica* **C153**, 75 (1988).

[6] B. Friedl, C. Thomsen, and M. Cardona, *Phys. Rev. Lett.* **65**, 915 (1990).

[7] B. Gozz Levi, *Phys. Today*, March 1990, p. 20.

[8] J.P. Carbotte, *Rev. Mod. Phys.* **62**, 1027 (1990).

[9] C.G. Olson *et al*, Science **245**, 731 (1989).

[10] R. Manzke *et al.*, *Europhys. Lett.* **9**, 477 (1989).

[11] T. Takahashi *et al.*, *Nature* **334**, 691 (1988).

[12] J.-M. Imer *et al.*, *Phys. Rev. Lett.* **62,** 336 (1989).

[13] A. Khurana, *Phys. Today*, September 1990, p. 17 and references therein.

[14] P.B. Allen and B. Mitrovic, *Sol. State Phys.* **37**, 1 (1982); Z.M. Galasiewicz, *Superconductivity and Quantum Fluids* (Pergamon, N.Y., 1970) p. 66.

[15] A.L. Fetter and J.D. Walecka, *Quantum Theory of Many-Particle Systems*, (McGraw-Hill, N.Y., 1971).

[16] A. Junod, in *Physical Properties of High-Temperature Superconductors*, vol. II, ed. by D.M. Ginsburg (World Scientific, Singapore, 1990).

[17] A.D. Kulkarni, F.W. de Wette, J. Prade, U. Schroeder, and W. Kress, *Phys. Rev.* **B43**, 5451 (1991).

[18] R. Micnas, J. Ranninger, and S. Robaszkiewicz, *Rev. Mod. Phys.* **62**, 113 (1990).

[19] J.M. Blatt, *Theory of Superconductivity* (Academic, N.Y., 1964), p. 206.

[20] Z. Schlesinger *et al.*, *Nature* **343**, 242 (1990).

[21] S.D. Brorson *et al*, *Sol. State Comm.* **74**, 1305 (1990).

[22] C. Thomsen *et al.*, *Sol. State Comm.*.**75**, 219 (1990).

3

Fermion Clustering in an Exactly-Soluble N-Fermion Model for Hadronic, Nuclear and Superconductivity Physics

J. Dukelsky, C. Esebbag[1] and M. de Llano[2]

ABSTRACT We highlight the dynamical similarities between quark interactions, effective nucleon-nucleon interactions and the $3D$ electron fluid "jellium" model, on the one hand, and the exactly-soluble $1D$ fermion fluid with attractive delta-fuction pairwise potentials, on the other. The latter model is characterized by clustering of particles at low density and strong coupling. This motivates solving the Cooper pair problem within the model for all couplings and/or densities, a task accomplished exactly in graphical form. For *weak* coupling the essential singularity characteristic of standard $3D$ low-temperature superconductivity emerges, while for *strong* coupling BCS theory reproduces that exact ground-state system energy. The transition from large-coherence-length (low-T_c-like) Cooper pairs to small ("bipolaron", high-T_c-like) ones is smooth.

3.1 Introduction

Marcos Moshinsky's impact in clarifying the role of symmetries in the many-body problem cannot be overstated. The soluble model to be applied in this paper to superconductivity, for example, was treated almost 25 years ago by C.N. Yang (Ref. [2], below) in a paper acknowledging Moshinsky.

An assembly of $N(\gg 1)$ fermions of mass m enclosed in a one-dimensional "box" of length L and interacting via an attractive pairwise delta-

---

[1]Universidad Autónoma de Madrid, Spain.
[2]North Dakota State University, Fargo, North Dakota, USA.

function potential is described by the Hamiltonian

$$H = -\frac{\hbar^2}{2m}\sum_{i=1}^{N}\frac{d^2}{dx_i^2} - v_0\sum_{i<j}^{N}\delta(x_i - x_j), \qquad v_0 > 0. \tag{1.1}$$

Given the number density $\rho \equiv N/L$, introducing the dimensionless coordinates $x_1' \equiv \rho x_i$ leads to the dimensionless Hamiltonian

$$H' \equiv \frac{mH}{\hbar^2\rho^2} = \frac{1}{2}\sum_{i=1}^{N}\frac{d^2}{dx_i'^2} - \lambda\sum_{i<j}^{N}\delta(x_i' - x_j'), \tag{1.2}$$

where $\lambda \equiv mv_0/\hbar^2\rho$ is a dimensionless coupling parameter, and is clearly the *only* variable upon which thhe ground-state properties of (1.2) can depend. Its range of variation is $0 \leq \lambda < \infty$, and the N-fermion system is seen to have the peculiar dynamics whereby high (low) particle density is associated with weak (strong) interaction. This is in marked contrast with many familiar $3D$ quantum fluids such as the liquid heliums (both ^{3}He and ^{4}He), nuclear and neutron matter, etc., where the pair interaction $V(r)$ is short-ranged attractive and repulsive at even shorter range. Figure 3.1, top panel, schematically displays the ground-state energy-per-particle of either liquid ^{3}He or nuclear matter as function of number density ρ. If the N-fermion system (1.1) consists of ν distinct species (*e.g.*, $\nu = 2$ if we have up and down spins possible), the Bethe *ansatz* [1] shows that (1.2) has only *one* bound state which at zero density has a total energy given [2] simply by

$$E_0(N) = [N/\nu]E_0(\nu), \tag{1.3}$$

where $[N/\nu]$ is the nearest integer, from below, to N/ν, and

$$E_0(\nu) \equiv -mv_0^2\,\nu(\nu^2 - 1)/24\hbar^2, \tag{1.4a}$$

$$\underset{\nu=2}{\longrightarrow} -mv_0^2/4\hbar^2 \equiv E_0. \tag{1.4b}$$

Equation (1.4b) gives the well-known elementary $1D$ quantum-mechanical result for the (sole) ground-state energy of two-particles of mass m (or one of reduced mass $m/2$) interacting with an attractive delta-function potential of strength v_0. The result (1.4a) is also the exact total ground-state energy of $\nu(\gg 1)$ bosons [3] with the Hamiltonian (1.1) where N is replaced by ν. The energy-per-boson is *not* extensive since it clearly collapses to $-\infty$ in the thermodynamic limit $\nu \to \infty$. Such a collapse is prevented if the particles are fermions by the Pauli principle. In $3D$, however, this is *not* the case. A determinant of plane waves (with wave vectors $\mathbf{k}$ occupying a Fermi sphere in k-space of radius k_F) used as a trial function readily gives for high density an expectation value of (1.1) in $3D$ of the form $N\{A\rho^{2/3} - \frac{1}{2}\rho v_0\}$, with A a positive constant. This being a rigorous upper bound for the exact $E_0(N)$ (with ρ in fact a variational parameter

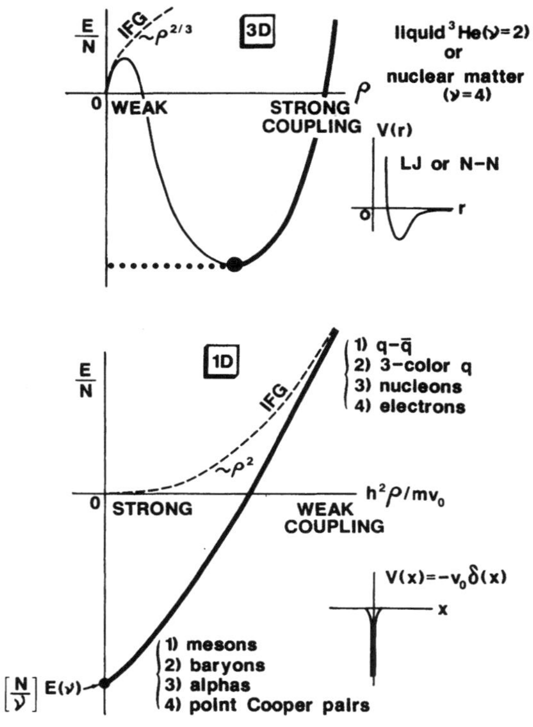

FIGURE 3.1. Comparision of ground-sate energy-per-particle *vs.* density curves for a tipycal $3D$ fermi liquid (like ^{3}He with 2 species or nuclear amtter with 4) [top panel] and for $1D$ many-fermion fluid defined by Eqs. (1.1) or (1.2) [bottom panel] IFG refers to the ideal Fermi gas.

that can be taken arbitrarily large) proves that the exact ground-state total energy collapses to negative infinity.

The bound N-fermion wave function is given [4] by

$$\begin{aligned} F &\quad (1,2,\ldots,\nu)F(\nu+1,\nu+2,\ldots,2\nu)\cdots F(N+1-\nu,\ldots,N-1,N) \\ \pm &\quad \text{(all permutations of indices belonging to different } F\text{'s)}, \end{aligned} \tag{1.5}$$

where

$$F(1,2,\ldots\nu) \equiv \prod_{i<j}^{\nu} \exp\left\{-\frac{mv_0}{2\hbar^2}|x_i-x_j|\right\} \chi(\sigma_1,\sigma_2,\ldots\sigma_\nu) \tag{1.6}$$

is totally-antisymmetric, since χ is an anti-symmetric "spin" function, and the space part is symmetric, each under the exchange of two particle indices from the set $1 \leq i,j \leq \nu$.

Thus, for $\rho = 0$ ($\lambda = \infty$) the system consists of an *ideal gas* of noninteracting $[N/\nu]$ "cluster" (composite) particles, each made up of ν distinct fermion species. It is remarkable that shortly before the word "soliton" was coined [5], McGuire showed [2] by using S-matrix techniques that the "clusters" of (1.1) have all the properties of solitons, namely: when scattered one from another their constituent fermions may rearrange and the clusters will be phase-shifted but they otherwise emerge intact (*i.e.*, there is no cluster "breakup" in collisions).

On the other hand, for $\lambda = 0$ ($\rho = \infty$) the ground-state N- particle energy of (1.1) is just the ideal Fermi gas limit

$$E_0(N) = \frac{N}{3}\frac{\hbar^2 k_F^2}{2m} = N\frac{\hbar^2\pi^2}{6m\nu^2}\rho^2, \tag{1.7}$$

since the "Fermi sea" consists of the interval in $1D$ k-space given by $[-k_F, k_F]$, and consequently $\rho = N/L = \nu k_F/\pi$. Figure 3.1, bottom panel, schematically summarizes the expected energy-per-particle as function of density for the $1D$ model Hamiltonian (1.1).

For $0 < \lambda < \infty$ (implying $\infty > \rho > 0$) the ground-state energy as a function of λ can be obtained numerically [6] by solving the so-called Gaudin equations [7]. In Figure 3.2 we plot the resulting dimensionless equation-of-state (EOS) for the model (thick curve) for $\nu = 2$ species. In Figure 3.2, $\hbar = m = 1$ was used, so that the ascissa refers to λ^{-1}. Dashed curves are rigorous *lower* bounds [8] to the energies both of the exact problem [Schrödinger equation (1.1)] and to the lowest possible energy state in the Hartree-Fock approximation. The three full thin curves refer to various mean-field results, both Thomas-Fermi and Hartree-Fock as discussed in more detail in Refs. [4] and [6]. Note that both rigorous bounds, the upper (HF) and the lower one, are unsatisfactorily far from the exact (full thick curve) EOS for intermediate values of λ. Before proceeding, we discuss all known applications of the model (1.1).

3.1.1 Hadronic and Nuclear Physics

Table I summarizes some recent applications of the $1D$ N-fermion model (1.1) to hadronic [9] and nuclear [10] physics. If the N-fermion assembly is composed of 2 species, quark and antiquark ($q, \overline{q}$) when $\lambda = 0$ ($\rho = \infty$), the resulting $\lambda = \infty$ ($\rho = 0$) clusters are *mesons*. If we begin with 3-color quarks (q_1, q_2, q_3) as ideal fermion gas constituents, color-neutral baryons form as the coupling increases. Finally, if the fermion gas is noninteracting nuclear matter with protons and neutrons, each of spin up or down, then $\nu = 4$ and *alpha particles* are formed as the coupling is increased. Alpha-particles clustering at low densities has also been found in *three* dimensions via variational studies [11] employing (effective) Skyrme nucleon-nucleon forces. The usefulness of the $1DN$-fermion model in hadronic physics becomes evident from the fact that quark interactions also share with the soluble model the same peculiar dynamics, namely, they interact weakly (strongly) when close together (far apart).

ν (# of species)	fermions	clusters
2	$q, \overline{q}$	mesons
3	3-color q's	baryons
4	nucleons	alphas

TABLE I. Constituent fermions ($\lambda = 0$: v_0 and/or $\rho = \infty$) and ν-fermion clusters ($\lambda = \infty$: $v_0 = \infty$ and/or $\rho = 0$) of the many-body Hamiltonian (1.1) or (1.2).

The field of hadronic physics has grown rapidly since the discovery of the European Muon Collaboration (EMC) effect in deep inelastic lepton scattering from nuclei. This effect is based on the experimental finding that quark momentum distributions in nuclei differ significantly from those in free nucleons. From several interpretations given, we emphasize here the modifications in quark structure of the nucleon due to quark exchange and quark-quark interactions between different nucleons, which may be qualitatively understood as nucleon "swelling". As one cannot derive even the nucleon from QCD, one must resort to *effective* quark interactions to describe hadrons. By means of these effective theories it is possible to describe hadron structure, hadron-hadron interactions, quark exchange effects in the nuclear medium, etc. But a complete description of nuclear systems from quark degrees of freedom is still only possible for very simple quark models. In spite of the simplicity of these models, it is instructive to investigate them as a guide to more elaborate and realistic approaches.

Recently, the $1D$ N-fermion Hamiltonian (1.1) has been used [9] as a model for nuclear matter in terms of quarks. One starts with N nonrelativistic fermions (quarks) with $SU(3)$ color as the only internal degree of freedom and moving in one dimension. The quark-quark interaction is the same as in (1.1) but with $v_0\delta(x_i - x_j)$ replaced by $\frac{3}{8}g\lambda_i\lambda_j\delta(x_i - x_j)$ where λ are the color $SU(3)$ matrices. Although the zero-range delta potential does not resemble too much the one gluon exchange potential, it produces the desired quark clustering effect. A system of N quarks interacting through (1.1) may then form bound clusters each of which are symmetric in configuration space and antisymmetric in color space. Color antisymmetry limits each cluster size to two or three particles. Three-quark clusters are not only energetically favorable over two-color ones but also colorless in agreement with QCD. The center of mass energy of each cluster is $E = -Mg^2$ with M the quark mass. The intrinsic wave function is given by (6) with $\nu = 3$, and χ the totally antisymmetric color wave function. The general solution for the ground state of a large N-quark system is again obtained by using the Bethe *ansatz*, imposing periodic boundary conditions in a length L which will finally tend to infinity but with finite density. A modified version of the Gaudin [7] equations then follows [9].

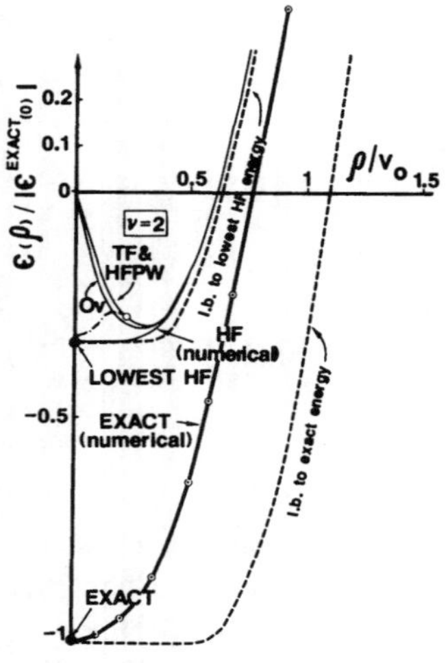

FIGURE 3.2. Exact ground-state EOS (in dimensionless units) of (1.1) or (1.2), full thick curve, obtained numerically. Dashed curves refer to rigorous lower bounds [8]; thin full curves to the Hartree-Fock approximation [4], [8], and to the Thomas-Fermi calculation discussed in Ref. [6].

There are two interesting limits where an analytic solution for the ground state energy can be obtained. At low density the system behaves as a Fermi gas of nucleons, and at high density as a Fermi gas of quarks. The numerical solution of the equations shows a smooth transition between the confined (low density) and the deconfined phase (high density). At low density the separation distance between three-quark clusters is much larger than their average size, and so the overlap is negligible. For this reason there is neither an effective repulsion due to quark exchange nor due to the direct delta interaction, and the clusters move freely inside the nuclear medium. At intermediate density the clusters partially overlap, resembling in a very crude manner the low-energy nuclear matter regime. An exchange between clusters, made possible by this non-zero overlap, modifies not only the nuclear properties but also the nuclear structure.

The Bethe *ansatz* is able to provide the exact lowest eigenvalue as a function of the density, but no additional information on either nuclear or nucleon structure. It would be very interesting to develop a many-body theory for three-body clustering to describe the intermediate density region which can be checked against the exact ground-state energy. Such an approximation has already been proposed within the context of the Petry model [12]. These approaches, when generalized to a realistic three-dimensional problem may be able to account f or the interplay of nucleon and nuclear structure, which is one of the most outstanding open problems in the borderline between nuclear and particle physics.

3.1.2 Solid State Physics

The simplest model of a $3D$ metal is the familiar "jellium" model [13] consisting of N point electrons of mass m moving in a rigid uniform background of equal but positive charge. The Hamiltonian is

$$H = -\frac{\hbar^2}{2m}\sum_{i=1}^{N}\nabla_i^2 + V_{bb} + V_{be} + V_{ee}, \tag{1.8}$$

where the first term is the N-electron kinetic energy and the three remaining terms are the (positive) background-background repulsive potential energy V_{bb}, the (negative) background-electron attraction V_{be}, and the (positive) electron-electron repulsion $V_{ee} \equiv \sum_{i<j}^{N} e^2/r_{ij}$. The Hamiltonian (1.8) can be reduced [14] to only *two* terms

$$H = -\frac{\hbar^2}{2m}\sum_{i=1}^{N}\nabla_i^2 + \sum_{i<j}^{N}{}'\frac{e^2}{r_{ij}}, \tag{1.9}$$

where the prime on the second summation means that in a Fourier decomposition of the interaction e^2/r_{ij}, the zero-momentum component [the (diverging) volume integral of the pair interaction] is put equal to zero. Letting $\rho \equiv N/L^3 \equiv (\frac{4}{3}\pi r_0^3)^{-1}$, we can define dimensionless coordinates $(x', y', z') \equiv (x/r_0, y/r_0, z/r_0)$ and $r' \equiv r/r_0$ to introduce the dimensionless Hamiltonian

$$H' \equiv a_0 r_s^2 H/e^2 = -\frac{1}{2}\sum_{i<j}^{N}\nabla_{r_i'}^2 + r_s\sum_{i<j}^{N}{}'\frac{1}{r_{ij}'}, \tag{1.10}$$

where $a_0 \equiv \hbar^2/me^2$ is the Bohr radius, and the dimensionless coupling parameter r_s is defined as

$$0 \le r_s \equiv r_0/a_0 \equiv \frac{me^2}{\hbar^2(\frac{4}{3}\pi\rho)^{1/3}} < \infty. \tag{1.11}$$

Evidently, we again encounter the same dynamics as in the $1DN$-fermion model, namely, strong (weak) coupling [large (small) e^2] is associated with low (high) density ρ. The $1DN$-fermion model may thus also be useful in tracing the evolution [15] from *large* weakly-coupled, overlapping "Cooper pairs" of electrons to the strong-coupling extreme of *small* tightly-bound "local pairs" (or "bipolarons") which appear to play an increaingly important role in recent theories [16] of high-temperature superconductivity where much smaller Cooper-like pairs are characteristic. Table II summarizes this application of the model, which was first considered by Takahashi [17] and which is the main focus of this paper. In the next section, we discuss this application in greater detail.

ν (# of species)	fermions	clusters	
2	electrons	dimers	Cooper pairs ($\lambda \to 0$)
			local pairs ($\lambda \to \infty$) ("bipolarons")

TABLE II. Same as Table I but for "electrons".

3.2 Cooper pairing

3.2.1 BCS Model Interaction in 3D

The Cooper fermion-pair problem [18] consists in solving the Schrödinger equation (in the momentum representation) for two fermions that interact via a potential $V(r)$ and that cannot scatter by phonon exchange into single-fermion states already occupied by the $N-2$ background fermions. The two fermions have opposite spins and zero center-of-mass motion. Thus, if $\mathbf{r} \equiv \mathbf{r}_1 - \mathbf{r}_2$, $\mathbf{K} \equiv \mathbf{k}_1 + \mathbf{k}_2 = 0$ so that $\mathbf{k}_1 = -\mathbf{k}_2 \equiv \mathbf{k}$, and the intrinsic space wave function is $\psi(\mathbf{r}) = \sum_k C_k e^{i\mathbf{k}\cdot\mathbf{r}}$. The coefficients $C_\mathbf{k}$ then satisfy

$$2\epsilon_k^0 C_k + \sum_k{}' C_k' V_{k',k} = E C_k, \quad C_k \equiv 0 \quad (k < k_F), \tag{2.1}$$

where $\epsilon_k^0 \equiv \hbar^2 k^2/2m$, E is the pair energy eigenvalue and, if L^D is the D-dimensional system volume,

$$V_{k',k} \equiv L^{-D} \int d^D r\, e^{-i\mathbf{k}'\cdot\mathbf{r}} V(\mathbf{r}) e^{i\mathbf{k}\cdot\mathbf{r}}. \tag{2.2}$$

Cooper used what is now called the "BCS model interaction": $V_{k',k} = -V$ if $E_F < \epsilon_k^0,\ \epsilon_{k'}^0 < E_F + \hbar\omega_D$ and $= 0$, otherwise, where $\hbar\omega_D$ is the maximum energy a phonon can have, and $V > 0$. This simplification lead him to the eigenvalue equation

$$1 = V \sum_k{}' \frac{1}{2\epsilon_k^0 - E} \tag{2.3}$$

where the prime on the summation sign means the $\mathbf{k}$-sum is restricted so that $E_F < \epsilon_k^0 < E_F + \hbar\omega_D$. The sum can be converted to an energy integral if the density of states $g(\epsilon)$ [$\equiv (L/2\pi)^D d^D k/d\epsilon$] is introduced, so that (2.14) becomes

$$1 = V \int_{E_F}^{E_F+\hbar\omega_D} d\epsilon \frac{g(\epsilon)}{2\epsilon - E} \simeq V g(E_F) \int_{E_F}^{E_F+\hbar\omega_D} \frac{d\epsilon}{2\epsilon - E}, \tag{2.4}$$

$$1 = V \int_{E_F}^{E_F+\hbar\omega_D} d\epsilon \frac{g(\epsilon)}{2\epsilon - E} \simeq V g(E_F) \int_{E_F}^{E_F+\hbar\omega_D} \frac{d\epsilon}{2\epsilon - E}, \tag{2.4}$$

where the last step follows in $3D$ from the empirical fact that in metals $\hbar\omega_D \ll E_F$. Note that this reduces the problem to a $2D$ one, where $g(\epsilon)$ is a constant independent of ϵ. The remaining integral is elementary and gives

$$E = 2E_F - \frac{2\hbar\omega_D}{e^{1/\lambda} - 1} \underset{\lambda\to 0}{\longrightarrow} 2E_F - 2\hbar\omega_D e^{-1/\lambda}, \qquad (3D) \tag{2.5}$$

where $\lambda \equiv g(E_F)V/2$.

Although in arriving at (2.6) we have started with the problem in 3D, it can be shown [19] that the essential singularity $e^{-1/\lambda}$ at $\lambda = 0$ *also emerges in* $1D$. This means that it is *unrelated* to the $2D$ property $\hbar\omega_D \ll E_F$ of the BCS interaction model —which restricts the interaction to the immediate vicinity of the Fermi surface— contrary to what might be expected from ordinary $2D$ quantum binding [20] where a similar essential singularity appears, and which seems to be the conventional wisdom.

3.2.2 Attractive Delta Potential Model in $1D$

We wish to solve the same problem in $1D$ for the interaction model (1.1), namely $V(r) = -v^0\delta(x)$, with $v_0 > 0$ and $x \equiv x_1 - x_2$. Then (2.3) gives $-v_0/L$ and instead of (2.4) we have

$$1 = \frac{v_0}{L} \sum_{k}{}' \frac{1}{2\epsilon_k^0 - E}, \tag{2.6}$$

with the prime now meaning *only* that $E_F < \epsilon_k^0 < \infty$. If this restriction were removed, then (2.7) becomes for $E < 0$

$$1 = \frac{v_0}{L}\frac{L}{2\pi} \int_{-\infty}^{\infty} \frac{dk}{\hbar^2 k^2/m + |E|} = \frac{v_0}{2\hbar}\sqrt{\frac{m}{|E|}}, \tag{2.7}$$

which on squaring gives the exact two-body Schrödinger result (1.4*b*) for the eigenvalue E, as expected.

But in the corresponding $1D$ Cooper problem, instead of (2.4) we now have exactly

$$1 = \frac{v_0}{L} C_1 \int_{E_F}^{\infty} d\varepsilon \frac{\varepsilon^{-1/2}}{2\varepsilon - E}, \tag{2.8}$$

where $g(\varepsilon) = C_1 \varepsilon^{-1/2}$ is the $1D$ density of states with $C_1 \equiv \sqrt{m/2}\, L/\pi\hbar$. Introducing the dimensionless $\epsilon \equiv E/2E_F$, the integral in (2.19) gives

$$\frac{1}{2\sqrt{E_F}} \frac{1}{\sqrt{\epsilon}} \ln\left(\frac{1+\sqrt{\epsilon}}{1-\sqrt{\epsilon}}\right) \quad \text{if } \epsilon > 0, \tag{2.9}$$

$$\frac{1}{\sqrt{E_F}}\frac{1}{\sqrt{|\epsilon|}}\left(\frac{\pi}{2}-\arctan\frac{1}{\sqrt{|\epsilon|}}\right) \quad \text{if } \epsilon < 0. \tag{2.10}$$

The Fermi energy $E_F \equiv \hbar^2 k_F^2/2m$ is related to the number density $\rho = N/L = 2k_F/\pi$, and since $\lambda \equiv mv_0/\hbar^2\rho$ the prefactor in (2.9) is then just $2\lambda\sqrt{E_F}/\pi^2$, so that (2.9), (2.10) and (2.11) lead to the *two* implicit transcendental equations for ϵ, namely

$$\pi^2\sqrt{\epsilon}/\lambda = \ln\left(\frac{1+\sqrt{\epsilon}}{1-\sqrt{\epsilon}}\right) \quad \text{if } \epsilon > 0, \tag{2.11}$$

$$\pi^2\sqrt{|\epsilon|}/\lambda = \pi - 2\arctan\frac{1}{\sqrt{|\epsilon|}} \quad \text{if } \epsilon < 0. \tag{2.12}$$

Both right-hand side members behave like $2\sqrt{|\epsilon|}+0(|\epsilon|^{3/2})$ for small $|\epsilon|$, but (2.12) is concave up in ϵ while (2.13) is concave down in $|\epsilon|$. Consequently, there will be *non trivial* ($|\epsilon| > 0$) *solutions* (NTS) whenever

$$0 < \lambda < \pi^2/2, \qquad \text{for } \epsilon > 0, \tag{2.13}$$

$$\pi^2/2 < \lambda \qquad \text{for } \epsilon < 0, \tag{2.14}$$

with the value $\lambda = \pi^2/2$ corresponding to $\epsilon = 0$. Figure 3.3 depicts the graphical solutions, where full curves refer to the right-hand side (thick curve) and left-hand side (thin curve) of (2.12), while dashed curves refer to the right-hand side (thick dashed curve) and left-hand side (thin dashed curve) of (2.13). The dot-dashed straight line is the asymptote $2\sqrt{|\epsilon|}$ of both right-hand side members of (2.12) and (2.13) for small $|\epsilon|$. Consider The two "Leggett extremes" [15] of $\lambda \to 0^+$ and $\lambda \to \infty$, wich are, respectively, the weakly-bound, strongly-overlapping Cooper pair limit and the tightly-bound (point boson) "dimer" limit. Let $\Delta \equiv 2E_F - E$ be the (positive) binding energy of the wearkly-bound Cooper pair. Defining $\delta \equiv \Delta/2E_F$, (2.12) can be expanded for $1-\epsilon \equiv \delta \to 0^+$ and yields $\pi^2/\lambda \simeq -\ln(\delta/4)$ or

$$\delta \underset{\lambda\to 0}{\simeq} 4e^{-\pi^2/\lambda}, \tag{2.15}$$

meaning that

$$\begin{aligned} 2E_F - E \equiv \Delta &\simeq 8E_F e^{-\pi^2\hbar^2\rho/mv_0} \\ &\neq \Delta_0 + \Delta_1 v_0 + \Delta_2 v_0^2 + \cdots. \end{aligned} \tag{2.16}$$

The same type essential singularity in coupling (2.15) also appears in the $3D$ Cooper pair problem [18], in the $3D$ many-electron BCS theory gap parameter and, finally, in the BCS superconductive transition temperature [14]. On the other hand, $\lambda \equiv mv_0\ \hbar^2\rho \to \infty$ is equivalent to finite v_0 and ρ (or $E_F \to 0$, and so it should correspond to treating the two fermions *in a vacuum*. In this case (2.13) becomes, since $|\epsilon| \equiv E/2E_F \to \infty$,

$$\frac{\pi^2\sqrt{|\epsilon|}}{\lambda} = \pi - \frac{2}{\sqrt{|\epsilon|}} + \cdots, \quad \sqrt{|\epsilon|} \simeq \lambda/\pi, \tag{2.17}$$

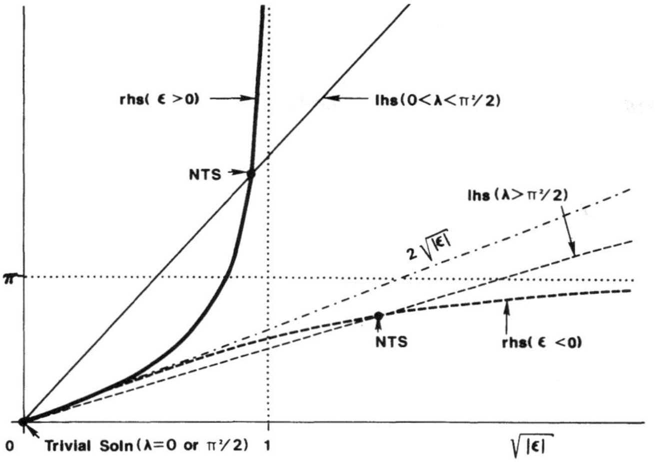

FIGURE 3.3. Right- and left-hand side of (2.12) (full curves) and of (2.13) (dashed curves), plotted against $\sqrt{|\epsilon|}$. Intersections denote nontrivial solutions for both positive and negative $\epsilon \equiv E/2E_F$ values. The dot-dashed line is the small $\sqrt{|\epsilon|}$ asymptote of the right-hand side of both (2.12) and (2.13).

which upon squaring again gives $E = E_0(2)$ of (4*b*), as it should.

The Leggett question [15] of "what happens to a Cooper pair as density is decreased" is here trivially but explicitly answered: it becomes a point bound "dimer" (or "bipolaron"). More significantly, this "dimerization" can also be seen to occur when *all N* fermions (not just Cooper pair partners) are treated on an equal footing, as in the BCS many-fermion theory to which we now turn.

3.2.3 BCS Gap and Number Equations

For the derivation of the two fundamental equations of the BCS theory of an interacting many-fermion system is any dimension we follow the notation of Fetter and Walecka's text [14]. That discussion is more general than most in that does not assume the chemical potential μ to coincide with the value E_F (which depends on ρ but not on v_0), a result occurring only for weak coupling. For the Hamiltonian (1.1) the interaction matrix elements $\langle k_1 k_2 | V | k_3 k_4 \rangle$ are simply $(v_0/L)\delta_{k_1+k_2,k_3+k_4}$, and the HF single-particle energies ϵ_k then become

$$\epsilon_k = \epsilon_k^0 - \tfrac{1}{2}\rho v_0, \tag{2.18}$$

where as before $\epsilon_k^0 \equiv \hbar^2 k^2/2m$. Putting $\xi_k \equiv \epsilon_k - \mu$, the gap parameter Δ is defined as

$$\Delta \equiv \frac{v_0}{L}\sum_k v_k \sqrt{1 - v_k^2}, \tag{2.19}$$

where the occupation probability v_k^2 in a state k (for *one* spin direction) is given by

$$v_k^2 \equiv \tfrac{1}{2}(1 - \xi_k/E_k), E_k \equiv \sqrt{\Delta^2 + \xi_k^2}. \tag{2.20}$$

Thus, (3.2) simplifies to the *gap equation*

$$1 = \frac{v_0}{2L}\sum_k \frac{1}{E_k} \longrightarrow \frac{v_0}{4\pi}\int_{-\infty}^{\infty} dk \left[\left(\frac{\hbar^2 k^2}{2m} - \frac{1}{2}\rho v_0 - \mu\right)^2 + \Delta^2\right]^{-1/2}. \tag{2.21}$$

Simultaneously, the *number equation* (which determines μ) becomes

$$N = 2\sum_k v_k^2 = \sum_k (1 - \xi_k/E_k) \longrightarrow \frac{L}{2\pi}\int_{-\infty}^{\infty} dk \left[1 - \frac{\epsilon_k - \mu}{\sqrt{(\epsilon_k - \mu)^2 + \Delta^2}}\right]. \tag{2.22}$$

The two coupled, implicit equations (3.5) and (3.6) must be solved (numerically, at worst) to determine both Δ and μ as functions of the dimensionless coupling parameter λ. This would ultimately allow the total energy-per-particle (which for $\lambda \to 0$ is well-known [14] to be quadratic in Δ) to be calculated in the BCS approximation for all $0 < \lambda < \infty$. This computation is in progress, but we report the result for the two extremes $\lambda = 0$ and $\lambda = \infty$ which is straightforward.

For $\rho \to 0$ (3.6) implies that $v_k \to 0$ which from (3.2) means that $\Delta \to 0$. The gap equation (3.5) then becomes an implicit equation for μ alone,

$$1 = \frac{v_0}{2\pi}\int_0^{\infty} dk \frac{1}{\hbar^2 k^2/2m - \mu}, \tag{2.23}$$

which, if $\mu = -|\mu|$, gives the same type of integral as in (2.8), so that we immediately find

$$\mu = -\frac{1}{2}|E| \equiv -\frac{1}{2}\frac{mv_0^2}{4\hbar^2} = \frac{1}{2}E_0(2), \tag{2.24}$$

i.e., μ in this limit is *one-half* the energy of a single cluster $E_0(2)$ as given by (1.4b). One can further see that the exact $\rho \to 0$ ($\lambda \to \infty$) total energy result (1.3) is fully reproduced in the BCS approximation. This comes from the fact that the BCS μ given by (3.8) is indeed the *exact* value of μ in this limit. The latter assertion follows from (1.3) and the Hugenholtz-van Hove theorem [21] stating that the ground-state energy-per-particle $E_0(N)/N$ of an interacting many-fermion system at zero pressure ($P \equiv \rho^2\partial[E_0(N)/N]/\partial\rho = 0$) is identical with its chemical potential μ. [A result equivalent to the first equality in (3.8) has been obtained [22] in the zero-density limit of the BCS theory of a $2D$ fermion gas with arbitrary short-ranged pair interactions.] Finally, in the other (trivial) extreme $\lambda \to 0$, or $v_0 \to 0$, from (3.2) Δ again vanishes. This reduces v_k^2 in (3.3) to the step function $\theta(\mu - \hbar^2 k_F^2/2m)$, which in (3.6) with $\rho \equiv N/L = 2k_F/\pi$ integrates to give $\mu = \hbar^2 k_F^2/2m \equiv E_F$.

3.3 Conclusions

The exactly-soluble many-fermion system with pairwise attractive delta-function interactions in $1D$ possesses the same qualitative dynamics as a quark fluid or an electron fluid "jellium" model in $3D$. As such, it may be extremely valuable in testing more realistic $3D$ models of quark confinement in hadronic, or of electron correlations in solid-state, physics.

The Cooper pair problem can be solved for the model exactly for all coupling and density in graphical form. The weak-coupling and/or high-density extreme is characteirzed by the same type essential singularity of the $3D$ Cooper problem with the BCS model interaction, of the many-electron BCS theory gap-parameter and of the BCS critical-temperature formula of low-temperature superconductivity.

It will be interesting to solve, numerically if necessary, the BCS theory equations to compute the ground-state energy and compare with the exact results for all values of coupling and/or density. However, one already notes that perfect coincidence occurs at *both* extremes, namely: *(a)* of weakly- bound, strongly-overlapping Cooper pairs, and *(b)* of tightly-bound, point ("bipolaronic") dimers. As extremes *(a)* and *(b)* seem to characterize low- and high-temperature superconductivity, respectively, the model should prove of great usefulness in the study of the highly-correlated electron fluid presumably responsible for these phenomena.

3.4 Acknowledgements

M. de Ll. thanks Professor S.A. Moszkowski for many discussions.

3.5 References

[1] H.A. Bethe, *Z. Phys.* **71**, 205 (1931).

[2] J.B. McGuire, *J. Math. Phys.* **5**, 622 (1964); C.N. Yang, *Phys. Rev.* **168**, 1920 (1968).

[3] F. Calogero, A. Degasperis, *Phys. Rev.* **A 11**, 265 (1975).

[4] V.C. Aguilera-Navarro, E. Ley-Koo, M. de Llano, S.M. Peltier, and A. Plastino, *J. Math. Phys.* **23**, 2439 (1982).

[5] N.J. Zabusky and M.D. Kruskal, *Phys. Rev. Lett.* **15**, 240 (1965).

[6] M. Casas, C. Esebbag, A. Extremera, J.M. Getino, M. de Llano, A. Plastino, and H. Rubio, *Phys. Rev.* **A** (in press).

[7] M. Gaudin, *Phys. Lett.* **A 24**, 55 (1967).

[8] E.H. Lieb and M. de Llano, *J. Math. Phys.* **19**, 860 (1978).

[9] D.S. Koltun, *Phys. Rev.* **C 36**, 2047 (1987); D.S. Koltun and S. Tosa, *Phys. Lett.* **B 172**, 267 (1986); S. Tosa, *Phys. Rev.* **C 34**, 2302 (1986).

[10] T. Kebukawa, *Phys. Rev.* **C 35**, 794 (1986).

[11] M. de Llano, *Nucl. Phys.* **A 317**, 183 (1979).

[12] S. Pittel, J. Engel, and J. Dukelsky, *Phys. Lett.* **B 247**, 185 (1990).

[13] E.P. Wigner, *Phys. Rev.* **46**, 1002 (1934).

[14] A.L. Fetter and J.D. Walecka, *Quantum Theory of Many-Particle Systems* (McGraw-Hill, New York, 1971) p. 21 and p. 330 ff.

[15] A.J. Leggett, *J. Phys. (Paris)* **41**, C7 (1980).

[16] N.F. Mott, *Contemp. Phys.* **31**, 373 (1990); R. Micnas, J. Ranninger, and S. Robaszkiewicz, *Rev. Mod. Phys.* **62**, 1 (1990); R. Friedberg and T.D. Lee, *Phys. Rev.* **B 40**, 6745 (1989); *Phys. Lett.* **A 152**, 417 (1991); R. Friedberg, R.D. Lee, and H.C. Ren, *Phys. Lett.* **152**, 423 (1991).

[17] M. Takahashi, *Prog. Theor. Phys.* **44**, 348 (1970).

[18] L.N. Cooper, *Phys. Rev.* **104**, 1189 (1956).

[19] C. Esebbag, J.M. Getino, M. de Llano, S.A. Moszkowski, U. Oseguera, A. Plastino, and H. Rubio, *The essential singularity in Cooper pairing in one-, two-, and three-dimensions* (to be published).

[20] L.D. Landau and E.M. Lifshitz, *Quantum Mechanics* (Pergamon, New York, 1977) p. 163.

[21] N.M. Hugenholtz and L. van Hove, *Physica* **24**, 363 (1958).

[22] K. Miyake, *Prog. Theor. Phys.* **69**, 1794 (1984); M. Randeria, J.-M. Duan, and L.-Y. Shiek, *Phys. Rev. Lett.* **62**, 981 (1989); *Phys. Rev.* **B. 41**, 327 (1990).

4

The Scattering Approach to Quantum Electronic Transport

Pier A. Mello[1]

ABSTRACT Some of the efforts to describe the problem of quantum electronic transport through a scattering approach are reviewed. Two-terminal devices are mainly considered. The key object in the scattering approach is the transfer matrix M, whose dimensionality is $2N$, N being the number of channels, or transverse modes; M is multiplicative, so that the M for the total system is the product of the M's of the various scatterers. The various statistical distributions are assumed *isotropic*, a multichannel generalization of the familiar random-phase model in one dimension. For a given distribution of the n individual M's, the total M has a distribution that, as n grows, tends to a limit, depending only on one parameter: the elastic mean free path ℓ; the limiting distribution is universal, in the sense that it has a unique form once ℓ is specified. One obtains excellent results for the weak-localization correction to the conductance, universal conductance fluctuations, the backscattering enhancement, long-range correlations of reflection or transmission coefficients and time-symmetry breaking by a magnetic field. An attempt to relax the assumption of isotropy, and also an extension to a three-terminal 1-dimensional device are briefly described.

4.1 Introduction

The last few years have seen spectacular experiments [1]-[6] realized with "mesoscopic" devices, as well as important efforts to provide a theoretical explanation [7]-[29] of the observed phenomena.

Quantum mechanics teaches us that a perfect crystal would be a perfect electric conductor. If a nonzero resistivity ρ is observed, it is because of deviations from a perfect crystal, of which we may consider two types: *(a)* **thermal lattice vibrations** —an *inelastic* energy exchange between the conduction electrons and the lattice phonons— that typically gives a dependence $\rho \sim T$ at room temperature and $\rho \sim T^5$ below a few °K; *(b)* **static lattice imperfections** which give rise to *elastic* scattering events and produces a temperature-independent contribution as $T \rightarrow 0$. For a

[1]Instituto de Física, UNAM.

macroscopic piece of metal [*i.e.* whose dimension L is large compared with the inelastic mean free path (mfp) ℓ_{in}], the electron may suffer an arbitrary number of elastic collisions (with mfp ℓ_{el}), but as long as it doesn't suffer an inelastic one, its wave function conserves its phase coherence. Roughly, phase memory is lost in every piece of wire of length $\sim \ell_{in}$. If $L \gg \ell_{in}$, the net result is that the resistance $R \sim L$, which we call "Ohmic behavior". On the other hand, if the system is mesoscopic ($L < \ell_{in}$), the wave function keeps its *phase coherence inside the whole sample*, and we may even see interference effects, which are certainly non-Ohmic! This can be achieved for sufficiently small systems ($\sim 1\mu$) at sufficiently low temperatures ($\sim$ 35 mK).

In the experiment of Ref. [3] the system was a ring, attached to external leads, through which Aharonov-Bohm type of interference oscillations could be observed!

On the theoretical side, we search for a quantum-transport theory. We can think of the system being connected, through perfect leads, to reservoirs with chemical potentials $\mu_1, \mu_2, \ldots$. Then the currents $I_1, I_2, \ldots$, in the various leads are related through Büttiker's equation [16], [17]

$$I_i = \frac{e}{h}\left[(N_i - R_{ii})\mu_i - \sum_{j(\neq i)} T_{ij}\mu_j\right], \tag{1.1}$$

where T_{ij} —the transmission coefficient from wire j to i, and R_{ii} —the reflection coefficient in wire i, have to be calculated quantum mechanically.

Because of the disorder present in our samples, T_{ij}, R_{ii} vary from sample to sample: a *statistical* study of these quantities is thus needed.

The simplest possibility is that of a two-terminal device, for which the conductance, in units of e^2/h, is $g \sim 2T$, where 2 is a spin factor. In this article we shall mainly present some of the efforts that have been done to describe quantum transport in two-terminal systems using the *scattering approach*. Only at the end of the paper we briefly mention an extension of the above techniques to multiterminal devices. We shall always assume our system to be at 0°K, so that our attention is focused on quantum interference effects in the electronic transport (only $\ell_{el} = \ell < \infty$).

One of the backbones of the presentation that follows is the concept of universality, an important notion in transport theory in disordered system. The idea is that probability distributions for quantities associated with a large (compared with ℓ) piece of a metal is insensitive to the microscopic details that specify completely the system. Consider, for instance, the probability distribution $P_L(g)$ for the conductance g in an ensemble of samples of size L. For small L this distribution is certainly not universal and can be practically any function of g. It depends on all the details specifying the ensemble, like the shape of a single impurity potential, the impurity concentration and their statistics. For large L, however, one can expect that the information about microscopic details will be washed out and the distribution $P_L(g)$ will assume some universal shape. Only few parameters will be needed then to specify the distribution completely. When such a situation occurs, we shall say that we have a *central-limit theorem* (CLT), in analogy with the familiar CLT of elementary statistics.

In the present paper we shall be able to show a CLT under various circumstances. We begin our analysis with two-terminal systems.

4.2 Two-terminal systems

4.2.1 A Central-Limit Theorem for Multiscattering Processes. The Isotropic Model

In this section we use the multichannel scattering approach to indicate how a CLT can be obtained [22].

In the scattering approach [21], the disordered system (a piece of wire) is sandwiched between two perfect leads, where the scattering states at the Fermi energy define N channels: $N \sim (k_F W)^{d-1}$, where k_F is the Fermi momentum, W the width of the system and d the dimensionality; N is thus proportional to the cross-sectional area. The $2N \times 2N$ *transfer matrix* M relates the $2N$-component wave function (N waves travel in either direction) on the right of the system to that on the left. *(T) Time-reversal symmetry* implies

$$M^* = \Sigma_x M \Sigma_x \tag{2.1}$$

and *flux conservation (FC)* requires

$$M \Sigma_z M^\dagger = \Sigma_z \ ; \tag{2.2}$$

here $\Sigma_x = \begin{pmatrix} 0 & 1 \\ 1 & 0 \end{pmatrix}, \Sigma_z = \begin{pmatrix} 1 & 0 \\ 0 & -1 \end{pmatrix}$, where $1, 0$ are the $N \times N$ unit and zero matrices, respectively.

The matrix M has $N(2N+1)$ independent parameters and can be represented in the *polar* form [21]

$$M = \begin{bmatrix} u & 0 \\ 0 & u^* \end{bmatrix} \begin{bmatrix} \sqrt{1+\lambda} & \sqrt{\lambda} \\ \sqrt{\lambda} & \sqrt{1+\lambda} \end{bmatrix} \begin{bmatrix} v & 0 \\ 0 & v^* \end{bmatrix}, \tag{2.3}$$

where u, v are $N \times N$ unitary matrices and λ is a real, diagonal matrix with N arbitrary positive elements $\lambda_1, ..., \lambda_N$. The transfer matrices form a group; with FC and T-symmetry, which is the case we are discussing here, this group can be identified as the simplectic group $Sp(2N, R)$ (Ref. [21]).

Quantities of interest can be written in terms of u, v, λ; for instance, the $N \times N$ reflection and transmission matrices (when incidence is from the left) are given by

$$r = -v^T \sqrt{\frac{\lambda}{1+\lambda}}\, v, \qquad (2.4)$$

$$t = u \frac{1}{\sqrt{1+\lambda}}\, v, \qquad (2.5)$$

and the total transmission coefficient T by

$$T = \sum_a \frac{1}{1+\lambda_a}, \qquad (2.6)$$

in terms of which the conductance is $g = 2T$, as indicated in the Introduction.

Let our system consist of n units, each being a slice of microscopic thickness, whose transfer matrix M is sampled from a collection (ensemble) with a differential probability

$$dP_1(M) = p_1(M)\, d\mu(M), \qquad (2.7)$$

where the index 1 indicates that we refer to one unit; $d\mu(M)$ is the invariant measure (that remains unchanged when all the M's are multiplied by a fixed one M_o), given by [21]

$$d\mu(M) = J(\boldsymbol{\lambda}) \left[\prod_{a=1}^{N} d\lambda_a\right] d\mu(u)\, d\mu(v), \qquad (2.8)$$

where $\boldsymbol{\lambda} = (\lambda_i, ..., \lambda_N)$,

$$J(\boldsymbol{\lambda}) = \prod_{a<b} |\lambda_a - \lambda_b|, \qquad (2.9)$$

and $d\mu(u), d\mu(v)$ are the unitary group $U(N)$ invariant measures associated with u, v.

Consider now two units in series, assuming that they are statistically independent and sampled from the same distribution (2.6). The resulting distribution $p_2(M)$ is given by the convolution [21]

$$p_2(M) = \int p_1(MM_1^{-1}) p_1(M_1)\, d\mu(M_1) \equiv p_1 * p_1. \qquad (2.10)$$

By successive convolutions we can get the distribution $p_n(M)$ corresponding to any number of units n. The question is whether, as $n \to \infty$, $p_n(M)$ tends to a limiting universal distribution.

In Ref. [22], $p_1(M)$ was assumed *isotropic; i.e.*, $p_1(M) = p_1(\boldsymbol{\lambda})$, independent of u, v. This is the multichannel generalization of the one-dimensional random-phase model; u, v are thus distributed, for one unit, according to the $U(N)$ invariant measure. This assumption is mathematically very convenient, because the convolution of two functions of $\boldsymbol{\lambda}$ only can be shown to remain a function of $\boldsymbol{\lambda}$. The physical limitations of the isotropy assumption will be discussed later on.

We have the recursion relation

$$p_{n+1}(M) = p_n * p_1. \tag{2.11}$$

In the isotropic model we can write (2.10) using the definition of convolution (2.9) as

$$p_{n+1}(\boldsymbol{\lambda}) = \langle p_n(\boldsymbol{\lambda} + \delta\boldsymbol{\lambda}) \rangle_1, \tag{2.12}$$

where the brackets on the right-hand side indicate an average over M_1, evaluated with the probability distribution $p_1(M_1)\, d\mu(M_1)$.

From now on we proceed in a way familiar in the derivation of the Fokker-Planck equation. We treat $\delta\boldsymbol{\lambda}$ in (2.11) as a small quantity (locally weak disorder) and expand the right-hand side of (2.11) in a Taylor series to obtain

$$w_{n+1}(\boldsymbol{\lambda}) = w_n(\boldsymbol{\lambda}) + \frac{1}{N}\langle \mathrm{tr}\lambda_1 \rangle_1 \hat{D} w_n(\boldsymbol{\lambda}) + \mathcal{O}(\langle (\lambda_1)_a (\lambda_1)_b \rangle). \tag{2.13}$$

Here

$$w(\boldsymbol{\lambda}) = p(\boldsymbol{\lambda}) J(\boldsymbol{\lambda}), \tag{2.14}$$

with $J(\boldsymbol{\lambda})$ given in (2.8), represents the joint probability density of the variables $\lambda_1, \ldots, \lambda_N$; $\hat{D}$ is the differential operator

$$\hat{D} = \frac{2}{N+1} \sum_a \frac{\partial}{\partial \lambda_a} \lambda_a (1 + \lambda_a) J(\boldsymbol{\lambda}) \frac{\partial}{\partial \lambda_a} \frac{1}{J(\boldsymbol{\lambda})}. \tag{2.15}$$

We now iterate (2.13) m times and take the *weak-scattering limit*: the density ν of units and their number $n \to \infty$, while the total reflection coefficient of one unit (summed over outgoing and averaged over incoming channels) tends to 0, so that $n/\nu = L$ is the fixed length of the sample and

$$\nu \frac{1}{N} \sum_{ab} \langle R_{ab} \rangle_1 = \frac{1}{\ell} \tag{2.16}$$

is finite, ℓ being the overall *mfp* for reflection. The probability density $w_L(\boldsymbol{\lambda})$ then obeys the diffusion equation [22]

$$\frac{\partial w_L(\boldsymbol{\lambda})}{\partial L} = \frac{1}{\ell} \frac{2}{N+1} \sum_{a=1}^{N} \frac{\partial}{\partial \lambda_a} \left[\lambda_a (1 + \lambda_a) J(\boldsymbol{\lambda}) \frac{\partial}{\partial \lambda_a} \frac{w_L(\boldsymbol{\lambda})}{J(\boldsymbol{\lambda})} \right] \tag{2.17}$$

with the initial condition $w_o(\boldsymbol{\lambda}) = \delta(\boldsymbol{\lambda})$, a one-sided δ-function.

In Eq. (2.16), the information about the microscopic distribution enters only via the parameter ℓ, other details being irrelevant. Thus, whatever is the solution of (2.16), it is *universal* once ℓ is specified. This constitutes a CLT .

Originally, the diffusion Eq. (2.15) was arrived at from a different route [21]; *building blocks* of length δL, small enough macroscopically, but still containing many of the above considered microscopic slices, were modelled by a *maximum entropy distribution*, constrained only by a fixed value of the average $mfp\ \ell$. Successive convolutions of building blocks can then be used to produce the distribution for an arbitrary length L, and this process can be represented by the diffusion equation (2.16). This seems a remarkable result. In the elementary theory of statistics, successive convolutions produce eventually a Gaussian which, in turn, is a maximum-entropy distribution constrained only by centroid and variance. Similarly, in the present more complicated context, very many successive convolutions generate, for a system of length δL, an M-distribution that is the one of maximum entropy for a given value of the average $mfp\ \ell$. For arbitrary lengths L the M-distribution is then the solution of the diffusion equation: one can show that that distribution is itself the one of maximum entropy, once the density of λ_a's is specified! [19].

Thus, even in the present highly non-trivial situations, successive convolutions generate distributions of maximum entropy. It would certainly be desirable to explain that connection on more general grounds: however, such a general framework is not available yet.

4.2.2 Results

We discuss some of the consequences of the above formalism.

For one-dimensional systems (one-dimensional) the number of channels is $N = 1$ and the above diffusion equation reduces to

$$\frac{\partial w_s(\lambda)}{\partial s} = \frac{\partial}{\partial \lambda}\left[\lambda(1+\lambda)\frac{\partial w_s(\lambda)}{\partial \lambda}\right], \tag{2.18}$$

where $s = L/\ell$ is the length measured in units of the mfp. The initial condition is $w_o(\lambda) = \delta(\lambda)$, a one-sided δ-function.

Equation (2.17) was first obtained by Melnikov [26]. Some time ago we discussed its solution as a group-theoretical problem [20]. In fact, the transfer matrices M are now 2×2 and form the group $Sp(2,R) \sim SU(1,1)$, whose unitary irreducible representations (unirreps) are well known. A statistical distribution $p(M)$ can be expanded in terms of those unirreps, just as the usual Fourier expansion in terms of e^{ikx} (unirreps of the translation group). One can then prove a convolution theorem, so that a convolution in the original M-space becomes a product in the dual space (of the unirreps). This was used to prove a CLT along precisely the same lines as the familiar CLT of statistics! The result appears as a linear combination of the unirreps of $SU(1,1)$. When the weak-scattering limit is taken, the solution satisfies

Eq. (2.17) and is given by [20]

$$w_s(\lambda) = \int_0^\infty f_t(\lambda) e^{-(t^2+\frac{1}{4})s} 2t \ \tanh \pi t \ dt, \tag{2.19}$$

where

$$f_t(\lambda) = D_{00}^{\frac{1}{2}+it}(\lambda, \mu, \nu) = {}_2F_1(\tfrac{1}{2}+it, \tfrac{1}{2}-it; 1; -\lambda), \tag{2.20}$$

is the $m = m' = 0$ matrix element of the unirrep belonging to the continuous class, for the integral case and in the principal interval; μ, ν are two angles that appear, for $N = 1$, instead of u, v of (2.3). This restriction occurs because of the isotropy assumption, that implies that all our distribution functions depend only on λ, and not on μ, ν.

The solution (2.18) becomes $\delta(\lambda)$ as $s \to 0$ and, for $s \gg 1$, it becomes lognormal in λ.

Certain averages can be obtained directly from (2.17). For instance, for the average of $\lambda = R/T$ (Landauer's resistance) one obtains

$$\langle \lambda \rangle_s = \tfrac{1}{2}(e^{2s} - 1), \tag{2.21}$$

that increases exponentially with length, as a quantum mechanical effect.

We now turn to the more exciting multichannel case.

As a direct consequence of the isotropy assumption, without making use of the diffusion equation, one finds, for the average reflection coefficients R_{ab} [the square of the ab matrix element of (2.4a)], the relation [24, 25]

$$\frac{\langle R_{aa} \rangle}{\langle R_{a \neq b} \rangle} = 2. \tag{2.22}$$

This equation means that backward scattering to the same channel is enhanced by a factor 2, as compared with the scattering to any other channel. This is precisely the prediction of *weak-localization* theory, where the argument is that the various paths contribute with random phases, except for a path and its time-reversed one, which contribute coherently and give rise to a factor 2 in the backward direction. However, in an optical experiment, let us say, backward enhancement is not infinitely sharp, but occurs inside a cone of width $\sim 1/(k_F \ell)$, which is very small, but nonzero. So, although the "gross structure" is correctly predicted, we miss the cone. This is a direct consequence of isotropy.

We have not been able to solve the diffusion equation (2.16) for $N > 1$. However, for $N \gg 1$ the expectation value of various physical quantities have been calculated to various orders. Generally speaking, one finds excellent results for systems much longer than wide [$L \gg W$, or quasi $1D(q1D)$ systems]. For instance, for the average transmission coefficient one finds [23], [25]

$$\langle T \rangle = \frac{N\ell}{L} - \frac{1}{3} + \cdots. \tag{2.23}$$

We recognize the first term as the classical Ohmic behaviour of the conductance: $g \sim 2T$ is directly proportional to the cross-sectional area and inversely proportional to the length of the system. The second term is a negative correction, just as it occurs in *weak-localization* theory: in fact, the above discussed backscattering enhancement must occur, by flux conservation, at the expense of the transmission coefficient, thus implying a negative correction. A diagrammatic calculation of the weak-localization correction gives precisely $-1/3$ for $q1D$ systems!

For a square we obtain basically the same result, instead of a logarithmic dependence on the size: this is a weakness of the model outside $q1D$.

For the variance of T, the first nonzero term in the expansion gives [23], [25]

$$\text{var}\, T = 2/15 + \cdots, \tag{2.24}$$

a result independent of the number N of channels, the length L of the conductor and the *mfp* ℓ (*universal conductance fluctuations*). From (2.23) we find

$$\text{rms}\, g = \sqrt{8/15} = 0.730, \tag{2.25}$$

which is precisely what emerges from perturbation expansions as well as numerical simulations for $q1D$ systems.

One can calculate more details of the statistical distribution, as correlations between pairs of transmission or reflection coefficients [24]. The basic structure of the result, with regards to its channel dependence is, in our model, a direct consequence of isotropy, just as we found in Eq. (2.21). That structure is similar to the one obtained by perturbative calculations for $q1D$, when the latter one is available. The corresponding coefficients are found from the diffusion equation and again check beautifully. Outside $q1D$ we again miss a cone, just as before, for which it seems that isotropy is to be blamed.

4.3 Beyond the isotropic model

For $q1D$ systems the isotropic model gives excellent results, in agreement with perturbative calculations and consistent with experiment, for the weak-localization correction to the conductance, universal conductance fluctuations, the backscattering enhancement, long-range correlations between pairs of transmission or reflection coefficients and time-reversal symmetry breaking by a magnetic field.

However, isotropy is *directly* responsible for "missing the cone" in the backscattering enhancement and in the correlation between transmission and reflection coefficients.

In the isotropic model the N channels are mixed randomly among themselves, in a completely "democratic" way, so that diffusion in the *transverse direction* occurs "instantaneously" and the *dimensionality* d of the system enters only through the number of channels $N \sim (k_F W)^{d-1}$.

Recently, an attempt has been made [27], [28] at constructing a more general approach not restricted by isotropy. In the model described above, the isotropy assumption for *one* scatterer goes "hand in hand" with the polar representation (2.3), that automatically satisfies FC and T-symmetry. Physically, however, we expect the transfer matrix for one scatterer to differ only little from the unit matrix: $M = I+\epsilon$; for such an M_1 we *cannot* assume an isotropic distribution! In this approach, the "Cartesian" representation (consisting of the actual matrix elements M_{ab} themselves) is more natural; however, FC is no longer automatically satisfied. One can show, though, that if FC is first imposed *on average* and then the *weak-scattering limit* is taken [described right below Eq. (2.14)], *exact* FC *is recovered!* One finds, in the end, a diffusion equation containing, as parameters, $\{\ell_{ab}\}$, $\{\ell'_{ab}\}$, *i.e.* the *mfp's* from channel b to a for backward and forward scattering, respectively. Once these parameters are given, the resulting distribution is a fixed, *universal* function. We thus have again a central-limit theorem! Now the structure of the ℓ, ℓ' matrices certainly depends on the dimensionality.

Preliminary results are encouraging. Consider physical quantities that depend only on $\{\lambda_a\}$ (*trace-like* quantities). If we adopt an *equivalent-channel model*, replacing all the ℓ's by a single one, the statistical properties of a trace become identical to those of the old isotropic model. Any improvement on the isotropic-model results must then arise from the spread among the *mfp's*. Quantities other than traces behave differently in the isotropic model as in the equivalent-channel model, the latter performing better and the *mfp's* spread still improving the results. For phases the situation is really dramatic. In the one-channel case, for instance, the transmission amplitude t has random phase for *any* length in the isotropic model: $\langle t/t^* \rangle = 0$, $0 \leq L < \infty$. In this new model, $\langle t/t^* \rangle = 1$ for $L = 0$ and decreases to zero within a "randomization distance" of about $\ell/3$, which is much more satisfactory.

The new diffusion equation has not been fully explored yet. We think that it offers physical possibilities not contained in the isotropic model, like the description of the physics in the transverse direction. These possibilities will be studied in the future.

4.4 A three-terminal system

As we saw in the Introduction, there is great interest in studying more than two-terminal devices. However, the extension of the two-terminal model of the previous section to more complicated systems is not a trivial one. Recently, an attempt has been made at providing such an extension for a three-terminal device consisting of one-dimensional wires. This turns out to be soluble, in the sense that it can be entirely reduced to quadratures.

In Ref. [29] the attention was focused on the voltage drop along a one-dimensional conductor, which is connected, via perfect leads, to reservoirs with chemical potentials μ_1, μ_2. A voltage probe is constructed as a third wire connecting the point of interest along the main conductor to a third reservoir of chemical potential μ_3, when the current I_3 along it vanishes. The point of departure is Buttiker's equation (1.1) which, for the case just described, gives

$$\mu_3 = \frac{T_{31}\mu_1 + T_{32}\mu_2}{T_{31} + T_{32}}. \tag{4.1}$$

As explained right after Eq. (1.1), T_{31}, and T_{32} are the transmission coefficients from wires 1 or 2 to wire 3. These transmission coefficients can be calculated from the 3×3 S-matrix of the full three-terminal device, that we designate by S. At the junction between wire 3 and the arms 1 and 2 of the main wire we imagine a splitter, characterized by a 3×3 S-matrix that we designate by S°. The matrix S can then be expressed in terms of S° and the transfer matrices $M^{(i)} (i = 1, 2, 3)$ of the three wires.

One case that was studied is that of a splitter modelled by a particular S° that couples wire 3 to wires 1 and 2 symmetrically. The three wires are assumed statistically independent, each one described by Eqs. (2.17)–(2.19). As the disorder in the main wire increases, the statistical distribution of the measured voltage μ_3 (constructed by running an experiment with a collection of wires of the same length that we imagine having in stock) becomes so wide that we can hardly speak of a macroscopic variable in the usual sense of the word! For instance, suppose we connect the "alligator clip" right in the middle of the main wire. If disorder is sufficiently strong, the largest probability is to find a voltage in the vicinity of μ_1 or μ_2 (with equal probability), instead of $(\mu_1 + \mu_2)/2$!

Another case that was studied is that of a "perfect" main conductor, the experiment being run with a collection of splitters that we imagine having in stock. The resulting μ_3-distribution may again be very wide, between μ_1 and μ_2!

These are surprising results. Of course they are expected to become less "violent" as we increase the number of channels. In any case, these results may be the starting point for a more complete application of random-matrix models to multichannel mesoscopic systems.

4.5 References

[1] C.P. Umbach, S. Washburn, R.B. Laibowitz, and R.A. Webb, *Phys. Rev.* **B 30**, 4048 (1984).

[2] S. Washburn, C.P. Umbach, R.B. Laibowitz, and R.A. Webb, *Phys. Rev.* **B 32**, 4789 (1985).

[3] R.A. Webb, S. Washburn, C.P. Umbach, F.P. Milliken, R.B. Laibowitz, and A.D. Benoit, *Physica* **140 A**, 175 (1986).

[4] A.D. Benoit, C.P. Umbach, R.B. Laibowitz, and R.A. Webb, *Phys. Rev. Lett.* **58**, 2343 (1987).

[5] W.J. Skocpol, P.M. Mankiewick, R.E. Howard, L.D. Jackel, D.M. Tennant, and A.D. Stone, *Phys. Rev. Lett.* **58**, 2347 (1987).

[6] C.P. Umbach, S. Santhanam, C. Van Hasendonck, and R.A. Webb, *Appl. Phys. Lett.* **50**, 1289 (1987).

[7] B.L. Al'tshuler, *Pis'ma Zh. Eksp. Teor. Fiz.* **41**, 530 (1985) [*JETP Lett.* **41** 648 (1985)].

[8] B.L. Al'tshuler and D.E. Khmel'nitskiĭ, *Pis'ma Zh. Exsp. Teor. Fiz.* **42**, 291 (1985) [*JETP Lett.* **42**, 359 (1986)].

[9] A.D. Stone, *Phys. Rev. Lett.* **54**, 2692 (1985).

[10] P.A. Lee and A.D. Stone, *Phys. Rev. Lett.* **55**, 1622 (1985).

[11] B.L. Al'tshuler and B.I. Shklovskiĭ, *Zh. Eksp. Teor. Fiz.* **91**, 220 (1986) [*Sov. Phys. JETP* **64**, 127 (1986)].

[12] P.A. Lee, A.D. Stone, and H. Fukuyanna, *Phys. Rev.* **B 35**, 1039 (1987).

[13] S. Feng, C. Kane, P.A. Lee, and A.D. Stone, *Phys. Rev. Lett.* **61**, 834 (1988).

[14] M. Büttiker, *Phys. Rev. Lett.* **57**, 1761 (1986).

[15] M. Büttiker, *Phys. Rev.* **B 35**, 4123 (1987).

[16] M. Büttiker, *IBM J. Res. Develop.* **32**, 317 (1988).

[17] A.D. Stone and A. Szafer, *IBM J. Res. Develop.* **32**, 384 (1988).

[18] H.U. Baranger, A.D. Stone, and D.P. Di Vincenzo, *Rap. Comm. Phys. Rev.* **B 37**, 6521 (1988).

[19] D.A. Stone, P.A. Mello, K.A. Muttalib, and J.L. Pichard, in *Mesoscopic Systems*, P. Lee and B. Al'tshuler Eds. (North-Holland), to appear.

[20] P.A. Mello, *J. Math. Phys.* **27**, 2876 (1986).

[21] P.A. Mello, P. Pereyra, and N. Kumar, *Ann. Phys. (N.Y.)* **181**, 290 (1988).

[22] P.A. Mello and B. Shapiro, *Phys. Rev.* **B 37**, 5860 (1988).

[23] P.A. Mello, *Phys. Rev. Lett.* **60**, 1089 (1988).

[24] P.A. Mello, E. Akkermans, and B. Shapiro, *Phys. Rev. Lett.* **61**, 459 (1988).

[25] P.A. Mello and A.D. Stone, *Phys. Rev.* **B**, in press.

[26] V.I. Melnikov, *Pis'ma Zh. Eksp. Teor. Fiz.* **32**, 244 (1980) [*JETP Lett.* **32** 225 (1980)].

[27] P.A. Mello and S. Tomsovic, *Phys. Rev. Lett.*, to be published.

[28] P.A. Mello and S. Tomsovic, *Phys. Rev.*, to be published.

[29] S. Godoy and P.A. Mello, *Phys. Rev. Lett.*, to be published.

5

Symmetry-Avoided Crossings and their Role in the Catalytic Activity of Transition Metals

Octavio Novaro[1]

ABSTRACT A series of transition metal atoms interacting with hydrogen, methane and other small molecules are studied at an *ab-initio* LCAO-MO-SCF level including intensive configuration Interaction, with a CI space of $\sim 10^6$ configurations. All the metals studied (Pt, Pd, Ni, Ru, Fe, Ag, Cu, Zn, Ga, Zn^+, Zn^{++}, Ga^+, Ga^{++}) show the importance of maximal d-occupation states, which usually are highly excited states for the isolated atom but become the ground states of the metal-molecule complex. This implies the existence of symmetry-avoided crossings. The importance of these avoided crossings for the catalytic phenomenon is analyzed.

5.1 A personal introduction

In the late 1960's Marcos Moshinsky was in one of the most fruitful and active periods of his life (but then again he always is), this being the time when he was reaping the fruits of his accomplishments [1] in Nuclear Physics [2], and becoming more deeply involved in his valuable contributions in group theoretical methods [3]. As the newest member of his group of students, I was therefore surprised when he suggested that I work with him in a quite different field, that of atomic and molecular physics. Marcos had evidently made a penetrating diagnosis; this was an important field in which no Mexican physicists or chemists had yet made any contributions. He thrust me, along with some other young students, in the desired direction. He did so in the Moshinsky fashion, with authority, confidence and an extremely original and personal approach. It is no surprise that the first work done in this field in Mexico had a peculiar flavor [4]-[7]; exact many-body solutions, model systems, hidden symmetries, global overviews on problems that all too often are treated on an atom-by-atom, molecule-by-molecule approach. There was an immediate interest among experts and groups in the field of atomic and molecular physics, as I learned person-

[1]Instituto de Física, UNAM.

ally from R. F. W. Bader, O. Sinanoğlu, C. Wulfman, and D. W. Davies. Manuel Berrondo had a similar experience with C. Bloch, C. Coulson, P. O. Löwdin, and I. Prigogine.

Marcos Moshinsky has gone to new problems since, and Berrondo, myself and many other of his students have taken our own separate ways within and outside the field of atomic and molecular physics. Still I like to recognize his mark in some of the work we all do, and even in the work of our own former students down the line. Because we not only learned to love and respect Marcos, we also hopefully learned his most valuable lesson (each one up to his own personal limitations); to establish your own questions, to seek your own way into truth and to do so with relentless passion. The present paper reflects a long line of work that I feel stems from my formation under Marcos Moshinsky.

5.2 General introduction

The symmetry avoided crossing principle was established early [8] in the application of quantum mechanics to molecular structure [9]. A very simple discussion for the case of diatomic molecules appears in the book of Coulson [10]; we reproduce it here for completeness.

Consider a system with two or more states described by independent eigenfunctions, say ψ_1 and ψ_2, which may be derived from a variational or perturbational calculation and which will be considered to be separately normalized. The expectation values of the energy of each component is given as follows:

$$\int \psi_1^* H \psi_1 dV = H_{11}, \tag{2.1a}$$

$$\int \psi_2^* H \psi_2 dV = H_{22}, \tag{2.1b}$$

where H is the quantum Hamiltonian of the system. The application of the variational principle introducing a trial function

$$\psi = a_1\psi_1 + b_2\psi_2 \tag{2.2}$$

leads to a set of secular equations

$$a_1(H_{11} - E) + a_2(H_{12} - ES_{12}) = 0, \tag{2.3a}$$

$$a_1(H_{12} - ES_{12}) + a_2(H_{22} - E) = 0, \tag{2.3b}$$

where E is the variational energy associated to the expansion (2.2) and S_{12} is the overlap integral involving ψ_1 and ψ_2. By eliminating the expansion coefficients a_1 and a_2 we obtain E as a root of the secular equation

$$(E - H_{11})(E - H_{22}) - (H_{12} - ES_{12}^2) = 0, \tag{2.4}$$

which defines a parabola that takes negative values at $E = H_{11}$ and $E = H_{22}$. The two roots are related to the combination of ψ_1 and ψ_2, so we have two energy values derived from the two possible combinations (*i.e.* two eigenvectors defined by two pairs of values for a_1 and a_2). Notice that because E is negative at H_{11} and H_{22} and positive for large positive or negative ordinates it must cross the axis at two points. This implies that the two energy values E_1 and E_2 of the system with states ψ_1 and ψ_2 are as follows: one is smaller than the smallest of the pair H_{11} and H_{22}, and the other larger than the largest of this pair. As H_{11} would be the eigenvalue for the system if only state ψ_1 existed and viceversa, we see that H_{11} and H_{22} have repelled each other through their interaction and reached the separated values E_1 and E_2.

This is the avoided crossing rule and evidently only applies to states of the same symmetry, because only in such a case can we combine them in Eq. (2.2), which implies equal symmetry restrictions. Different symmetry states may cross freely. This non-crossing rule applies on the Born-Oppenheimer approximation. It can also apply when a perturbative term is added to the Hamiltonian; in this case $H_{12} = V_{12}$ makes H_{11} and H_{22} repel each other.

Diatomic molecules rigorously obey this non-crossing rule, with the exception [11] of a one-electron system like H_2^+. For larger molecules, intersections may be allowed, due to the growing number of degrees of freedom. For three-dimensional potential surfaces, they may intersect at a point joining two conical sections, for n-dimensions, contacts are at $n - 2$ dimensional hypersurfaces.

We should therefore explain that in the following, when applying the non-crossing rule to polyatomic systems, we are really scanning a particular sector of the hypersurface, associated to a single degree of freedom at a time, say the separation between a transition metal atom and a small molecule, without the simultaneous optimization of all other degrees of freedom.

Before discussing how symmetry avoided crossings may play a role in the activity of transition metal catalytic centers, let us schematically review some of the theoretical efforts to understand the catalytic phenomenon. Even as most catalysis experts and technicians were still quite satisfied to call this field of research more an art than a science, the first pioneering efforts [12], [13] to apply quantum-mechanical concepts, albeit in a purely qualitative way, were being done. In the early 70's semiempirical calculations were being addressed to different aspects, most of them to the structure of the heterogeneous catalysts, as in the school of Baetzold [14], [15] for example. A smaller group of researchers were more interested in understanding the chemical changes induced at a heterogeneous active center or an organometallic homogeneous catalyst [16]-[18].

Ab-initio calculations on these same questions [19] soon followed, showing that the transition metal atoms active in processes like polymerization were very often in excited states, stabilized by the surrounding atmosphere of the ligands forming the organometallic complex or the surface structures of the solid catalysts. In their studies on the oxidative addition of hydrogen onto biphosphine platinum complexes, for example, several authors [20]-[22] coincide that the Pt center is not in its isolated atom $^3D(5d^96s^1)$ ground state, but rather in a $^1S(5d^{10}6s^0)$ closed-shell excited state. We shall attempt to show through successive examples that this is not an exceptional case and that apparently it is a recurring situation that, if confirmed, may shed new light to understand the ability of transition metals to be such versatile catalysts.

The first and (to the best knowledge of the present author) only paper explicitly showing that the active sites of a catalyst are associated to metallic-atom excited states, was based on studies of a single metal: copper [23]. The idea originated from studies on the photochemical activation of small molecules (H_2^{24-26}, N_2^{27}, CH_4^{28}) by selectively photoexcited Cu atoms and clusters. Our present aim here is to extend this study to several other transition metal atoms (including, Pt, Ru, Pd, Ag, Fe, Ni, Zn, and Ga and their ions, as well as Be and Li) and test the conclusions of Ref. [23] in all these metal-molecule systems.

We shall show that for each metal atom (with the exception of Pd) the excited states are always responsible for the capture and activation of reactants like the H_2 molecule. In effect, excited states, generally those having maximal d-shell occupancy, are strongly attractive and through one or a series of avoided crossings with the lower-lying states which originally were either repulsive or nonbinding, induce the formation of deep potential energy wells. Thus the stability of, say, ground-state RuH_2, is originally due to a Ru excited state which, through a symmetry avoided crossing reverts the tendency of ground-state Ru to repel H_2, reaching the deep well after surmounting an activation barrier. The role played by the avoided crossing leaves its mark in the molecular orbital nature of the ground state, a fact that had puzzled some researchers studying only the latter. All the metal atoms studied here show this behaviour, with the exception of Pd whose own ground state is a closed shell $^1S(4d^{10})$. As Pd is the only transition metal with this characteristic, its behavior is in reality a confirmation of the general trend, even if in this case the maximal d-shell occupancy state cannot suffer avoided crossings in its descending potential curve (with ethylene [29] or H_2^{30}), being the lowest state. It is on he other hand striking that so different metals as Fe, Ru, Cu, Ag, Ni, Pt, Zn, or Ga, all show avoided crossings as the origin of their reactivity.

5.3 Method

Since we will be reporting here diverse sets of calculations, addressed at different systems (usually formed by a transition metal atom and a small molecule) and carried out by one or another of my students and collaborators, it is impractical to give a detailed description of all the basis sets, contraction schemes, model potentials, etc. It would require dozens of tables, so the reader is kindly asked to go to the original papers and theses. Here we shall only describe the common general features of all these studies.

The approach begins with SCF calculations followed by extensive variational and perturbational configuration interaction (CI). The SCF parts were done either using the PSHONDO program [31] or PSHF [32], a derivation thereof which includes the pseudopotential method of Barthelat *et al.*, [33]. For many of the metal atoms studied here the pseudopotentials of the PSHF manual [32] were used, an exception was Ru for which the LaJohn *et al.* [34] pseudopotential and basis sets were used. In essentially all cases the basis sets were always of double-zeta quality plus polarization (but not including *f*-orbitals) and treating the *d*-shells explicitly. More details can be found in the original theses and publications. The CI part was obtained using the CIPSI algorithm [35] where usually of the order of one million configurations were included in a Møller-Plesset second order perturbation scheme (for Pd, where only the ground state was studied, about one tenth of this space was used [29]). In every case where it is necessary, the relativistic effects were taken into account through the use of relativistic pseudopotentials [36]. No use of spin-orbit corrections was made.

The great advantage of the PSHONDO-CIPSI sequence of programs over other alternative methods, is its great capacity and stability for the simultaneous study of several excited states. The standard procedure used in practically all of the systems studied here implied the simultaneous optimization of the basis set for all states to be studied (ground and excited), this included ionic states in the cases of Zn and Ga. The calculations were calibrated by obtaining reasonably well the energy separations between the ground and excited states of the isolated metal atom and also usually studying the metal hydrides to compare with theoretical and experimental results. The general procedure is exemplified by the series of papers on CuH_2 [23]-[26] to which the reader should refer as the one followed in our yet unpublished new studies. In the latter, a modification of this general procedure was the use of multireference CI when necessary to correct the SCF results. This was systematically used for instance in the cases of Zn, Zn^+, Zn^{++}, Ga, Ga^+, and Ga^{++} for which three or more states of each species were simultaneously studied [37].

5.4 Results

As mentioned in the introduction, we shall describe here the behavior of several metal atoms in their ground and excited states as they interact with hydrogen and other small molecules. The metal atoms studied are: Pt, Ru, Fe, Pd, Ag, Cu, Zn, Ga, Ni, Be, and Li. Most of the results presented here are new and perhaps would be interesting *per se*, as they involve excited

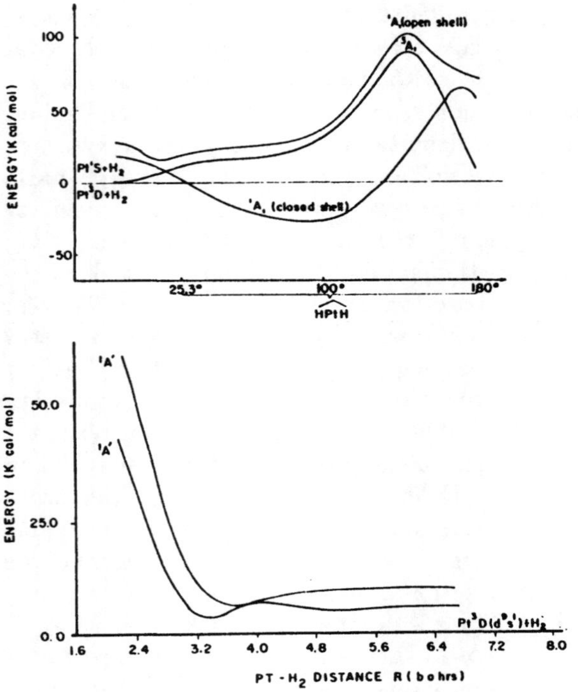

FIGURE 5.1. Potential energy curves for the PtH_2 system. *(a)* The case of the C_{2v} symmetry of PtH_2 for a triplet and two singlet states. *(b)* The linear PtH_2 system (C_s symmetry) for the two lowest singlet states.

and ionic states not studied elsewhere and in some cases they give insight about the way a particular transition metal species is capable of activating very important chemical bonds (*e.g.* the H-H, the C-H, the C=C and N≡N bonds, etc.) Our main interest here is different however, we want to test a few novel and intriguing proposals: *(a)* that it is generally the maximal *d*-shell occupancy states that produce the spontaneous attraction of the metals to otherwise rather inert, closed-shell molecules like methane or hydrogen, *(b)* that through a series of symmetry avoided crossings it is the descent of these usually (with the exception of Pd) high-lying states of the metal-molecule complex which provoke the lower curves to become attractive after their initial repulsions to the reactants and generates a change of curvature (at the apex of an activation barrier) and *(c)* that these avoided crossings leave a trace in the nature of the ground state of the metal-molecule system, which is usually recognizable from a population analysis (*e.g.* the ground state of the complex shows that the metal atom dominantly has maximal *d*-shell occupancy character). The fact that metal atom after atom apparently behave the same way gives great strength to these proposals [23].

5.4.1 Pt-H_2 Interactions

We shall first discuss the case of a third-row transition metal atom, namely platinum, interacting with a hydrogen molecule. The Pt atom has a ground state with a $^3D(5d^9 6s^1)$ configuration and many low-lying excited states with relatively small energy separations, among these the lowest singlet state has a $^1S(5d^{10})$ configuration and lies only 18 kcal/mol above the ground state energy. Therefore we must optimize our basis set for these states simultaneously and use a pseudopotential which naturally has to contain relativistic effects, very crucial for a heavy atom like Pt. The reader can see the details in Ref. [38], where the approach of the Pt equidistintantly to both H moieties was studied. We have added another geometry; Pt approaching linearly towards the H_2 molecule. In other words two different symmetries of the Pt H_2 system, C_{2v} and C_s, are studied here. The results are given in Fig. 1. In Fig. 5.1*a* the C_{2v} curves are given, showing that the ground-state Pt atom simply repels the H_2 molecule and only when the H-Pt-H angle is near to 180° does the energy go down, although it never reaches the asymptotic Pt+H_2 value. This is valid for all triplet C_{2v} representations although only the $^3A_1$ curve is depicted in Fig. 5.1*a*. The singlet states on the other hand show a couple of attractive curves which suffer an early avoided crossing making the upper curve [associated to the $^1A_1(d^9)s^1$ state] to present a relative minimum and then goes up descending again only at angles near 180°. Meanwhile the lower $^1A_1(d^{10})$ curve is repelled downwards reaching a deep minimum. Evidently the C_{2v} capture of H_2 by Pt is done through this excited-state of $^1A_1$ symmetry and with a close *d*-shells d^{10} configuration (maximal *d*-shell occupancy) and its stabilization is preceded by a symmetry-avoided crossing. Many other curves were calculated and reported in Ref. [38] but for simplicity we shall omit then here. The important point to note is that the $^1A_1(d^{10})$ closed-shell state is responsible for the stable form of PtH_2, with Pt-H separations of 3.2 bohrs and a H-Pt-H angle of 180°.

The Fig. 5.1*b* shows the linear Pt-H-H energy curves. The two singlet states of the $^1A_1$ representation of C_s symmetry suffer an avoided crossing as in the C_{2v} symmetry, but in this case both curves only have shallow minima located respectively at a Pt-nearest H distance of 3.5 bohrs for state $^1A_1(d^9s^1)$, and 3.2 for $^1A_1(d^{10})$. The fact that the linear approach is less effective for H_2 capture by transition metal atoms is apparently true in general, at least for most cases studied here (the case of N_2 capture is quite the opposite, the linear approach being much more favorable than the C_{2v} one [27]). In consequence, in most of the following we shall concentrate mainly in the C_{2v} approach of H_2 to metals.

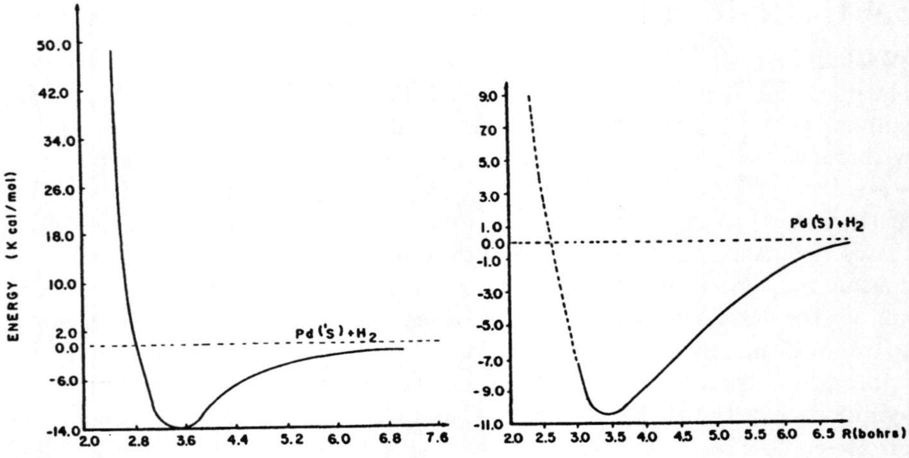

FIGURE 5.2. Potential energy curves for the PdH_2 system. *(a)* C_{2v} symmetry of the PdH_2 ground state. *(b)* Same as before but now for the C_s (linear structure) symmetry.

5.4.2 Pd-H_2 Interactions

As mentioned before, palladium is an exception among transition metal atoms in the sense that its ground state is a closed-shell $^1S(4d^{10})$ state. In a previous study [29] on the Pd-ethylene interactions, this state was shown to be quite capable of capturing C_2H_4 molecules even though its complete *d*-shell could not receive charge donation from the olefin. Pd did "retrodonate" charge to the latter in the sense of the classical Chatt and Duncanson model [12], [13], [17]. A similar situation is found [30] for Pd-H_2 interactions. In Fig. 5.2 two attractive curves corresponding to the C_{2v} and linear approaches of Pd to H_2 are given. Contrary to the case of Pt, the Pd ground state does capture H_2 (notice however that this ground state corresponds to the Pt *excited* state which does capture H_2, both having a d^{10} configuration), and besides it does so only slightly less efficiently for the linear than for the C_{2v} configuration, as Fig. 5.2 shows.

From the above discussion it is evident that Pd does not present symmetry avoided crossings associated to its potential energy well PdH_2 minimum. Yet in a sense these results fit quite well in the general scheme proposed here; it is the maximal *d*-shell lowest-multiplicity state (d^{10}) which is responsible for H_2 capture and obviously its configuration will be dominant in the lowest Pd H_2 energy minimum. If this is confirmed for other transition metals it will immediately imply avoided crossings because in no other case do we have a closed-shell ground state.

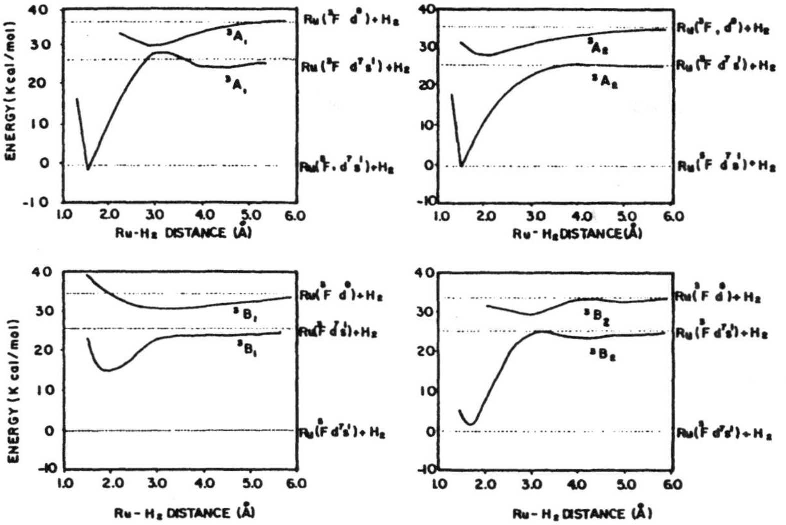

FIGURE 5.3. Potential energy curves for the triplet states of RuH_2, in the C_{2v} symmetry. All four representations of C_{2v} are depicted showing avoided crossings between the lowest triplet and the maximal d-shell occupancy excited triplet. These curves correspond to an unrelaxed H_2 intromolecular distance of 0.74Å.

5.4.3 Ru-H_2 Interactions

Recent results [39] on the interactions of the Ru excited states with H_2 show very interesting features. The quintuplet (spin 2) states of Ru, including the $^5F(4d^75s^1)$ ground state lead to repulsive C_{2v} curves in their interaction with H_2. The first excited state is $^3F(4d^75s^1)$, but a higher triplet state $^3F(4d^85s^0)$ that is usually ignored in analyzing Ru-atom reactions, will be shown to play a crucial role in the process. In Fig. 5.3 all of the representations ($^3A_1$,$^3A_2$,$^3B_1$, and $^3B_2$) of the C_{2v} Ru H_2 curves stemming from both Ru^3F states are presented. In all of the curves originating from the maximal d-shell occupancy state $^3F(d^8)$ an initial attraction is evident. The four curves stemming from $^3F(d^7s^1)$ are essentially nonbinding at first. At a Ru-H_2 center of mass separation of $\sim$ 3Å the descending curves suffer avoided crossings with the lower curves of the same representation, so the $^3A_1$ curves repel and so do the $^3A_2$ curves, etc. In a word four independent symmetry avoided crossings are evident in Fig. 5.3. The originally non-binding curves immediately descend steeply, all of them (with the exception of $^3B_1$) reaching or transcending the asymptotic ground state $Ru^5F(d^7s^1)$+H_2 reference value. Further relaxation of the H-Ru-H angle makes for even deeper wells for all four curves, as depicted in Fig. 5.4. We see that the first excited state of $Ru^3F(d^7s^1)$ produces important C_{2v} attractions of H_2 for all representations, with wells lying between 15 and 25

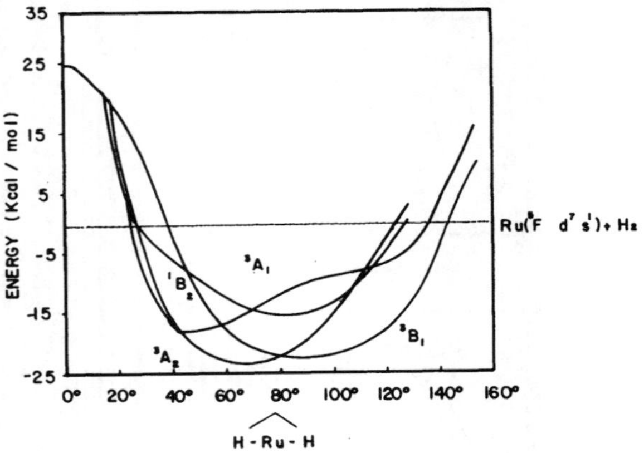

FIGURE 5.4. Same system as in Fig. 5.3 but now allowing the relaxation of the H–H distance.

kcal/mol below the ground-state Ru quintuplet $+H_2$ asymptotic value. But remember that these $Ru^3F(d^7s^1)$ curves became attractive *only* after being subjected to the symmetry avoided crossings of Fig. 5.3, where each curve has been repelled by its counterpart, an equal symmetry curve coming from the higher-lying $Ru^3F(d^8)$ state. Thus we may say that this higher triplet originates the stabilization of the RuH_2 system. A consequence of this is the character of the wave function in the potential energy wells of Fig. 5.4. In effect a very clear-cut d^8 character of the molecular orbitals has been reported [39]. For instance, the lowest minimum curve in Fig. 5.4, with $^3A_2$ symmetry, has a reference SCF configuration; $1a_1^2, 1b_1^2 2a_1^2 1b_2^2 1a_2^1 3a_1^1$. At long separations, the $1a_1$, orbital is associated to the H_2 σ-bond and $2a_1, 3a_1, 1a_2, 1b_1$ and $1b_2$ come from the Ru d-subsell while $4a_1$ from the Ru 5s orbital. (The latter is absent from the $^3A_2$ reference configuration.) The $^1b_2$ orbital is associated to the d_{yz} of Ru and interacts with the $H_2\sigma^*$ antibonding orbital from symmetry considerations. The configurations of the other representations also correspond to the $Ru^3F(d^8)$ state [39]. Considering the large weight that the reference configuration carries in the CI scheme, the CI does not alter this fact.

5.4.4 Ag-H_2 Interactions

The nine potential energy curves of the Ag+H_2 reaction with Ag in its ground (^{2}S) and first two excited states (^{2}D) and (^{2}P) have been analyzed here in complete analogy with the case of the Cu+H_2 reaction [24]-[26]. Both linear and C_{2v} approaches of Ag to H_2 have been considered but the latter is substantially more favorable and we shall concentrate on this C_{2v} symmetry here. In fact only the ground-state curve (stemming from the Ag(^{2}S) state interacting with $^1\Sigma_g^+$ H_2) with a $^2A_1$ representation, and the

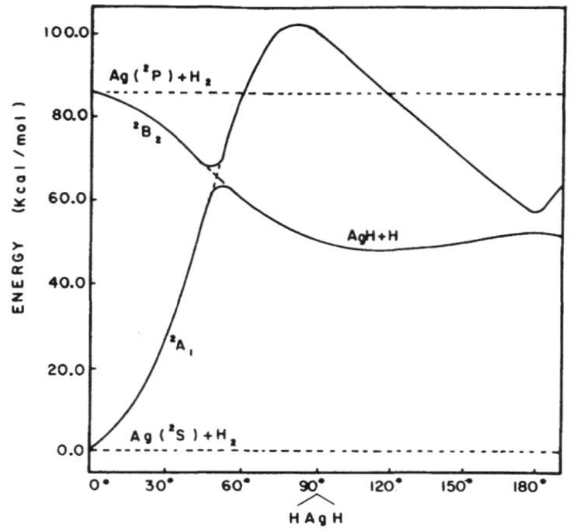

FIGURE 5.5. Potential energy curves for the AgH_2 system in the C_{2v} symmetry. Two states are depicted, the $^2A_1$ state correlacted with the silver ground state and the $^2B_2$ state correlated to the first excited state of Ag(^{2}P).

most favorable $^2B_2$ representations stemming from the first excited state (^{2}P) of Ag are represented in Fig. 5.5. They are by far the most interesting ones.

We should mention that the spectra of Ag and Cu differ mainly in the ordering of their first two excited states, the ground state for both is ^{2}S but in Cu the first excited state is ^{2}D followed by ^{2}P while in Ag the first excited state corresponds to a ^{2}P($d^{10}s^1p^1$) configuration. This ^{2}P state of Ag spontaneously captures H_2 in a $^2B_2$ representation of C_{2v}, as shown in Fig. 5.5. The curve consequently goes down until it meets the repulsive $^2A_1$ curve originating in the ground-state Ag+H_2 interaction. Belonging to different representations, these two curves would cross freely if it were not by the Herzberg-Teller vibronic coupling [40] that first was demonstrated by our group [25] to occur for two similar curves in the case of the CuH_2 system. The vibrational modes of AgH_2 include: a nonsymmetrical stretching mode b_2 and a symmetrical breathing mode a_1. If either one of them is associated to curve $^2A_1$ and the other to $^2B_2$ of Fig. 5.5 or viceversa, the vibronic coupling implies mutual perturbation between the curves which therefore avoid each other at the crossing point region (dotted lines in Fig. 5.5).

The validity of our AgH_2 curves may be established by comparison with photochemical matrix-isolation experiments [41] as was done for the CuH_2 system [24]-[26]. In fact the discussion of the similarities between both systems will allow us to keep the discussion here short. The reader is kindly asked to refer to the CuH_2 series of papers [23]-[26]. On the other hand, we shall emphasize here the main differences and their experimental consequences. In reference [41] a mixture of Ag atoms, H_2 molecules, and Kr gas were condensed at low temperatures (12°K) and a solid was formed where Ag and H_2 are trapped. No reaction between them is observed unless a photoexcitation with exactly the wavelength that corresponds to the $^2P \leftarrow ^2S$ transition is used to irradiate the mixture. Immediately the atomic spectrum is depleted and a reaction product AgH is obtained (all this verified by emission and absorption *uv*-visible and infrared spectroscopies, Raman, EPR, etc [41]).

Our Fig. 5.5 explains this process. The $^2P \leftarrow ^2S$ transition implies the immediate capture of H_2 (in the Ag ground state $^2A_1$ it was repelled) reaching the avoided crossing. A nonradiative transition [25] to the lower curve will lead either to the reactants Ag (^{2}S)+ H_2^*, or to the products AgH+H. The energy gain from the photon is enough to overcome the activation barrier. All this has been thoroughly discussed in the literature [23]-[26] for CuH_2 and no more discussion will be needed here.

Rather let us emphasize now the main differences between CuH_2 and AgH_2 as described in Fig. 5.5. One observed difference is that for the latter [41] a competing reaction, Ag dimerization, is more notable than for copper. More importantly the AgH+H products (specially the latter; H atom vibrations in the Kr lattice are hardly observed [41] in the absorption band spectra, although EPR confirms their existence) are less evident than the case of CuH + H.

As the preparation of both systems was identical, we must conclude that the Ag reaction is intrinsically less efficient. Furthermore, the thermal reaction CuH+H→ Cu + H_2 was both experimentally [42] and theoretically [26] very efficient. AgH + H trapped in a krypton matrix and heated (20° K) also allows some thermal recovery of ground state H_2 and Ag. However this recovery [41] is only 20% while for copper it was 50%. If we compare Fig. 5.5 with the results for Cu [25], [26] we can understand these differences. The valley of the AgH + H products lies substantially higher in energy than for CuH + H and the highest activation barrier separating the reactants and the products valleys is about 25 kcal/mol for copper, 65 kcal/mol for silver. In fact the barrier for silver can be barely surmounted thanks to the large energy gain provided by the photon, but its efficiency cannot be expected to be as large as for copper, well in accord with experimental facts.

We should warn that these new results for AgH_2 are preliminary. Their main limitations are that localization effects and other important corrections proposed by Ramírez [43] and found to be crucial in the study of silver clusters [44] were not included and spin-orbit couplings are also absent. These effects may diminish the Ag+H_2 barrier of Fig. 5.5. The present results however comprise the first attempt at a very important and challenging problem and the above discussed conclusions derived from them must be at least qualitatively correct, as they fully reproduce the main experimental facts [41].

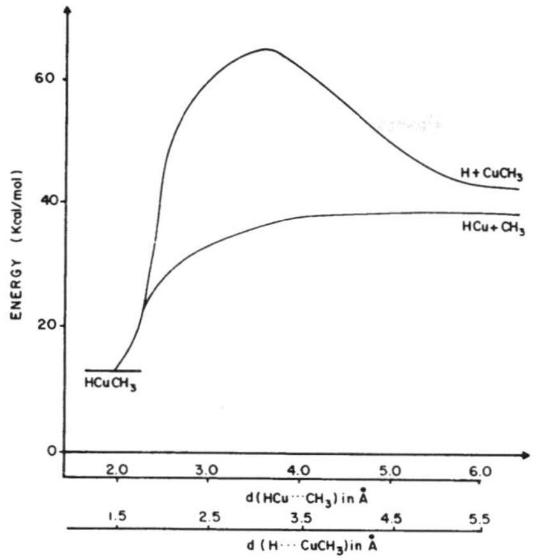

FIGURE 5.6. Potential energy curves relating the $HCuCH_3$ intermediate to the products of the reaction $Cu+CH_4$. Potential energy curves relating the $HCuCH_3$ intermediate to the products of the reaction $Cu+CH_4$. Potential energy curves relating the $HCuCH_3$ intermediate to the products of the reaction $Cu+CH_4$.

5.4.5 Cu-CH$_4$ Interactions

While the Cu + H_2^{24-26} and the Cu + N_2^{27} reaction coordinates have been thoroughly studied theoretically, the technologically very promising [45] problem of the activation of natural gas by copper needs more attention. Enough experimental work has been done to establish its viability [28], [46]-[47]. Theoretically only the relative energies of the reactants and products were obtained at an SCF level [48] and the structure of the intermediate complex at a CI level [49]. For our present purposes, however, the evolution of the chemical reaction coordinates including the copper excited states (after all it is a photochemical process including two separate photoactivations [46]-[47]) is the most important question. Some preliminary [50] results will be presented here. The first photoactivation implies a very similar process to that discussed in the previous subsection and involving the same copper states as for the Cu + H_2 reaction. [24]-[26]. The reaction with methane differs from the one with H_2, because a well identified and stable intermediate ($HCuCH_3$) is observed [28]. It is only by a second-photon process [46]-[47] that the two observed sets of final products (CuH + CH_3; H + $CuCH_3$) are obtained. In Fig. 5.6 our theoretical curves for this second process (from intermediate to final products) is presented. A first set of products (CuH + CH_3, by far the most important from a technological [45] point of view) present no activation barrier, a second set (H + $CuCH_3$) has a high barrier. The second photon provides enough energy for these endothermic products and even to overcome the barrier. The latter however greatly acts to discriminate against the second set of products, a fact known experimentally: the H + $CuCH_3$ products are eighty times more abundant[28].

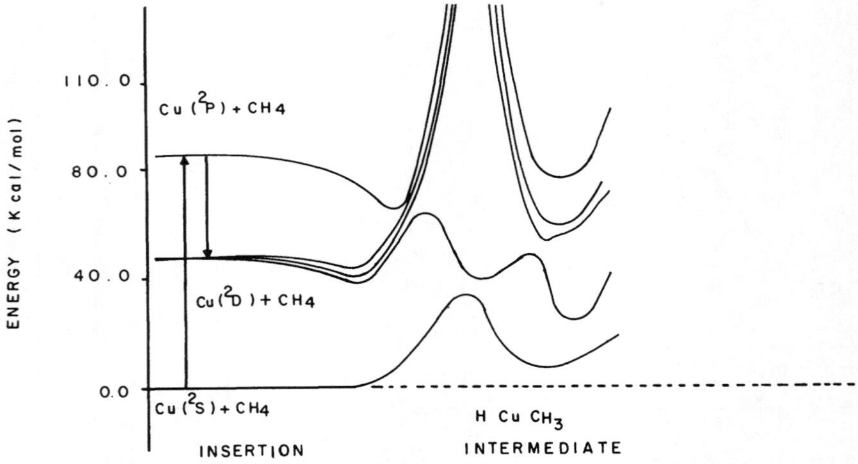

FIGURE 5.7. Potential energy curves relating the reactants of the methane + copper reaction (the latter in its ^{2}S, ^{2}P and ^{2}D states) with the $HCuCH_3$ intermediate. Only the states of the A' representation of C_s are depicted.

The first photoactivation, from ground-state Cu(^{2}S) to (^{2}P) leads from the Cu + CH_4 reactant valley to the intermediate complex $HCuCH_3$. The 9 curves related to the (^{2}S), (^{2}D), and (^{2}P) states of Cu have been studied [50]. The most relevant ones are depicted in Fig. 5.7. These curves are obtained at a C_s symmetry. The ground-state has $^2A'$ representation as do three curves from the (^{2}D) state and one from (^{2}P). Only the latter is initially attractive, as in the case of Cu + H_2. The three (^{2}D)A' curves initially are repulsive, but as Fig. 5.7 shows they meet the (^{2}P)A' descending curve in a region where multiple symmetry-avoided crossings make the lowest ^{2}D-derived curve to shift down and the ^{2}P-derived curve to go up. This now descending (^{2}D) curve eventually meets the ground-state A', also initially repulsive and, through a final avoided crossing creates the deep minimum responsible for the remarkable stability [28] of the $HCuCH_3$ complex.

5.4.6 The Cases of Ni and Fe

The first row of transition metal atoms has been thoroughly studied, often using all-electron, extensive CI methods. However, a few specific questions raised in such studies merit further theoretical efforts to obtain at least qualitative answers. One such case is NiH_2, for which Blomberg and Siegbahn [51] reported singlet and triplet energy surfaces. The ground state of Ni is a ^{3}F state which has a repulsive $^3A_1$ curve for NiH_2, reaching a minimum only at a angle of 180° after surmounting a high (80 kcal/mol) activation barrier. The Ni excited ^{1}D($3d^9 4s^1$) state on the other hand is highly attractive to H_2 in $^1A_1$ symmetry, reaching a deep potential energy

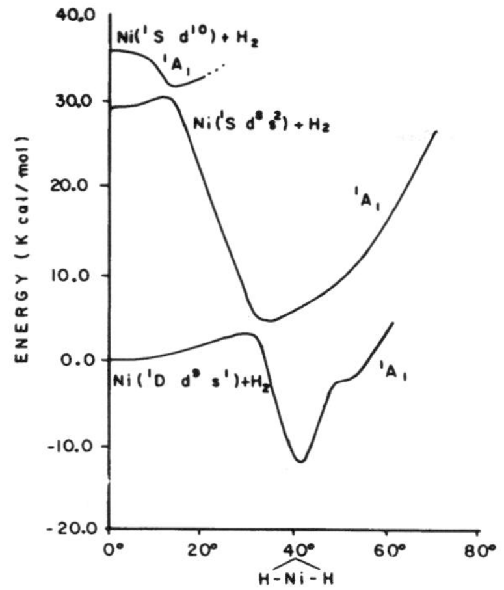

FIGURE 5.8. Potential energy curves for the NiH_2 system. The lowest and the two first excited singlets are represented, all in the ${}^1A_1$ representation of C_{2v}.

well, much lower than any other state and having a bent (C_{2v}) geometry. Interestingly these authors note that this well corresponds to a closed-shell configuration which makes one immediately think of the highly excited $(3d^{10}4s^0)^1S$ state of Ni. Will the latter capture H_2 spontaneously? Does it provoke avoided crossings with lower states of NiH_2 with the same ${}^1A_1$ symmetry? This question has been answered in Ref. [52], although not all of the relevant calculations have yet been completed. The lowest singlet of Ni as mentioned above is ${}^1D(3d^94s^1)$; the first two excited singlets are ${}^1D(3d^84s^2)$ and ${}^1S(3d^{10})$. The corresponding $NiH_2^1A_1$ curves are given in Fig. 5.8. We see that for the maximal d-shell occupancy state ${}^1S(3d^{10})$, the ${}^1A_1$ curve is originally attractive, suffering an avoided crossing with the ${}^1A_1$ from the ${}^1D(3d^84s^2)$ state which originally repells H_2. After the avoided crossing this last curve descends and suffers a second symmetry avoided crossing with the lowest ${}^1A_1$ curve of a ${}^1D(3d^94s^1)$ configuration which afterwards reaches its minimum, first reported by Blomberg and Siegbahn [51]. We have now evidence [52] that this minimum derives from a series of avoided crossings stemming from the ${}^1S(3d^{10})$ excited state of Ni. This would explain the closed-shell character of the ground state singlet reported in Ref. [51].

Iron was studied by Rubinovitz and Nixon [53] experimentally and they report that it is an excited state ${}^5P(3d^74s)$ which captures H_2. On the other hand, theoretical results [54] on the lowest triplet of the FeH_2 system, ${}^3F(3d^74s^1)$ also show two shallow attractive wells for the ${}^3B_1$ and ${}^3B_2$ representations of C_{2v}. Our own calculations [37] show that these two wells show important contributions of a $(3d^84s^0)$ configuration, hinting that a symmetry avoided crossing with a higher ${}^3F$ state takes place in the forma-

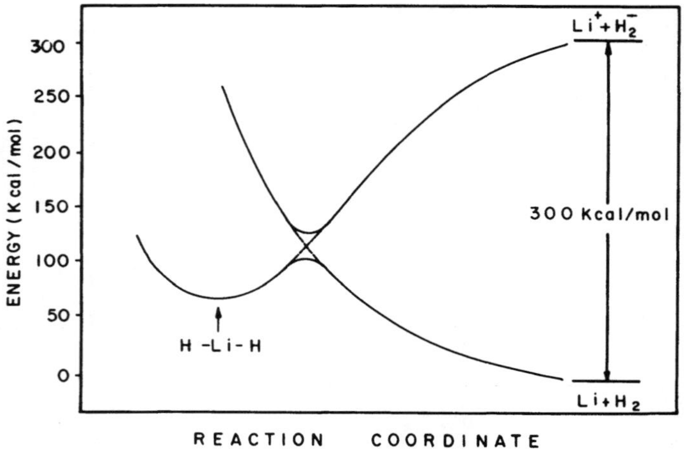

FIGURE 5.9. Covalent and ionic potential energy curves of the LiH_2 system at C_{2v} symmetry. The avoided crossing here implies a charge transfer process.

tion of these $^3B_1$ and $^3B_2$ wells. At the present moment our calculations [37] on the behaviour of these excited ^{5}P and ^{3}F states of FeH_2 show that indeed symmetry avoided crossings are important for this system also.

5.4.7 The Cases of Li and Be

Now let us look at nontransition metals, as Li and Be, to see their differences and similarities to the cases so far studied. The Li + H_2 reaction was studied by our group [55]. The ground states of both the linear form and the C_{2v} symmetry LiH_2 structures are highly ionic. In Fig. 5.9 we show the avoided crossings between the ground state and ionic state of LiH_2 at the C_{2v} configuration. So we see that even in this simple case the nature of the ground state stems from on avoided crossing. Here the absence of an inner *d*-shell obviously makes for a very different behavior than transition-metal atoms and Li lacks the versatility of the other metals studied thus far. It only can interact through its high ionicity, but even in this case we can formally describe its interactions with H_2 using a similar image as the one used for transition metal atoms in the previous subsections.

The case of beryllium is quite intriguing: Be_2 is (with He_2) the weakest and hardest to detect elementary molecule in nature, its equilibrium distance is almost 4Å, an enormous separation for such a small atom. However, bulk berylium is a semimetal with internuclear distances of 2Å. Theoretical calculations [56]-[57] show that Be_4 is already a quite stable system. The cause is more or less evident, after all, the approach of a third Be atom to (weakly bound) Be_2 necessarily implies the participation of the empty 2*p*-orbitals, so as not to violate Pauli's principle. This produces enormous

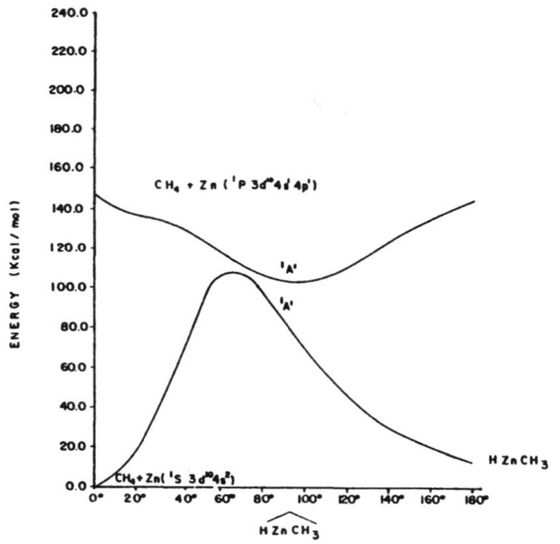

FIGURE 5.10. Potential energy curves for the $ZnCH_4$ system. The two lowest singlets with $^1A'$ representations of C_s are reported.

three-body nonadditive attractions which are solely responsible for the stability of Be_4 and beyond [57]. Again, excited states are responsible for the binding.

This overly simple examples are only brought to give a more general perspective to the main aspect of the previous results; the role of excited states.

5.4.8 The Cases of Zn and Ga

For Zn and Ga we shall study [37] their interactions with H_2 when the metal atoms are in their neutral state and in their ionized and doubly-ionized states. For each of these cases the ground state and first excited states are simultaneously optimized at a multiconfiguration plus extensive CI level [37]. Independently, the interaction of neutral Zn with methane was studied [50]. This particular system is depicted in Fig. 5.10 where we see that the neutral ground state of Zn $^1S(3d^{10}4s^2)$ is initially repulsive, while the first excited state $^1P(3d^{10}4s^14p^1)$ is attractive towards the CH_4 molecule. Both curves belong to the $^1A'$ representation of C_s and they meet at an opening angle H-Zn-CH_3 of 60°, for which an avoided crossing inevitably occurs causing the formation of two potential energy well minima, a shallow one for the upper curve at an angle of 100° and a deep well for the linear (C_{3v} symmetry) form of ground-state $HZnCH_3$.

Thus this system is a perfect example of our main proposal; the ground state of Zn only changes its initial methane repulsion to an attraction due to a symmetry-avoided crossing with a descending curve from an excited Zn state (see Fig. 5.10). The population analysis of the C_{3v} $HCuCH_3$ system

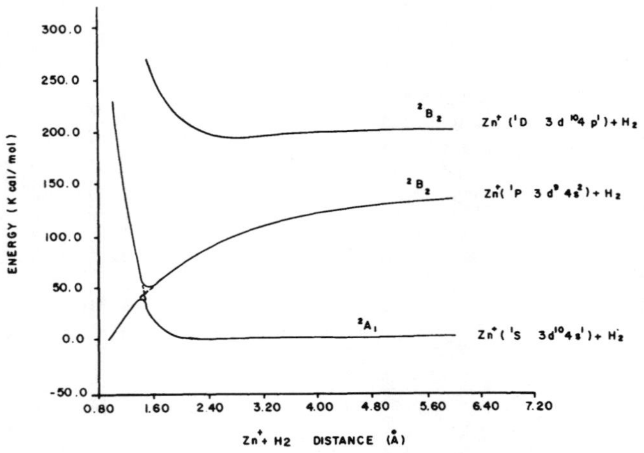

FIGURE 5.11. Potential energy curves of the Zn^+ H_2 system. The three lowest doublets are reported.

shows the participation of the excited ^{2}P state in its formation [50].

Now we pass to the interactions with H_2. Neutral Zn was studied in its two lowest singlet states already discussed in Fig. 5.10 as well as its lowest triplet $^3P(3d^{10}4s^1 4p^1)$ which lies between them. The interaction of these three states with H_2 will not be presented here however, considering that we have already discussed the Zn-CH_4 interactions and it is not worth devoting more space to them. Suffice it to say that again the Zn ground state is originally repulsive (in C_{2v} symmetry it belongs to the $^1A_1$ representation), while the $^3B_2$ curve is non-binding. Just as in the Zn + CH_4 case it is the ^{1}P state that captures H_2, specially in its $^1B_2$ representation. See Ref. [37] for details. More interesting to us here is the case of $Zn^+ + H_2$ for which we studied the three lowest doublets, the ground state $^2A_1$ curve associated to $Zn^{+2}S(3d^{10}4s^1)$ and the two lowest $^2B_2$ curves associated to the first excited state of $Zn^{+2}P(3d^{10}4s^0 4p^1)$ and its second excited state $^2D(3d^9 4s^2)$. Notice that this Zn^+ ordering coincides qualitatively with the case of silver discussed before. The Zn^+H_2 curves also have a behavior similar to those of Ag as Fig. 5.11 shows. The $^2B_2$ (^{2}D) curve is nonbinding, $^2B_2(^2P)$ is quite attractive while $^2A_1(^2S)$ is repulsive. The two last curves cross when the Zn^+-H_2 center-of-mass separation is ~ 1.7 Å. At this crossing the situation is identical to that discussed in Subsection 5.4.4 for silver. A vibronic coupling represented by the dashed lines in Fig. 5.11 makes $^2A_1$ revert its ascent and reaches a shallow minimum for a linear Zn^+H_2 ground state. The similarities with Ag are evident and we shall not dwell on this any longer.

The case of Zn^{++} interacting with H_2 is also interesting. Its ground state has a $^1A_1$ curve stemming from the Zn^{++} ground state $^1S(3d^{10})$. This im-

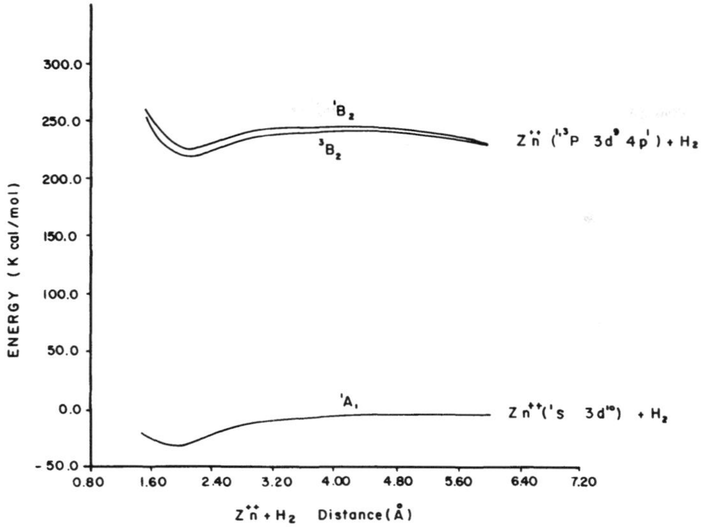

FIGURE 5.12. Potential energy curves of the $Zn^{++}H_2$ system. The two lowest singlets and a triplet are reported.

plies a situation similar to the case of Pd, discussed in Subsection 5.4.2. Figure 12 shows the mild attraction between H_2 and ground-state Zn^{++}. Two other states, the lowest triplet and first excited singlet of Zn^{++} are also weakly attractive in their B_2 representations of C_{2v} for $Zn^{++}H_2$, as Fig. 5.12 shows. Their separation from the ground state $^1A_1$ is very large ($\sim$ 250 kcal/mol) so they do not play a role in its stabilization. In any case the behavior of Zn^{++} is so similar to the case of Pd that no further discussion will be given here [37]. The case of Ga will be discussed only very briefly here, due to limitations of space. Details about the interactions of H_2 with Ga, Ga^+ and Ga^{++} in C_{2v} symmetry will be given elsewhere [37]. Here we shall mention only that a similar behavior to the other metals discussed above is observed. Neutral Ga is interesting because its first excited states involve the $n = 5$ shell orbitals, in the states $^2S(3d^{10}4s^25s^1)$ and $^2P(3d^{10}4s^25s^05p^1)$ which lie 70 and 93 kcal/mol above the $^2P(3d^{10}4s^24p^1)$ ground state. The latter strongly repells H_2 until its $^2B_2$ curve meets the first excited $^2B_2$ attractive curve stemming from the Ga $^2P(3d^{10}4s^25p^1)$ state. After that, the ground state $^2B_2$ curve goes down to a well which is slightly above the asymptotic energy of the Ga and H_2 reactants. This well again stems from an avoided crossing [37].

Ga^+ was studied in its ground state singlet and the first triplet and singlet excited states. The latter is the one responsible for the H_2 capture [37] much like the case of neutral Zn. Ga^{++} was studied in its ground and first three excited doublets. The first excited state $^2P(3d^{10}4s^04p^1)$ is responsible for the C_{2v} capture of H_2 in a $^2B_2$ configuration, and via

Herzberg-Teller couplings generates the avoided crossing responsible for the ground state $Ga^{++}H_2$ potential energy well. This again is similar to the discussion of Ag and Zn^+ given above.

We have now presented enough examples of the important role played by symmetry avoided crossings and we now proceed to our final conclusions.

5.5 Conclusions

We now will attempt to correlate the main result repeatedly obtained here —the existence of multiple avoided crossings among the transition-metal excited-state interaction curves with different adsorbates— to the catalytic phenomenon. The transition metal atoms have a notable characteristic: they have a large number of low-lying excited states that are only separated by a few kcal/mol in energy from each other. This is specially true for the heavier transition metals (*i.e.* those that are the most versatile and powerful catalysts). It has repeatedly been observed [16]-[22] that when these atoms are in organometallic complexes (typically the case of homogeneous catalysts) their electronic states do not correspond to their ground states but to excited states of maximal *d*-shell occupancy. For pure metal catalysts, on the other hand, it has been proposed [23] that in the irregularities of the surfaces of the catalyst (and we must remember that it is in these irregularities that the active centers of heterogeneous catalysts are generally located [58]), the metal atoms are also in excited states. This is related to the insaturation of multibody nonadditive contributions which precisely occurs at such irregularities [57].

What we have added here to the above picture concerning the role of excited metal atom states in catalysis, is a quite substantial ammount of evidence for the systematic presence of avoided crossings in metal-adsorbate interactions. We have used a couple of examples of small adsorbates: hydrogen, methane (in previous work also nitrogen [27], ethylene [29], etc.) On the other hand many different metals and metal ions were tested here including: Pt, Pd, Ni, Ru, Fe, Ag, Cu, Zn, Zn^+, Zn^{++}, Ga, etc. From all of them a systematic picture has developed. Symmetry avoided crossings are the cause of H_2 or CH_4 capture by most of these metals. The ground state of the metal H_2 complex has a population compatible with a maximal occupancy *d*-subsell which corresponds to a metal atom excited state that descends and interacts with lower, initially inactive states. Thus the excited states of maximal *d*-occupancy apparently play a major role in the initiation of the catalytic process. If this is confirmed in future work it would have interesting consequences for the understanding of the catalytic phenomenon.

5.6 Acknowledgements

All of the new results presented here were obtained by my students presently (or recently) finishing their Ph.D. or M.Sc. theses. Among them I thank: J.M. Martínez-Magadan, S. Castillo, F. Colmenares, C. Jarque, A. Ramírez, R. Santamaría, and G. Casillas.

Another, long overdue acknowledgement involves an incident that occurred almost 25 years ago. I had decided to leave my studies. Marcos Moshinsky thought I had the potential to do research. I told him he was wrong. He didn't listen and urged me to do research in Molecular Physics. Perhaps he may have been wrong, but this shows that even so, he usually gets what he wants.

5.7 References

[1] T. Brody and M. Moshinsky, *Tables of Transformation Brackets* (Instituto de Física, UNAM, 1960).

[2] M. Moshinsky, *Nucl. Phys.* **A107**, 481 (1968) and references therein.

[3] M. Moshinsky, *J. Math. Phys.* **7**, 691 (1966) and references therein.

[4] M. Moshinsky and O. Novaro, *J. Chem. Phys.* **48**, 4162 (1968).

[5] E. Chacón, M. Moshinsky, O. Novaro, and C. Wulfman, *Phys. Rev.* **A3**, 166 (1971).

[6] M. Moshinsky, A. Calles, and O. Novaro, *J. Physique* **31**, 125 (1970).

[7] A. Calles, M. Moshinsky, O. Novaro, and O. Sinanoğlu, *Phys. Lett.* **40A**, 79 (1972).

[8] J. von Neumann and E. Wigner, *Zeits. Phys.* **30**, 467 (1929).

[9] C. Coulson, *Proc. Camb. Phil. Soc.* **33**, 111 (1937).

[10] C. Coulson, *Valence* (Oxford Univ. Press, London, 1972).

[11] C. Coulson and A. Joseph, *Int. J. Quantum Chem.* **1**, 337 (1967).

[12] J. Chatt and L.A. Duncanson, *J. Chem. Soc.* **2939**, (1953).

[13] P. Cossee, *J. Catal* **3**, 80 (1964).

[14] R.C. Baetzold, *J. Chem. Phys.* **55**, 4363 (1971).

[15] R.C. Baetzold, *J. Am. Chem.* **82**, 738 (1975).

[16] D.R. Armstrong, P.G. Perkins, and J.J.P. Stewart, *J. Chem. Soc.* (Dalton) (1972).

[17] O. Novaro, S. Chow, and P. Magnouat, *J. Cat.* **42**, 131 (1976).

[18] D.R. Armstrong, O. Novaro, M.E. Ruiz, and R. Linarte, *J. Cat.* **51**, 108 (1978).

[19] O. Novaro, E. Blaisten, E. Clementi, G. Giunchi, and M.E. Ruiz, *J. Chem. Phys.* **68**, 2337 (1978).

[20] J.O. Noell and P.J. Hay, *J. Am. Chem. Soc.* **104**, 4578 (1982).

[21] S. Obara, K. Kitaura, and K. Morokuma, *J. Am. Chem. Soc.* **107**, 5653 (1985).

[22] J.J. Low and W.A. Goddard, *J. Am. Chem. Soc.* **106**, 6928 (1984).

[23] J. García-Prieto, M.E. Ruiz, and O. Novaro, *J. Chem. Soc.* **107**, 5653 (1985).

[24] M.E. Ruiz, J. García-Prieto, and O. Novaro, *J. Chem. Phys* **50**, 1529 (1984).

[25] J. García-Prieto, M.E. Ruiz, E. Poulain, G.A. Ozin, and O. Novaro **81**, 5920 (1984).

[26] M.E. Ruiz, J. García-Prieto, E. Poulain, G.A. Ozin, R.A. Poirier, S.M. Mattar, I.G. Csizmadia, C. Gracie, and O. Novaro, *J. Phys. Chem.* **90**, 279 (1986).

[27] M. Sánchez-Zamora, O. Novaro, and M.E. Ruiz, *J. Phys. Chem.* **94**, 2746 (1990).

[28] G.A. Ozin, D.F. Mclntosh, S.A. Mitchel, and J. García-Prieto, *J. Am. Chem. Soc.* **103**, 1575 (1981).

[29] C. Jarque, O. Novaro, and M. E. Ruiz, *Mol. Phys.* **62**, 129 (1987).

[30] C. Jarque, Ph.D. Thesis, Cambridge University (in preparation).

[31] M. Dupuis, J. Rys, and H.F. King, *J. Chem. Phys.* **65**, 111 (1976).

[32] J.P. Daudey, Laboratoire de Physique Quantique Internal Publication, Université Paul Sabatier (Toulouse, France 1990), available on request.

[33] J. C. Barthelat, P. Durand, and A. Serafini, *Mol. Phys.* **33**, 179 (1977).

[34] L.A. LaJohn, P.A. Christiansen, R.B. Ross, T. Asharoo, and W.C. Ermler, *J. Chem. Phys.* **36**, 2812 (1987).

[35] B. Huron, P. Rancurel, and J.P. Malrieu, *J. Chem. Phys* **58**, 5745 (1973).

[36] M. Pelissier, *J. Chem Phys.* **58**, 5745 (1983).

[37] J.M. Martínez-Magadán, M.Sc. Thesis, Fac. de Química, UNAM, 1991 (in preparation).

[38] E. Poulain, J. García-Prieto, M.E. Ruiz, and O. Novaro, *Internat. J. Quantum Chem.* **29**, 1181 (1986).

[39] F. Colmenares, M.Sc. Thesis, Fac. de Química, UNAM, Dec. 1990.

[40] G. Herzberg, *Electronic Spectra of Polyatomic Molecules* (Van Nostrand, New York, 1966) p.67.

[41] S.A. Mitchell, J. Farrell, G. A. Kenney, and G.A. Ozin, *J. Am. Chem. Soc.* **102**, 7702 (1980).

[42] A. Ramírez-Solís, Ph.D. Thesis, Université P. Sabatier (Toulouse, France) August 1990.

[43] A. Ramírez-Solís, J.P. Daudey, O. Novaro, and M.E. Ruíz, *Zeits. Phys.* **D15**, 71 (1990).

[44] G.A. Ozin and C. Gracie, *J. Phys. Chem.* **84**, 643 (1984).

[45] E. Garfield, *Current Contents* **12**, 3 (1985).

[46] G.A. Ozin and J.G. Mc Caffrey, *J. Am. Chem. Soc.* **104**, 7351 (1982).

[47] J.M. Parniss, S.A. Mitchell, J. García-Prieto, and G.A. Ozin, *J. Am. Chem. Soc.* **107**, 8169 (1985).

[48] R.A. Poirier, G.A. Ozin, D.F. Mclntosh, I.G. Czismadia, and R. Daudel, *Chem. Phys. Lett.* **101**, 221 (1983).

[49] G.E. Quelch and I.H. Hillier, *Chem. Phys.* **121**, 183 (1988).

[50] S. Castillo, Ph.D. Thesis, Fac. de Ciencias, UNAM (in preparation).

[51] M.R.A. Blomberg and P.E.M. Siegbahn, *J. Chem. Phys.* **78**, 5682 (1983).

[52] G. Casillas, B.Sc. Thesis, Fac. de Ciencias, UNAM (in preparation).

[53] R.L. Rubinowitz and E.R. Nixon, *J. Phys. Chem.* **90**, 1942 (1986).

[54] P.E.M. Siegbahn, M.R.A. Blomberg, and C.W. Bauschlicher, *J. Chem. Phys.* **81**, 1373 (1984).

[55] J. García-Prieto, W.L. Feng, and O. Novaro, *Chem. Phys. Lett.* **119**, 128 (1985).

[56] C.W. Bauschlicher, D.H. Liskow, C. F. Bender, and H.F. Schaefer, *J. Chem. Phys.* **62**, 4815 (1975).

[57] O. Novaro and W. Kołos, *J. Chem. Phys.* **67**, 5066 (1977).

[58] J.F. Benson, H.S. Hwang, and M. Boundart *J. Catal.* **30**, 146 (1973).

6

The Symplectic Model and Potential-Energy-Surfaces

Octavio Castaños[1] and Peter O. Hess[2]

ABSTRACT We give a brief summary of the (pseudo-)symplectic model. We then show how to construct a microscopically justified potential-energy-surface (PES) for this model, using coherent states. Two applications are presented, one to the light nucleus $^{24}_{12}$Mg and the other to the heavy nucleus $^{238}_{92}$U.

6.1 Introduction

We want to thank the Organizing Committee for having invited us to this Symposium in honor to Marcos Moshinsky. The authors were former students of his and have collaborated with him on many research topics. Between them, the work related to the symplectic model of nuclei plays a prominent role [1]. There are other groups in the world which also have made important contributions to this subject [2]-[5].

The symplectic model is an extension of the SU(3) model of Elliott [6]. The SU(3) model can only be applied to $2s$–$1d$ light nuclei because for higher shell nuclei, $N > 3$, the spin-orbit-coupling pushes the highest spin-orbital $j = N + \frac{1}{2}$ to the next lowest shell. In the seventies, Hecht *et al.* [7] proposed the pseudo-SU(3) ($\widetilde{\mathrm{SU}}(3)$) model, which takes effectively into account the spin-orbit interaction and therefore can be applied to heavy nuclei.

But both SU(3) and $\widetilde{\mathrm{SU}}(3)$ suffer from the fact that they need effective charges which essentially have to be adjusted. These effective charges are due to the restriction of calculations to a single shell. In reality, a microscopic hamiltonian with the physical quadrupole-quadrupole interaction already mixes different shells. This mixture is taken into account by the symplectic model. Recently, in Ref. [5] a simplified, contracted, version of this model was constructed and applied to light nuclei. Unfortunately, due to the spin-orbit interaction this version can only be applied to light nuclei.

[1]Instituto de Ciencias Nucleares, UNAM, México.

[2]*id.* and fellow of the *Deutscher Akademischer Austauschdienst.*

In order to describe heavy nuclei the pseudo-symplectic ($\widetilde{\text{Sp}}(3,R)$) model was developed [8].

The Sp(3,R) and $\widetilde{\text{Sp}}(3,R)$ models, which have been effective in describing low energy collective spectra and $B(E2)$-transitions, can be related to the geometrical model following the procedure indicated in Ref. [9].

First, we will give in Section 2 a short review of the symplectic model Sp(3,R), the $\widetilde{\text{SU}}(3)$ and $\widetilde{\text{Sp}}(3,R)$ models. In Section 3, we derive the PES for the $\widetilde{\text{Sp}}(3,R)$ model using coherent states [10]. In Section 4 we will apply the method to the light nucleus $^{24}_{12}$Mg and to the heavy nucleus $^{238}_{92}$U. In Section 5 we will discuss the results obtained and give an outlook to possible applications.

6.2 The pseudo-symplectic model

As is well known, the SU(3) symmetry is badly broken for heavy nuclei due to the spin-orbit interaction. In each shell N the highest spin member, $j_{\max} = N + \frac{1}{2}$, is pushed down among the orbitals of the next lower shell. Nevertheless, the remaining orbitals can be mapped to states of a pseudo oscillator ($\widetilde{\text{SU}}(3)$). To achieve this, one maps the orbital, $j = l \pm \frac{1}{2}$ to the pseudo orbitals $j = \tilde{l} \mp \frac{1}{2}$ (excluding $j_{\max} = N + \frac{1}{2}$) where $\tilde{l}$ is the pseudo-orbital momentum and $\tilde{s} = \frac{1}{2}$ the pseudo-spin. One obtains then the structure of the $\tilde{N} = N - 1$ pseudo-oscillator shell. The pseudo spin-orbit partners, with the same $\tilde{l}$, are experimentally nearly degenerate [7, 11, 12]. The separation of these doublets, which can be reproduced by an $\tilde{l}^2$ term, is washed out in deformed systems by the quadrupole-quadrupole interaction [13]. For increasing deformation, the classification according to $\widetilde{\text{SU}}(3)$ is even better realized, as shown in Ref. [14]. The near degeneracy of the pseudo-spin-orbit doublets implies a small pseudo spin-orbit coupling. The $\widetilde{\text{SU}}(3)$ scheme has been used with success in calculations of a variety of physical phenomena, such as backbending for heavy deformed nuclei [15, 16] and the structure of super deformed bands [17, 18].

For heavy nuclei, one has to distinguish the proton shells from the neutron shells. In each space one has the "normal" parity orbital, which can be identified with a pseudo shell with $\tilde{N}_\alpha = N_\alpha - 1$ ($\alpha = \pi$ for protons and ν for neutrons), and the intruder orbital from the shell above. A general state can then be written as a direct product of a state classifying the "normal" parity states by a $\widetilde{\text{SU}}(3)$ irreducible representation (irrep) and a state for the "abnormal" parity orbital classified by the one use a seniority scheme. For low lying collective states, the nucleons in the "abnormal" parity orbitals can be considered to be in a seniority zero state. Only when the backbending region is reached do other seniorities other than zero become important. Therefore, below the backbending region, states of a many-particle basis can be written as a coupled product of proton and neutron states each classified in the $\widetilde{\mathrm{SU}}(3)$ symmetry

$$|\psi\rangle = \left[|(\lambda_\pi,\mu_\pi)\rangle \times |(\lambda_\nu,\mu_\nu)\rangle\right]^{\rho(\lambda\mu)}_{KLM}, \tag{2.1}$$

where we have used the usual notation of Elliott [6] and the index "π" refers to protons and "ν" to neutrons.

The pseudo symplectic model ($\widetilde{\mathrm{Sp}}(3,R)$) extends $\widetilde{\mathrm{SU}}(3)$ in a similar way as the Sp(3,R) model does for SU(3). We will therefore first review the Sp(3,R) model, which is applicable only to light nuclei. (For more details see Ref. [5] and references therein). As a microscopic hamiltonian we use

$$H = \hbar\omega N - \tfrac{1}{2}\chi[(Q^c\cdot Q^c) - (Q^c\cdot Q^c)^{TE}] + H_r. \tag{2.2}$$

In (2.2) N is the harmonic oscillator part and counts the total number of oscillator quanta. It is written as a sum $N = N_\pi + N_\nu$ where N_π counts the quanta carried by protons and N_ν by neutrons. The operator Q^c_{2m} is the mass quadrupole operator, *i.e.*

$$Q^c_{2m} = \sqrt{\frac{16\pi}{5}}\sum_{s=1}^{A} r_s^2 Y_{2m}(\Omega_s) \tag{2.3}$$

which, in contrast to the Elliott quadrupole operator, mixes shells. The operator H_r is a residual interaction defined in Ref. [16] and is essentially the SU(3) counter part of the rigid rotor. It serves to lift the degeneracy of equal angular momentum states within the same irrep (irreducible representation) of SU(3). This interaction will play no significant role for the construction of the PES and we refer the reader to Ref. [16]. The interaction $(Q^c\cdot Q^c)^{TE}$ in Eq. (2.2) refers to the "trace equivalent" part. It is just the diagonal contribution of $(Q^c\cdot Q^c)$ averaged over each shell. Substracting this from $(Q^c\cdot Q^c)$ results in a quadrupole-quadrupole interaction which does preserve the oscillator structure, with spacing of $2\hbar\omega$, when averaged over all states within each shell. The explicit form of $(Q^c\cdot Q^c)^{TE}$ is [5]

$$(Q^c\cdot Q^c)^{TE} = \tfrac{65}{7}N_c^2 + \tfrac{250}{7}N_c + \tfrac{10}{3}N_sN_c + \tfrac{5}{2}C_2(\lambda_s,\mu_s) + 10N_s, \tag{2.4}$$

where N_s is the total number of quanta in the lowest SU(3)-irrep (λ_s,μ_s) and $C_2(\lambda_s,\mu_s) = \lambda_s^2 + \lambda_s\mu_s + \mu_s^2 + 3\lambda_s + 3\mu_s$, while $2N_c$ counts the number of collective excitation quanta. Note that the operators appearing in (2.2) are generators of the Sp(3,R) group. This group consists of the generators [$\eta_{sm}(m = -1,0,+1, s = 1,\cdots,A-1)$ is a creation operator of one oscillator quanta and ξ^{sm} is an annihilation operator with $[\xi_1^{m'}\eta_m] = \delta_m^{m'}$, see Ref. [5])]

$$\begin{aligned} B^+_{lm} &= [\eta\times\eta]^l_m, \\ B_{lm} &= [\xi\times\xi]^l_m, \qquad l = 0,2, \\ C_{lm} &= \tfrac{1}{2}\left([\eta\times\xi]^l_m + [\xi\times\eta]^l_m\right), \qquad l = 0,1,2. \end{aligned} \tag{2.5}$$

Alternatively, we can define the generators $Q^a_m = \sqrt{6}C_{2m}$ and $L^a_m = -\sqrt{2}C_{1m}$, the latter being the angular momentum. (The index "a" refers to operators which only act within a shell.) The physical quadrupole operator is [5]

$$Q^c_m = Q^a_m + \tfrac{\sqrt{6}}{2}(B^+_{2m} + B_{2m}). \tag{2.6}$$

With H being a function of the Sp(3,R) generators we can divide the shell model space into irreps of Sp(3,R) and restrict our considerations to within the lowest irrep in energy or the lowest ones. A irrep is defined by the lowest weight state

$$|N_s(\lambda_s,\mu_s)KL-L\rangle \equiv |\text{min}\rangle \tag{2.7}$$

which satisfies $B_{lm}|\text{min}\rangle = 0$ and $Q_{2-2}|\text{min}\rangle = Q_{2-1}|\text{min}\rangle = L_{-1}|\text{min}\rangle = 0$ [Q_{2-2}, Q_{2-1}, L_{-1} and B_{lm} ($l = 0,2$) are the lowering operators of the Sp(3,R)-algebra]. In (2.7) we use the usual notation to denote an SU(3)-state, "s" refering to "shell". In (2.7) (λ_s,μ_s) denotes the SU(3)-irrep, N_s the total number of quanta plus $\frac{1}{2}n$ ($n = A-1$), K is a multiplicity index, l the orbital angular momentum and j the total spin. The states of a given $Sp(3,R)$ irrep, defined by the minimum weight state (2.7), are finally obtained by applying an arbitrary polynomial in the pair creation operators B^+_{lm} and an appropriate coupling. This irrep is still infinite, but in practical calculations one truncates in the total number of pair excitations.

Instead of working in the full Sp(3,R) model it is easier to go to the contracted limit. What essentially is done is to approximate the pair operators B^+_{lm} and B_{lm} by ideal boson operators via [5]

$$B^+_{lm} = \sqrt{\frac{4N_s}{3}}\,b^+_{lm} \quad \text{and} \quad B_{lm} = (B^+_{lm})^+. \tag{2.8a}$$

Observing that N_s is a big number (*e.g.* 62.5 for $^{24}_{12}$Mg and 574.5 for $^{238}_{92}$U) the operators b^+_{lm} and b_{lm} fulfill, within a good approximation, the usual boson commutation relations. This property can be related to a technique called "group contraction" and the interesting reader is refered to Ref. [5] and references therein. In this procedure the generators Q^a_m and L^a_m are simple sums of operators acting only within the SU(3) irrep given by the minimum weight state and of operators acting on the "bosonic" part without changing quanta, *e.g.*, $Q^a_m = Q^s_m + Q^b_m$ with "s" refering to the Elliott operator acting only within the SU(3)-irrep of the valence shell and

$$Q^b_m = -2\sqrt{2}\left((b^+_{2m}b_{00} + b^+_{00}b_{2m}) + \tfrac{\sqrt{7}}{2}\left[b^+ \times b\right]^2_m\right). \tag{2.8b}$$

The important advantage of working in the contracted model is that a general state can now be constructed as the product of the states of the SU(3) irrep (λ_s,μ_s), defined by the lowest one in a Sp(3,R) irrep, and a polynomial in b^+_{lm} ($l = 0,2$) creating an U(6) state. Calculations of matrix elements are greatly simplified. The corrections to the exact result are of order $1/N_s$. Within the contracted symplectic model, the microscopic hamiltonian (2.2) acquires the form

$$\begin{aligned} H^c &= 2\hbar\omega N_b + \chi\Big[\big(\tfrac{5}{3}N_s + \tfrac{75}{7}\big)N_b + \tfrac{25}{14}N_b^2 \\ &\quad -\tfrac{1}{2}\mathbf{Q}^a\cdot\mathbf{Q}^a - \sqrt{2N_s}(\mathbf{b}^+\cdot\mathbf{Q}^a + \mathbf{Q}^a\cdot\mathbf{b}) \\ &\quad +10\sqrt{N_s}(b_{00}^+ + b_{00}) - N_s(\mathbf{b}^+\cdot\mathbf{b}^+ + \mathbf{b}\cdot\mathbf{b}) \\ &\quad -2N_sN_d + \tfrac{5}{4}C_2(\lambda_s,\mu_s)\Big] + H_r. \end{aligned} \tag{2.9}$$

In Ref. [5] and [19], the contracted Sp(3,R) model was applied to nuclei in the s-d-shell.

What is then the $\widetilde{\text{Sp}}(3,R)$ model? First one has to rewrite the generators of Sp(3,R) into tensors of $\widetilde{\text{SU}}(3)$. This is done in Ref. [19] and [20]. A given operator $\hat{\mathcal{O}}$ (standing for a generator of Sp(3,R)) is a sum of different tensors of $\widetilde{\text{SU}}(3)$ whose leading term transforms to $\widetilde{\text{SU}}(3)$ as $\hat{\mathcal{O}}$ does for SU(3), *i.e.*

$$\hat{\mathcal{O}} = \kappa\tilde{\hat{\mathcal{O}}} + \cdots \tag{2.10}$$

The other terms give contributions of less then ten percent [8]. The κ-factor is not exactly one and can be deduced using the tables of Ref. [19] and [20]. Rewriting the hamiltonian (2.2) in terms of the pseudo-operators leads then to a similar structure

$$\tilde{H} = \hbar\omega\tilde{N} - \tfrac{1}{2}\chi\big((\tilde{Q}^c\cdot\tilde{Q}^c) - (\tilde{Q}^c\cdot\tilde{Q}^c)^{TE}\big) + \tilde{H}_r$$

where the κ–factor of $\tilde{N}$ is one and for $\tilde{Q}^c_{2m}$ it is absorbed into the quadrupole strength parameter χ. Now we can proceed in the same fashion as before, *i.e.* the extension of $\widetilde{\text{SU}}(3)$ to $\widetilde{\text{Sp}}(3,R)$ is done in the same way as SU(3) to Sp(3,R), when we stay within the "normal" parity states.

What happens with mixtures of the "abnormal" states? If we consider multiple $2\hbar\omega$ excitations in the "normal" parity space, should we not also consider excitations within the "abnormal" parity states, as well as mixing with the "normal" parity states? The answer is that this is not necessary for states in even-even nuclei that lie below the backbending region, because the "abnormal" parity parts of these wave functions will be dominated by pairing correlations which differ from quadrupole correlations in that they do require couplings to higher shells for strength enhancement. Only for states immediately below and above the pairing gap, which is coincident with the onset of backbending, or for odd-A nuclei are "abnormal" parity configurations expected to play an important role in the dynamics.

In the next section we will show how an effective potential can be derived from the above microscopic hamiltonian using coherent states. There we will also indicate how to determine the parameters of the theory relevant for the construction of the PES. (The relevant parameters are $\hbar\omega$, N_s, χ and κ.)

6.3 A procedure to construct a PES

A usual definition of an effective potential is

$$V_{\text{eff}}(\alpha) = \langle \psi(\alpha)|H|\psi(\alpha)\rangle, \tag{3.1}$$

where $\psi(\alpha)$ is a state depending on parameters which in our case are related to collective coordinates. In order to probe the entire microscopic space, $\psi(\alpha)$ should contain contributions from all the Hilbert space. One possible choice, in our problem, is to use a direct product of a combination of states in the SU(3) irrep (λ_s, μ_s) and a coherent state in the operators b^+_{lm}, b_{lm} (*i.e.* we have to vary over contributions given by shell excitations and over all Sp(3,R)-irreps). Let us be more specific:

(i) The SU(3) states are denoted by $|N_s(\lambda_s\mu_s)\sigma\rangle \equiv |q\rangle$ and the expectation value of the operator Q^s_m is given by

$$\langle q|Q^s_m|q\rangle = q^s_m, \tag{3.2}$$

where q^s_m is a c-number. In fact, it can be shown that the state $|N_s(\lambda_s\mu_s)\sigma\rangle$, in the limit of small angular momentum compared to the size of the irrep (λ_s, μ_s), becomes an eigenstate of Q^s_m (see Refs. [16, 21]–[23]). We define

$$q^s_m = \sqrt{\frac{5}{\pi}} N_s \alpha^s_m, \tag{3.3}$$

because then $(\alpha^s \cdot \alpha^s)$ is related to the deformation squared β^2_s of the state $|q\rangle$. In fact, β_s, and γ_s can be related to λ_s and μ_s [16].

(ii) For the coherent state, we use

$$|\alpha\rangle \equiv \exp(\alpha_2 \cdot (b^+_2 - b_2) + \alpha_0(b^+_0 - b_0))|0\rangle = e^x|0\rangle, \tag{3.4}$$

where $|0\rangle$ is the boson vacuum defined via $b_{lm}|0\rangle = 0$ and α_{2m}, α_{00} are parameters which we have to relate to the variables α^g_m of the geometrical model [9].

The full state is then given by $|\alpha, q\rangle = |\alpha\rangle|q\rangle$. In order to include all the microscopic states, we allow changes to a different Sp(3,R)-irrep when the resulting value for V_{eff} is lower. This is perfectly allowed, even though H does not change dynamically to different irreps once a given irrep is chosen. Using the relations

$$\begin{aligned}
\left\langle 0 \left| e^{-x} \begin{pmatrix} b^+_{lm} \\ b_{lm} \end{pmatrix} e^x \right| 0 \right\rangle &= \left\langle 0 \left| \begin{pmatrix} b^+_{lm} + \alpha_{lm} \\ b_{lm} + (-1)^{l-m}\alpha_{lm} \end{pmatrix} \right| 0 \right\rangle \\
&= \begin{pmatrix} \alpha_{lm} \\ (-1)^{l-m}\alpha_{l-m} \end{pmatrix}, \\
N_b &= N_d + b^+_{00} b_{00} = b^+_2 \cdot b_2 + b^+_{00} b_{00},
\end{aligned}$$

and

$$Q_m^a = Q_m^s - 2\sqrt{2}(b_{2m}^+ b_{00} + b_{00}^+ b_{2m} + \frac{\sqrt{7}}{2}[b^+ \times b]_m^2), \tag{3.5}$$

it is straightforward to evaluate the expectation value of the hamiltonian. We then obtain an analytic expression for the microscopically justified energy surface [23]

$$\begin{aligned}\langle \alpha; q|H|\alpha; q\rangle &= \{2\hbar\omega + \chi[\tfrac{3}{2} - \tfrac{7}{3}N_s - \tfrac{87}{7}\alpha_{00}^2 + 16\sqrt{N_s}\alpha_{00}]\}\boldsymbol{\alpha}\cdot\boldsymbol{\alpha} \\ &+\{2\hbar\omega + \tfrac{5}{3}\chi N_s - \tfrac{15}{2}\chi\}\alpha_{00}^2 + 20\chi\sqrt{N_s}\alpha_{00} \\ &+2\sqrt{2}\chi(2\alpha_{00} - \sqrt{N_s})\boldsymbol{\alpha}\cdot\mathbf{q}^s + \tfrac{25}{14}\chi\alpha_{00}^4 \\ &+\sqrt{14}\chi\mathbf{q}^s\cdot[\boldsymbol{\alpha}\times\boldsymbol{\alpha}]^2 + \sqrt{7}(4\sqrt{N_s} - 8\alpha_{00})\chi\boldsymbol{\alpha}\cdot[\boldsymbol{\alpha}\times\boldsymbol{\alpha}]^2 \\ &-\tfrac{3}{14}\chi(\boldsymbol{\alpha}\cdot\boldsymbol{\alpha})^2 - \tfrac{3}{16}\chi(\mathbf{q}^s\cdot\mathbf{q}^s),\end{aligned} \tag{3.6}$$

where $\langle\alpha, q|4C_2(\lambda_s, \mu_s)|\alpha, q\rangle \cong \mathbf{q}^s\cdot\mathbf{q}^s$ has been used.

Note that *two* types of deformation parameters appear: (i) α_m^s, which is related to the deformation of the lowest SU(3)-irrep within a given Sp(3,R)-irrep, and (ii) α_m which gives the contribution to the deformation resulting from mixtures of higher shells. This will become clearer when we relate α_m^g, the geometric deformation variable, to α_m^s and α_m.

For this purpose, we compare the expectation values of the microscopic monopole and quadrupole operator (b is the oscillator length, $b^2 = \hbar/(m\omega)$)

$$\begin{aligned}\langle\alpha; q|Q_{00}^c|\alpha; q\rangle &= b^2\langle\alpha; q|\{N_s + 2N_b - \sqrt{N_s}(b_{00}^+ + b_{00}\}|\alpha; q\rangle \\ &= b^2\{N_s + 2[\boldsymbol{\alpha}\cdot\boldsymbol{\alpha} + \alpha_{00}^2] - 2\sqrt{N_s}\alpha_{00}\},\end{aligned} \tag{3.7a}$$

$$\begin{aligned}\langle\alpha; q|Q_{2m}^c|\alpha; q\rangle &= b^2\langle\alpha; q|\{Q_m^s + Q_m^b + \sqrt{2N_s}(b_{2m}^+ + b_{2m})\}|\alpha; q\rangle \\ &= b^2\Big\{\sqrt{\tfrac{5}{\pi}}N_s\alpha_m^s - \sqrt{2}[4\alpha_{00}\alpha_{2m} \\ &\qquad +\sqrt{7}[\boldsymbol{\alpha}\times\boldsymbol{\alpha}]_m^2] + 2\sqrt{2N_s}\alpha_{2m}\Big\},\end{aligned} \tag{3.7b}$$

to the expressions of their geometrical counter-parts

$$Q_{00}^c \rightarrow b^2N_s(1 + F(\alpha^g)) \simeq b^2N_s(1 + \tfrac{5}{4\pi}(\alpha^g\cdot\alpha^g)) \tag{3.8a}$$

$$Q_{2m}^c \rightarrow b^2N_s\sqrt{\tfrac{5}{\pi}}G_m(\alpha^g) \simeq b^2N_s\sqrt{\tfrac{5}{\pi}}\left(\alpha_m^g - \sqrt{\tfrac{10}{7\pi}}[\alpha^g\times\alpha^g]_m^2\right). \tag{3.8b}$$

In Eqs. (3.8) we give the form of F and G up to second order in $(\alpha^g\cdot\alpha^g)$ where we used volume conservation, but in general they have more complicated expressions. Now we proceed to compare Eqs. (3.8) and (3.7). First we express α_m and α_{00} in (3.7) by

$$\alpha_m = \sum_{K=0}\alpha_m^{(K)}, \quad \alpha_{00} = \sum_K \alpha_{00}^{(K)}, \tag{3.9}$$

where the upper index (K) refers to the order in α_m^g and α_m^s. Then a Taylor series expansion of F and G in terms of α_m^g is made. Afterwards we compare equations (3.7) with (3.8). We have done this up to second order yielding the results

$$\begin{aligned} \alpha_{2m}^{(0)} &= \alpha_{00}^{(0)} = 0, \\ \alpha_{2m}^{(1)} &= \sqrt{\frac{5}{8\pi}}\sqrt{N_s}(\alpha_m^g - \alpha_m^s), \qquad \alpha_{00}^{(1)} = 0, \end{aligned}$$

and

$$\begin{aligned} \alpha_m^{(2)} &= \frac{5}{16\pi}\sqrt{\frac{N_s}{7}}([\alpha^g \times \alpha^g]_m^2 + 14[\alpha^g \times \alpha^s]_m^2 - 7[\alpha^s \times \alpha^s]_m^2), \\ \alpha_{00}^{(2)} &= \frac{5}{8\pi}\sqrt{N_s}((\alpha^s \cdot \alpha^s) - 2(\alpha^g \cdot \alpha^s)). \end{aligned} \quad (3.10)$$

Note that the geometrical deformation variable is to lowest order a *sum* of the deformation variables α_{2m}^s and α_{2m}, *i.e.* the deformation has a contribution from the lowest SU(3)-irrep in a Sp(3,R)-irrep and from the excitation to higher shells. On substituting this result into Eq. (3.6) and neglecting contributions of order $(1/N_s)$ we are led to the geometrical potential energy form

$$\langle\alpha, q|H|\alpha, q\rangle = V = V_0 + V_1,$$

where the subindices 0 and 1 denote the approximations up to second and the third order in deformation, respectively. The resulting expressions are

$$\begin{aligned} V_0 &= \tfrac{65}{16\pi}\chi N_s^2\beta_s^2 + \tfrac{5}{4\pi}N_s[\hbar\omega - \tfrac{7}{6}\chi N_s](\beta_g - \beta_s)^2 \\ &\quad -\chi N_s^2\tfrac{5}{\pi}\beta_s\beta_g + \tfrac{5}{2\pi}N_s[\hbar\omega + \tfrac{5}{6}\chi N_s]\beta_s\beta_g[1 - \cos(\gamma_g - \gamma_s)] \quad (3.11a) \\ V_1 &= \tfrac{10}{112\pi}\sqrt{\tfrac{5}{\pi}}N_s[\hbar\omega - \tfrac{91}{6}\chi N_s]\beta_g^2\cos^3\gamma_g \\ &\quad + \tfrac{10}{112\pi}\sqrt{\tfrac{5}{\pi}}N_s[\hbar\omega + \tfrac{5}{6}\chi N_s]\beta_s \\ &\quad \times \ [13\beta_g^2\cos(2\gamma_g + \gamma_s) - 21\beta_s\beta_g\cos(\gamma_g + 2\gamma_s) \\ &\quad + \ 7\beta_s^2\cos 3\gamma_s], \end{aligned} \quad (3.11b)$$

where we assumed for convenience that the body fixed system of the two deformation variables α_{2m}^g and α_{2m}^s coincide, *i.e.*

$$\begin{aligned} \beta_s^2 &= (\alpha^s \cdot \alpha^s), \\ \beta_g^2 &= (\alpha^g \cdot \alpha^g), \\ (\alpha^g \cdot \alpha^s) &= \beta_s\beta_g\cos(\gamma_g - \gamma_s), \\ (\alpha^g \cdot [\alpha^g \times \alpha^g]^2) &= -\sqrt{\tfrac{2}{7}}\beta_g^3\cos(3\gamma_g), \\ (\alpha^s \cdot [\alpha^g \times \alpha^g]^2) &= -\sqrt{\tfrac{2}{7}}\beta_s\beta_g^2\cos(2\gamma_g + \gamma_s), \\ (\alpha^g \cdot [\alpha^s \times \alpha^s]^2) &= -\sqrt{\tfrac{2}{7}}\beta_s^2\beta_g\cos(\gamma_g + 2\gamma_s), \end{aligned}$$

and

$$(\alpha^s \cdot [\alpha^s \times \alpha^s]^2) \;=\; -\sqrt{\tfrac{2}{7}}\beta_s^3 \cos(3\gamma_s).$$

(Note that under the symmetry transformations leading from one body fixed frame to another [9] we have to transform simultaneously γ_g and γ_s, *i.e.* if $\gamma_g \to \gamma_g - \frac{2\pi}{3}$ then also $\gamma_s \to \gamma_s - \frac{2\pi}{3}$ and if $\gamma_g \to -\gamma_g$ also $\gamma_s \to -\gamma_s$. This is because we assumed both body fixed systems equal.) It turns out that for heavy nuclei the contribution given by V_1 is of order less than five percent and can therefore safely be neglected. For light nuclei, these corrections are of the order of ten percent. We will therefore restrict to V_0 only when we apply to specific nuclei.

From (3.11a), we can estimate the position of the minimum yielding

$$(\beta_g)_0 = \left[1 + \frac{2\chi N_s}{(\hbar\omega - \frac{7}{6}\chi N_s)}\right]\beta_s, \tag{3.12}$$

where the subscript "0" denotes values at the minimum. Eq. (3.12) implies that with $\chi \neq 0$ the deformation increases, which reflects the contribution to the deformation of the higher shell excitations.

The PES given by (3.11) is restricted to one irrep of Sp(3,R), or equivalently to a fixed β_s, γ_s which are related to the Sp(3,R)-irrep. However, as pointed out earlier in this section, we have also to vary the parameter β_s, γ_s as long as it corresponds to a permitted Sp(3,R)-irrep. This leads to a set of PES's from which another can be constructed that is a function of four variables $(\beta_g, \gamma_g, \beta_s, \gamma_s)$. In order to obtain a simpler theory which only depends on β_g, γ_g as independent variables, we have to relate β_s, γ_s to β_g, γ_g. We propose a procedure which, for convenience, consists of three distinct steps:

(a) In the first step, we divide the $\gamma-$range in a grid of steplength of 5° (this, of course, can be changed). In each sector we look for the dominant Sp(3,R)-irrep, *i.e.* with largest eigenvalue of the second order Casimir operator of SU(3). This gives us 12 values $((\beta_s)_{\max}, (\gamma_s)_{\max})$.

(b) In the second step, we construct a PES associated with a particular set $((\beta_s)_{\max}, (\gamma_s)_{\max})$. First we fix γ_g and $\gamma_s = (\gamma_s)_{\max}$, so we are left with a two dimensional problem in (β_g, β_s). We will constrain the system to a minimal energy path along the two dimensional surface defined by the potential. This gives us a relation for β_s in terms of β_g. It is done for *fixed but arbitrary* γ_g. As a result we are left with 12 different PES.

(c) In the last step we fix (β_g, γ_g) and ask which of the above PES acquires the lowest value. Having done so in the whole $(\beta_g, \gamma_g)-$plane we obtain finally the desired PES.

The result of the first step depends greatly on the specific nucleus and how to get the correct $SU(3)$ irreps is explained in Ref. [8]. The second step results in [23]

$$\beta_s = \beta_g \tan(\varphi), \tag{3.13a}$$

with

$$\varphi = \tfrac{1}{2}\arctan\left\{\frac{8}{13}\frac{[\hbar\omega + \frac{5}{6}\chi N_s]}{\chi N_s}\cos(\gamma_g - \gamma_s)\right\}. \tag{3.13b}$$

The third step depends on *(a)* and *(b)*. Eq. (3.13a) gives the dependence of β_s and β_g which is, however, restricted by the Pauli principle, *i.e.* when the maximal possible β_s value is reached ($\beta_s = (\beta_s)_{\max}$) it is trapped to this value. Then for higher deformations, $\beta_g \geq (\tan(\varphi))^{-1}(\beta_s)_{\max}$, the potential acquires the form (3.11).

We must still determine the parameters $\hbar\omega, N_s$ and χ. The first one is given by $\hbar\omega = 41A^{-\frac{1}{3}}$ for heavy nuclei [9] and by $\hbar\omega = 45A^{-\frac{1}{3}} - 25A^{-\frac{2}{3}}$ for light nuclei [24]. The number N_s is given by the total number of quanta carried by the SU(3) state (λ_s, μ_s) plus $n/2$ with $n = A - 1$. This is determined by filling at a given, experimentally measured, deformation (β_g) the nucleons (protons and neutrons separately) into the Nilsson orbitals and counting the quanta. The parameter χ is either determined by fitting the $B(E2; 0_1^+ \to 2_1^+)$ transition in a microscopic calculation or by inverting Eq. (3.12) and using the measured $(\beta_g)_0$ deformation.

With this we can now proceed to apply our proposed method to specific nuclei.

6.4 Application to $^{24}_{12}$Mg and $^{238}_{92}$U

6.4.1 $^{24}_{12}$Mg:

Being a *s*-*d*-shell nucleus, where the spin-orbit interaction is small, no distinction between protons and neutrons is made. The nucleus $^{24}_{12}$Mg has 8 valence nucleons in the *s*-*d*-shell. Using the supermultiplet scheme (U(4)), we have

$$\begin{array}{ccccc} \mathrm{U}(24) & \supset & \mathrm{U}(6) & \times & \mathrm{U}(4) \\ [1^8] & \to & [4^2] & \times & [2^4] \end{array} \tag{4.1}$$

where we consider only the dominant irrep in the reduction of $[1^8]$. Using the tables of Ref. [25], which give the reduction of U(6) ⊃ SU(3), we get the allowed SU(3)-irreps $(\lambda_s\mu_s)$, which are listed in Table I with their corresponding β_s and γ_s–values.

For the parameters $\hbar\omega$, N_s, and χ, we obtain respectively $\hbar\omega = 12.6$ MeV, $N_s = 62.5$ and $\chi = 0.0415$ MeV where the value of χ has been taken from Ref. [19].

With these values, Eq. (3.11a) and the procedure proposed in the last section we obtain the results shown in Figs. 6.1 and 6.2. In Fig. 6.1 we give

TABLE I. Allowed SU(3)-irreps.

λ_s, μ_s	β_s	γ_s [degrees]
(8,4)	0.309	20.63
(4,6)	0.261	35.50
(0,8)	0.238	54.79
(6,2)	0.221	17.00
(2,4)	0.172	38.2
(4,0)	0.134	8.95
(0,2)	0.080	46.10

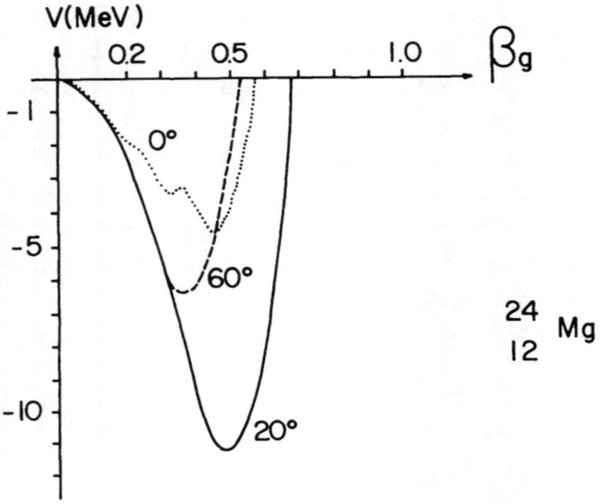

FIGURE 6.1. Cuts in β_g at $\gamma_g = 0°, 20°, 60°$ through the PES of $^{24}_{12}$Mg.

three cuts through the PES at $\gamma_g = 0°, 20°, 60°$. The value 20° corresponds approximately to the γ_s value 20.6° of the dominant irrep (8,4). The cut through 0° shows some deviation from a smooth behaviour which comes from changing the parameters β_s, γ_s to a different irrep corresponding to another range in the γ_g–grid. A similar behaviour is shown in Fig. 6.2 where the letters a, b, c indicate the irrep that contribute in this range. We see that (8,4) dominates over a wide range in $\gamma_g(\beta_g = \text{fixed})$ and that only three irreps, namely $a = (8,4)$, $b = (4,6)$ and $c = (0,8)$, really contribute in this cut. No comparison to the geometrical model can be made because of non existing calculations and because light nuclei are generally considered to be bad examples for the application of a phenomenological, geometrical collective model.

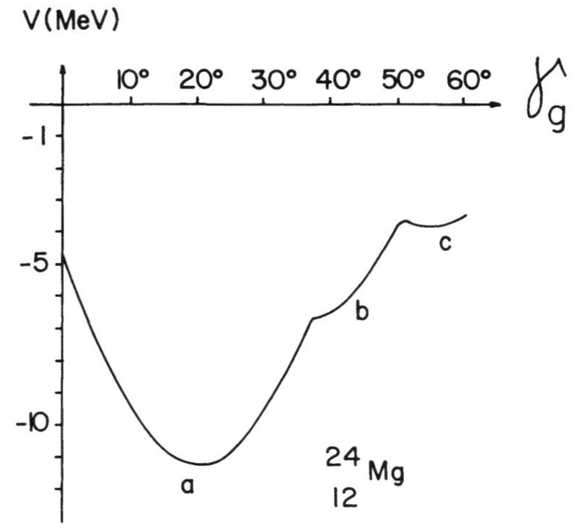

FIGURE 6.2. Cut in γ_g through the PES of $^{24}_{12}$Mg for $\beta_g = 0.476$. The letters a, b, c indicate from which irrep of Sp(3,R), denoted by the lowest irrep (λ_s, μ_s) of SU(3), the contribution to the PES comes from ($a = (8,4)$, $b = (4,6)$, $c = (0,8)$).

6.4.2 $^{238}_{92}$U:

These results can be compared with existing calculations in the geometrical model [9, 26]. In the case of a heavy nucleus, we have to work in the contracted pseudo-$\widetilde{\text{Sp}}(3,R)$ model, *i.e.* we have to distinguish "normal" and "abnormal" nucleons, as was explained in the second section. The Nilsson orbitals have to be filled separately for neutrons and protons. Using the experimental deformation of 0.25, we get 6 valence protons in the $\tilde{N} = 4$ pseudo oscillator shell and 12 valence nucleons in the $\tilde{N}_\nu = 5$ pseudo shell. The valence nucleons in the "abnormal" states are 4 and 8 for protons and neutrons respectively.

For the parameters, we find $\hbar\omega = 6.62$ MeV, $N_s = 574.5$ and $\chi = 0.00135$ MeV, where the latter is taken from Ref. [8], where it was adjusted to the $B(E2; 0_1^+ - 2_1^+)$ transitions in a microscopic calculation within the contracted pseudo-$Sp(3, R)$ model.

In Figs. 6.3 and 6.4, cuts through the resulting PES are presented. In the case of the contracted pseudo-Sp(3,R) model, we have to substitute in (3.11a) the variable β_g by $\tilde{\beta}_g$ whose relation to β_g is given by $\beta_g = \kappa\tilde{\beta}_g$. The reason lies in the mapping of the Sp(3,R)-generators into their pseudo-counterparts where this factor appears and is similar, but not equal, for all generators in the same shell. For ^{238}U, we get $\kappa = 1.14$ [8]. In Fig. 6.3 the cuts are in β_g for $\gamma_g = 0.9°, 30°$ and $59°$, the γ_s–values for the irreps (54,0) (the dominant one) (26,26) and (0,42), respectively. The comparison to the geometrical calculation [26] is also given. As it is seen, the agreement is, considering the simplicity of our procedure, rather good. The differences lie in the position of the absolute minimum and in its stiffness.

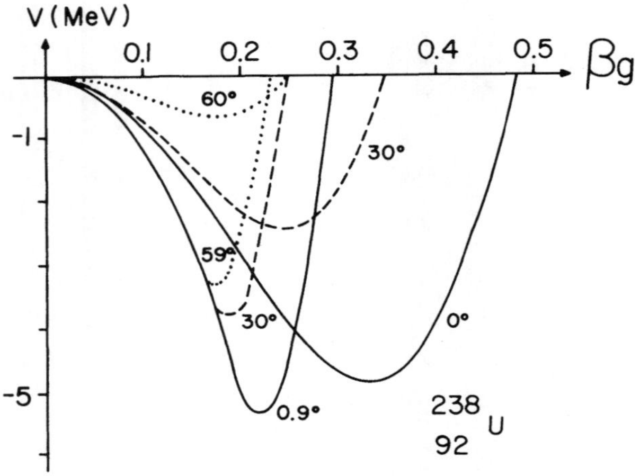

FIGURE 6.3. Cut in β_g at $\gamma_g = 0.9°$, $30°$, $59°$ (corresponding to the γ_s-values of (54,0), (26,26) and (0,42)) through the PES of $^{238}_{92}$U. The results are compared to calculations within the General Collective Model (GCM) [9, 26].

In Fig. 6.4 we show a cut through γ_g for $\beta_g = 0.22$, *i.e.* the position of the minimum. We see that eight irreps contribute in this cut and that the dominant irrep (54,0) only covers a small range, in contrast to the case $^{24}_{12}$Mg. The discontinuities may be an artifact, considering that we do not include Sp(3,R)-symmetry breaking interactions as, for example, the pairing interaction. This is currently a topic of our investigations.

6.5 Conclusions

In this contribution, we gave a summary of a recent development in nuclear structure physics. In Section 2, we gave a review of the pseudo-Sp(3,R) model and its contracted version. This model is an analogous extension of the pseudo-SU(3) model as is the Sp(3,R) model for SU(3). The latter is only applicable to light nuclei, *e.g.* in the *s*-*d*-shell, while the pseudo extension can be applied to heavy nuclei. With this we are able to describe the low lying collective spectra of well deformed heavy nuclei.

In Section 3, we presented a possible procedure for deducing a Potential-Energy-Surface (PES) starting from a microscopic hamiltonian and evaluating its expectation value with respect to a certain trial wave function. This function depends on several parameters, where some have to be related to the geometrical deformation variable and others are related to the irreps of Sp(3,R) with which we are working. The latter parameters, we could relate to the deformation of the lowest SU(3)-irrep in a Sp(3,R) irrep, while the former were related to the contribution to deformation coming from

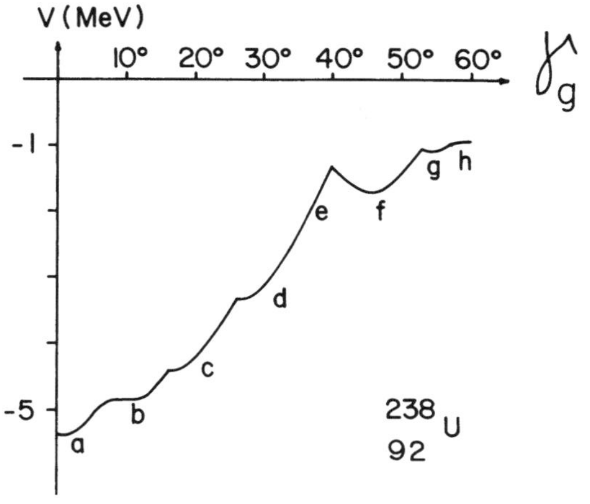

FIGURE 6.4. Cut in γ_g through the PES of $^{238}_{92}$U for $\beta_g = 0.22$. The letters a-h denote the irreps Sp(3,R), given by the lowest SU(3) irrep (λ_s, μ_s), *i.e.* $a = (54,0)$, $b = (46,10)$, $c = (40,16)$, $d = (30,24)$, $e = (26,26)$, $f = (12,36)$, $g = (4,40)$, $h = (0,42)$.

admixtures of higher shells. The geometrical deformation parameter can then be related, to lowest order, to a sum of these deformations. As the part which describes the admixtures of higher shells, we used a coherent state.

Having evaluated the expectation value and substituted the dependence on α^g_m, we were left with a rather simple potential in structure when restricting to only one irrep of $Sp(3,R)$, *i.e.* the parameters β_s, γ_s were fixed. Allowing the latter to vary modified the global structure, leading to complex forms. One essential feature was that, due to the Pauli principle at a given angle γ_g, β_s was restricted by the largest $SU(3)-$irrep with $\gamma_s \simeq \gamma_g$. As a consequence we got the stabilizing raising part in β_g.

Finally we applied our procedure to a light s-d-shell nucleus, *i.e.* $^{24}_{12}$Mg and to a heavy nucleus, *i.e.* $^{238}_{92}$U. Comparing the latter to calculations within a geometrical model led to a rather good agreement. This is somewhat surprising considering the simplicity of our crude procedure.

In the future, we intend to investigate the predictions of the PES when varying from one nucleus to the other. A question addressed will be, for example, when $\gamma-$unstable nuclei arise or when a nucleus is prolate, oblate or triaxial. This has then to be compared to current models. Another question is when a second minimum at high deformation (superdeformation [17, 27]) appears.

6.6 Acknowledgements

We acknowledge the collaboration with J.P. Draayer and P. Rochford with whom parts of the results where obtained.

6.7 References

[1] L. Weaver, L.C. Biedenharn, and R.Y. Cusson, *Ann. Phys.* **77**, 250 (1973); *ibid.,* **112**, 493 (1976).

[2] V. Vanagas, *Sov. J. Part. Nucl.* **7**, 118 (1976); *ibid.,* **11**, 172 (1980).

[3] D.J. Rowe, *Rep. Prog. Phys.* **48**, 1419 (1985).

[4] O. Castaños, A. Frank, E. Chacón, P.O. Hess, and M. Moshinsky, *Journ. of Math. Phys.* **23**, 2537 (1982); M. Moshinsky; *J. Math. Phys.* **25**, 1555 (1984); E. Chacón, P.O. Hess, and M. Moshinsky, *J. Math. Phys.* **25**, 1565 (1984); O. Castaños, E. Chacón, and M. Moshinsky; *J. Math. Phys.* **25**, 2815 (1984); M. Moshinsky, M. Nicolescu, and R.T. Sharp, *J. Math. Phys.* **26**, 2995 (1985); E. Chacón, P.O. Hess, and M. Moshinsky, *J. Math. Phys.* **28**, 2223 (1987).

[5] O. Castaños, and J.P. Draayer, *Nucl. Phys.* **A491**, 349 (1989).

[6] J.P. Elliott, *Proc. Roy. Soc. London* **A245**, 128 (1958); *ibid.,* **A245**, 562 (1958).

[7] R.D. Ratna Raju, J.P. Draayer, and K.T. Hecht, *Nucl. Phys.* **A202**, 433 (1973).

[8] O. Castaños, P.O. Hess, J.P. Draayer, and P. Rochford, *Nucl. Phys.* **A524**, 471 (1991).

[9] J.M. Eisenberg, and W. Greiner, *Nuclear Theory I: Nuclear Models,* 3rd edition (North-Holland, 1987).

[10] K.T. Hecht, *The Vector Coherent State Method and its Application to Problems of Higher Symmetries*, Springer-Heidelberg (1987).

[11] A. Arima, M. Harvey, and K. Shimizu, *Phys. Lett.* **B30**, 517 (1969).

[12] A. Bohr, I. Hamamoto, and B.R. Mottelson, *Phys. Scr.* **26**, 267 (1982).

[13] J.P. Draayer, Pseudo Spin, Pseudo SU(3) and Superdeformation, preprint, Louisiana State University (1990).

[14] J.P. Draayer, H.A. Naqvi, Pseudo-Spin Symmetry and High-Spin Physics, preprint, Louisiana State University (1991).

[15] J.P. Draayer, and K.J. Weeks, *Ann. Phys.* **156**, 41 (1984).

[16] O. Castaños, J.P. Draayer, and Y. Leschber, *Z. Phys.* **A329**, 33 (1988).

[17] W. Nazarewiz, P.J. Twin, P. Fallon, and J.D. Garrett, *Phys. Rev. Lett.* **64**, 654 (1990).

[18] F.S. Stephens *et al., Phys. Rev. Lett.* **64**, 2623 (1990); *ibid.,* **65**, 301 (1990).

[19] C. Bahri, J.P. Draayer, O. Castaños, and G. Rosensteel, *Phys. Lett.* **B234**, 430 (1990).

[20] O. Castaños, J.P. Draayer, and Y. Leschber, *Nucl. Phys.* **A473**, 494 (1987).

[21] H. Ui, *Prog. Theor. Phys.* **44**, 153 (1970).

[22] D.J. Rowe, R. Le Blanc, and J. Repka, *J. Phys.* **A22**, L309 (1989).

[23] O. Castaños, P.O. Hess, J.P. Draayer, and P. Rochford, Microscopic Interpretation of Potential Energy Surfaces, submitted for publication.

[24] J. Blomquist, and A. Molinari, *Nucl. Phys.* **A106**, 545 (1968).

[25] J.P. Draayer, and Y. Leschber, Branching Rules for Unitary Symmetries of the Nuclear Shell Model, LSU-Report (1987).

[26] P.O. Hess, M. Seiwert, J.A. Maruhn, and W. Greiner, *Z. Phys.* **A296**, 147 (1980).

[27] W. Nazarewicz, in: *International Conference on Contemporary Topics in Nuclear Structure Physics*, Cocoyoc, México (World Scientific, 1988), p.467; K. Schiffer, J. Gascon, B. Herskind, *ibid.*, p.467; P.J. Twin, *ibid.*, p.445.

7

The SU(3) Generalization of Racah's $SU(2j+1) \supset SU(2)$ Group-Subgroup Embedding

James D. Louck[1]
Lawrence C. Biedenharn[2]

ABSTRACT Racah showed how to embed the symmetry group, SU(2) or SO(3), of a physical system in the general unitary group $SU(2j+1)$, where the latter group is the most general group of linear transformations of determinant 1 that leaves invariant the inner product structure of an arbitrary state space H_j of the physical system. This state space H_j is at the same time the carrier space of an irreducible representation (irrep) $[j]$ of the symmetry group. This embedding is achieved by classifying the vector space of mappings $H_j \rightarrow H_j$ as irreducible tensor operators with respect to the underlying symmetry group. These irreducible tensors are the generators of the Lie algebra of $SU(2j+1)$. Racah's method is reviewed within the framework of unit tensor operators. The generalization of this technique to the symmetry group $U(3)$ to obtain the embedding $U(3) \subset U(n)$, where $n = \dim[m]$ is the dimension of an arbitrary irrep of U(3). As in the SU(2) case, the group U(3) is the symmetry group of a physical system, and $U(\dim[m])$ is the most general group of linear transformations that preserves the inner product structure of an arbitrary state space $H_{[m]}$ of the system. This state space $H_{[m]}$ is at the same time the carrier space of irrep $[m]$ of the symmetry group U(3). Preliminary results on the Lie algebraic vanishings of U(3) Racah coefficients in consequence of the embedding $SU(3) \subset E_6 \subset SU(27)$ are given.

7.1 Introduction

Symmetry techniques have played a large role in the development of quantum physics and chemistry, from the indispensable applications of angular momentum and permutational symmetry, to applications of higher symmetries originating from the families of orthogonal, unitary, symplectic groups, etc. Marcos Moshinsky and his students have contributed to this field for

---

[1]Los Alamos National Laboratory, New Mexico, USA.
[2]Duke University, Durham, North Carolina, USA.

over forty years, especially in exploiting boson operator methods for realizing the carrier spaces of irreducible representations (irreps) of groups of physical interest. Contributions most relevant to the subject of this paper include Refs. [1]–[7].

One well-known example of these techniques is Racah's embedding of SU(2), resp., SO(3), in the general unitary group $\mathrm{SU}(2j+1)$:

$$\mathrm{SU}(2) \subset \mathrm{SU}(2j+1), \text{ resp.}, \mathrm{SO}(3) \subset \mathrm{SU}(2j+1), \; j = 1, 3/2, \ldots. \tag{1.1}$$

Mathematically, the embeddings in Eq. (1.1) can be realized in many ways, most of them trivially. The essential point of Racah's method, however, is that the SU(2) [resp., SO(3)] is to be identified as the ***symmetry group*** of a given physical system, while the group $\mathrm{SU}(2j+1)$ has quite a different origin: It is not usually a symmetry of the physical system as a whole (global symmetry), but rather is realized as a group of unitary transformations that leave invariant a given energy eigenspace of the Hamiltonian, or a model Hamiltonian, being used to describe the system. The problem posed by the embedding (1.1), and solved by Racah, is to effect the embedding is such a way that the symmetry group structure fits into the general unitary transformations of the degenerate energy eigenspace as a subgroup of transformations. It is for this reason that the method finds many applications to physical problems, sometimes unexpected.

Racah [8, 9] solved this physical embedding problem by the technique of classifying the Lie algebraic generators of the group $\mathrm{SU}(2j+1)$ as irreducible tensor operators with respect to the transformations of the energy eigenspace corresponding to the underlying symmetry group, SU(2) or SO(3). In this manner, one obtains a basis of the Lie algebra of $\mathrm{SU}(2j+1)$ in which the structure constants are given in terms of the 3-j and 6-j coefficients of the symmetry group. The applications of the Lie algebra of $\mathrm{SU}(2j+1)$, presented in this way, to atomic and nuclear spectroscopy are well-known.

We review in Section 2 a slight generalization of the embedding (1.1) to the form

$$\mathrm{U}(2) \subset \mathrm{U}(2j+1), \; j = 1, 3/2, \ldots. \tag{1.2}$$

The methods employed point the way to the U(3) generalization, which is the subject of this paper. With slight modifications, one can also obtain the embedding

$$\mathrm{SO}(3) \subset \mathrm{U}(2j+1). \tag{1.3}$$

The generalization to U(3) is the embedding expressed by

$$\mathrm{U}(3) \subset \mathrm{U}(\dim[m]), \tag{1.4}$$

where

$$\dim[m] = (m_{13} - m_{23} + 1)(m_{13} - m_{33} + 2)(m_{23} - m_{33} + 1)/2 \tag{1.5}$$

denotes the dimension of an arbitrary (integral) irrep

$$[m] = [m_{13}, m_{23}, m_{33}] \tag{1.6}$$

of U(3). The m_{i3} ($i = 1, 2, 3$) are arbitrary integers (positive, zero, or negative) satisfying

$$m_{13} \geq m_{23} \geq m_{33}. \tag{1.7}$$

In analogy with Racah's approach, the U(3) group in Eq. (1.4) is to be the symmetry group of the physical system, while the group U(n), with $n = \dim[m]$, is to be a group of transformations of a degenerate energy eigenspace.

The major deterrent to a straightforward generalization of the embedding (1.2) to that of relation (1.4) is the resolution of the multiplicity problem, which afflicts all unitary groups beyond U(2). This involves additional structure, *operator pattern labelling,* which makes the associated Wigner (3-j) and Racah (6-j) coefficient structure of U(3) decidedly more intricate than that of U(2). Nonetheless, these additional operator labellings and their structural relations are the essential key for developing comprehensive applications of higher unitary symmetry to problems such as that posed by the embedding (1.4).

It is convenient to present in Section II Racah's method for U(2) by using directly the concept of a U(2) Wigner operator. This approach places clearly in evidence the parallel route to be followed in the U(3) generalization. The underlying algebra of the U(2) Wigner (unit tensor) operators is an essential ingredient of this approach.

The generalization to U(3) is developed in Section 3 with emphasis on the role of operator patterns and the U(3) Wigner operator algebra.

In Racah's embedding of SU(2) [or SO(3)] in U($2j+1$), the structure constants of the Lie algebra factorize into a product of WCG and Racah coefficients. The existence of a group G such that, for example, SO(3) $\subset G \subset$ U($2j+1$) can lead to the vanishing of a Racah coefficient, even though all the triangle and symmetry conditions are fulfilled. The first example was given by Racah [9]. (See Vanden Berghe *et al.* [10] and also Beyer *et al.* [11] for a summary of such Lie algebraic zeroes as of 1986.) It is, therefore, natural to ask: Can nontrivial zeroes of the SU(3) 6-j coefficients arise in consequence of the existence of a group G satisfying

$$\mathrm{SU}(3) \subset G \subset \mathrm{SU}(\dim[m])\ ? \tag{1.8}$$

The general method of effecting the embedding (1.8) is given in Section 4, and the question of zeroes for $G = \mathrm{E}_6$ is addressed in Section 5.

7.2 Resumé of Racah's method

We summarize in subsections 1–10 below the vector space and algebraic structures underlying Racah's SU(2) $\subset$ U($2j+1$) embedding. The emphasis is on the algebraic properties of unit tensor operators, since this makes transparent the generalization to U(3).

7.2.1 Vector Space

H_j denotes a finite-dimensional inner product space with orthonormal basis given by

$$B_j = \{|\, jm\rangle \,|\, m = -j, -j+1, \ldots, j\} \tag{2.1}$$

of dimension

$$\dim H_j = \dim[j] = 2j + 1. \tag{2.2}$$

Here j may be any integer or half-integer in the set

$$j \in \{0, 1/2, 1, \ldots\}. \tag{2.3}$$

7.2.2 Action of the SU(2) Lie algebra on H_j

The angular momentum operators $J_+ = J_1 + iJ_2$, J_3, $J_- = J_1 - iJ_2$, where $\mathbf{J} = (J_1, J_2, J_3)$ is the total angular momentum of the physical system in question (generators of the quantal rotation group, $SU(2)$) have the standard action on the space H_j:

$$J_\pm|jm\rangle = [(j \mp m)(j \pm m + 1)]^{1/2}|jm \pm 1\rangle, \tag{2.4}$$

$$J_3|jm\rangle = m|jm\rangle, \tag{2.5}$$

$$\mathbf{J}^2|jm\rangle = j(j+1)|jm\rangle, \tag{2.6}$$

where $\mathbf{J}^2 = J_1^2 + J_2^2 + J_3^2$ denotes the total (squared) angular momentum.

7.2.3 Action of SU(2) on H_j

The action of the group SU(2) generated by the Lie algebra basis $\mathbf{J} = (J_1, J_2, J_3)$ on the basis H_j is given by

$$T_U : H_j \rightarrow H_j, \text{ each } U \in \mathrm{SU}(2), \tag{2.7}$$

where

$$T_U|jm\rangle = \sum_{m'} D^j_{m'm}(U)|jm'\rangle, \tag{2.8}$$

in which $D^j(U)$ denotes the Wigner D-matrix (see Eq. (3.86) of Ref. [12] (Vol. 8) for their explicit definition, as used here); that is, the correspondence

$$U \rightarrow D^j(U), \text{ each } U \in \mathrm{SU}(2), \tag{2.9}$$

is a representation of the group SU(2) by unitary matrices of dimension, $\dim[j] = 2j + 1$. This unitary property is also expressed by

$$T_{U^{-1}} = T_{U^\dagger}, \tag{2.10}$$

where $\dagger$ denotes Hermitian conjugation of U.

7.2.4 Irreducible Unit Tensor Operator Maps $H_j \to H_j$

The operator maps

$$\left\langle \begin{matrix} & 0 & \\ a & & -a \\ & \alpha & \end{matrix} \right\rangle : H_j \to H_j \tag{2.11}$$

defined by

$$\left\langle \begin{matrix} & 0 & \\ a & & -a \\ & \alpha & \end{matrix} \right\rangle |jm\rangle = \begin{cases} C^{jaj}_{m,\alpha,m+\alpha}|jm+\alpha\rangle, & a = 0, 1, \ldots, 2j \\ 0, & a = 2j+1, 2j+2, \ldots, \end{cases} \tag{2.12}$$

with $\alpha = -a, -a+1, \ldots, a$, are a basis of all mappings of the vector space H_j into itself. The C-coefficients in the definition (2.12) are Wigner-Clebsch-Gordon (WCG) coefficients of SU(2); that is, these coefficients are the elements of the real orthogonal matrix that reduces the Kronecker product representation $D^j(U) \times D^j(U)$ into a direct sum of the irreps $D^a(U)$, which we denote symbolically by

$$[j] \times [j] = \sum_{a=0}^{2j} \oplus [a] . \tag{2.13}$$

There are, of course, exactly $(\dim[j])^2 = (2j+1)^2$ operators defined by Eq. (2.12), including the identity map

$$\left\langle \begin{matrix} & 0 & \\ 0 & & 0 \\ & 0 & \end{matrix} \right\rangle = \mathbf{1},$$

and in accord with Eq. (2.13), we have

$$(2j+1)^2 = \sum_{a=0}^{2j} (2a+1). \tag{2.14}$$

The WCG coefficients are related to the 3-j coefficients by

$$C^{abc}_{\alpha\beta\gamma} = (-1)^{a-b+\gamma} \begin{pmatrix} a & b & c \\ \alpha & \beta & -\gamma \end{pmatrix} \Big/ (2c+1)^{1/2} . \tag{2.15}$$

The operators defined by Eq. (2.12) obey the following relations:

$$\sum_{\alpha} \left\langle \begin{matrix} & 0 & \\ a & & -a \\ & \alpha & \end{matrix} \right\rangle \left\langle \begin{matrix} & 0 & \\ b & & -b \\ & \alpha & \end{matrix} \right\rangle^{\dagger} = \mathbf{1}, \text{ on } H_j; \tag{2.16}$$

$$\sum_{m} \left\langle jm \left| \left\langle \begin{matrix} & 0 & \\ a & & -a \\ & \alpha & \end{matrix} \right\rangle \left\langle \begin{matrix} & 0 & \\ b & & -b \\ & \beta & \end{matrix} \right\rangle^{\dagger} \right| jm \right\rangle = \delta_{ab}\delta_{\alpha\beta}\frac{(2j+1)}{(2a+1)}. \tag{2.17}$$

In consequence of property (2.16), the operators defined by Eq. (2.12) are called *unit tensor operators* or *Wigner operators*.

For each $a \in \{0, 1, \ldots, 2j\}$, the set of $2a+1$ maps (2.11), corresponding to $\alpha = -a, -a+1, \ldots$, a, transform under the action of the unitary operator T_U defined by Eq. (2.8) according to

$$T_U \left\langle \begin{matrix} & 0 & \\ a & & -a \\ & \alpha & \end{matrix} \right\rangle T_{U^{-1}} = \sum_{\alpha'} D^a_{\alpha'\alpha}(U) \left\langle \begin{matrix} & 0 & \\ a & & -a \\ & \alpha' & \end{matrix} \right\rangle, \text{ each } U \in \mathrm{SU}(2). \tag{2.18}$$

This *equivariance relation* expresses the property that the set of operators

$$\left\{ \left\langle \begin{matrix} & 0 & \\ a & & -a \\ & \alpha & \end{matrix} \right\rangle \middle| \alpha = -a, -a+1, \ldots, a \right\} \tag{2.19}$$

is an irreducible tensor operator of SU(2).

7.2.5 Algebra of Unit Tensor Operators

The set of $(2j+1)^2$ operator maps $H_j \to H_j$ defined by Eq. (2.12) are a basis of the vector space of all linear maps $H_j \to H_j$. The scalars of this vector space are invariant operators with respect to SU(2). In addition, these unit tensor operators obey the following product law:

$$\left\langle \begin{matrix} & 0 & \\ b & & -b \\ & \beta & \end{matrix} \right\rangle \left\langle \begin{matrix} & 0 & \\ a & & -a \\ & \alpha & \end{matrix} \right\rangle = \sum_c \mathbf{W}^{abc}_{000} C^{abc}_{\alpha,\beta,\alpha+\beta} \left\langle \begin{matrix} & 0 & \\ c & & -c \\ & \alpha+\beta & \end{matrix} \right\rangle, \tag{2.20}$$

where $\mathbf{W}^{abc}_{000}$ denotes a *Racah invariant operator*. Its eigenvalue on the space H_j is given by

$$W^{abc}_{000}(j) = [(2c+1)(2j+1)]^{1/2}\, W(jajb; jc)\,; \tag{2.21}$$

that is,

$$\mathbf{W}^{abc}_{000}|jm\rangle = W^{abc}_{000}(j)|jm\rangle. \tag{2.22}$$

The W-coefficient in the right-hand side of Eq. (2.21) is a Racah coefficient, which is given in terms of the 6-j coefficient by

$$W(jajb : jc) = (-1)^{2j+a+b} \begin{Bmatrix} j & a & j \\ b & j & c \end{Bmatrix}. \tag{2.23}$$

7.2.6 Weyl Basis of the Lie Algebra of $\mathrm{U}(2j+1)$

The set of n^2 Weyl basis elements of the Lie algebra of $\mathrm{U}(n)$ is given by

$$\{E_{ik} \mid \; i, k = 1, 2, \ldots, n\}. \tag{2.24}$$

These generators of the general unitary group U(n) (indeed, of the general linear group) obey the commutation relations

$$[E_{ik}, E_{i'k'}] = \delta_{ki'} E_{ik'} - \delta_{ik'} E_{i'k}. \tag{2.25}$$

For $n = 2j+1$, we denote the set (2.24) of generators of U($2j+1$) by

$$\{E_{j-m'+1,j-m+1} | m', m = -j, -j+1, \ldots, j\}. \tag{2.26}$$

7.2.7 Racah Basis of the Lie Algebra of U($2j+1$)

The real orthogonal matrix R of dimension $(2j+1)^2$ is defined by

$$(R)_{a\alpha;m'm} = [(2a+1)/(2j+1)]^{1/2} \left\langle jm' \middle| \left\langle \begin{matrix} & 0 & \\ a & & -a \\ & \alpha & \end{matrix} \right\rangle \middle| jm \right\rangle . \tag{2.27}$$

Property (2.17) expresses the fact that the matrix R with rows enumerated by the index pairs (a, α) and columns by the index pairs (m', m) is real orthogonal. This matrix is now used to define a new basis, ***the Racah basis***, of the Lie algebra of U($2j+1$):

$$E^a_\alpha = \sum_{m',m} (R)_{a\alpha;m'm} E_{j-m'+1,j-m+1}, \tag{2.28}$$

where we suppress j in the notation E^a_α for the Racah basis set:

$$\{E^a_\alpha \mid \ a = 0, 1, \ldots, 2j; \quad \alpha = -a, \ -a+1, \ldots, a\}. \tag{2.29}$$

Since the matrix R is real orthogonal, relation (2.28) can be inverted to give the Weyl generators in terms of the Racah generators:

$$E_{j-m'+1,j-m+1} = \sum_{a,\alpha} (R)_{a\alpha;m'm} E^a_\alpha. \tag{2.30}$$

The embedding of the quantal rotation group SU(2) in U($2j+1$) is now obtained in the following way: The relation between SU(2) angular momentum operators (J_+, J_3, J_-) and Wigner operators is given by

$$\begin{aligned} J_+ &= \left(2\mathbf{J}^2\right)^{1/2} \left\langle \begin{matrix} & 0 & \\ 1 & & -1 \\ & 1 & \end{matrix} \right\rangle, \quad J_3 = \left(\mathbf{J}^2\right)^{1/2} \left\langle \begin{matrix} & 0 & \\ 1 & & -1 \\ & 0 & \end{matrix} \right\rangle, \\ J_- &= \left(2\mathbf{J}^2\right)^{1/2} \left\langle \begin{matrix} & 0 & \\ 1 & & -1 \\ & -1 & \end{matrix} \right\rangle. \end{aligned} \tag{2.31}$$

From these relations and Eqs. (2.28), we obtain the following operator identities on H_j:

$$\begin{aligned} J_+ &= -[2(2j+1)j(j+1)/3]^{1/2} E^1_{+1}, \\ J_3 &= [(2j+1)j(j+1)/3]^{1/2} E^1_0, \\ J_- &= [2(2j+1)j(j+1)/3]^{1/2} E^1_{-1}. \end{aligned} \tag{2.32}$$

These relations give explicitly the embedding of the physical SU(2) group in U(2j + 1).

The commutators of the angular momentum operators (J_+, J_3, J_-) with the elements (2.28) of the Racah basis are given by

$$[J_\pm, E^a_\alpha] = [(a \mp \alpha)(a \pm \alpha + 1)]^{1/2} E^a_{\alpha\pm1}, \tag{2.33}$$

$$[J_3, E^a_\alpha] = E^a_\alpha. \tag{2.34}$$

These relations show that *the Racah basis of the Lie algebra of* U(2j + 1) *consists of irreducible tensor operators with respect to* SU(2). (The results, Eqs. (2.33)–(2.34), are most easily derived by using relations (2.37) below.) Relations (2.33)–(2.34), in turn, imply the transformation property:

$$T_U E^a_\alpha T_{U^{-1}} = \sum_{\alpha'} D^a_{\alpha'\alpha}(U) E^a_{\alpha'}, \quad \text{each} \quad U \in \mathrm{SU}(2). \tag{2.35}$$

7.2.8 Structure Constants in the Racah Basis

If $X = (X_{ik})$ and $Y = (Y_{ik})$ are arbitrary matrices of dimension n (over the complex numbers), we define the (extended) elements of the Lie algebra of U(n) by

$$L_X = \sum_{i,k=1}^{n} X_{ik} E_{ik}, \quad L_Y = \sum_{i,k=1}^{n} Y_{ik} E_{ik}. \tag{2.36}$$

It is then easily proved that

$$[L_X, L_Y] = L_{[X,Y]}. \tag{2.37}$$

Application of this relation to any pair of elements E^a_α and E^b_β in the Racah basis and use of the product law (2.20) for unit tensor operators gives

$$[E^a_\alpha, E^b_\beta] = \sum_{c,\gamma} A^{abc}_{\alpha\beta\gamma} E^c_\gamma, \tag{2.38}$$

where the *structure constants* are given by

$$A^{abc}_{\alpha\beta\gamma} = [(-1)^{a+b-c} - 1] C^{abc}_{\alpha\beta\gamma} [(2a+1)(2b+1)]^{1/2} W(abjj; cj). \tag{2.39}$$

[We have used well-known symmetries of the WCG and Racah coefficients to bring the structure constants to the form (2.39).]

7.2.9 Action of U(2j + 1) on H$_j$

Let $V \in U(2j+1)$, and let the elements of V be enumerated by $(V_{m'm})$ with $m' = j, j-1, \ldots, -j$ denoting rows (read from left to right), and

$m = j, j-1, \ldots, -j$ denoting columns (read from top to bottom) in the matrix V. Then the action S_V of $\mathrm{U}(2j+1)$ of H_j is given by

$$S_V|jm\rangle = \sum_{m'} V_{m'm}|jm'\rangle, \quad \text{each} \quad V \in \mathrm{U}(2j+1). \tag{2.40}$$

In particular, since $D^j(U) \in \mathrm{U}(2j+1)$, we find that

$$S_{D^j(U)} = T_U. \tag{2.41}$$

This result shows clearly that the space H_j carries the fundamental representation

$$D^{[10\cdots0]}(V) = V \tag{2.42}$$

of $\mathrm{U}(2j+1)$ and that this representation, when restricted to the quantal rotation subgroup SU(2), as embedded in $\mathrm{U}(2j+1)$ in the Racah basis, reduces to the irreps of this SU(2):

$$D^{[10\cdots0]}(D^j(U)) = D^j(U), \quad \text{each} \quad U \in \mathrm{SU}(2). \tag{2.43}$$

7.2.10 Expansion to U(2) ⊂ U(2j + 1)

The generators of U(2) are obtained by adjoining the invariant operator

$$E_0^0 = (2j+1)^{-1/2} \sum_{i=1}^{2j+1} E_{ii} \tag{2.44}$$

to the SU(2) generators (J_+, J_3, J_-). The transition to U(2) is then best made by using the full U(2) Gel'fand patterns as explained in detail in Ref. [12].

7.2.11 The Modification to SO(3) ⊂ U(2j + 1)

The results given in subsections 1–8 above apply equally well to the group SO(3) of 3×3 real, proper orthogonal matrices corresponding to the rotations of a physical system in $\Re^3$. It is customary for this case to write $j = \ell$, $\mathbf{J} = \mathbf{L} = (L_1, L_2, L_3)$, and to replace $D^\ell(U)$ by $\mathcal{D}^\ell(R) = D^\ell(U(R))$, where $\pm U(R) \to R$ in the two-to-one homomorphism of SU(2) onto SO(3).

We conclude this summary of Racah's embedding as described in subsections 1–8 above by again pointing out that SU(2) [resp., SO(3)] is the group under whose action the physical system is invariant (symmetry group). The group $\mathrm{U}(2j+1)$ [resp., $\mathrm{U}(2\ell+1)$] need not be a symmetry group of the full physical system, but only of special quantum states of the system. This general unitary group always has the significance of being the most general group of linear transformations of the space H_j that preserves the inner product structure of this space.

7.3 The $U(3) \subset U(\dim[m])$ embedding

We can now give the

$$U(3) \subset U(\dim[m]) \tag{3.1}$$

generalization of Racah's $U(2) \subset U(2j+1)$ embedding by following, step by step, the procedures given in 1–9 in Section 2. The group U(3), or SU(3), is taken as the symmetry group of a physical system, or the model of such a system. The meaning of the group $U(\dim[m])$ is that it is the most general group of linear transformations of the carrier space $H_{[m]}$ of an irrep of U(3), which preserves the inner product. The basic problem is to classify the set of all maps $H_{[m]} \to H_{[m]}$ as irreducible tensor operators with respect to the underlying symmetry group U(3). We proceed as in Section 2.

7.3.1 Vector Space

$H_{[m]}$ denotes a finite-dimensional inner product space with orthonormal basis given by

$$B_m = \left\{ \left| \begin{matrix} m_{13} & & m_{23} & & m_{33} \\ & m_{12} & & m_{22} & \\ & & m_{11} & & \end{matrix} \right\rangle \middle| \begin{matrix} \text{this array is a} \\ \text{Gel'fand pattern} \end{matrix} \right\}. \tag{3.2}$$

We recall that a U(3) Gel'fand pattern is any triangular array of integers, positive, zero, or negative, satisfying the betweenness relations:

$$\begin{matrix} m_{13} \geq m_{12} \geq m_{23} \geq m_{22} \geq m_{33} \\ m_{12} \geq m_{11} \geq m_{22}. \end{matrix} \tag{3.3}$$

We denote such a Gel'fand pattern by

$$\begin{pmatrix} [m] \\ m \end{pmatrix}, \quad [m] = [m_{13} m_{23} m_{33}], \quad m = \begin{pmatrix} m_{12} & & m_{22} \\ & m_{11} & \end{pmatrix}. \tag{3.4}$$

At times we use a notation less encumbered with subscripts. The dimension of the space $H_{[m]}$ is $\dim[m]$, as given by Eq. (1.5).

7.3.2 Action of the U(3) Lie Algebra on $H_{[m]}$

We denote the set of Weyl basis elements of the Lie algebra of the symmetry group U(3) of a physical system by

$$\{K_{ij} | i, j = 1, 2, 3\}. \tag{3.5}$$

These operators are, by definition, linear mappings $H_{[m]} \to H_{[m]}$ satisfying the commutation relations

$$[K_{ij}, K_{k\ell}] = \delta_{jk} K_{i\ell} - \delta_{i\ell} K_{kj}, \tag{3.6}$$

and the Hermitian conjugate relations $K_{ij}^{\dagger} = K_{ji}$. The action of these generators of U(3) on $H_{[m]}$ is given by

$$K_{ij}\left|\begin{matrix}[m]\\ m\end{matrix}\right\rangle = \sum_{m'}\left\langle\begin{matrix}[m]\\ m'\end{matrix}\right| K_{ij}\left|\begin{matrix}[m]\\ m\end{matrix}\right\rangle\left|\begin{matrix}[m]\\ m'\end{matrix}\right\rangle, \tag{3.7}$$

where the matrix elements $\langle\ldots|K_{ij}|\ldots\rangle$ are the standard ones [13].

The state space of a physical system having U(3) symmetry can then be decomposed as a direct sum $\sum\oplus H_{[m]}$ (including multiplicities, as required). The vector space $H_{[m]}$ used throughout this section has this significance of a subspace of states of such a physical system.

7.3.3 Action of U(3) on $H_{[m]}$

For each $U \in \mathrm{U}(3)$, the Lie algebra $\{K_{ij}\}$ generates a unitary transformation T_U of the space $H_{[m]}$:

$$T_U\left|\begin{matrix}[m]\\ m\end{matrix}\right\rangle = \sum_{m'} D^{[m]}_{m'm}(U)\left|\begin{matrix}[m]\\ m'\end{matrix}\right\rangle, \tag{3.8}$$

where $D^{[m]}(U)$ denotes a unitary matrix with rows and columns enumerated by the U(2) Gel'fand patterns m' and m, respectively. Here the irrep label $[m]$ occurs as the top row in both patterns:

$$\begin{pmatrix}[m]\\ m'\end{pmatrix} \quad \text{and} \quad \begin{pmatrix}[m]\\ m\end{pmatrix}. \tag{3.9}$$

The unitary irreducible representation matrices

$$\{D^{m}(U) \mid U \in \mathrm{U}(3)\} \tag{3.10}$$

are much-studied objects [5, 6], [14]–[24] and are known explicitly, but are quite complicated. The only property we need here is that the correspondence

$$U \to D^{[m]}(U), \quad \text{each} \quad U \in \mathrm{U}(3) \tag{3.11}$$

is a representation of U(3) by unitary matrices of dimension, dim$[m]$. The unitary property is also expressed as

$$T_{U^{-1}} = T_{U^{\dagger}}. \tag{3.12}$$

7.3.4 Irreducible Unit Tensor Operator Maps $\mathbf{H}_{[m]} \to \mathbf{H}_{[m]}$

It is well-known how to determine the irrep labels $[k] = [k_1k_2k_3]$ of the irreducible U(3) tensor operator maps:

$$\mathbf{T}([k]) : H_{[m]} \to H_{[m]}. \tag{3.13}$$

The U(3) irrep labels are just those occurring in the reduction of the Kronecker product

$$[\overline{m}] \times [m] = \sum \oplus [k]. \tag{3.14}$$

Here $[\overline{m}]$ denotes the irrep of U(3) that is the complex conjugate of $[m]$:

$$[\overline{m}] = [\overline{m}_{13}\overline{m}_{23}\overline{m}_{33}] = [-m_{33}, -m_{23}, -m_{13}]. \tag{3.15}$$

(See, for example, Ref. [14].)

The explicit decomposition of the Kronecker product $[\overline{m}] \times [m]$ into irreps of U(3) is given by

$$[\overline{m}] \times [m] = \sum_{[k]} I([\overline{m}] \times [m]; [k])[k], \tag{3.16}$$

where $I([\overline{m}] \times [m]; [k])$ denotes the (intertwining) number of times irrep $[k]$ is contained in the representation $[\overline{m}] \times [m]$. (See Refs. [25]–[28] for a discussion of these numbers and their relation to the Littlewood-Richardson numbers.) The summation in Eq. (3.16) is over all irreps

$$[k] = [k_1 k_2 k_3] \tag{3.17}$$

such that the k_i are integers satisfying the following relations:

$$\begin{array}{ll} \text{i)} & k_1 + k_2 + k_3 = 0, \\ \text{ii)} & 0 \le k_1 \le m_{13} - m_{33}, \\ \text{iii)} & -(m_{13} - m_{33}) \le k_3 \le 0. \end{array} \tag{3.18}$$

The interwining number has the following explicit values:

$$I([\overline{m}] \times [m]; [k]) = k_1 - k_2 + 1, \tag{3.19}$$

for $k_2 \ge 0$ and $k_1 \le \min(m_{13} - m_{23}, m_{23} - m_{33})$;

$$I([\overline{m}] \times [m]; [k]) = \min(m_{13} - m_{23} - k_2, m_{23} - m_{33} - k_2), \tag{3.20}$$

for $k_2 \ge 0$ and $k_1 > \min(m_{13} - m_{23}, m_{23} - m_{33})$.

For $k_2 < 0$, we have the relation

$$I([\overline{m}] \times [m]; [k]) = I([\overline{m}] \times [m]; [\overline{k}]), \tag{3.21}$$

so that together Eqs. (3.19), (3.20) and (3.21) give explicitly all multiplicity numbers appearing in the Kronecker product reduction, Eq. (3.16).

The explicit set of unit tensor operators, which constitute a basis for all maps $H_{[m]} \to H_{[m]}$, is given by

$$\left\langle \begin{matrix} \lambda \\ [k] \\ \alpha \end{matrix} \right\rangle = \left\langle \begin{matrix} & & 0 & & \\ & \ell & & -\ell & \\ k_1 & & k_2 & & k_3 \\ & \alpha_{12} & & \alpha_{22} & \\ & & \alpha_{11} & & \end{matrix} \right\rangle, \tag{3.22}$$

where the irrep label $[k]$ runs over all values satisfying the conditions (i)-(iii) above. For each such $[k]$, the entry in the *operator pattern*

$$\lambda = \begin{pmatrix} & 0 & \\ \ell & & -\ell \end{pmatrix}, \tag{3.23}$$

and the α_{ij} in the Gel'fand pattern

$$\alpha = \begin{pmatrix} \alpha_{12} & & \alpha_{22} \\ & \alpha_{11} & \end{pmatrix}, \tag{3.24}$$

assume the following values:

$$\ell = k_2, k_2+1, \ldots, k_2 - 1 + I([\overline{m} \times [m]; [k]); \tag{3.25}$$

where

$$k_1 \geq \alpha_{12} \geq k_2 \geq \alpha_{22} \geq k_3; \alpha_{12} \geq \alpha_{11} \geq \alpha_{22}. \tag{3.26}$$

The restriction of the ℓ-values to the domain given by Eq. (3.25), rather than the full domain admitted by the betweenness conditions, is a consequence of the canonical splitting conditions for the U(3) WCG coefficients as determined by null space (see Refs. [29]–[35]).

That the counting is correct for the reduction given by Eqs. (3.16)–(3.26) may be verified from the equation

$$(\dim[m])^2 = \sum_{[k]} I([\overline{m}] \times [m]; [k]) \dim[k], \tag{3.27}$$

where the summation is over all $[k]$ satisfying conditions (3.18).

It is useful to give an example illustrating the above rules:

$$\begin{aligned} [0,-1,-2] \times [2\ 1\ 0] &= [0\ 0\ 0] \oplus 2[1,0,-1] \oplus [1,1,-2] \\ &\quad \oplus [2,-1,-1] \oplus [2,0,-2]. \end{aligned} \tag{3.28}$$

The set of patterns labelling the unit tensor operator maps $H_{[2\ 1\ 0]} \rightarrow H_{[2\ 1\ 0]}$ are

$$\begin{gathered} \begin{pmatrix} & & 0 & & \\ & 0 & & 0 & \\ 0 & & 0 & & 0 \\ & 0 & & 0 & \\ & & 0 & & \end{pmatrix}, \begin{pmatrix} & & 0 & & \\ & 0 & & 0 & \\ 1 & & 0 & & -1 \\ & & \alpha & & \end{pmatrix}, \begin{pmatrix} & & 0 & & \\ & -1 & & -1 & \\ 1 & & 0 & & -1 \\ & & \alpha & & \end{pmatrix}, \\ \begin{pmatrix} & & 0 & & \\ & 1 & & -1 & \\ 1 & & 1 & & -2 \\ & & \alpha' & & \end{pmatrix}, \begin{pmatrix} & & 0 & & \\ & 1 & & -1 & \\ 2 & & -1 & & -1 \\ & & \alpha'' & & \end{pmatrix}, \begin{pmatrix} & & 0 & & \\ & 0 & & 0 & \\ 2 & & 0 & & -2 \\ & & \alpha''' & & \end{pmatrix}, \end{gathered} \tag{3.29}$$

in which α, α', α'', α''' runs over all patterns satisfying betweenness. This provides the $1 + 2(8) + 10 + 10 + 27 = 64$ mappings of the space $H_{[2\ 1\ 0]}$, with $\dim[2\ 1\ 0] = 8$, onto itself.

The action of the unit tensor operator (3.22) on the space $H_{[m]}$ is given by

$$\left\langle \begin{matrix} \lambda \\ [k] \\ \alpha \end{matrix} \right\rangle \left| \begin{matrix} [m] \\ m \end{matrix} \right\rangle = \sum_{m'} \left\langle \begin{matrix} [m] \\ m' \end{matrix} \right| \left\langle \begin{matrix} \lambda \\ [k] \\ \alpha \end{matrix} \right\rangle \left| \begin{matrix} [m] \\ m \end{matrix} \right\rangle \left| \begin{matrix} [m] \\ m' \end{matrix} \right\rangle, \qquad (3.30)$$

where the notation $\langle \cdots | \langle \ \rangle | \cdots \rangle$ denotes a WCG coefficient of U(3). The operators defined by Eq. (3.30) obey the following relations:

$$\sum_{\alpha} \left\langle \begin{matrix} \lambda \\ [k] \\ \alpha \end{matrix} \right\rangle \left\langle \begin{matrix} \lambda' \\ [k] \\ \alpha \end{matrix} \right\rangle^{\dagger} = \delta_{\lambda\lambda'} \mathbf{1}, \quad \text{on} \quad H_{[m]}, \qquad (3.31)$$

$$\sum_{\lambda} \left\langle \begin{matrix} \lambda \\ [k] \\ \alpha \end{matrix} \right\rangle^{\dagger} \left\langle \begin{matrix} \lambda \\ [k] \\ \alpha' \end{matrix} \right\rangle = \delta_{\alpha\alpha'} \mathbf{1}, \quad \text{on} \quad H_{[m]}, \qquad (3.32)$$

$$\sum_{m} \left\langle \begin{matrix} [m] \\ m \end{matrix} \right| \left\langle \begin{matrix} \lambda \\ [k] \\ \alpha \end{matrix} \right\rangle \left\langle \begin{matrix} \lambda' \\ [k] \\ \alpha' \end{matrix} \right\rangle^{\dagger} \left| \begin{matrix} [m] \\ m \end{matrix} \right\rangle = \delta_{\lambda\lambda'} \delta_{\alpha\alpha'} \frac{(\dim[m])}{(\dim[k])}. \qquad (3.33)$$

In consequence of relation (3.31), these operators are called unit tensor operators.

For each irrep label $[k]$, the set of $\dim[m]$ maps (3.22) corresponding to all Gel'fand patterns α transforms under the action of the unitary operator T_U defined by Eqs. (3.8)–(3.9) according to

$$T_U \left\langle \begin{matrix} \lambda \\ [k] \\ \alpha \end{matrix} \right\rangle T_{U^{-1}} = \sum_{\alpha'} D^{k}_{\alpha'\alpha}(U) \left\langle \begin{matrix} \lambda \\ [k] \\ \alpha' \end{matrix} \right\rangle. \qquad (3.34)$$

This equivariance relation expresses the property that the set of operators

$$\left\{ \left\langle \begin{matrix} \lambda \\ [k] \\ \alpha \end{matrix} \right\rangle \middle| \, \alpha \text{ is a Gel'fand pattern} \right\} \qquad (3.35)$$

is an irreducible tensor operator of U(3) for each operator pattern λ.

7.3.5 Algebra of U(3) Unit Tensor Operators

The set of $(\dim[m])^2$ operator maps $H_{[m]} \to H_{[m]}$ defined by Eqs. (3.30) is a basis of the vector space of all linear maps $H_{[m]} \to H_{[m]}$. The scalars of this vector space are invariant operators with respect to U(3). In addition, these unit tensor operators obey the following *product law*:

$$\left\langle \begin{matrix} \lambda' \\ [k'] \\ \alpha' \end{matrix} \right\rangle \left\langle \begin{matrix} \lambda \\ [k] \\ \alpha \end{matrix} \right\rangle = \sum \left\{ \begin{matrix} \lambda'' & \lambda' & \lambda \\ [k''] & [k'] & [k] \\ \alpha'' & \alpha' & \alpha \end{matrix} \right\} \left\langle \begin{matrix} \lambda'' \\ [k''] \\ \alpha'' \end{matrix} \right\rangle, \qquad (3.36)$$

where the quantity $\left\{\begin{smallmatrix} \cdot & \cdot & \cdot \\ \cdot & \cdot & \cdot \end{smallmatrix}\right\}$ is an invariant operator; it is defined in terms of U(3) WCG coefficients and U(3) Racah invariant operators by

$$\begin{Bmatrix} \lambda'' & \lambda' & \lambda \\ [k''] & [k'] & [k] \\ \alpha'' & \alpha' & \alpha \end{Bmatrix} = \sum_{\gamma'} \left\langle \begin{matrix} [k''] \\ \alpha'' \end{matrix} \middle| \left\langle \begin{matrix} \gamma' \\ [k'] \\ \alpha' \end{matrix} \right| \begin{matrix} [k] \\ \alpha \end{matrix} \right\rangle \times \left\{ \begin{pmatrix} [k''] \\ \lambda'' \end{pmatrix} \begin{pmatrix} \gamma' \\ [k'] \\ \lambda' \end{pmatrix} \begin{pmatrix} [k] \\ \lambda \end{pmatrix} \right\}. \tag{3.37}$$

The summation in Eq. (3.36) is over all irreps $[k'']$ contained in $[k'] \times [k]$, and also over all Gel'fand patterns α'' and operator patterns λ'', as described in Eq. (3.41) below. In Eq. (3.37), the symbol $\langle \cdots | \langle \, \rangle | \cdots \rangle$ denotes a U(3) WCG coefficient, and $\{(\cdots)(\cdots)(\cdots)\}$ denotes a U(3) Racah invariant operator. The eigenvalue of this U(3) Racah invariant on the space $H_{[m]}$ is denoted

$$\left\{ \begin{pmatrix} [k''] \\ \lambda'' \end{pmatrix} \begin{pmatrix} \gamma' \\ [k'] \\ \lambda' \end{pmatrix} \begin{pmatrix} [k] \\ \lambda \end{pmatrix} \right\} ([m]), \tag{3.38}$$

and is defined by the action of the invariant operator on an arbitrary vector of $H_{[m]}$, which may be taken to be a basis vector:

$$\left\{ \begin{pmatrix} [k''] \\ \lambda'' \end{pmatrix} \begin{pmatrix} \gamma' \\ [k'] \\ \lambda' \end{pmatrix} \begin{pmatrix} [k] \\ \lambda \end{pmatrix} \right\} \left| \begin{matrix} [m] \\ m \end{matrix} \right\rangle = \left\{ \begin{pmatrix} [k''] \\ \lambda'' \end{pmatrix} \begin{pmatrix} \gamma' \\ [k'] \\ \lambda' \end{pmatrix} \begin{pmatrix} [k] \\ \lambda \end{pmatrix} \right\} ([m]) \left| \begin{matrix} [m] \\ m \end{matrix} \right\rangle. \tag{3.39}$$

The U(3) Racah coefficient given by the notation (3.38) is a real number of considerable complexity in its dependence on the integer entries entering its definition. For clarity, we display the symbol fully, in unabbreviated notation:

$$\left\{ \begin{pmatrix} k_1'' & & k_2'' & & k_3'' \\ & \ell'' & & -\ell'' & \\ & & 0 & & \end{pmatrix} \begin{pmatrix} & & \gamma_{11}' & & \\ & \gamma_{12}' & & \gamma_{22}' & \\ k_1' & & k_2' & & k_3' \\ & \ell' & & -\ell' & \\ & & 0 & & \end{pmatrix} \begin{pmatrix} k_1 & & k_2 & & k_3 \\ & \ell & & -\ell & \\ & & 0 & & \end{pmatrix} \right\} ([m_{13} m_{23} m_{33}]) \tag{3.40}$$

By construction, this coefficient is zero unless all of the following *generalized triangle conditions* are fulfilled:

$$\begin{aligned}
&\text{(i) } [k] \in [\overline{m}] \times [m] \text{ and } [k'] \in [\overline{m}] \times [m];\\
&\text{(ii) } [k''] \in [k'] \times [k];\\
&\text{(iii) } [k''] = [k] + \Delta(\gamma'), \text{ where } \Delta(\gamma') \text{ denotes the } \textit{shift} \text{ of the operator pattern } \gamma' \text{ defined by } \Delta(\gamma') = \\
&\qquad [\Delta_1(\gamma'), \Delta_2(\gamma'), \Delta_3(\gamma')] = [\gamma'_{11}, \gamma'_{12} + \gamma'_{22} - \gamma'_{11}, k'_1 + k'_2 + k'_3 - \gamma'_{12} - \gamma'_{22}];\\
&\text{(iv) all patterns satisfy the betweenness relations;}\\
&\text{(v) } [m] \text{ is an arbitrary U(3) irrep label.}
\end{aligned} \tag{3.41}$$

Here we use $\in$ to denote, for example, that $[k]$ occurs in the Kronecker product $[\overline{m}] \times [m]$ reduction.

The above notation for a U(3) Racah coefficient is explained in detail in Refs. [14, 36]; indeed, the notation extends to U(n). It is useful to remark that only operator patterns and the irrep labels of the irrep space $H_{[m]}$ appear in this notation, the latter reflecting that the coefficients have their origin as invariant operators, and the former reflecting that operator patterns are structual elements arising in the resolution of the multiplicity of irreps in the Kronecker product for general U(n).

The notation described above applies equally well to U(2) Racah coefficients, which are related in the following way to the standard W or 6-j notation:

$$\left\{ \begin{pmatrix} k''_1 & & k''_2 \\ & \lambda'' & \end{pmatrix} \begin{pmatrix} & \gamma' & \\ k'_1 & & k'_2 \\ & \lambda' & \end{pmatrix} \begin{pmatrix} k_1 & & k_2 \\ & \lambda & \end{pmatrix} \right\} (m_{12}, m_{22})$$
$$= [(2e+1)(2f+1)]^{1/2} W(abcd; ef), \tag{3.42}$$

where

$$\begin{aligned}
a &= \tfrac{1}{2}(m_{12} - m_{22}) - [\lambda'' - \tfrac{1}{2}(k''_1 + k''_2)],\\
b &= \tfrac{1}{2}(k_1 - k_2),\\
c &= \tfrac{1}{2}(m_{12} - m_{22}),\\
d &= \tfrac{1}{2}(k'_1 - k'_2),\\
e &= \tfrac{1}{2}(m_{12} - m_{22}) - [\lambda' - \tfrac{1}{2}(k'_1 + k'_2)],\\
f &= \tfrac{1}{2}(k''_1 - k''_2),
\end{aligned} \tag{3.43}$$

with $k''_1 = k_1 + \gamma'$, $k''_2 = k_2 + (k'_1 + k'_2 - \gamma')$ and $\lambda'' = \lambda + \lambda'$. It is a fact that U($n$) Racah coefficients may always be taken (by a phase choice) to be SU(n) Racah coefficients.

The U(2) notation above is redundant, but nonetheless is indispensible for the U(3) Racah coefficients and the general U(n) case in exhibiting clearly the structure features of these coefficients. The notation shows unambiguously how Racah coefficients inherit the operator pattern from those of the Wigner coefficients, these patterns themselves being the structural elements on which is based the resolution of the multiplicity problem. This is most clearly brought out by the following identities (see Refs. [36]–[38]) for the case at hand:

$$\left\{\begin{matrix} \lambda'' & \lambda' & \lambda \\ [k''] & [k'] & [k] \\ \alpha'' & \alpha' & \alpha \end{matrix}\right\} = \left\langle\begin{matrix} \lambda' \\ [k'] \\ \alpha' \end{matrix}\right\rangle \left\langle\begin{matrix} \lambda \\ [k] \\ \alpha \end{matrix}\right\rangle \left\langle\begin{matrix} \lambda'' \\ [k''] \\ \alpha'' \end{matrix}\right\rangle^{\dagger}, \tag{3.44}$$

$$\left\{\begin{pmatrix} [k''] \\ \lambda'' \end{pmatrix} \begin{pmatrix} \gamma' \\ [k'] \\ \lambda' \end{pmatrix} \begin{pmatrix} [k] \\ \lambda \end{pmatrix}\right\} = \sum_{\alpha,\alpha',\alpha''} \left\langle\begin{matrix} [k''] \\ \alpha'' \end{matrix}\right| \left\langle\begin{matrix} \gamma' \\ [k'] \\ \alpha' \end{matrix}\right\rangle \left|\begin{matrix} [k] \\ \alpha \end{matrix}\right\rangle$$
$$\times \left\langle\begin{matrix} \lambda' \\ [k'] \\ \alpha' \end{matrix}\right\rangle \left\langle\begin{matrix} \lambda \\ [k] \\ \alpha \end{matrix}\right\rangle \left\langle\begin{matrix} \lambda'' \\ [k''] \\ \alpha'' \end{matrix}\right\rangle^{\dagger}. \tag{3.45}$$

It is worth remarking here that the general algebra of unit tensor operators is infinite-dimensional (countably), but since all the operators introduced in this section are maps $H_{[m]} \to H_{[m]}$, the algebra (3.36) is finite-dimensional, there being altogether $(\dim[m])^2$ elements in the algebra, which closes. Indeed, all these unit tensor operators are represented on $H_{[m]}$ by real orthogonal matrices of dimension, $\dim[m]$, a key fact that we utilize in Subsection 7 below.

7.3.6 Weyl Basis of the Lie Algebra of U(dim[m])

The set of $(\dim[m])^2$ Weyl basis elements of U(dim[m]) is given by

$$\left\{ E_{m'm} \left| \begin{pmatrix} [m] \\ m' \end{pmatrix} \text{ and } \begin{pmatrix} [m] \\ m \end{pmatrix} \text{ are Gel'fand patterns} \right.\right\}. \tag{3.46}$$

The subscripts m' and m are U(2) Gel'fand patterns that share the same U(3) irrep label $[m]$. These operators obey the commutation relations

$$[E_{m'm}, E_{m'''m''}] = \delta_{m\,m'''} E_{m'm''} - \delta_{m'm''} E_{m'''m}, \tag{3.47}$$

where for the evaluation of the Kronecker δ's two U(2) Gel'fand patterns are defined to be equal when their corresponding entries are equal, and otherwise are defined to be unequal.

A bijection of the Weyl basis elements (3.46) to the standard set

$$\{E_{ij} \,| i,j = 1,2,\ldots,\dim[m]\} \tag{3.48}$$

may be obtained from the following rules: For given U(3) irrep labels $[m]$, we introduce a total ordering on the set of all U(2) patterns by writing

$$\begin{pmatrix} m'_{12} & & m'_{22} \\ & m'_{11} & \end{pmatrix} > \begin{pmatrix} m_{12} & & m_{22} \\ & m_{11} & \end{pmatrix} \tag{3.49}$$

whenever the first nonzero entry in the 3-tuple

$$(m'_{12} - m_{12}, m'_{22} - m_{22}, m'_{11} - m_{11}) \tag{3.50}$$

is non-negative; otherwise, we write $m > m'$. The bijection is then given by

$$\begin{pmatrix} m_{13} & & m_{23} \\ & m_{13} & \end{pmatrix} \to 1, \ldots, \begin{pmatrix} m_{23} & & m_{33} \\ & m_{33} & \end{pmatrix} \to \dim[m], \tag{3.51}$$

with all intermediate patterns being mapped in turn to the integer given by the rule: if $m' > m$ with $m' \to n'$ and $m \to n$, then $n' < n$. This rule accords with the one used in Eq. (2.26), where the index $j - m + 1$ is mapped to 1 for the highest weight ($m = j$) and to $2j + 1$ for the lowest weight ($m = -j$).

7.3.7 Racah Basis of the Lie Algebra of U(dim[m])

The real orthogonal matrix R of dimension $(\dim[m])^2$ is defined by

$$R\left[\begin{pmatrix} \lambda \\ [k] \\ \alpha \end{pmatrix}\begin{pmatrix} m \\ [m] \\ m' \end{pmatrix}\right] = \left[\frac{\dim[k]}{\dim[m]}\right]^{1/2}\left\langle \begin{matrix}[m]\\ m'\end{matrix} \middle| \left\langle \begin{matrix}\lambda \\ [k] \\ \alpha\end{matrix}\right\rangle \middle| \begin{matrix}[m]\\ m\end{matrix}\right\rangle. \tag{3.52}$$

Property (3.33) expresses the fact that the matrix R with rows and columns enumerated by the patterns

$$\begin{pmatrix} \lambda \\ [k] \\ \alpha \end{pmatrix} \text{ and } \begin{pmatrix} m \\ [m] \\ m' \end{pmatrix}, \tag{3.53}$$

respectively, is real orthogonal. This matrix is now used to define a new basis, *the Racah basis*, of the Lie algebra of U(dim[m]):

$$E\begin{pmatrix} \lambda \\ [k] \\ \alpha \end{pmatrix} = \sum_{m',m} R\left[\begin{pmatrix} \lambda \\ [k] \\ \alpha \end{pmatrix}\begin{pmatrix} m \\ [m] \\ m' \end{pmatrix}\right] E_{m'm}. \tag{3.54}$$

Since the matrix R is real orthogonal, relation (3.54) can be inverted to given the Weyl generators in terms of the Racah generators:

$$E_{m'm} = \sum_{\begin{pmatrix} \lambda \\ [k] \\ \alpha \end{pmatrix}} R\left[\begin{pmatrix} \lambda \\ [k] \\ \alpha \end{pmatrix}\begin{pmatrix} m \\ [m] \\ m' \end{pmatrix}\right] E\begin{pmatrix} \lambda \\ [k] \\ \alpha \end{pmatrix}. \tag{3.55}$$

The embedding of the symmetry group U(3) in the U(dim$[m]$) is now obtained in the following way: The relation between the K_{ij} generators of U(3) given by Eqs. (3.5)–(3.7) and U(3) Wigner operators is given by (see Refs. [14, 39, 40])

$$K_{ij} = (-1)^j (\mathbf{I})^{1/2} \left\langle \begin{array}{ccccc} & & 0 & & \\ & 0 & & 0 & \\ 1 & & 0 & & -1 \\ & & (i,j) & & \end{array} \right\rangle, \quad i \neq j, \tag{3.56}$$

$$\frac{(K_{11} - K_{22})}{\sqrt{2}} = -(\mathbf{I})^{1/2} \left\langle \begin{array}{ccccc} & & 0 & & \\ & 0 & & 0 & \\ 1 & & 0 & & -1 \\ & 1 & & -1 & \\ & & 0 & & \end{array} \right\rangle, \tag{3.57}$$

$$\frac{(K_{11} + K_{22} - 2K_{33})}{\sqrt{6}} = (\mathbf{I})^{1/2} \left\langle \begin{array}{ccccc} & & 0 & & \\ & 0 & & 0 & \\ 1 & & 0 & & -1 \\ & 0 & & 0 & \\ & & 0 & & \end{array} \right\rangle, \tag{3.58}$$

$$K_{11} + K_{22} + K_{33} = \mathbf{I}_1 \left\langle \begin{array}{ccccc} & & 0 & & \\ & 0 & & 0 & \\ 0 & & 0 & & 0 \\ & 0 & & 0 & \\ & & 0 & & \end{array} \right\rangle. \tag{3.59}$$

In Eq. (3.56), the index pair (i,j) denotes the unique Gel'fand pattern having weight $e_i - e_j$, where $e_1 = (1\ 0\ 0), e_2 = (0\ 1\ 0), e_3 = (0\ 0\ 1)$. The invariant operators appearing in these relations have the following definitions:

$$\mathbf{I} = (3\mathbf{I}_2 - \mathbf{I}_1^2)/27, \tag{3.60}$$

with

$$\mathbf{I}_1 = \sum_i K_{ii}, \quad \mathbf{I}_2 = \sum_{i,j} K_{ij}K_{ji}. \tag{3.61}$$

The eigenvalues of $\mathbf{I}_1$ and $\mathbf{I}_2$ on $H_{[m]}$ are

$$I_1([m]) = m_{13} + m_{23} + m_{33}, \tag{3.62}$$

$$I_2([m]) = \sum_{i=1}^{3}(m_{i3} + 3 - i)^2 - 2\sum_{i=1}^{3}(m_{i3} + 3 - i) + 1. \tag{3.63}$$

We thus obtain the following relations on $H_{[m]}$:

$$K_{ij} = (-1)^j A([m]) E \left\langle \begin{array}{ccccc} & & 0 & & \\ & 0 & & 0 & \\ 1 & & 0 & & -1 \\ & & (i,j) & & \end{array} \right\rangle, \quad i \neq j, \tag{3.64}$$

$$K_{ii} = A_{ii}([m])E\left\langle \begin{array}{ccccc} & & 0 & & \\ & 0 & & 0 & \\ 0 & & 0 & & 0 \\ & 0 & & 0 & \\ & & 0 & & \end{array} \right\rangle + B_{ii}([m])E\left\langle \begin{array}{ccccc} & & 0 & & \\ & 0 & & 0 & \\ 1 & & 0 & & -1 \\ & 1 & & -1 & \\ & & 0 & & \end{array} \right\rangle$$

$$+C_{ii}([m])E\left\langle \begin{array}{ccccc} & & 0 & & \\ & 0 & & 0 & \\ 1 & & 0 & & -1 \\ & 0 & & 0 & \\ & & 0 & & \end{array} \right\rangle, \tag{3.65}$$

where

$$\begin{aligned} A([m]) &= [(\dim[m])I([m])/8]^{1/2}, \\ A_{11}([m]) &= A_{22}([m]) = A_{33}([m]) = [(\dim[m])/8]^{1/2} I_1([m])/3, \\ -B_{11}([m]) &= B_{22}([m]) = A([m])/\sqrt{2}, \quad B_{33}([m]) = 0, \\ C_{11}([m]) &= C_{22}([m]) = -C_{33}([m])/2 = A([m])/\sqrt{6}. \end{aligned} \tag{3.66}$$

These relations give explicitly the embedding of the symmetry group U(3) in U(dim$[m]$).

The commutators of the U(3) generators K_{ij} with the general elements of the Racah basis (3.54) are given by

$$\left[K_{ij}, E\begin{pmatrix}\lambda\\ {[k]}\\ \alpha\end{pmatrix}\right] = \sum_{\alpha'} \left\langle \begin{matrix}[k]\\ \alpha'\end{matrix} \middle| K_{ij} \middle| \begin{matrix}[k]\\ \alpha\end{matrix}\right\rangle E\begin{pmatrix}\lambda\\ {[k]}\\ \alpha'\end{pmatrix}, \tag{3.67}$$

where $\langle \cdots | K_{ij} | \cdots \rangle$ denotes the standard Gel'fand-Zetlin matrix elements of the U(3) generators K_{ij}. These relations show that *the Racah basis of the Lie algebra of* U(dim$[m]$) *consists of irreducible tensor operators with respect to* U(3). Globally, this relation is expressed by

$$T_U E\begin{pmatrix}\lambda\\ {[k]}\\ \alpha\end{pmatrix} T_{U^{-1}} = \sum_{\alpha'} D^{k}_{\alpha'\alpha}(U) E\begin{pmatrix}\lambda\\ {[k]}\\ \alpha'\end{pmatrix}, \text{ each } U \in \mathrm{U}(3). \tag{3.68}$$

7.3.8 Structure Constants in the Racah Basis of the Lie Algebra of U(dim$[m]$)

We now apply relation (2.37) and use the product law of U(3) unit tensor operators to obtain the following relations for the Racah basis of the Lie algebra of $U(\dim[m])$:

$$\left[E\begin{pmatrix}\lambda'\\ [k']\\ \alpha'\end{pmatrix}, E\begin{pmatrix}\lambda\\ [k]\\ \alpha\end{pmatrix}\right] = \sum_{\begin{pmatrix}\lambda''\\ [k'']\\ \alpha''\end{pmatrix}} A\begin{bmatrix}\lambda'' & \lambda' & \lambda\\ [k''] & [k'] & [k]\\ \alpha'' & \alpha' & \alpha\end{bmatrix} E\begin{pmatrix}\lambda''\\ k''\\ \alpha''\end{pmatrix}, \tag{3.69}$$

where the *structure constants* in the relation are given by

$$A\begin{bmatrix}\lambda'' & \lambda' & \lambda\\ [k''] & [k'] & [k]\\ \alpha'' & \alpha' & \alpha\end{bmatrix} = \left[\frac{\dim[k]\ \dim[k']\ \dim[k'']}{\dim[m]\ \dim[m]\ \dim[m]}\right]^{1/2}$$

$$\times\left(\begin{Bmatrix}\lambda'' & \lambda' & \lambda\\ [k''] & [k'] & [k]\\ \alpha'' & \alpha' & \alpha\end{Bmatrix}([m]) - \begin{Bmatrix}\lambda'' & \lambda' & \lambda\\ [k''] & [k'] & [k]\\ \alpha'' & \alpha' & \alpha\end{Bmatrix}([m])\right) \tag{3.70}$$

with [see Eqs. (3.37)–(3.39)]

$$\begin{Bmatrix}\lambda'' & \lambda' & \lambda\\ [k''] & [k'] & [k]\\ \alpha'' & \alpha' & \alpha\end{Bmatrix}([m]) = \sum_{\gamma}\left\langle\begin{matrix}[k'']\\ \alpha''\end{matrix}\right|\left\langle\begin{matrix}\gamma'\\ [k']\\ \alpha'\end{matrix}\right\rangle\left|\begin{matrix}[k]\\ \alpha\end{matrix}\right\rangle$$

$$\times\left\{\begin{pmatrix}[k'']\\ \lambda''\end{pmatrix}\begin{pmatrix}\gamma'\\ [k']\\ \lambda'\end{pmatrix}\begin{pmatrix}[k]\\ \lambda\end{pmatrix}\right\}([m]). \tag{3.71}$$

This result for the structure constants is the U(3) analogue of the SU(2) result given by Eq. (2.39). It is to be noted, however, that the U(dim[m]) structure constants given by Eqs. (3.70)–(3.71) entail a summation over the operator patterns γ'; that is, *the structure constants do not in the general case assume a factored form into a product of a WCG coefficient and a Racah coefficient.* This result complicates the situation for determining Lie algebraic zeroes of the U(3) Racah coefficients, as we discuss in Section 5.

7.3.9 Action of U(dim[m]) on $\mathbf{H}_{[m]}$

Let $V \in$ U(dim[m]), and let the elements of V be enumerated by the U(2) Gel'fand patterns m' and m described in Eqs. (3.46)–(3.51). Then the action of U(dim[m]) on $H_{[m]}$ is given by

$$S_V\left|\begin{matrix}[m]\\ m\end{matrix}\right\rangle = \sum_{m'} V_{m'm}\left|\begin{matrix}[m]\\ m'\end{matrix}\right\rangle, \quad \text{each } V \in \text{U}(\dim[m]). \tag{3.72}$$

In particular, since $D^{[m]}(U) \in$ U(dim[m]), each $U \in$ U(3), we find that

$$S_{D^{[m]}(U)} = T_U. \tag{3.73}$$

This result shows that the space $H_{[m]}$ carries the fundamental representation

$$D^{[10\cdots 0]}(V) = V \tag{3.74}$$

of U(dim$[m]$) and that this representation, when restricted to the symmetry group U(3) as embedded in U(dim$[m]$) in the Racah basis, reduces to the irreps of this U(3):

$$D^{[10\cdots 0]}(D^{[m]}(U)) = D^{[m]}(U), \quad \text{each} U \in \mathrm{U}(3). \tag{3.75}$$

7.4 Racah basis for the Lie algebra of any subgroup $G \subset \mathrm{U}(\dim[m])$

Let G be an arbitrary subgroup of U(dim$[m]$):

$$G \subset \mathrm{U}(\dim[m]). \tag{4.1}$$

Let the Lie algebra of G have basis

$$\{X_1, X_2, \ldots, X_p\} \tag{4.2}$$

with commutation relations

$$[X_r, X_s] = \sum_t C^t_{rs} X_t, \quad r, s = 1, 2, \ldots, p, \tag{4.3}$$

where $\{C^t_{rs}\}$ denote the structure constants.

We may restrict the fundamental $[1\ 0\ \cdots\ 0]$ irrep of U(dim$[m]$) to G and obtain a representation of G by matrices of dimension, dim$[m]$. We denote the corresponding representation of the basis elements $\{X_r\}$ of the Lie algebra by

$$\{M_1, M_2, \ldots, M_p\}. \tag{4.4}$$

These (dim$[m]$) $\times$ (dim$[m]$) matrices then satisfy the commutation relations (4.3):

$$[M_r, M_s] = \sum_t C^t_{rs} M_t, \quad r, s = 1, 2, \ldots, p. \tag{4.5}$$

We next obtain the realization of the Lie algebra of G on the vector space $H_{[m]}$ in the following way: Define the linear map

$$L_r : H_{[m]} \rightarrow H_{[m]} \tag{4.6}$$

by

$$L_r = \sum_{m', m} (M_r)_{m'm} E_{m'm}, \tag{4.7}$$

where $\{E_{m'm}\}$ is the set of basis elements of the $U(\dim[m])$ Lie algebra, as described by Eqs. (3.46)–(3.51). The maps in the set

$$\{L_1, L_2, \ldots, L_p\} \tag{4.8}$$

then also satisfy the commutation relations:

$$[L_r, L_s] = \sum_t C^t_{rs} L_t, \quad r,s = 1,2,\ldots,p. \tag{4.9}$$

The operators L_r in the set (4.6)–(4.7) can be expressed in terms of the Racah basis of the Lie algebra of U(dim[m]) by using Eq. (3.55), which relates the Weyl and Racah basis:

$$L_r = \sum_{\left(\begin{smallmatrix}\lambda\\ [k]\\ \alpha\end{smallmatrix}\right)} R_r \begin{pmatrix}\lambda\\ [k]\\ \alpha\end{pmatrix} E \begin{pmatrix}\lambda\\ [k]\\ \alpha\end{pmatrix}, \tag{4.10}$$

where the coefficients $R_r(\,\cdot\,)$ are defined by

$$R_r \begin{pmatrix}\lambda\\ [k]\\ \alpha\end{pmatrix} = \sum_{m',m} (M_r)_{m'm} R \left[\begin{pmatrix}\lambda\\ [k]\\ \alpha\end{pmatrix} \begin{pmatrix}m\\ [m]\\ m'\end{pmatrix} \right], \quad r = 1,2,\ldots,p. \tag{4.11}$$

Using identity (4.10) for the L_r in the commutation relations (4.8), and Eq. (3.69) for the commutator of Racah basis elements, we obtain the following relation between structure constants:

$$\sum_{\left(\begin{smallmatrix}\lambda'\\ [k']\\ \alpha'\end{smallmatrix}\right)\left(\begin{smallmatrix}\lambda\\ [k]\\ \alpha\end{smallmatrix}\right)} R_r \begin{pmatrix}\lambda'\\ [k']\\ \alpha'\end{pmatrix} R_s \begin{pmatrix}\lambda\\ [k]\\ \alpha\end{pmatrix} A \begin{bmatrix}\lambda'' & \lambda' & \lambda\\ [k''] & [k'] & [k]\\ \alpha'' & \alpha' & \alpha\end{bmatrix}$$

$$= \sum_t C^t_{rs} R_t \begin{pmatrix}\lambda''\\ [k'']\\ \alpha''\end{pmatrix}. \tag{4.12}$$

An important point to note here is that, once the matrix realization

$$X_r \to M_r, \quad r = 1,2,\ldots,p \tag{4.13}$$

of the Lie algebra of G is given, the operator realization of this algebra on the space $H_{[m]}$,

$$X_r \to L_r, \quad r = 1,2\ldots,p, \tag{4.14}$$

in the Racah basis, as given by Eqs. (4.10)–(4.11), is *uniquely* determined.

7.5 Zeroes of U(3) Racah coefficients

We have shown in the Section 3 how the U(3) Racah coefficients enter into the structure constants of the Lie algebra of U(dim[m]) in the Racah basis of this algebra. The symbol for these coefficients and some of its properties are given in Eqs. (3.36)–(3.45). The coefficient is zero, by definition, whenever the integers entering into the arrays occurring in the symbol fall outside their domains as described by the generalized triangle conditions in (i)-(v) given in Eq. (3.41).

It is a well-known fact that the SU(2) Racah coefficients possess infinite families of zeroes, even when all the triangle conditions are fulfilled and all symmetries taken into account. A few of these zeroes (nine, in fact, as of 1988) have found explanation in terms of a group G lying between SU(2) [or SO(3)] and $U(2j+1)$:

$$SU(2) \subset G \subset U(2j+1), \quad \text{for some } j,$$

or

$$SO(3) \subset G \subset U(2j+1), \quad \text{for some } j. \tag{5.1}$$

(See Ref. [11] for summaries as of 1986; see also Refs. [10], [41]–[46]); since the Lie algebra of G is a sub-Lie algebra of that of $U(2j+1)$, containing also the SU(2) [or SO(3)] Lie algebra, the Lie algebra of G can be realized in terms of the Racah basis. This then implies relations on the structure constants given by Eq. (2.39), leading in some instances to a required vanishing of a Racah coefficient. We call zeroes of this type *Lie algebraic zeroes.*

Infinite families of zeroes of the Racah coefficients, including the Lie algebraic ones, have also been obtained as the solutions of classical Diophantine equations that arise directly from the explicit form of these coefficients. Brudno [47] initiated these Diophantine equation studies, and, since, a number of publications [48]–[57] along these lines have followed. (No comprehensive theory of all zeroes of SU(2) Racah coefficients has yet appeared, to our knowledge.) (Aside from the discovery of several Lie algebraic zeroes, the occurrence of numerous zeroes was first shown numerically in Ref. [58]; see also Ref. [59].)

The question of zeroes of the SU(3) Racah coefficients is complicated beyond that of SU(2) by another structural feature: The structure constants in the Racah basis given by Eq. (3.69) no longer appear in a factored form as they do in the SU(2) case of Eq. (2.39). This situation is unavoidable, since it is a direct consequence of the multiplicity structure of the Kronecker product reduction. On the other hand, if G is a group such that

$$SU(3) \subset G \subset U(\dim m), \tag{5.2}$$

then the Racah basis of the Lie algebra of G must close with attendant relations between the structure constants. Does this imply also some SU(3)

Racah coefficient zeroes? We examine this question more closely below for the exceptional Lie group E_6, which has the embedding

$$SU(3) \subset E_6 \subset SU(27). \tag{5.3}$$

Still another feature enters into the discussion of U(3) Racah coefficient vanishings. Certain zeroes are implied by the canonical splitting conditions that resolve the U(3) multiplicity. These zeroes are, in fact, a consequence of the fact that a certain class of Racah invariants must be the null invariant operator **0** in consequence of null space structural properties of operator patterns (see Ref. [34]). These zeroes go beyond the triangle rules of Eq. (3.41). It is proved in Ref. [34] (see Eqs. (3.15) of this reference) that these structural zeroes include the following ones among the coefficients described in Eqs. (3.38)–(3.41):

$$\left\{ \begin{pmatrix} [k''] \\ \lambda'' \end{pmatrix} \begin{pmatrix} \gamma' \\ [k'] \\ \lambda' \end{pmatrix} \begin{pmatrix} [k] \\ \lambda \end{pmatrix} \right\} = \mathbf{0}, \tag{5.4}$$

whenever the following condition is fulfilled:

$$i + i' - i'' \geq [k_1' - \Delta_1(\gamma')] - [k_3' - \Delta_3(\gamma')] + 2, \tag{5.5}$$

where the indices i, i', i'' have the following definitions in terms of the operator patterns appearing in the symbol (3.40):

$$i \;=\; \begin{cases} k_1 - \ell + 1, & \text{for } k_2 \geq 0 \\ -k_3 - \ell + 1, & \text{for } k_2 < 0 \end{cases}; \tag{5.6}$$

$$i' \;=\; \begin{cases} k_1' - \ell' + 1, & \text{for } k_2' \geq 0 \\ -k_3' - \ell' + 1, & \text{for } k_2' < 0 \end{cases}; \tag{5.7}$$

$$i'' \;=\; \begin{cases} k_1'' - \ell'' + 1, & \text{for } k_2'' \geq 0 \\ -k_3'' - \ell'' + 1, & \text{for } k_2'' < 0 \end{cases}. \tag{5.8}$$

The indices i, i', i'' arise in the canonical labelling of the operator patterns in a given multiplicity set in the following way (see Refs. [32]–[34]). It is sufficient to illustrate the rule for the index i, since the others are obtained similarly. U(3) operator patterns may be ordered by two different rules, which, it turns out, are compatible. In the first rule, we write for $k_2 \geq 0$:

$$\begin{pmatrix} k_1 \quad k_2 \quad k_3 \\ \Gamma_{k_1-\ell+1} \end{pmatrix} = \begin{pmatrix} k_1 \quad\; k_2 \quad\; k_3 \\ \ell \quad -\ell \\ 0 \end{pmatrix}; \tag{5.9}$$

that is,

$$\Gamma_{k_1-\ell+1} = \lambda = \begin{pmatrix} \ell \quad -\ell \\ 0 \end{pmatrix}, \quad \ell = k_1, k_1 - 1, \ldots, k_2. \tag{5.10}$$

We now apply the order rule on such patterns given by Eqs. (3.49)–(3.50) to obtain

$$\Gamma_1 > \Gamma_2 > \cdots > \Gamma_M, \quad M = k_1 - k_2 + 1. \tag{5.11}$$

Similarly, for $k_2 < 0$, we write

$$\Gamma_{-k_3-\ell+1} = \lambda = \begin{pmatrix} \ell & & -\ell \\ & 0 & \end{pmatrix}, \quad \ell = -k_2, -k_2 - 1, \ldots, -k_3, \tag{5.12}$$

with again

$$\Gamma_1 > \Gamma_2 > \cdots > \Gamma_M, \quad M = k_2 - k_3 + 1. \tag{5.13}$$

These inequalities on operator patterns then accord exactly with the nested properties of the null space of the unit tensor operators

$$\left\langle \begin{matrix} & \Gamma_t & \\ k_1 & k_2 & k_3 \\ & \bullet & \end{matrix} \right\rangle, \quad t = 1, 2, \ldots, M, \tag{5.14}$$

as given by

$$N(\Gamma_1) \supset N(\Gamma_2) \supset \cdots \supset N(\Gamma_M). \tag{5.15}$$

Operator patterns were, of course, introduced precisely to accommodate this nested null space structure of the associated unit tensor operators.

The results given in Eqs. (5.4)–(5.15) allows us to identify ***structural zeroes*** of the Racah coefficient associated with the canonical resolution of the multiplicity. We use this result below

Let us return now to the problem, for Racah coefficients, originating from the existence of the group-subgroup chain (5.3). Our first problem is to obtain the generators of E_6 in terms of the Racah basis of SU(27). The 27-dimensional irrep of E_6 is complex [60], which implies that the generators of this representation are antihermitian. Thus, for the E_6 Lie algebra, we may take the basis elements

$$\{M_1, M_2, \ldots, M_p\}, \quad \text{with } p = 78, \tag{5.16}$$

of Eq. (4.4) to be antihermitian

$$M_r^\dagger = -M_r, \quad r = 1, 2, \ldots, 78. \tag{5.17}$$

Using this result in Eq. (4.7), we obtain

$$L_r^\dagger = -L_r, \quad r = 1, \ldots, 78, \text{ on } H_{[4\ 2\ 0]}. \tag{5.18}$$

Thus, the Lie algebra of E_6 is realized on the space $H_{[4\ 2\ 0]}$ as antihermitian operators. For the consistency of Eq. (5.3), let us note that the SU(3) Lie algebra can also be realized by 27×27 real, skew-symmetric matrices. This representation is equivalent to the [4 2 0] irrep, and can be obtained by using the Gell-Mann [61] generators of SU(3).

We thus obtain eight of the E_6 generators as the eight SU(3) generators, as given in the Racah basis by Eqs. (3.56)–(3.66). The remaining seventy must come in conjugate pairs to realize the conjugation property (5.18). All maps $H_{[4\ 2\ 0]} \to H_{[4\ 2\ 0]}$ are labelled by the irreps occurring in the reduction of the Kronecker product $[0\ -2\ -4] \times [4\ 2\ 0]$. This reduction is given by Eqs. (3.16)–(3.21) to be

$$\begin{aligned}
&[0\ -2\ -4] \times [4\ 2\ 0] = \\
&\underset{125}{[4\ \ 0\ -4]} \oplus \underset{81}{[4\ -1\ -3]} \oplus \underset{81}{[3\ \ 1\ -4]} \oplus \\
&\underset{28}{[4\ -2\ -2]} \oplus \underset{28}{[2\ \ 2\ -4]} \oplus 2\underset{64}{[3\ \ 0\ -3]} \oplus 2\underset{35}{[3\ -1\ -2]} \oplus 2\underset{35}{[2\ \ 1\ -3]} \oplus \\
&3\underset{27}{[2\ \ 0\ -2]} \oplus \underset{10}{[2\ -1\ -1]} \oplus \underset{10}{[1\ \ 1\ -2]} \oplus 2\underset{8}{[1\ \ 0\ -1]} + \underset{1,}{[0\ 0\ 0]}
\end{aligned} \tag{5.19}$$

where we have written the dimension below each irrep label. We see from this result that the only irrep labels that qualify for labelling the remaining generators of E_6 in the Racah basis are the thirty-five dimensional conjugate irreps

$$[2\ \ 1\ -3] \text{ and } [3\ -1\ -2]. \tag{5.20}$$

The explicit conjugation relation for the corresponding unit tensor operators is

$$\begin{aligned}
\left\langle \begin{matrix} & \lambda & \\ 2 & 1 & -3 \\ & \alpha & \end{matrix} \right\rangle^{\dagger} &= (-1)^{\phi(\alpha)} \left\langle \begin{matrix} & \lambda & \\ 3 & -1 & -2 \\ & \overline{\alpha} & \end{matrix} \right\rangle, \\
\lambda &= \begin{pmatrix} & 0 & \\ 1 & & -1 \end{pmatrix} \text{ or } \begin{pmatrix} & 0 & \\ 2 & & -2 \end{pmatrix},
\end{aligned} \tag{5.21}$$

where

$$\phi(\alpha) = \alpha_{11} + \alpha_{12} + \alpha_{22}, \quad \overline{\alpha} = \begin{pmatrix} -\alpha_{22} & & -\alpha_{12} \\ & -\alpha_{11} & \end{pmatrix}. \tag{5.22}$$

We conclude from these relations that the orthogonal matrix defined by Eq. (3.52) has the property

$$R\left[\begin{pmatrix} & \lambda & \\ 2 & 1 & -3 \\ & \alpha & \end{pmatrix} \begin{pmatrix} m \\ [m] \\ m' \end{pmatrix}\right] = (-1)^{\phi(\alpha)} R\left[\begin{pmatrix} & \overline{\lambda} & \\ 3 & -1 & -2 \\ & \overline{\alpha} & \end{pmatrix} \begin{pmatrix} m' \\ [m] \\ m \end{pmatrix}\right]. \tag{5.23}$$

Using this result in Eq. (4.11), together with the antihermitian property of the matrix M_r, now gives

$$R_r^* \begin{pmatrix} & \lambda & \\ 2 & 1 & -3 \\ & \alpha & \end{pmatrix} = -(-1)^{\phi(\alpha)} R_r \begin{pmatrix} & \overline{\lambda} & \\ 3 & -1 & -2 \\ & \overline{\alpha} & \end{pmatrix} \tag{5.24}$$

for each of the operator patterns λ in Eq. (5.21) (note that $\overline{\lambda} = \lambda$).

We now use relation (5.23) in Eq. (4.10), selecting only the terms $[2 \ \ 1 - 3]$ and $[3 \ -1 \ -2]$ from the sum, to conclude that the remaining generators of E_6 in the Racah basis are given by

$$\begin{aligned} L_r &= \sum_{\ell=1}^{2}\sum_{\alpha} R_r \begin{pmatrix} & & 0 & & \\ & \ell & & -\ell & \\ 2 & & 1 & & -3 \\ & & \alpha & & \end{pmatrix} E \begin{pmatrix} & & 0 & & \\ & \ell & & -\ell & \\ 2 & & 1 & & -3 \\ & & \alpha & & \end{pmatrix} \\ &\quad - \sum_{\ell=1}^{2}\sum_{\alpha} R_r^* \begin{pmatrix} & & 0 & & \\ & \ell & & -\ell & \\ 2 & & 1 & & -3 \\ & & \alpha & & \end{pmatrix} (-1)^{\phi(\alpha)} E \begin{pmatrix} & & 0 & & \\ & \ell & & -\ell & \\ 3 & & -1 & & -2 \\ & & \overline{\alpha} & & \end{pmatrix}, \end{aligned} \tag{5.25}$$

where $r = 9, 10, \ldots, 78$, since we can choose the generators

$$\{L_1, L_2, \ldots, L_8\} \tag{5.26}$$

to be those of the SU(3) subalgebra. This set of operators then satisfies the antihermitian property:

$$L_r^\dagger = -L_r, \quad r = 1, 2, \ldots, 78. \tag{5.27}$$

We observe again that once the 27-dimensional antihermitian generators in (5.16) of the 27-dimensional representation of E_6 are specified, there is no freedom left in expressing the generators L_r in terms of the Racah basis, since the R_r-coefficients in Eqs. (5.25), (5.26) are uniquely obtained from Eq. (4.11).

It follows from Eq. (5.25) that the following operators are a basis for these generators:

$$L_\alpha^{(1)} = \sum_{\ell=1}^{2} \left[E \begin{pmatrix} & & 0 & & \\ & \ell & & -\ell & \\ 2 & & 1 & & -3 \\ & & \alpha & & \end{pmatrix} - (-1)^{\phi(\alpha)} E \begin{pmatrix} & & 0 & & \\ & \ell & & -\ell & \\ 3 & & -1 & & -2 \\ & & \overline{\alpha} & & \end{pmatrix} \right], \tag{5.28}$$

$$L_\alpha^{(2)} = i \sum_{\ell=1}^{2} \left[E \begin{pmatrix} & & 0 & & \\ & \ell & & -\ell & \\ 2 & & 1 & & -3 \\ & & \alpha & & \end{pmatrix} + (-1)^{\phi(\alpha)} E \begin{pmatrix} & & 0 & & \\ & \ell & & -\ell & \\ 3 & & -1 & & -2 \\ & & \overline{\alpha} & & \end{pmatrix} \right], \tag{5.29}$$

where α runs over all thirty-five Gel'fand patterns for the given irrep labels. These operators then satisfy the following identities on the space $H_{[4\ 2\ 0]}$:

$$L_\alpha^{(1)\dagger} = -L_\alpha^{(1)}, \quad L_\alpha^{(2)\dagger} = -L_\alpha^{(2)}. \tag{5.30}$$

Let us now consider the consequences of the E_6 embedding (5.3). We have, first of all, that the commutators

$$\left[E\left(\begin{array}{ccccc} & & 0 & & \\ & 0 & & 0 & \\ 1 & & 0 & & -1 \\ & & (i,j) & & \end{array}\right), L_\alpha^{(a)}\right], \quad a = 1, 2, \quad \left\{\begin{array}{l} i \neq j = 1,2,3 \\ i = j = 1,2, \end{array}\right. \tag{5.31}$$

must close on the generators of E_6. These commutators must, by construction of the Racah basis, close automatically on the generators of E_6 because of the irreducible tensor property expressed by Eq. (3.67) [see Eqs. (3.56)–(3.59) and (3.64)–(3.66)]. That is, the closure of the commutators (5.31) is a property of the Racah basis, and *not* of the E_6 embedding. Indeed, in consequence of the irreducible tensor operator property (3.67) and the generator relations (3.56)–(3.59), the structure constants for the Racah basis must obey the relations:

$$\begin{aligned} &A\left[\begin{array}{ccc} \lambda'' & \begin{array}{ccccc} & & 0 & & \\ & 0 & & 0 & \\ {[1} & & 0 & & -1] \\ & & (i,j) & & \end{array} & \lambda \\ {[k'']} & & {[k]} \\ \alpha'' & & \alpha \end{array}\right] \\ &\quad = \delta_{[k''][k]} \left\langle \begin{array}{c} [k] \\ \alpha'' \end{array} \right| \left\langle \begin{array}{ccccc} & & 0 & & \\ & 0 & & 0 & \\ 1 & & 0 & & -1 \\ & & (i,j) & & \end{array} \right\rangle \left| \begin{array}{c} [k] \\ \alpha \end{array} \right\rangle, \end{aligned} \tag{5.32}$$

for $i \neq j = 1,2,3;\ i = j = 1,2;$

$$A\left[\begin{array}{ccc} & \begin{array}{ccccc} & & 0 & & \\ & 0 & & 0 & \\ {[0} & & 0 & & 0] \\ & 0 & & 0 & \\ & & 0 & & \end{array} & \\ \lambda'' & & \lambda \\ {[k'']} & & {[k]} \\ \alpha'' & & \alpha \end{array}\right] = \delta_{[k''][k]}\delta_{\lambda''\lambda}\delta_{\alpha''\alpha}. \tag{5.33}$$

These relations already lead to interesting and nontrivial identities between WCG and Racah coefficients when the structure constants given by Eqs. (3.70), (3.71) are substituted for the left-hand side in Eq. (5.32). These relations are, as pointed out above, *not* properties of the E_6 embedding, but rather of the canoncial splitting and null space conditions defining the solution of the multiplicity problem.

For the possibility of a SU(3) Racah coefficient vanishing in consequence of the E_6 embedding (5.3), one requires the closure of the commutators

$$\left[L_\alpha^{(a)}, L_{\alpha'}^{(a')}\right], \quad a, a' = 1, 2, \tag{5.34}$$

on the E_6 generators. In the next step, we substitute relations (5.28)–(5.29) for the generators into the commutator (5.31), using relation (3.69) to obtain a linear combination of the Racah basis elements of the form

$$E\begin{pmatrix}\lambda'' \\ [k''] \\ \alpha''\end{pmatrix} \quad \text{with} \quad \begin{cases}[k''] \in [2 \quad 1 \ -3] \times [2 \quad 1 \ -3], \\ [k''] \in [2 \quad 1 \ -3] \times [3 \ -1 \ -2], \\ [k''] \in [3 \ -1 \ -2] \times [3 \ -1 \ -2].\end{cases} \tag{5.35}$$

For closure of the Lie algebra of E_6, the coefficients of each operator (5.35) must vanish for $[k''] \neq [1 \quad 0 \ -1], [2 \quad 1 \ -3], [3 \ -1 \ -2]$. This leads to linear relations between the structure constants.

To determine if there are any SU(3) Racah coefficient zeroes associated with a linear relation between structure constant described above, one expresses these coefficients in terms of WCG and Racah coefficients using Eqs. (3.70), (3.71). One must at the same time account for the null space zeroes as described by Eqs. (5.5)-(5.15). The result of this is a set of relations between WCG and Racah coefficients. Finally, one must take into account the symmetry relations for WCG coefficients, as well as those for the Racah coefficients. Since such symmetry relations are not, at this time, known generally for generic irrep labels, it is a sizeable task to implement this procedure. It may, indeed, require numerical calculation of the relevant coefficients.

We have formulated here the rather intricate structural relations that underly the problem of zeroes of SU(3) Racah coefficients. We hope to implement this process for particular coefficients in a future paper, using either more detailed knowledge of the required symmetries than presently available, or with the help of numerical calculations.

7.6 Acknowledgements

Work performed under the auspices of the U.S. Department of Energy. We wish to acknowledge the help of Dr. Joris van der Jeugt in our initial work on this problem. We also thank the organizers of this conference, honoring Marcos Moshinsky, for the invitation to present this work.

7.7 References

[1] V. Bargmann and M. Moshinsky, "Group theory of harmonic oscillators (I). The collective modes," Nucl. Phys. **18** (1960), 697–712; "(II). The integrals of motion for the quadrupole-quadrupole interaction," *ibid.* **23** (1961), 177–199.

[2] M. Moshinsky, "Wigner coefficients for the SU_3 group and some applications," Rev. Mod. Phys. **34** (1962), 813-828.

[3] M. Moshinsky, "Bases for the irreducible representations of the unitary groups and some applications," J. Math. Phys. **4** (1963), 1128–1139.

J.G. Nagel and M. Moshinsky, "Operators that raise or lower the irreducible vector spaces of U_{n-1} contained in an irreducible vector space of U_n, J. Math. Phys. **6** (1965), 682–694.

[4] T.A. Brody, M. Moshinsky, and I. Renero, "Recursion relations for the Wigner coefficients of unitary groups," J. Math. Phys. **6** (1965), 1540–1546.

[5] E. Chacón and M. Moshinsky, "Representations of finite U_3 transformations," Physics Letters **23** (1966), 567–569.

[6] E. Chacon, "Representation coefficients for the SU(3) group," Rev. Mex. Fś. **17** (1968), 315–325.

[7] P. Kramer and M. Moshinsky, "Group theory of harmonic oscillators and nuclear structure," in *Group Theory and Its Applications* (E.M. Loebl, Ed.), Vol. I, pp. 339–468. Academic Press, New York, 1968.

[8] G. Racah, "Theory of complex spectra I," Phys. Rev. **61** (1942), 186–197; II. *ibid.* **62** (1942), 438–462; III. *ibid* **63** (1943), 368–382; IV, *ibid.* **76** (1949), 1352–1365.

[9] G. Racah, *Group Theory and Spectroscopy.* Lectures at the Institute for Advanced Study, Princeton, New Jersey, 1951; published in Ergeb. Exakt. Naturw. **37** (1965), 28–84.

[10] G. Vanden Berghe, H. DeMeyer, and J. Van der Jeugt, "Tensor operator realizations of E_6 and structural zeros of the $6j$-symbol," J. Math. Phys. **25** (1984), 2585–2589.

[11] W.A. Beyer, J.D. Louck, and P.R. Stein, "Zeroes of Racah coefficients and the Pell equation," Acta Appl. Math. **7** (1986), 257–311.

[12] L.C. Biedenharn and J.D. Louck, *Angular Momentum in Quantum Physics*, Encyclopedia of Mathematics and Its Applications, Vol. 8; *The Racah-Wigner Algebra in Quantum Theory*, Vol. 9; edited by G.-C. Rota, Addison-Wesley, Reading, MA, 1981. (Reissued: Cambridge University Press, London and New York, 1985).

[13] I.M. Gel'fand and M.L. Zetlin, "Finite representations of the group of unimodular matrices," Doklady Akad. Nauk **71** (1980), 825–28. (Appears in translation in I.M. Gel'fand, R.A. Minlos, and Z.Ya. Shapiro, *Representations of the Rotations and Lorentz Groups and Their Applications*, Pergamon, New York, 1963. Translated from the Russian by G. Cummins and T. Boddington); I. M. Gel'fand and M. I. Graev, "Finite-dimensional irreducible representations of the unitary and full linear groups, and related special functions," Izv. Akad. Nauk SSSR Ser. Mat. **29** (1965), 1329–1356 [Am. Math. Soc. Transl. **64** (1967), Ser. 2, 116–146].

[14] J.D. Louck, "Recent progress toward a theory of tensor operators in the unitary groups," Amer. J. Phys. **38** (1970), 3–42.

[15] W.J. Holman and L.C. Biedenharn, "The representations and tensor operators of the unitary groups U(n)," in *Group Theory and Its Applications* (E.M. Loebl, Ed.), Vol. II, pp. 1–73. Academic Press, New York, 1971.

[16] W.J. Holman, "On the general boson states of $U_n * U_n$ and $Sp_4 * Sp_4$," Nuovo Cimento **4A** (1971), 904–931.

[17] J.D. Louck and L.C. Biedenharn, "The structure of the canonical tensor operators in the unitary groups. III. Further development of the boson polynomials and their implications," J. Math. Phys. **14** (1973), 1336–1357.

[18] C.J. Henrich, "The Gel'fand states of certain representations of U(n) and the decomposition of products of representations of U(2)," J. Math. Phys. **16** (1975), 2271–2288.

[19] P. Kramer, "Finite representations of the unitary group and their applications in many body physics," in *Group Theoretical Methods in Physics: Proceedings of the Fifth International Colloquium* (R.T. Sharp and B. Kolman, Eds.), pp. 173–179. Academic Press, New York, 1977.

[20] J.-P. Gazeau, M.-Cl. Dumont-Lepage, and A. Ronveaux, "Gelfand lattice polynominals and irreducible representations of U(n)," J. Math. Phys. **19** (1978), 734–748.

[21] A. Antillón and T.H. Seligman, "Explicit expressions for double Gel'fand states," J. Math. Phys. **23** (1982), 473–480.

[22] Y. Fujiwara and H. Horiuchi, "Properties of double Gel'fand polynomials and their application to multiplicity-free problems", Memoirs of the Faculty of Science of Kyoto University (Japan) 1982.

[23] J.P.S. Kung and G.-C. Rota, "The invariant theory of binary forms," Bull. Am. Math. Soc. **10** (1984), 27–85.

[24] J. Grabmeier and A. Kerber, "The evaluation of irreducible polynomial representations of the general linear groups and of the unitary groups over fields of characteristic 0," Acta Appl. Math. **8** (1987), 271–291.

[25] J.D. Louck and L.C. Biedenharn, "Some properties of the interwining numers of the general linear group," in *Science and Computers, Advances in Supplementary Studies* (G.-C. Rota, Ed.). Academic Press, New York, 1984.

[26] K. Baclawski, "A new rule for computing Clebsch-Gordan series," Adv. Appl. Math. **5** (1984), 418–432.

[27] B.G. Wybourne, *Symmetry Principles and Atomic Spectroscopy*, Wiley (Interscience), New York, 1970.

[28] B.G. Wybourne, "Properties of half-filled shells in *d*- and *f*-configurations," in *Symmetry and Structural Properties of Condensed Matter* (W. Florek, T. Lulek, M. Mucha, Eds.), pp. 155–186. World Scientific, 1991.

[29] L.C. Biedenharn, A. Giovannini, and J.D. Louck, "Canonical definition of Wigner operators in U_n," J. Math. Phys. **8** (1967), 691–700.

[30] L.C. Biedenharn and J.D. Louck, "A pattern calculus for tensor operators in the unitary groups," Commun. Math. Phys. **8** (1968), 80–131.

[31] E. Chacón, M. Ciftan, and L.C. Biedenharn, "On the evaluation of the multiplicity-free Wigner coefficients of U(n)," J. Math. Phys. **13** (1972), 577–589.

[32] L.C. Biedenharn, J.D. Louck, E. Chacón, and M. Ciftan, "On the structure of the canonical tensor operators in the unitary groups. I. An extension of the pattern calculus rules and the canonical splitting in U(3)," J. Math. Phys. **13** (1972), 1957–1984.

[33] L.C. Biedenharn and J.D. Louck, "On the structure of the canonical tensor operators in the unitary groups. II. The tensor operators in $U(3)$ characterized by maximal null space," J. Math. Phys. **13** (1972), 1985–2001.

[34] J.D. Louck, M.A. Lohe, and L.C. Biedenharn, "Structure of the canonical U(3) Racah functions and the U(3):U(2) projective functions," J. Math. Phys. **16** (1975), 2408–2426.

[35] M.A. Lohe, L.C. Biedenharn, and J.D. Louck, "Structural properties of the self-conjugate SU(3) tensor operators," J. Math. Phys. **18** (1977), 1883–1891.

[36] J.D. Louck and L.C. Biedenharn, "Canonical unit adjoint tensor operators in U(n)," J. Math. Phys. **11** (1970), 2368–2414.

[37] J.D. Louck and L.C. Biedenharn, "Special functions associated with SU(3) Wigner-Clebsch-Gordon coefficients," in *Symmetry and Structural Properties of Condensed Matter* (W. Florek, T. Lulek, and M. Mucha, Eds.), pp. 207–229. World Scientific, 1991.

[38] J.D. Louck and L.C. Biedenharn, "Canonical and noncanonical methods in applications of group theory to physical problems (to appear in Adv. uan t. Chem., (1991).

[39] G.E. Baird and L.C. Biedenharn, "On the representations of semisimple Lie groups. II," J. Math. Phys. **4** (1963), 1449–1466; "III. The explicit conjugation operation for SU(n)," *ibid.* **5** (1964), 1723–1730; "IV, A canonical classification for tensor operators in SU_3," *ibid.* **5** (1965) 1730–1747.

[40] J.D. Louck, "Group theory of harmonic oscillators in n-dimensional space," J. Math. Phys. **6** (1965), 1786–1804.

[41] B.R. Judd, *Operator techniques in Atomic Spectroscopy* (McGraw-Hill, New York, 1963), p. 151.

[42] H.T. Wadzinski, "The group F_4 and its generators," Nuovo Cimento **B62** (1969), 247–257.

[43] B.R. Judd, "Topics in atomic theory," in *Topics in Atomic and Nuclear Theory,* Caxton Press, Christchurch, New Zealand, 1970, pp. 1–60.

[44] A. de-Shalit and I. Talmi, *Nuclear Shell Theory,* Academic Press, New York, 1963.

[45] J. Van der Jeugt, G. Vanden Berghe, and H. De Meyer, "Boson realization of the Lie algebra F_4 and non-trivial zeroes of the $6j$ symbol," J. Phys. A.: Math. Gen. **16** (1983), 1377–1382.

[46] H. De Meyer, G. Vanden Berghe, and J. Van der Jeugt, "On Lie algebras built from SO(3)," J. Math. Phys. **25** (1984), 751–754.

[47] S. Brudno, "Nontrivial zeroes of the Wigner (3-j) and Racah (6-j) coefficients. I. Linear solutions," J. Math. Phys. **26** (1985), 434–435.

[48] K.S. Rao, "A note on the classification of the zeroes of angular momentum coefficients," J. Math. Phys. **26** (1985), 2260–2261.

[49] S. Brudno and J.D. Louck, "Nontrivial zeroes of weight 1 $3j$- and $6j$-coefficients: Relation to Diophantine equations of equal sums of like powers," J. Math. Phys. **26** (1985), 2092–2094.

[50] S. Brudno and J.D. Louck, "Nontrivial zeroes of the Racah quadrupole invariant," J. Math. Phys. **26** (1985), 1125–1127.

[51] A. Lindner, "Non-trivial zeroes of the Wigner ($3j$) and Racah ($6j$) coefficients," J. Phys. A: Math. Gen. **18** (1985), 3071–3072.

[52] K.S. Rao, "A note on the classification of the zeroes of angular momemtum coefficients," J. Math. Phys. **26** (1985), 2260–2261.

[53] A. Bremner, "On Diophantine equations and nontrivial Racah coefficients," J. Math. Phys. **27** (1986), 1181–1184.

[54] A. Bremner and S. Brudno, "A complete determination of the zeroes of weight-1 $6j$ coefficients," J. Math. Phys. **27** (1986), 2613–2615.

[55] J.J. Labarthe, "Parametrization of the linear zeroes of $6j$ coefficients," J. Math. Phys. **27** (1986), 2964–2965.

[56] K. Srinivasa Rao and V. Rajeswari, "An algorithm to generate the polynomial zeroes of degree one of the Racah coefficients," J. Phys. A: Math. Gen. **20** (1987), 507–509.

[57] S. Brudno, "Nontrivial zeroes of the Wigner ($3j$) and Racah ($6j$) coefficients, II. Some nonlinear solutions," J. Math. Phys. **28** (1987), 124–127.

[58] S.H. Koozekanani and L.C. Biedenharn, "Non-trivial zeroes of the Racah (6-j) coefficient," Rev. Mex. Fś. **23** (1974), 327–340.

[59] M.J. Bowick, "Regge symmetries and null 3-j and 6-j symbols," Thesis, University of Canterbury, Christchurch, New Zealand, 1976.

[60] J. Tits, *Tabellen zu den einfachen Lie Gruppen und ihren Darstellungen.* Lecture Notes in Mathematics **40** (1967), Springer-Verlag, Berlin.

[61] M. Gell-Mann, "Symmetries of baryons and mesons," Phys. Rev. **125** (1962), 1067–1084.

8

Scaling and Universality in the Shock Compression of Condensed Matter

Alfonso Mondragón and Mariblanca Astiz[1]

ABSTRACT From universality and scale invariance we derive a set of thermodynamic constraints on the equation of state of those materials for which the linear relation between shock and particle velocities is valid with non-vanishing intercept. These constraints are sufficient to determine the universal form which the incomplete equation of state of those materials takes at very high pressures and large values of the compression.

8.1 Introduction

For many years it has been known that under strong shock compression many different materials behave in a similar way [1]. The shock wave velocity U_s and the particle velocity U_p appear to be related by the linear equation

$$U_s = A + BU_p, \tag{1.1}$$

where A and B are constants characteristic of the material. When combined with the Rankine-Hugoniot equations for conservation of mass and momentum [2], the linear relationship between the velocities gives an empirical expression for the pressure on the Hugoniot as function of the compression which is singular at the maximum value of the compression [2]. It has also been established that the linear relation between shock and particle velocities is a consequence of the conservation of mass across the shock front [3]. Since no specific thermodynamic information about the physical system is required to determine the functional relation (1.1), it might be said that, in the regime of strong shock compression, the conservation of mass dominates over the microscopic dynamics of the system. It gives rise to scale invariance and a law of corresponding states [4]. However, not all the information about the thermodynamic properties of the system is absent. For those physical systems which obey the linear relation with $A \neq 0$, the singularity in the pressure on the Hugoniot is a double pole while for

---

[1]Instituto de Física, UNAM.

other systems, like the perfect gas, the linear relation is obeyed with $A = 0$, and the singularity is a single pole or a branch point. In analogy with the thermodynamics of a second order phase transition near a critical point [5], we might say that we have two classes of universality characterized by the critical exponents 2 and ν, with $-1 < \nu < 2$, respectively.

The aim of this work is to translate the statement "a physical system which obeys the linear relationship with $A \neq 0$" into precisely worded thermodynamic expressions in the form of thermodynamic constraints on the equation of state. It will be shown that the thermodynamic characterization obtained in this way is sufficient to determine the universal form which the incomplete equation of state of such a system takes in the limit of very high pressures and large values of the compression. A complete equation of state may also be derived introducing the temperature as the inverse of the integrating factor of the entropy. In a particular case, we give analytical expressions for the thermodynamic coefficients, the specific heat at constant volume and the Grüneisen parameter. Although scale invariance in the strong shock compression of condensed materials is a consequence of mass conservation across the shock front, our results show that the specific form it takes for a particular physical system, that is, the numerical value of the critical exponent is a manifestation of a property of the specific internal energy which is a state function. Therefore, the consequences following from this property should show up also in static compression experiments when the values of the pressure and compression are comparable to those attained in strong shock compression.

8.2 Rankine-Hugoniot equations

Let us consider a steady shock wave travelling in an infinite material medium in thermodynamic equilibrium on each side of the shock front. For sufficiently strong shocks, the rigidity effects are negligible so that a solid may be treated as a fluid. The motion of the shock front is related to the thermodynamic properties of the fluid by the laws of conservation of mass, momentum and energy. Let U_s be the velocity of propagation of the shock front. Let ρ_o and U_{po} be the mass density and particle velocity (mean velocity of the matter) in the region which has not yet been reached by the shock front, and ρ and U_p be the corresponding quantities in the fluid just behind the front. Then, the Rankine-Hugoniot equations for conservation of mass and momentum across a steady shock wave are

$$\rho_o(U_s - U_{po}) = \rho(U_s - U_p), \tag{2.1}$$

and

$$P - P_o = \rho_o(U_s - U_{po})(U_p - U_{po}), \tag{2.2}$$

respectively. These equations may be combined with the equation for conservation of energy to give a third Rankine-Hugoniot relation [2]

$$E - E_o = \tfrac{1}{2}(V_o - V)(P + P_o), \tag{2.3}$$

where P_o, P and E_o, E are the fluid pressure and specific internal energy in front of and behind the shock front. We have also introduced the specific volumes $V_o = 1/\rho_o$ and $V = 1/\rho$.

8.3 Universality

It has been observed that under strong shock compression many different materials behave in a similar way. The shock velocity and the particle velocity behind the shock are related by the linear equation

$$U_s = A + BU_p, \tag{3.1}$$

where A and B are constants characteristic of the material. This relation applies to all kinds of condensed materials, liquids as well as solids. These include elemental metals, metallic compounds, other chemical compounds and mixtures (*e.g.* alkali-metal hallides unreacted explosives, rocks, polymers and plastics) [6]. Departures from linearity can usually be attributed to the porosity of the material or to phase transitions [7].

The universality of the behaviour of shock compressed materials may be understood as a consequence of the condition of conservation of mass, equation (2.1), which may also be written as

$$\frac{U_s}{U_p} = \frac{\rho}{\rho - \rho_o} + \frac{\rho_o}{\rho_o - \rho}\,\frac{U_{po}}{U_p}. \tag{3.2}$$

For strong shocks, $U_s \gg U_p$ and $U_p \gg U_{po}$. Hence, when there is a limit to the shock compression such that the compressibility κ_H on the Hugoniot vanishes at the maximum value of the compression ρ_L/ρ_o, we may write

$$\frac{U_s}{U_p} \underset{\rho\to\rho_L}{\longrightarrow} \frac{\rho_L}{\rho_L - \rho_o} = B. \tag{3.3}$$

Then, for strong shock compression, we have that

$$U_s \simeq A + BU_p, \tag{3.4}$$

where A and B are constants. Prieto and Renero [3] have shown that under the conditions stated above, in the limit of strong shock compression, the general form of the functional relation between U_s and U_p takes the form

$$U_s = A + BU_p + F(U_p), \tag{3.5}$$

where A and B are constants, and $F(U_p)$ and its first derivative dF/dU_p are rapidly vanishing functions of U_p when $\rho/\rho_o \to \rho_L/\rho_o$.

8.4 The empirical expressions for the pressure and internal energy on the shock Hugoniot

The linear relation (3.1) and the Rankine-Hugoniot equations for conservation of mass and momentum may be combined to obtain the well known empirical expression for the pressure as function of the compression on the shock Hugoniot

$$P_H = \frac{\left(\frac{A}{B}\right)^2 \frac{1}{V_o} 1 - \frac{V}{V_o}}{\left(\frac{V}{V_o} - \frac{B-1}{B}\right)^2 + P_o.} \tag{4.1}$$

This expression is valid for any material obeying the linear relationship between shock and particle velocities with $A \neq 0$. Substitution of (4.1) for P_H in the Rankine-Hugoniot equation for conservation of energy, equation (2.3), gives the empirical expression for the specific internal energy on the shock Hugoniot

$$E_H = \frac{1}{2}\left(\frac{A}{B}\right)^2 \frac{\left(1 - \frac{V}{V_o}\right)^2}{\left(\frac{V}{V_o} - \frac{B-1}{B}\right)^2} + (V_o - V)P_o + E_o. \tag{4.2}$$

8.5 A law of corresponding states: scaling

Introducing the concept of similarity of equations, Prieto [4] has shown that when scaled or reduced dimensionless variables p and x are defined as

$$p = P_H / P_C, \tag{5.1}$$

with

$$P_C = A^2 / V_o B, \tag{5.2}$$

and

$$x = \frac{B}{V_o}(V_o - V), \tag{5.3}$$

then, the empirical expression for the pressure on the Hugoniot, Eq. (3.7), takes the universal form

$$p = \frac{x}{(1-x)^2} + p_o. \tag{5.4}$$

Similarly, the reduced specific internal energy may be written as

$$e = \frac{x^2}{(1-x)^2} + 2p_o x + e_o, \tag{5.5}$$

where

$$e = 2\left(\frac{B}{A}\right)^2 E. \tag{5.6}$$

States of shock compression of different materials will be said to be in corresponding states if their relative compressions and their pressures are in the ratio

$$\left(1 - \frac{V_1}{V_{10}}\right)\left(1 - \frac{V_2}{V_{20}}\right)^{-1} = \frac{B_1}{B_2} \tag{5.7}$$

and

$$\frac{P_2}{P_1} = \frac{P_{2C}}{P_{1C}}. \tag{5.8}$$

In the plane (p, x), the reduced of all materials obeying the linear relation coincide in one single curve, represented by the plot of equation (5.4). Corresponding states of shock compression for different materials are represented by the same point on this curve. As we approach the limiting value of the compression, the singular term $(1-x)^{-2}$ dominates. The form of the pressure or the internal energy on the Hugoniot does not change, only the *scale* changes. Thus we might say that in the neighbourhood of the limiting compression there are no natural length scales left.

This situation is somewhat reminiscent of the behaviour of a thermodynamic system undergoing a second order phase transition near a critical point [5]. In this latter case the susceptibilities, specific heats and other thermodynamic properties of the system are singular functions of the temperature at the critical temperature. The singularities are characterized by the critical exponents. Different sets of values of the critical exponents define different universality classes. All physical systems in the same universality class obey a law of corresponding states. From the knowledge of the critical exponents and scale invariance, that is, the law of corresponding states, it is possible to derive explicit expressions for the analytical form of the free energy and the equation of state of the system near the critical point valid for all systems in the same universality class.

In the case of the asymptotic regime of strong shock compression, the pressure and the internal energy on the Hugoniot are singular functions of the compression near the limiting value of the compression. For those physical systems that obey the linear relation (3.1) with $A \neq 0$, the singularity is characterized by the "critical" exponent 2. For ideal systems, like the perfect gas, that obey the linear relationship with $A = 0$, the value of the "critical" exponent is 1. Therefore, keeping in mind the analogy with the thermodynamics of the critical point, it makes sense to look for explicit expressions for the analytical form of the thermodynamic potentials and the equation of state valid in the asymptotic regime of strong shock compression for all physical systems in the same universality class.

8.6 Formal implicit solution for the pressure on the Hugoniot

In order to proceed with the discussion of our problem we need a thermodynamic expression for the specific internal energy of the system which combined with the Rankine-Hugoniot equation for conservation of energy gives the pressure on the Hugoniot as function of compression.

It is convenient to refer both the internal energy and pressure to their zero temperature values at the given density

$$\hat{E}(V) \equiv E(V, T=0), \tag{6.1}$$

$$\hat{P}(V) \equiv P(V, T=0) = -\frac{d\hat{E}}{dV}. \tag{6.2}$$

To obtain the values of E and P, at a finite temperature T, we can express E and P as functions of V and T. Then E and P are obtained by integrating along a path in the (V,T) plane from the reference state $(V_o, T=0)$ to the state (V,T): first along the cold isotherm $T=0$ from V_o to V, then from $T=0$ to T at constant V. The result is

$$E = \hat{E}(V) + \int_0^T dT' \left(\frac{\partial E}{\partial T'}\right)_V, \tag{6.3}$$

$$P = \hat{P}(V) + \int_0^T dT' \left(\frac{\partial P}{\partial T}\right)_V, \tag{6.4}$$

or, introducing the entropy S,

$$E = \hat{E}(V) + \int_0^T dT' \left(\frac{\partial S}{\partial T'}\right)_V, \tag{6.5}$$

and

$$P = \hat{P}(V) + \int_0^T dT' \left(\frac{\partial S}{\partial V}\right)_{T'}. \tag{6.6}$$

The effect of the heat content on the equation of state may conveniently be expressed by the dimensionless ratio

$$R \equiv \frac{(E-\hat{E})}{V(P-\hat{P})} = \frac{\displaystyle\int_0^T dT\, T \left(\frac{\partial S}{\partial T}\right)_V}{\displaystyle\int_0^T dT\, V \left(\frac{\partial S}{\partial V}\right)_T}. \tag{6.7}$$

According to this definition, the ratio R is specified as a function of T and V. However, since it is a state function, with the help of the equation of state we may express R as function of P and V. In this way, we obtain

$$E = \hat{E} + R\, V(P-\hat{P}) \tag{6.8}$$

which, when combined with the Rankine-Hugoniot equation for conservation of energy (2.3) gives

$$P_H = \frac{(2R_H+1)^{-1}\left[(1-\frac{V}{V_o})P_o + \frac{2}{V_o}(E_o - \hat{E}) + 2R_H\hat{P}\frac{V}{V_o}\right]}{\left(\frac{V}{V_o} - \frac{1}{2R_H+1}\right)}. \qquad (6.9)$$

The ratio R_H is, of course, still specified as a function of $P_H(V)$ and V, so equation (6.9) is no particular simplification on the general case. However, in the following, it will be shown that written in this form it is particularly appropriate for the discussion of the asymptotic regime of strong shock compression.

In the empirical expressions for the pressure and the specific internal energy on the Hugoniot, (4.1) and (4.2), the singular term $(V/V_o - V_L/V_o)^{-2}$ dominates over all other terms when V/V_o is close to V_L/V_o. If $R(P,V)$ is such that the solution $P_H(V)$ to (6.9) has the same functional dependence on V/V_o as the empirical expression (4.1) in the regime of strong shock compression, then the specific internal energy, obtained by substitution of this solution for P_H in the Rankine-Hugoniot equation for conservation of energy, Eq. (2.3), also reproduces the functional dependence on V/V_o of the empirical expression (4.2). Both grow as $(V/V_o - V_L/V_o)^{-2}$ when V approaches V_L. The ratio $R_H(P_H(V),V)$ remains finite for all $V \geq V_L$. It follows that the numerator in the right hand side of (6.9) is non-singular for $V \geq V_L$. Therefore, the singular behaviour of the empirical expression for P_H is reproduced by the thermodynamic expression (6.9) only if the denominator in the right hand side of this equation has a double zero at $V = V_L$.

Hence, the thermodynamic expression for the pressure on the Hugoniot, Eq. (6.9), and the corresponding one for the specific internal energy will give a linear relation between U_s and U_p with $A \neq 0$, when

$$\frac{V}{V_o} - \frac{1}{2R_H(P_H(V),V)+1} \underset{V\to V_L}{\longrightarrow} \left(\frac{V}{V_o} - \frac{V_L}{V_o}\right)^2 g\left(\frac{V_L}{V_o}\right). \qquad (6.10)$$

8.7 Conditions on $R(P,V)$ for a double pole in $P_H(V)$

The denominator in the implicit equation for $P_H(V)$ may be written as

$$\frac{V}{V_o} - \frac{1}{2R_H+1} = \frac{V_L}{V_o} + \left(\frac{V}{V_o} - \frac{V_L}{V_o}\right) - \frac{1}{2R_H+1}. \qquad (7.1)$$

Since the state of the system to which (6.10) applies is constrained to be on the Hugoniot, the arguments (P,V) in R_H are linked by a functional relation. In Eq. (6.9) we chose V as the independent variable, then R_H is a function of V only. We will assume that R_H is a regular function of V at V_L and make a Taylor expansion of $(2R_H+1)^{-1}$ about V_L/V_o, in this way we obtain

$$\begin{aligned}\frac{V}{V_o}-\frac{1}{2R_H+1} = & \left(\frac{V_L}{V_o}-\frac{1}{2R_L+1}\right) \\ & +\left[1+\frac{2V_o}{(2R_L+1)^2}\left(\frac{dR_H}{dV}\right)_{V_L}\right]\left(\frac{V}{V_o}-\frac{V_L}{V_o}\right) \\ & +\frac{1}{(2R_L+1)^2}\left[\frac{2}{2R_L+1}\left(\frac{dR_H}{dV}\right)^2_{V_L}\right. \\ & \left.-\left(\frac{d^2R_H}{dV^2}\right)\right]\left(\frac{V}{V_o}-\frac{V_L}{V}\right)^2+\cdots, \end{aligned} \tag{7.2}$$

where

$$R_L = \lim_{V\to V_L} R_H\left(P_H(V),V\right). \tag{7.3}$$

The condition (6.10) will be satisfied if the two first terms in the right hand side of Eq.(7.2) vanish when $V \to V_L$, that is

$$\lim_{V\to V_L}\frac{1}{2R_H(P_H(V),V)+1}=\frac{V_L}{V_o}, \tag{7.4}$$

and

$$\lim_{V\to V_L}\left(\frac{dR_H}{dV}\right)_H = -\frac{(2R_L+1)^2}{2V_o}, \tag{7.5}$$

and the coefficient of $(V/V_o - V_L/V_o)^2$ has a finite non–vanishing value in the limit $V \to V_L$

$$0<\left|\frac{2}{2R_L+1}\left(\frac{dR_H}{dV}\right)^2_{V_L}-\left(\frac{d^2R_H}{dV^2}\right)_{V_L}\right|<\infty. \tag{7.6}$$

These conditions make sense only if the first and second derivatives of R_H with respect to V exist and do not vanish at $V = V_L$. Now, since R is a function of two variables, the pressure P and specific volume V, then, its first and second derivatives with respect to V on the Hugoniot will depend on both the pressure on the Hugoniot $P_H(V)$ and its derivatives dP_H/dV and d^2P_H/dV^2. Thus, if the conditions under which the pressure on the Hugoniot will have a pole of second order, Eqs. (7.4) and (7.5), are to be satisfied, some consistency conditions should also be satisfied.

8.8 Consistency conditions

The conditions on $R(P,V)$ for a double pole in $P_H(V)$, Eqs. (7.5) and (7.6), may be satisfied only if dR_H/dV and d^2R_H/dV^2 are finite and non-vanishing. By imposing these conditions, we are requiring that $P_H(V)$ and its first and second derivatives be singular and simultaneously we are demanding that $R_H, dR_H/dV$ and d^2R_H/dV^2 be regular at $V = V_L$. Since R_H and its derivatives are explicit functions of P_H and its derivatives, some additional conditions should be imposed on R to make these two requirements consistent with each other.

Let us consider first the derivative of R_H with respect to V on the Hugoniot, which occurs in (7.5). It may be written as

$$\frac{dR_H}{dV} = \left(\frac{\partial R}{\partial P}\right)_V \frac{dP_H}{dV} + \left(\frac{\partial R}{\partial V}\right)_P . \tag{8.1}$$

If the conditions (7.4) and (7.5) are satisfied and d^2R_H/dV^2 is finite and non-vanishing, then according to (6.9), (7.1) and (7.2), the thermodynamic expression for $P_H(V)$ is a singular function of V with a dominant singular term proportional to $(V/V_o - V_L/V_o)^{-2}$, and its derivative is proportional to $(V/V_o - V_L/V_o)^{-3}$. In terms of the pressure

$$\frac{dP_H}{dV} \simeq -2(V_o - V_L)^{-1/2} P_H^{3/2} \underset{V \to V_L}{\longrightarrow} \infty. \tag{8.2}$$

Therefore, the term $(\partial R/\partial P)_V \, dP_H/dV$ will be finite and non-vanishing when V approaches V_L only if the derivative $(\partial R/\partial P)_V$ vanishes as $P^{-3/2}$ for large values of P,

$$\left(\frac{\partial R}{\partial P}\right)_V \sim -\frac{1}{2} r(V) P^{-3/2}, \tag{8.3}$$

where $r(V)$ is a function of V but not P. Integrating (8.3) with respect to P, we obtain

$$R(P,V) \sim R_L + r(V) P^{-1/2}. \tag{8.4}$$

This result suggests that $(P^{-1/2}, V)$ is a natural and convenient choice of independent variables to represent the functional dependence of the ratio R on the pressure and compression for materials obeying the linear relationship between U_s and U_p with $A \neq 0$,

$$R = R(P^{-1/2}, V). \tag{8.5}$$

Furthermore, the condition (8.3) on $(\partial R/\partial P)_V$ leads us to expect that, when expressed as function of $P^{-1/2}$ and V, R has a Taylor expansion

$$R = R_L + r(V) P^{-1/2} + w(V) P^{-1} + \cdots \tag{8.6}$$

which is rapidly convergent for small $P^{-1/2}$ and $V < V_o$.

Then, the term $(\partial R/\partial P)_V \, dP_H/dV$ in the right hand side of (8.1) is finite and non-vanishing if

$$r(V_L) \neq 0. \tag{8.7}$$

This, together with (8.5) and (8.6) is the first consistency condition.

In order to be more specific about dR_H/dV, we must express the derivatives $(\partial R/\partial P)_V$ and $(\partial R/\partial V)_P$ appearing in (8.1) in terms of thermodynamic coefficients.

From the identity (6.8), a straightforward calculation gives

$$\left(\frac{\partial R}{\partial V}\right)_P = \left[\frac{1}{P\kappa_T}\left(\frac{1}{\Gamma} + T\alpha_P\right) - R - 1\right]\frac{1}{V}, \tag{8.8}$$

and

$$\left(\frac{\partial R}{\partial P}\right)_V = \frac{1}{P}\left[\frac{1}{\Gamma} - R\right], \tag{8.9}$$

where κ_T is the isothermal compressibility, α_P is the coefficient of thermal expansion, Γ is the Grüneisen parameter

$$\Gamma = V\alpha_P/\kappa_T C_V, \tag{8.10}$$

and C_V is the specific heat at constant volume.

Substitution of (8.8) and (8.9) in (8.1) gives

$$\frac{dR_H}{dV} = \frac{1}{P_H}\left(\frac{dP_H}{dV}\right)\left[\frac{1}{\Gamma} - R_H\right] + \frac{1}{V}\frac{T\alpha_P}{P_H\kappa_T} + \frac{1}{V\Gamma P\kappa_T} - \frac{1}{V}(R_H + 1). \tag{8.11}$$

From (8.2),

$$\frac{1}{P_H}\frac{dP_H}{dV} \sim -2(V_o - V_L)^{-1/2}P_H^{1/2} \underset{V\to V_L}{\longrightarrow} \infty. \tag{8.12}$$

Therefore, its coefficient in (8.10) should vanish

$$\lim_{V\to V_L}\left(\frac{1}{\Gamma} - R\right) = 0, \tag{8.13}$$

that is

$$R_L = \lim_{V\to V_L}\frac{1}{\Gamma} = \frac{1}{\Gamma_L}. \tag{8.14}$$

This is the second consistency condition; when combined with (7.4) it gives the well known thermodynamic expression for the limiting value of the compression [8], [9]

$$\frac{V_L}{V_o} = \frac{\Gamma_L}{2 + \Gamma_L}, \tag{8.15}$$

and, from here, the coefficient B in the linear relation (3.1)

$$B = 1 + \tfrac{1}{2}\Gamma_L. \tag{8.16}$$

The two consistency conditions derived above are necessary but not sufficient to guarantee that dR_H/dV is finite in the limit $V \to V_L$. We still have to verify that the terms $T\alpha_P/P\kappa_T$ and $1/V\Gamma P\kappa_T$ are finite in that limit. This can not be done without the explicit knowledge of the equation of state. However, in the next section, it will be shown that a complete equation of state may be obtained from the first consistency condition. Then, it will be shown that the terms just mentioned are also finite in the limit $V \to V_L$.

Now, let us turn our attention to the coefficient of $(V/V_o - V_L/V_o)^2$ in (7.2). It may be written as

$$\left[\left(\frac{dR_H}{dV}\right)^2 - \frac{1}{2}(2R_H+1)\left(\frac{d^2R_H}{dV^2}\right)\right] = \left(\frac{1}{P_H}\frac{dP_H}{dV}\right)^2 C_1 - \left(\frac{1}{P_H}\frac{d^2P_H}{dV^2}\right) C_2 + \left(\frac{1}{P_H}\frac{dP_H}{dV}\right) C_3 + C_4, \tag{8.17}$$

where

$$C_1 = \left(\frac{1}{\Gamma} - R\right)^2 + \left(\frac{1}{\Gamma_L} + \frac{1}{2}\right)\left[\frac{P}{\Gamma^2}\left(\frac{\partial\Gamma}{\partial P}\right)_V + 2\left(\frac{1}{\Gamma} - R\right)\right], \tag{8.18}$$

$$C_2 = \left(\frac{1}{\Gamma_L} + \frac{1}{2}\right)\left(\frac{1}{\Gamma} - R\right), \tag{8.19}$$

$$C_3 = \frac{2}{V}\left\{\left(\frac{1}{P\kappa_T}\frac{1}{\Gamma} - R + \frac{T\alpha_P}{P\kappa_T} - 1\right)\left(\frac{1}{\Gamma} - R + \frac{1}{\Gamma_L} + \frac{1}{2}\right) + \left(\frac{1}{\Gamma_L} + \frac{1}{2}\right)\frac{V}{\Gamma^2}\left(\frac{\partial\Gamma}{\partial V}\right)_P\right\}, \tag{8.20}$$

and

$$C_4 = \frac{1}{V^2}\left\{2\left(1 + \frac{1}{\Gamma_L}\right)\left(\frac{1}{P\kappa_T}\frac{1}{\Gamma} - R + \frac{T\alpha_P}{P\kappa_T} - 1\right) - \left(\frac{1}{\Gamma_L} + \frac{1}{2}\right)\left(\frac{1}{P\kappa_T}\frac{1}{\Gamma} + \frac{T\alpha_P}{P\kappa_T}\right)\left(1 - \frac{V}{\Gamma}\left(\frac{\partial\Gamma}{\partial V}\right)_P\right)\right\} \tag{8.21}$$

The first three terms in the right hand side of Eq. (8.17) have a divergent factor, namely, the product $P_H^{-1}dP_H/dV$ or $P_H^{-1}d^2P_H/dV^2$. Therefore, the left hand side of Eq. (8.17) can not be finite when $V \to V_L$ unless the coefficients C_i, $i = 1, 2, 3$, vanish in that limit:

$$\lim_{V \to V_L} C_i(P_H, V) = 0, \qquad i = 1, 2, 3. \tag{8.22}$$

These requirements lead to further consistency conditions

$$\lim_{V \to V_L} \left[\left(\frac{\partial \Gamma}{\partial V} \right)_P \right]_H = 0, \tag{8.23}$$

$$\lim_{V \to V_L} \left[P_H \left(\frac{\partial \Gamma}{\partial P} \right) \right]_H = 1, \tag{8.24}$$

$$\lim_{V \to V_L} (P \kappa_T)_H = 1, \tag{8.25}$$

$$\lim_{V \to V_L} (T \alpha_P)_H = 1. \tag{8.26}$$

The subscript H indicates that the corresponding function is constrained to be on the Hugoniot.

These consistency conditions and the ones previously found, Eqs. (8.5), (8.6), (8.7) and (8.13), may be interpreted as strong constraints on the functional form that the equation of state of the physical system takes when it is strongly shock compressed. In the analysis leading to the derivation of the consistency conditions the only assumption made about the physical system was that, in the limit of strong shock compression, it is such that the linear relation between the shock velocity U_s and the particle velocity U_p is obeyed with $A \neq 0$. Since the linear relationship is a consequence of mass conservation across the shock front, the only particular physical assumption about the system is that the constant A is non–vanishing. What we have, in fact, accomplished in this section is the translation of this characterization into thermodynamical constraints on the equation of state. Furthermore, since the equations of state of all such systems are all subject to the same constraining conditions, it is reasonable to expect that in the regime of very high pressure and compression they all take a functional form compatible with these constraints. That this is, in fact, the case will be shown in the next section.

8.9 The complete equation of state in the strong shock regime

A complete thermodynamic description of the states of the shock compressed system in the vicinity of and along the Hugoniot curve is obtained only when a complete equation of state linking the temperature with the pressure and specific volume is given. In this section we shall derive from the first consistency condition an explicit expression for the universal form taken by the incomplete equation of state of the systems which obey the linear relationship with $A \neq 0$ in the regime of strong shock compression. A complete equation of state may then be obtained introducing the temperature as the integrating factor of the entropy.

The first consistency condition, Eqs. (8.5), (8.6) and (8.7), gives the ratio

R as function of P and V in the form of an expansion in powers of $P^{-1/2}$,

$$R = \frac{1}{\Gamma_L} + r(V)P^{-1/2} + w(V)P^{-1} + \cdots, \tag{9.1}$$

where $r(V)$ and $w(V)$ are unknown functions of V. From the identity (6.8), we obtain a similar expression for the specific internal energy

$$\begin{aligned}(E - \hat{E}) &= \frac{V}{\Gamma_L}P\left(1 - \frac{\hat{P}}{P}\right) + r(V)VP^{1/2}\left(1 - \frac{\hat{P}}{P}\right) \\ &\quad + w(V)V\left(1 - \frac{\hat{P}}{P}\right) + \cdots. \end{aligned} \tag{9.2}$$

Since we are interested in describing the states close to or on the Hugoniot, P^{-1} is of the same order of magnitude as $(V - V_L)^2$, therefore, we may set $V = V_L$ in $r(V)$ and $w(V)$ and neglect the terms $\hat{P}/P$ without any real loss of generality. In this approximation, we get

$$E - \hat{E} = \frac{1}{\Gamma_L}VP + r_L VP^{1/2} + \tilde{w}_L V, \tag{9.3}$$

in $\tilde{w}$ we have collected the coefficients of all terms independent of $P^{-1/2}$ and $(V - V_L)$.

For an infinitesimal change in the specific internal energy, (9.3) gives

$$d(E - \hat{E}) = \left(\frac{1}{\Gamma_L} + \frac{1}{2}r_L P^{-1/2}\right)VdP + \left(\frac{1}{\Gamma_L}P + r_L P^{1/2} + \tilde{w}\right)dV. \tag{9.4}$$

Now, from the fundamental thermodynamic identity, we have

$$d(E - \hat{E}) = TdS - P\left(1 - \frac{\tilde{P}}{P}\right)dV, \tag{9.5}$$

hence, in the same approximation as before, we obtain for the entropy

$$\begin{aligned} dS &= \frac{1}{T}\left\{\left(\frac{1}{\Gamma_L} + \frac{1}{2}r_L P^{-1/2}\right)VdP \right. \\ &\quad \left. + \left[\left(\frac{1}{\Gamma_L} + 1\right)P + r_L P^{1/2} + \tilde{w}\right]dV\right\}, \end{aligned} \tag{9.6}$$

where S is the specific entropy and T is the temperature.

Recalling that $1/T$ is the integrating factor of the entropy, we finally arrive at

$$T = \tau\left[\left(\frac{1}{\Gamma_L} + 1\right)P + r_L P^{1/2} + \tilde{w}\right]V \tag{9.7}$$

where τ is a constant. Since the integrating factor of (9.6) is not unique, (9.7) is a particular solution to the problem of finding a complete equation of state for linear systems with $A \neq 0$. However, it is general enough to illustrate the essential features of the problem, while being simple enough to allow for the explicit verification of the consistency conditions. In this work, we will treat it as representative of the class of functional forms that the equation of state of a physical system which obeys the linear relation (3.1) with $A \neq 0$ might take at very high pressures and large values of the compression. It is only in this very restricted sense that we might call it a "universal" form of the equation of state in the regime of strong shock compression.

The equation for an infinitesimal change of the entropy now reads

$$dS = \frac{1}{\tau}\left\{\frac{\frac{1}{\Gamma_L} + \frac{1}{2} r_L P^{-1/2}}{\left(\frac{1}{\Gamma_L}+1\right) P + r_L P^{1/2} + \tilde{w}\, dP + \frac{1}{V} dV}\right\}. \tag{9.8}$$

The equation may readily be integrated to obtain S as function of pressure and specific volume. Once S is known, the thermodynamic potentials may be obtained from S and the equation of state (9.7).

8.10 The thermodynamic coefficients, the specific heat and the Grüneisen parameter

Now that we have obtained analytical expressions for the entropy and the equation of state, we may proceed to verify that these functions sátisfy the consistency conditions. First, we shall derive analytical expressions for the thermodynamic coefficients, the specific heat and the Grüneisen parameter.

The thermodynamic coefficients α_P and κ_T are readily obtained from the equation of state, Eq. (9.7). The result is

$$\alpha_P = 1 \Big/ \tau \left[\left(\frac{1}{\Gamma_L}+1\right) P + r_L P^{1/2} + \tilde{w}\right] V, \tag{10.1}$$

or, in terms of the temperature

$$\alpha_P = 1/T, \tag{10.2}$$

and

$$\kappa_T = \frac{1}{P}\left[\frac{1 - \frac{1}{2} r_L P^{-1/2} + \tilde{w} P^{-1}}{\frac{1}{\Gamma_L} + 1 + r_L P^{-1/2} + \tilde{w} P^{-1}}\right]. \tag{10.3}$$

The specific heat C_V may be obtained from the expression for an infinitesimal change of the entropy, Eq. (9.8), the equation of state (9.7) and the thermodynamic relation

$$C_V = T\left(\frac{\partial S}{\partial T}\right)_V = T\left(\frac{\partial S}{\partial P}\right)_V \left(\frac{\partial P}{\partial T}\right)_V. \tag{10.4}$$

the thermodynamic relation

$$C_V = T\left(\frac{\partial S}{\partial T}\right)_V = T\left(\frac{\partial S}{\partial P}\right)_V\left(\frac{\partial P}{\partial T}\right)_V. \tag{10.4}$$

The result is

$$C_V = 1/\tau. \tag{10.5}$$

The Grüneisen parameter Γ may be derived from the thermodynamic relation

$$\left(\frac{\partial(E-\hat{E})}{\partial(P-\hat{P})}\right)_V = \left(\frac{\partial(E-\hat{E})}{\partial P}\right)_V = \frac{V}{\Gamma}, \tag{10.6}$$

and the expression for $(E-\hat{E})$, Eq. (9.2), we obtain

$$\frac{1}{\Gamma} = \frac{1}{\Gamma_L} + \frac{1}{2}r_L P^{-1/2} + \tilde{w}P^{-1}. \tag{10.7}$$

Hence,

$$\left(\frac{\partial\Gamma}{\partial V}\right)_P = 0, \tag{10.8}$$

and

$$\left(\frac{\partial\Gamma}{\partial P}\right)_V = \Gamma^2 P^{-1}\left[\frac{1}{4}r_L P^{-1/2} + \tilde{w}P^{-1}\right]. \tag{10.9}$$

In order to compute the limits implicit in the conditions for a double pole in $P_H(V)$, Eqs. (7.5) and (7.6), we still have to express the derivatives of $P_H(V)$ in terms of P_H. If the conditions for a double pole in the thermodynamic expression for $P_H(V)$ are satisfied, we may write

$$P_H(V) = N(V) \Big/ \left(\frac{V}{V_o} - \frac{V_L}{V_o}\right)^2 g(V), \tag{10.10}$$

where $N(V)$ is a shorthand for the numerator in the right hand side of (6.9), and $g(V)$ is the coefficient of $(V/V_o - V_L/V_o)^2$ in (6.10) and (7.2). Then, the derivatives of $P_H(V)$ may be written as

$$\frac{dP_H}{dV} = -\frac{2}{V_o}\left(\frac{g(V)}{N(V)}\right)^{1/2} P_H^{3/2} + \left(\frac{N'}{N} - \frac{g'}{g}\right)P_H, \tag{10.11}$$

and

$$\begin{aligned}\frac{d^2P_H}{dV^2} &= \frac{6}{V_o^2}\frac{g(V)}{N(V)}P_H^2(V) \\ &+ \left\{\frac{NN''-(N')^2}{N^2} - \frac{gg''-(g')^2}{g^2} - \left(\frac{N'}{N} - \frac{g'}{g}\right)\right\}P_H \end{aligned} \tag{10.12}$$

$N'(V)$, $g'(V)$ and $N''(V)$, $g''(V)$ are the first and second derivatives of N and $g(V)$ with respect to V.

Combining these results with (8.18), (8.19), (8.20) and the expressions for the thermodynamic coefficients, we obtain

$$\lim_{V\to V_L}\left[\left(\frac{1}{P_H}\frac{dP_H}{dV}\right)^2 C_1 - \frac{1}{P_H}\frac{d^2P_H}{dV^2}C_2\right]_H = \frac{1}{V_o^2}\frac{g(V_L)}{N(V_L)}\left(r_L^2 + 4\left(\frac{1}{\Gamma_L}+\frac{1}{2}\right)\tilde{w}\right),$$
$$\lim_{V\to V_L}\left[\left(\frac{1}{P_H}\frac{dP_H}{dV}\right)C_3\right] = 0, \tag{10.13}$$

and

$$\lim_{V\to V_L} C_4(V) = -\frac{1}{2V_o^2}\left(\frac{2+\Gamma_L}{\Gamma_L}\right)^3\frac{1+\Gamma_L}{\Gamma_L}. \tag{10.14}$$

With these results and Eqs. (8.17), (7.2), we obtain an equation for $g(V_L)$

$$g(V_L) = 2\left(\frac{\Gamma_L}{2+\Gamma_L}\right)^3\frac{1}{V_o^2}\left\{\frac{g(V_L)}{N(V_L)}\left(r_L^2+4\left(\Gamma_L+\frac{1}{2}\right)\tilde{w}\right) - \frac{1}{2}\left(\frac{2+\Gamma_L}{\Gamma_L}\right)^3\frac{1+\Gamma_L}{\Gamma_L}\right\}. \tag{10.15}$$

From this expression, it is fairly obvious that $g(V_L)$ is finite and non-vanishing, thus verifying the consistency of our method.

Now, we shall compute the limit implicit in the second condition for a double pole in $P_H(V)$, Eq. (7.5). From (8.11) and the results previously found in this section

$$\left(\frac{dR_H}{dV}\right)_{V_L} = \lim_{V\to V_L}\left[\left(\frac{1}{P_H}\frac{dP_H}{dV}\right)\left(\frac{1}{\Gamma}-R_H\right)\right] = \lim_{V\to V_L}\left[\left(\frac{1}{P_H}\frac{dP_H}{dV}\right)P_H^{-1/2}\left(-\frac{1}{2}r_L\right)\right]. \tag{10.16}$$

Hence,

$$\left(\frac{dR_H}{dV}\right)_{V_L} = \frac{1}{V_o}\left(\frac{g(V_L)}{N(V_L)}\right)^{1/2} r_L, \tag{10.17}$$

which is just a verification that $(dR_H/dV)_{V_L}$ is indeed finite and non-vanishing when $r_L \neq 0$.

Making use of this result, the second condition for a double pole in $P_H(V)$, Eq. (7.5), takes the form

$$r_L = -\frac{1}{2}\left(\frac{2+\Gamma_L}{\Gamma_L}\right)^2\left(\frac{N(V_L)}{g(V_L)}\right)^{1/2} \tag{10.18}$$

which gives r_L in terms of $N(V_L)$ and $g(V_L)$.

Now, we substitute the expression for $V_L/V_o, (dR_H/dV)_{V_L}$ and r_L given in Eqs. (8.15), (10.18) and (10.19) respectively, and we verify that

$$\frac{V}{V_o} - \frac{1}{2R_H + 1} = \left(\frac{V}{V_o} - \frac{V_L}{V_o}\right)^2 g(V) \tag{10.19}$$

in the regime of strong shock compression when V is close to V_L. In this way we demonstrate that the thermodynamic expression for $P_H(V)$, Eq. (10.10) is indeed correct.

In this section, we have explicitly verified that all the consistency conditions are satisfied. We have also expressed the conditions on R for a double pole in P_H in terms of the parameters appearing in the "universal" form, Eq. (9.3), taken by the incomplete equation of state in the regime of strong shock compression. These results, Eqs. (7.5), (10.8) and (10.19), are valid as long as the consistency conditions are satisfied, independently of the particular form of the integrating factor of (9.6) which is chosen to define the temperature.

8.11 A thermodynamic expression for the constant A

In order to derive a thermodynamic expression for the constant A occurring in the linear relationship between the shock and particle velocities we will compare the thermodynamic and empiric expressions for $P_H(V)$ in the limit $V \to V_L$.

From (10.10)

$$\lim_{V \to V_L} \left[\left(\frac{V}{V_o} - \frac{V_L}{V_o}\right)^2 P_H^{(th)}(V)\right] = \frac{N(V_L)}{g(V_L)}, \tag{11.1}$$

and from (4.1)

$$\lim_{V \to V_L} \left[\left(\frac{V}{V_o} - \frac{V_L}{V_o}\right)^2 P_H^{(e)}(V)\right] = \frac{A^2}{B^3}\frac{1}{V_o}. \tag{11.2}$$

Then,

$$A = (V_o B^3)^{1/2} \left(\frac{N(V_L)}{g(V_L)}\right)^{1/2} \tag{11.3}$$

with $B = 1 + \frac{1}{2}\Gamma_L$.

The factor $(N/g)^{1/2}$ may be written in terms of r_L, Eq. (10.19)

$$A = -\left(\frac{V_o}{2(2+\Gamma_L)}\right)^{1/2} \Gamma_L^2 r_L \tag{11.4}$$

Now, from the definition of r_L, Eqs. (8.3) and (10.7), we have

$$r_L = 2\left(\frac{\partial\Gamma^{-1}}{\partial P^{-1/2}}\right)_{V_L} = \frac{4}{\Gamma_L^2}\left[P^{3/2}\left(\frac{\partial\Gamma}{\partial P}\right)\right]_{V_L} . \tag{11.5}$$

Substitution of (11.5) in (11.4) gives

$$A = -2\left(\frac{2V_o}{2+\Gamma_L}\right)^{1/2} P^{3/2}\left(\frac{\partial\Gamma}{\partial P}\right)_{V_L} , \tag{11.6}$$

which is the desired result.

The expression for A in terms of r_L, Eq. (11.4), could also have been derived from the expression for $P_H(V)$ obtained by substitution of the specific energy $E - \hat{E}$ in terms of pressure, Eq. (9.3), in the Rankine-Hugoniot equation for conservation of energy, Eq. (2.3).

8.12 Summary of results and conclusions

The starting point of our analysis is a formal implicit, but exact solution for the pressure on the Hugoniot as function of the compression. By requiring that this exact thermodynamic expression reproduces the scaling behavior of the well known empirical expression for $P_H(V)$ we are led to impose three conditions on the ratio $R(P,V)$ of the specific internal energy to the product of pressure and specific volume. These conditions on R for a double pole in $P_H(V)$ at $V = V_L$ are satisfied only if $R_H(P_H(V),V)$ and its first and second derivatives with respect to V are finite and non-vanishing at $V = V_L$, while, at the same time, $P_H(V)$ and its first and second derivatives with respect to V are singular at V_L. Since $R(P,V)$ and its derivatives are explicit functions of the pressure P and its derivatives, some additional constraining conditions are imposed on R to make the two requirements compatible with each other. The consistency conditions translate into precise thermodynamic language the rather vague statement: "a physical system which obeys the linear relationship between shock and particle velocities with $A \neq 0$ when subject to very strong shock compression".

The thermodynamic characterization obtained in this way is sufficient to determine the functional form of the specific internal energy as function of $P^{-1/2}$ and V, for large values of the pressure P and values of the specific volume V close to the limiting value V_L. From this result we derive a universal form which the incomplete equation of state of such a system takes for large values of the pressure and compression. The temperature is introduced through its proper thermodynamic definition, namely, as the inverse of the integrating factor of the entropy in the equation of an infinitesimal change in the internal energy. Since the integrating factor is not unique, the complete equation of state is not uniquely defined. In this work, we chose the simplest form which allowed us to show that the consistency condition are satisfied. These results give the differential of the entropy as function of the pressure and the specific volume, from which the entropy and the thermodynamic potentials may be obtained by integration. Analytical expressions for the thermodynamic coefficients, the specific heat at constant volume and the Grüneisen parameter are also obtained which allow us to verify that all the consistency conditions are satisfied and, hence, that the analysis we have made is self-consistent. We end our thermodynamic description of the asymptotic regime of strong shock compression of condensed materials giving a thermodynamic expression for the constant A which occurs in the linear relation between shock and particle velocities.

We would like to stress that, although scale invariance in the strong shock compression of condensed materials is a consequence of mass conservation across the shock front, the specific form it takes for a particular physical system, that is, the numerical value of the critical exponent in the denominator of the empirical expression for $P_H(V)$, is a manifestation of a property of a state function of the system, namely, the ratio R as function of pressure and compression. In the case discussed in this paper, R is a regular function of $P^{-1/2}$ and V for large values of the pressure and compression. Since R is a state function, the thermodynamic consequences following from this property will show up whenever the system is in a state of thermodynamic equilibrium characterized by values of the pressure and compression close to those reached in strong shock compression. Therefore, it must be possible to observe them also in experiments of static compression of the same material when the pressure and compression reached is comparable to those attained in strong shock compression.

Finally, in most works on the thermodynamic properties of shock compressed materials, the thermodynamic coefficients and the Grünneisen parameter are given as functions of the compression or some other equivalent variable. In this work, we have given our results in terms of two variables, $P^{-1/2}$ and V. In order to compare our results with those of other authors which are specific for shock compression P must be on the Hugoniot, which means that we may replace $A\,B^{-3/2}V_o^{-1/2}\,(V/V_o - V_L/V_o)$ for $P_H^{-1/2}$ in our formulae.

8.13 References

[1] M.H. Rice, R.G. McQueen, and J.M. Walsh, *Solid State Phys.* **6**, 1 (1958); L.V. Al'tshuler, K.K. Krupnikov, and M.I. Brazhnik *Zh. Eksp. Teor. Fiz.* **34**, 886 (1958) [*Sov. Phys. JETP* **7**, 614 (1958)].

[2] Ya.B. Zel'dovich and Yu.P. Raizer, *Physics of Shock Waves and High-Temperature Hydrodynamic Phenomena* (Academic Press, New York, 1967).

[3] F.E. Prieto and C. Renero, Au sujet de la relation entre la vitesse de choc et la vitesse de particule. *Sciences et Techniques de l'Armement* **49**, 325 (1975).

[4] F.E. Prieto, A law of corresponding states for materials at shock pressures, *J. Phys. Chem. Solids* **35**, 279 (1974).

[5] L.E. Reichl, *A Modern Course in Statistical Physics* (University of Texas Press, Austin, 1980), Chapter 10.

[6] *Compendium of Shock Wave Data*, Ed. by M. van Thiel UCRL-50108 Vols. **1-3** (Lawrence Livermore National Laboratory 1977); see also *Shock Hugoniot Data*, Ed. by S. Marsh (University of California, Berkeley, 1980).

[7] R.G. McQueen, S.P. Marsh, J.W. Taylor, J.N. Fritz, and W.J. Carter, in *High-Velocity Impact Phenomena*, Ed. by R. Kinslow (Academic Press, New York, 1970).

[8] G.E. Duvall and G.R. Fowles *Shock Waves in High-Pressure Physics and Chemistry,* Ed. by R.S. Bradley (Academic Press, London and New York, 1963), Vol. **2**.

[9] R.G. McQueen and S.P. Marsh, *J. Appl. Phys.* **31**, 1253 (1960).

9

Deriving Nuclei from Quarks

S. Pittel, J. Engel[1], J. Dukelsky[2], and P. Ring[3]

ABSTRACT
Starting with an effective Hamiltonian for multi-quark systems, we propose a mapping of the quark degrees of freedom onto triplet fermions that represent colorless three-quark clusters. We end up with an effective Hamiltonian for the triplet fermions, which can be treated by traditional nuclear many-body techniques. In this approach, the nuclear ground state represents a quantum liquid of composite particles, the nucleons. Higher-lying states involve both excitations within the space of nucleons and excitations of the nucleons themselves. The method is tested in the context of a schematic model of quarks interacting through an angular momentum and isospin pairing interaction and is shown to be very promising.

9.1 Introduction

Traditionally, the nucleus is treated as a nonrelativistic system of pointlike structureless nucleons. Recently, experiments [1] by the European Muon Collaboration (EMC) have called this simple picture into question. There is now evidence that the structure of a nucleon in a nuclear medium is different from that in free space, suggesting that the nucleon and nuclear scales do not completely decouple.

Thus, there is currently great interest in trying to describe nuclei directly in terms of their quark and gluon constituents. Ideally, such a description would start from QCD, but little progress has been made along these lines. Somewhat less ambitious is to start with constituent quark models [2], in which the gluon and antiquark degrees of freedom have been suppressed. In these models, the quarks have effective masses of roughly 1/3 the nucleon mass and interact with one another through effective interactions that implicitly incorporate effects of the suppressed degrees of freedom.

[1]Bartol Research Institute, University of Delaware, Newark, USA.
[2]Departamento de Física Teórica, Universidad Autónoma de Madrid, Spain. On leave from the Comision Nacional de Energía Atómica of Argentina. Fellow of the Consejo de Investigaciones Científicas y Técnicas of Argentina.
[3]Physikdepartment, Technische Universität München, Germany.

Such models have been used with impressive success to describe one- and two-baryon systems; however, it has not yet proven feasible to apply them to multi-nucleon systems. The major obstacle has been the inability to incorporate the strong colorless clustering of three quarks into nucleons in traditional many-body variational methods. We have recently developed a method [3] for incorporating colorless three-quark clustering in a microscopic treatment of many-nucleon systems through the use of iterative mapping techniques [4].

The reason that three-body clustering is so difficult to treat is that three-fermion clusters are not exact fermions. Our method maps colorless three-quark clusters onto triplet fermions *that satisfy exact anticommutation rules.* The mapping leads from the original multi-quark Hamiltonian to an effective Hamiltonian for the triplet fermions, which incorporates quark Pauli effects. Since we now have *exact fermions*, we can treat their dynamics with standard fermion many-body techniques [5].

Our proposed method involves two steps. The first is a standard Dyson boson expansion [6], wherein pairs of fermions are mapped onto two-index bosons. We briefly review this method in Sect. 2. The second step is a mapping of boson-fermion pairs onto triplet fermions [3], [7]. This method is new and is described in Sect. 3. In Sect. 4, we describe an algebraic model of interacting quarks proposed by Petry and coworkers [8], which we have used to test our proposed method. In Sect. 5 we describe the results of our test calculations and in Sect. 6 present some concluding remarks and some issues for future consideration.

9.2 Boson expansions

In this section, we describe the method of boson expansions, which until recently was the only mapping technique in use. We start with a system of fermions, with creation and annihilation operators $q_i^\dagger$ and q_j, respectively. The basic idea of a boson expansion (or boson mapping) is to replace this problem by an equivalent problem of interacting bosons and (if necessary) additional ideal fermions. The bosons replace pairs of the original fermions and thus have two indices; we denote their creation and annihilation operators by $B_{ij}^\dagger$ and B_{ij}, respectively. The creation and annihilation operators of the ideal fermion are denoted by $a_i^\dagger$ and a_j, respectively. By assumption the ideal fermion operators commute with the boson operators.

The dynamics of the original fermion problem will be exactly reproduced in the ideal boson-fermion space if we require that all commutation relations be preserved by the mapping. In the Dyson method [6], this is accomplished by mapping operators according to

$$q_i^\dagger q_k^\dagger \Rightarrow B_{ik}^\dagger - \sum_{ll'} B_{il}^\dagger B_{kl'}^\dagger B_{ll'} - \sum_l (B_{il}^\dagger a_k^\dagger a_l - B_{kl}^\dagger a_i^\dagger a_l) ,$$

$$q_k q_i \Rightarrow B_{ik} , \qquad q_i^\dagger q_k \Rightarrow \sum_l B_{il}^\dagger B_{kl} + a_i^\dagger a_k . \tag{2.1}$$

There are several features of the Dyson boson mapping worth noting. The mapping is inherently non-Hermitian, as is reflected by the fact that the boson images of $q_i^\dagger q_k^\dagger$ and $q_k q_i$ are not hermitian adjoints of one another. It is also finite and does not involve infinite series expansions as would arise e.g. in the Generalized Holstein Primakoff boson expansion [6]. When we treat a system with an even number of particles and introduce bosons to represent *all* fermion pairs, we need not introduce ideal fermions. For systems with an odd number of particles, however, there will always be an unpaired particle and the ideal fermion must be included. Finally, one often wishes to introduce bosons to represent the dominant collective pair degrees of freedom only. In such cases, ideal fermions must be included, even for systems with an even number of particles.

9.3 Iterative mappings of quark systems

As noted earlier, our proposed method involves two successive mappings. In the first, pairs of quarks are mapped onto two-index (diquark) bosons, using the Dyson method described in Sect. 2. However, we only map pairs of quarks within a nucleon, since these are the dominant collective pairs; thus, we must include ideal fermions to represent the third quark of each nucleon. At this stage, the nucleon can be viewed as a (diquark) boson plus an ideal (quark) fermion and the nucleus has an equal number of diquark bosons and ideal quark fermions.

Next, we map boson-fermion pairs $B_{ik}^\dagger a_l^\dagger$ onto triplet fermions $c_{ik,l}^\dagger$, which obey the anticommutation relation

$$\{ c_{ik,l} , c_{i'k',l'}^\dagger \} = \delta_{ll'} (\delta_{ii'}\delta_{kk'} - \delta_{ik'}\delta_{ki'}) . \tag{3.1}$$

The Dyson mapping of general boson-fermion pairs to triplet fermions is [3], [7]

$$B_{ik}^\dagger a_l^\dagger \Rightarrow c_{ik,l}^\dagger - \sum_{p<q,r} c_{ik,r}^\dagger \, c_{pq,l}^\dagger \, c_{pq,r} , \qquad B_{ik} a_l \Rightarrow c_{ik,l} ,$$

$$a_i^\dagger a_k \Rightarrow \sum_{p<q} c_{pq,i}^\dagger \, c_{pq,k} , \qquad B_{ii'}^\dagger B_{kk'} \Rightarrow \sum_l c_{ii',l}^\dagger c_{kk',l} , \tag{3.2}$$

which preserves the relevant bilinear commutation algebra. Since we are mapping a system with an equal number of bosons and fermions, additional ideal degrees of freedom need not be included in this step.

Since the mapping is carried out in two steps, we end up with triplet fermions, $c^\dagger_{ij,k}$, that are not fully antisymmetric in their three indices. Thus, we must still impose a truncation to fully antisymmetric triplet fermions,

$$c^\dagger_{ijk} = \frac{1}{\sqrt{3}}(c^\dagger_{ij,k} + c^\dagger_{jk,i} + c^\dagger_{ki,j}) \; . \tag{3.3}$$

9.4 The Bonn quark shell model

We now briefly describe the model of interacting quarks proposed by Petry and coworkers [8] (often referred to as the Bonn Quark Shell Model or BQSM), which we have used to test our method. In this model, quarks interact via an angular momentum and isospin pairing interaction. An attractive feature of this model is that under special conditions (the single quark levels assumed to be degenerate) it can be solved exactly with group theoretical methods. Petry and his coworkers have suggested that with an appropriate choice of its parameters the BQSM is a semi-realistic quark model of nuclei. They propose taking the single quark energies from a confining oscillator potential whose size is that of the entire nucleus and choosing the pairing strength to give the nucleon excited states at the correct energy. However, several groups have taken exception to this claim [9], [10]. Suzuki and Hecht [9], for example, have shown that the model does not produce spatially localized nucleons, a necessary criterion for a truly realistic quark model of nuclei. We have used this model to test our method (even though it is not realistic) because it is in some cases solvable and because it contains the proper triplet color correlations.

The Hamiltonian for this model is

$$H \;=\; \sum_{1jmt} \varepsilon_j \; q^\dagger_{1jmt} q_{1jmt} \;-\; \frac{G}{4} \sum_{1jj'} A^\dagger_{1j} \, A_{1j'} \; , \tag{4.1}$$

where $q^\dagger_{1jmt}$ creates a quark with color quantum number 1, angular momentum quantum numbers (j, m) and isospin projection t, $A^\dagger_{1j} \;=\; \sum_{23mt} \epsilon_{123} \times q^\dagger_{2jmt} \; q^\dagger_{3j\tilde{m}\tilde{t}}$ is the collective quark pair for orbit j, ε_j is the single-quark energy for orbit j and G is the strength of the pairing interaction.

Using the iterative mapping method described above and restricting to a single collective ($J = 0$, $T = 0$) pair for each orbit, we end up with fully antisymmetric colorless triplet fermions, $c^\dagger_{j',jmt}$, in which two particles in orbit j' are fully paired and the third in orbit j is unpaired. The resulting Hamiltonian for these effective triplet fermions has the form

$$
\begin{aligned}
H = & \sum_{j_1 j_2 j m t} \Big\{ (\varepsilon_{j_1} + \varepsilon_{j_2} + \varepsilon_j)\delta_{j_1 j_2} \\
& - 2G\sqrt{\Omega_{j_1}\Omega_{j_2}} - 2G\sqrt{\frac{\Omega_{j_2}}{\Omega_{j_1}}}\delta_{j j_1} \Big\} c^{\dagger}_{j_1,jmt} c_{j_2,jmt} \\
& - G\sum_{j j_1}\sqrt{\frac{\Omega_{j_1}}{\Omega_j}} \sum_{j_2 j_3} \sum_{mm'tt'} \Big\{ c^{\dagger}_{j,j_2mt} c^{\dagger}_{j,j_3m't'} c_{j_1,j_2mt} c_{j,j_3m't'} \\
& + \frac{2}{3} c^{\dagger}_{j_2,jmt} c^{\dagger}_{j,j_3m't'} c_{j_2,j_3m't'} c_{j_1,jmt} \\
& - \frac{2}{3} c^{\dagger}_{j_2,jmt} c^{\dagger}_{j,j_3m't'} c_{j_1,j_3m't'} c_{j_2,jmt} \Big\} .
\end{aligned}
\qquad (4.2)
$$

The effective Hamiltonian (6) is non-Hermitian, reflecting the manner in which quark Pauli effects are transmitted by the Dyson mapping. However, it contains only one-body and two-body terms and thus can be readily treated with standard fermion many-body techniques, albeit modified for multi-index fermions and non-Hermitian Hamiltonians. As always, the choice of approximation is dictated by the physics of the system under investigation.

9.5 Results of test calculations for ^{16}O

We have carried out test calculations for the nucleus ^{16}O. For such a doubly-magic system, it is natural to use the (non-unitary) spherical Hartree Fock (HF) method, wherein the ground state is described by Slater determinants built up from single-particle states with definite angular momenta. For a non-Hermitian Hamiltonian, the bra and ket trial state vectors need not be adjoints of one another. Thus, we introduce separate bra and ket states:

$$
|\Psi\rangle = \prod_{i=1,\ldots,N} \Gamma^{\dagger}_{j_i m_i t_i}|0\rangle \qquad \text{and} \qquad \langle\Phi| = \langle 0| \prod_{i=1,\ldots,N} \gamma_{j_i m_i t_i} . \qquad (5.1)
$$

The correlated creation and annihilation operators, $\Gamma^{\dagger}_{jmt}$ and γ_{jmt}, are linear combinations of the fundamental creation and annihilation operators:

$$
\Gamma^{\dagger}_{jmt} = \sum_{j'} X_{j'j}\, c^{\dagger}_{j',jmt} \qquad \text{and} \qquad \gamma_{jmt} = \sum_{j'} Y_{j'j}\, c_{j',jmt} . \qquad (5.2)
$$

Their structure coefficients are determined by the variational condition,

$$
\delta\left(\langle\Phi|H|\Psi\rangle / \langle\Phi|\Psi\rangle\right) = 0 , \qquad (5.3)
$$

$\hbar\omega$	G	Quark Occupation Numbers						E_{GS}
		$1s_{1/2}$	$1p_{3/2}$	$1p_{1/2}$	$1d_{5/2}$	$2s_{1/2}$	$1d_{3/2}$	(MeV)
0	15	7.43	14.86	7.43	9.14	3.05	6.10	−6480.00
16	0	36	8	4	0	0	0	192.00
16	1.5	10.52	17.07	8.54	5.93	1.98	3.96	216.11
16	7.5	8.08	15.28	7.64	8.50	2.83	5.66	−2313.78
16	15	7.75	15.07	7.54	8.82	2.94	5.88	−5545.99

TABLE 9.1. Results of spherical HF test calculations for ^{16}O based on the BQSM. Shown are the number of quarks in the active quark orbits and the ground state energy. Values of $\hbar\omega$ and G are in MeV.

which can be solved with standard iterative techniques. Note that the structure coefficients of the various correlated triplet fermions (*i.e.* nucleons) need not be the same, nor need they be the same as the structure coefficents for a *free* nucleon ($N = 1$).

The results of several test calculations are summarized in Table 9.1. In each, we restricted the quarks to the $1s$, $1p$ and $2s - 1d$ shells of a confining oscillator potential and assumed that the unpaired quarks fill the $1s$ and $1p$ shells only. The calculations differ only in the choice of the oscillator energy $\hbar\omega$ and the pairing strength G. We show in the table both the calculated ground state energies and the occupation numbers for the various quark orbits.

The first calculation assumed a pairing strength of $G = 15$ MeV and degenerate single-particle orbits. With this choice of G, the lowest non-nucleonic excitations occur at a reasonable energy of 300 MeV [8]. With degenerate orbits, the BQSM can be solved exactly using group theoretical techniques. The ground state energy is given analytically by [8]

$$E_{GS} = -2\,G\,N\,(\Omega + 1) + G\,N\,(N - 1)\,, \tag{5.4}$$

where N is the number of collective triplets (nucleons) and $\Omega = \sum_j 2j+1$ is the total pair degeneracy of the active space. For $N = 16$, $G = 15$MeV and $\Omega = 20$, the analytic result is $E_{\mathrm{GS}} = -6480$ MeV, which is reproduced exactly by the spherical Hartree Fock calculation. The occupation numbers reflect the assumption that the unpaired quarks fill the $1s$ and $1p$ shells. Since the single-quark levels were degenerate, the unpaired quarks could have been distributed in any other fashion with no change in energy.

The other four calculations were carried out for nondegenerate single-quark energies appropriate to $\hbar\omega = 16$ MeV, which is reasonable for ^{16}O. The principal purpose of these calculations was to assess the extent to which Pauli violations occur within our method.

It is useful to clarify why and how Pauli violations can arise in methods based on the use of mapping techniques. As stated earlier, mappings carry one from the original space to an ideal space. In the ideal space, there are a subset of states (called physical states) that properly reflect the symmetry character of the original space and another subset (called unphysical states) that do not. Were it possible to diagonalize the Hamiltonian that arises from the mapping *exactly*, there would be a complete separation between the physical states (which are in one-to-one correspondence with the exact eigenstates of the original problem) and the unphysical states. It is in this sense that mappings fully transmit effects of the Pauli principle. However, once approximations are made (as invariably they must), physical and unphysical states can mix and Pauli violations can occur in the resulting states.

Indeed, for $G = 0$, the HF solution shown in the table has 36 quarks in the lowest $1s_{1/2}$ quark orbit, whereas only 12 are permitted by the Pauli principle. This is a clear violation of the quark Pauli principle, and is a reflection of strong unphysical state admixtures in the HF variational solution. In the realistic ($G = 15$ MeV) scenario, however, this is not the case. In fact, even when we reduce the pairing strength from this value by a factor of ten, we still find no Pauli violations at the quark level in the HF solution.

These results can be understood as follows: There are two scales in the problem, the nuclear scale (which is governed by the single-particle term in the Hamiltonian and has characteristic energies of the order of a few MeV), and the nucleon scale (which is governed by the pairing interaction $G\Omega$ and for which the characteristic energies are of the order of hundreds of MeV). The quark pairing interaction produces di-quark clustering at the nucleon scale, and it does this by scattering quarks strongly over a large number of single-quark orbits (which were produced at the nuclear scale). It is this strong scattering over a large number of quark orbits that precludes Pauli violations. Any *reasonable* quark interaction *must* produce strong clustering at the nucleon scale. As such, we do not expect Pauli violations to pose a problem in subsequent more realistic applications of our method.

9.6 Concluding remarks

There are several important issues still to be investigated. We need to consider the application of our method to more general quark Hamiltonians, which include not only more realistic two-quark interactions but perhaps three-quark forces as well. For more general Hamiltonians, our mapping equations may require some modification, since they were derived for one- plus two-body Hamiltonians only. We also need to consider the generalization of other fermion many-body techniques, such as TDA, RPA and

Brueckner theory, to multi-index fermions and non-Hermitian Hamiltonians. Such methods will be required if we wish to treat excited states of nuclei with our method. In addition, there are computational issues that must be addressed. To build strong spatial clustering in a spherical basis, we must mix a large number of single-quark orbits (even at the level of HF approximation) and it is still to be determined whether this will be computationally feasible.

Assuming that we can find an appropriate quark Hamiltonian and that we can treat it with our method, we should be able to address a number of interesting and important questions. In quark models, a nucleon (in medium) is the lowest-lying triplet fermion determined via some kind of Hartree-like procedure. Whether this physical nucleon is modified from the bare three-quark nucleon, as suggested by the EMC effect, is an important issue. We will be able to address this question by explicitly comparing *e.g.* the electromagnetic form factors of the bare and physical nucleons that emerge from quark models. Intimately related to the structure of nucleons in medium is the role of nucleon excited states (*e.g* the delta) in nuclear structure; we can also investigate this issue with our method. Put simply, since we are dealing with quark degrees of freedom from the outset, we have in principle a consistent framework with which to examine quark effects in nuclei.

9.7 Acknowledgements

This work was supported by the National Science Foundation under grant PHY-8901558, by the U.S. Department of Energy under contract DE-AC-03-76SF00098, by the Dirección General de Investigaciones Científicas y Técnicas of Spain, and by the Deutsche Forschungsgemeinschaft and the Bundesministerium für Forschung and Technologie.

9.8 References

[1] See *e.g.* European Muon Collaboration, J. Ashman *et al*, "A Measurement of the Spin Asymmetry and Determination of the Structure Function g_1 in Deep Inleastic Muon-Proton Scattering", *Phys. Lett.* **206B**, 364 (1988).

[2] M. Oka and K. Yazaki, "Baryon-Baryon Interaction from Quark-Model Viewpoint", in: *Quarks and Nuclei*, ed: W. Weise, (World Scientific, 1985); A. Faessler, F. Fernandez, C. Lübeck, and K. Shimizu, "The Nucleon-Nucleon Interaction and the Role of the [42] Orbital Six-Quark Symmetry", *Nucl. Phys.* **A402**, 555 (1983).

[3] S. Pittel, J. Engel, J. Dukelsky, and P. Ring, "The Nucleus as a Condensate of Collective Quark Triplets", *Phys. Lett.* **247B**, 185 (1990).

[4] S. Pittel, R. Bijker, and J. Dukelsky, "Boson Expansions for Systems of Interacting Bosons", *Phys. Lett.* **219B**, 5 (1989).

[5] P. Ring and P. Schuck, *The Nuclear Many-Body Problem* (Springer, New York, 1980).

[6] E.G. Marshalek, "How Microscopic Boson Models Work", *Nucl. Phys.* **A347**, 253 (1980).

[7] J. Dukelsky and S. Pittel, "Fermion Mapping of Boson-Fermion Pairs", *Phys. Rev.* **C42**, 2030 (1990).

[8] H.R. Petry, H. Hofestädt, S. Merk, K. Bleuler, H. Bohr, and K.S. Narain, "An Application of QCD in Nuclear Structure", *Phys. Lett.* **159B**, 363 (1985); H. Hofestädt, S. Merk, and H.R. Petry, "The Nucleus as a Multiquark System", *Z. Phys.* **A326**, 391 (1987).

[9] Y. Suzuki and K.T. Hecht, "Can a Quark Shell Model Lead to Nucleon Clusters?", *Phys. Lett.* **232B**, 195 (1989).

[10] A. Arima, K. Yazaki, and H. Bohr, "A Quark Shell Model Calculation of Nuclear Magnetic Moments", *Phys. Lett.* **183B**, 131 (1987); I. Talmi, "The Nuclear Shell Model —Of Nucleons or Quarks?", *Phys. Lett.* **205B**, 140 (1988).

10

Binding Energies of Nuclei and Atoms

Igal Talmi[1]

ABSTRACT Marcos Moshinsky made significant contributions to many fields of theoretical physics. These include nuclear and atomic physics. The subject to be described below is a link between these two fields.

In the early days of the shell model of nuclei, physicists were guided by the analogy with the structure of electron shells in atoms. A dramatic demonstration of the latter is offered by the plot of ionization energies as a function of atomic number. In it we see a characteristic large increase in ionization energy before a closed shell is reached and sharp drop beyond noble gas configurations.

Naturally, nuclear physicists were eager to demonstrate similar behaviour of nuclear separation energies. It took some manipulation of the data to produce something which bears feeble resemblance to the atomic case. Still, physicists wrote about anomalously high binding energies associated with closed shells. In the following, the correct behavior of separation energies of nuclei will be described and compared with theoretical predictions. We shall then return to atomic energies which will turn out to possess a deeper analogy with the nuclear case.

A big difference between atoms and nuclei is that the latter are made of two kinds of fermions —protons and neutrons. The structure of eigenstates is simpler when outside closed shells there are only nucleons of one kind (valence nucleons). Nucleons of the other kind are in closed shells as is the case for semi-magic nuclei. Therefore, we consider next the case of identical valence nucleons.

A most convenient formalism to describe states of n identical particles in a j-orbit is the seniority scheme introduced by Racah [1] in 1943. It is based on the idea of particle pairs coupled to states with $J = 0$ To define seniority it is convenient to use a quasi-spin formalism due to Kerman [2]. The operator

$$S_j^+ = \frac{1}{2}\sum_m (-1)^{j-m} a_{jm}^+ a_{j,-m}^+ \qquad (1)$$

[1]The Weizmann Institute of Science, Rehovot, Israel.

creates a pair of identical j-fermions in a state with $J = 0$. The operators a^+_{jm} are the usual creation operators of a fermion in the state with spin j and $j_3 = m$. the corresponding pair annihilation operator is

$$S_j^- = (S_j^+)^\dagger = \frac{1}{2}\sum_m (-1)^{j-m} a_{j,-m} a_{jm}. \tag{2}$$

The commutator of S_j^+ and S_j^- can be directly calculated to be

$$[S_j^+, S_j^-] = \sum_m a^+_{jm} a_{jm} - \tfrac{1}{2}(2j+1) = 2S_j^0. \tag{3}$$

It is equal to the number operator plus a constant from which follow the commutation relations

$$\left[S_j^0, S_j^+\right] = S_j^+, \qquad \left[S_j^0, S_j^-\right] = -S_j^-. \tag{4}$$

Thus, S_j^+, S_j^-, and S_j^0 satisfy the commutation relations of components of the angular momentum and are generators of the $SU(2)$ Lie algebra. All the knowledge about angular momentum may be used in the present case.

Racah introduced the pairing interaction operator which may be expressed as

$$S_j^+ S_j^- = \mathbf{S}_j^2 - (S_j^0)^2 + S_j^0. \tag{5}$$

Its eigenvalues are then given by

$$s(s+1) - \frac{1}{4}\left(\frac{2j+1}{2} - n\right)\left(\frac{2j+5}{2} - n\right), \tag{6}$$

where $s \geq 0$ is an integer or half integer. The seniority quantum number v introduced by Racah is simply related to s. the number v in a state of the j^v configuration is, loosely speaking, the number of unpaired particles. More precisely, a state with seniority v in the j^v configuration contains no pairs coupled to $J = 0$ and hence, is annihilated by S_j^-. Applying the operator (5) to such a state, the eigenvalue (6) with $n = v$ vanishes, from which follows

$$s = \frac{1}{2}\left(\frac{2j+1}{2} - v\right). \tag{7}$$

In terms of v, the eigenstates (6) are given by

$$\frac{1}{2}\frac{n-v}{2}(2j+3-n-v). \tag{8}$$

For describing ground states of semi-magic nuclei, states with lowest seniority (maximum pairing) are important. Such states are given by

$$v = 0: \qquad (S_j^+)^N \mid 0\rangle \qquad J = 0 \qquad n = 2N \tag{9}$$

$$v = 1: \qquad (S_j^+)^N a^+_{jm} \mid 0\rangle \qquad J = j,\ J_3 = m \qquad n = 2N+1 \tag{10}$$

In (9) and (10) the vacuum state $|0\rangle$ is the state of closed shells with no j-particles. As follows from the $SU(2)$ algebra, the action of S_j^+ on a state with given s increases n by two but does not change s (or v). In (9) the $v = 0$ state with no j-particles is just $v|0\rangle$. In (10) the state with $N = 0$ is $a_{jm}^+|0\rangle$ which definitely contains no $J = 0$ pairs.

The seniority scheme is the set of eigenstates of the operator (5). Thus, the pairing interaction (5) is, by definition, diagonal in the seniority scheme. It is often not realized that the pairing interactions is only a very special case of such interactions. There is a large variety of interactions which are diagonal in the seniority scheme and yet do not have the degeneracies of the pairing interaction. It is therefore impossible to express by a simple formula like (6) or (8) the eigenvalues of a general two-body interaction which is diagonal in the seniority scheme. Even in the j^2 configuration the eigenvalues

$$V_j = \langle j^2 JM|V|j^2 JM\rangle \tag{11}$$

depend not only on $n = 2$ and v. The $v = 2$ states with $J = 2, 4, \cdots, 2j-1$ may be very far from degenerate.

Still, for the states with lowest seniorities (9) and (10) a certain property of dynamical symmetry holds. Their eigenvalues for *any* two-body interaction which is diagonal in the seniority scheme may be obtained by the eigenvalues of a linear combination of the pairing interaction (5) and a constant interaction [whose eigenvalues in any state are proportional to $n(n-1)/2$]. We thus obtain for the $v = 0$ state and the $v = 1$ states the eigenvalues [3]

$$\tfrac{1}{2}n(n-1)\alpha + \tfrac{1}{2}n - v\beta. \tag{12}$$

The expression (12) is equal to the expectation values of *any* two-body interaction in the states (9) and (10). The coefficients α and β which depend on the interaction are linear combinations of V_0 and

$$\bar{V}_2 = \sum_{J>0,\,\text{even}} (2J+1)V_J \Big/ \sum_{J>0,\,\text{even}} (2J+1). \tag{13}$$

To determine α and β we may use the case of $n = 2, v = 0$ for which

$$V_0 = \alpha + \beta, \tag{14}$$

and the case of the closed j-orbit, $n = 2j+1, v = 0$ for which

$$\sum_{j\ \text{even}} (2J+1)V_J = V_0 + (j+1)(2j-1)\bar{V}_2 = \frac{2j(2j+1)}{2}\alpha + \frac{2j-1}{2}\beta. \tag{15}$$

From (14) and (15) we obtain

$$\alpha = \frac{2(j+1)\bar{V}_2 - V_0}{2j+1}, \qquad \beta = \frac{2(j+1)}{2j+1}(V_0 - \bar{V}_2). \tag{16}$$

In the case of the pairing interaction (5) $[V_0 = (2j+1)/2, \bar{V}_2 = 0]$ the values (16) are $\alpha = -\frac{1}{2}, \beta = j+1$.

In applying the expression (12) to nuclear binding energies we should add the single nuclear energy ϵ (the expectation value of the kinetic energy of a j-nucleon and its interaction with the closed shells) multiplied by n and the binding energy (B.E.) of the closed shells nucleus with $n = 0$. We thus obtain

$$\begin{aligned} & \text{B.E. } (j^n, v = 0 \text{ or } v = 1) \\ = \; & \text{B.E. } (n = 0) + n\epsilon + \tfrac{1}{2}n(n-1)\alpha + \tfrac{1}{2}n - v\beta. \end{aligned} \tag{17}$$

The mass formula (17) contains a constant term, a linear term, a quadratic term and a pairing term. The pairing term which gives rise to the well-known odd-even variation of nuclear binding energies is a direct result of the seniority scheme. In no way is it due to the two-body interaction being equal to the pairing interaction. To calculate the total binding energy from the interaction between free nucleons the latter should be well defined and the many body problem (approximately) solved. This, however, is far from being possible and the only practical way is to determine ϵ, α and β from experimental data in a consistent way. One of the consistency checks is to see whether (17) gives a good description of the experimental binding energies. A simple way of doing this is to plot nucleon separation energies as a function of n. From (17) we obtain

$$\begin{aligned} \text{B.E. } (j^n, v = 0 \text{ or } v = 1) \; - \; & \text{B.E. } (j^{n-1}, v' = 1 \text{ or } v' = 0) \\ & = \epsilon + (n-1)\alpha + (1-v)\beta. \end{aligned} \tag{18}$$

If the effective two-body interaction between identical nucleons is diagonal in the seniority scheme to a sufficient degree of approximation, separation energies plotted as a function of n should lie on two straight and parallel lines. There are several cases where this behavior can be checked experimentally and in all of them there is very good agreement with the theoretical expression (18) (see Fig. 9.1). The values of ϵ and α may be obtained only form binding energies. They may be due not only to the two-body interaction of j-nucleons but also to polarization of the core (the $n = 0$ nucleus) by valence nucleons. Any linear and quadratic changes as a function of n in B.E. $(n = 0)$ due to such polarization contribute to the values of ϵ and α. Also linear changes in ϵ contribute to the value of α. The value of β determined from binding energies may be compared with its expression (16) in terms of energy level difference. The agreement obtained in all cases is another consistency check on the determination of matrix elements of the effective interaction from experiment.

It is worthwhile to mention that for $j \leq 5/2$ there is only one state with given J, allowed by the Pauli principle, in all j^n configurations. In these cases, *any* interaction is trivially diagonal in the seniority scheme. In the

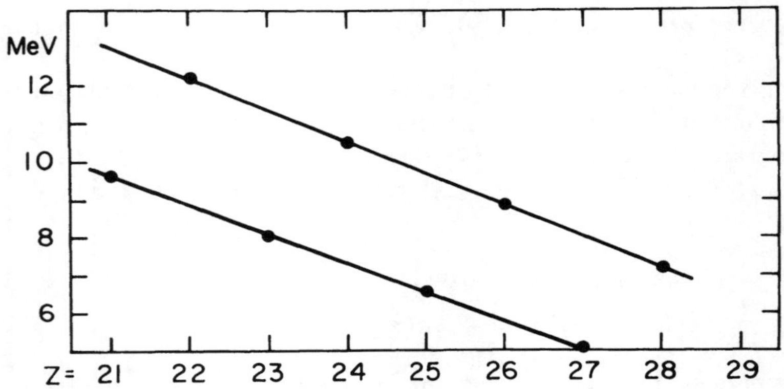

FIGURE 10.1. Proton separation energies of $1f_{7/2}$ nuclei.

$j = 7/2$ case there are two $J = 2$ states and two $J = 4$ states in the $(7/2)^4$ configuration. Still, it may be shown that *any two-body* interaction is seniority in the seniority scheme. Cases with $j \geq 9/2$ must be studied in order to see whether the effective interaction between identical nucleons is indeed diagonal in the seniority scheme. Analysis of spectra of $(1g_{9/2})^n$ and $(1h_{11/2})^n$ proton configurations shows that also in those cases the effective interaction is diagonal in the seniority scheme.

The values of the coefficient β determined from experiment are *large* and *attractive*. This gives rise to the pairing term in nuclear energies which increases the binding energies of even nuclei. It follows, according to (16), from the fact that $\bar{V}_2$, the center of mass of the $v = 2$ levels in the j^2 configuration is higher than V_0. The values of α determined from binding energies are small and *repulsive*. The separation energies of semi-magic nuclei go down (binding energies are usually expressed a positive numbers) as a magic number is approached! As a result, doubly magic nuclei are *less* bound (have smaller values of binding energy per nucleon) than some of their even-even neighbors. The fact that apart form the attractive pairing term there is average repulsion between identical nucleons in the same j-orbit should not come as a surprise. It follows form general features of *saturation* of nuclear energies. The *symmetry energy* term in binding energies is proportional to $T(T+1)$ or simply to $(N-Z)^2 = N^2 + Z^2 - 2NZ$ and its coefficient is *repulsive*. Hence, there is a repulsion which is linear and quadratic in n for $N = N_0 + n$ (Z fixed) or for $Z = Z_0 + n$ (N fixed).

This average repulsion must be present also between identical nucleons in different j-orbits. Surely, the features of nuclear saturation do not change at shell closures. Analysis of nuclear energies yields results consistent with this expectation. For example, the interaction between a proton in the $2p_{1/2}$ orbit and a proton in the $1g_{9/2}$ orbit is found to be 0.129 MeV in the state with $J = 5$ and 0.549 MeV in the $J = 4$ state. Thus, the average interaction is given by

$$\bar{V}(2p_{1/2}1g_{9/2}) = \sum_4^5 (2J+1)V_J \Big/ \sum_4^5 (2J+1) = 0.318 \text{ MeV}.$$

In the case of protons the interaction includes also the Coulomb repulsion. Still, there is an average repulsion also between neutrons in different orbits.

The average repulsion between identical valence nucleons should hold also in cases where strong mixing between different configurations takes place. Indeed, the average repulsion between identical nucleons in the same j-orbit together with the fact that the effective interaction between identical nucleons in a j-orbit together is diagonal in the seniority scheme lead to the scheme lead to the scheme of ***generalized seniority*** [4]. In that scheme a very special kind of configuration mixing is described. A generalized pair creation operator is defined as the linear combination

$$S^+ = \sum_j \alpha_j S_j^+ . \tag{19}$$

In (19) the operators S_j^+ are defined by (1) and the summation is extended over the j-orbits in a major shell. If all coefficients α_j are equal, (19) is a total quasi-spin operator and gives rise to the $SU(2)$ algebra and the seniority scheme as in a single j-orbit $[2j+1$ replaced by $2\Omega = \sum_j (2j+1)]$. Experimental data, however, indicate that the α_j coefficients for actual nuclei cannot be all equal.

In analogy with $v = 0$ states in j^n configurations we can define a state with generalized seniority zero by

$$(S^+)^N |0\rangle . \tag{20}$$

We now consider shell model Hamiltonians which have the states (20) as eigenstates. We normalize the Hamiltonian by subtracting the binding energy of the closed shell nucleons so that $H|0\rangle = 0$ and then study conditions that H must satisfy in order to obtain

$$H(S^+)^N|0\rangle = E_N (S^+)^N |0\rangle . \tag{21}$$

For $N = 1$ we obtain the condition

$$HS^+|0\rangle = V_0 S^+|0\rangle , \tag{22}$$

which states that the pair state created by S^+ is an eigenstate of H with eigenvalue V_0. From (22) the coefficients α_j may be determined. From the condition (21) with $N = 2$ we obtain by using (22)

$$\Big[[H, S^+], S^+\Big] = W(S^+)^2 . \tag{23}$$

The condition (23) is written as an operator equation since H is taken to include single nucleon energies and two-body interactions only. This is the reason that all higher order commutators of H with S^+ operators vanish. As a result, if (22) and (23) are satisfied it follows that (21) hold for *any* value of N and the eigenvalues are given by

$$H(S^+)^N|0\rangle = \left(NV_0 + \tfrac{1}{2}N(N-1)W\right)(S^+)|0\rangle. \tag{24}$$

Thus, binding energies of even-even nuclei are exact linear and quadratic functions of N. Due to the special configuration mixing expressed by (19) and (20) there are no breaks of the binding energy, or separation energy, at possible sub-shell closures. The ***pair separation*** energies are obtained from (24) to be

$$\text{B.E.}\,(N) - \text{B.E.}\,(N-1) = V_0 + (N-1)W. \tag{25}$$

The linear dependence of separation energies is verified in several regions. The coefficient W is repulsive indicating average repulsion between identical valence nucleons. Equation (17) for n even is a special case of (24) for $\alpha_{j'} = 0$, $j' \neq j$ and $V_0 = 2\epsilon + \alpha + \beta$, $W = 4\alpha$. For even-even nuclei, (24) and (25) are significant generalizations of the results obtained in the seniority scheme where the $SU(2)$ Lie algebra leads to simple results.

Two questions may now be raised. One concerns the origin of the extra stability of magic nuclei. As explained above, those are not more tightly bound than their preceding even-even neighbors. The extra stability is due to the fact that nuclei with nucleon numbers *beyond* shell closures are *less* bound. Beyond a magic number there is a sharp drop in separation energy, considerably bigger than the extrapolation of the linear decrease (Fig. 9.2). The other question is which interactions hold nuclei together. The $T = 1$ part of the effective interaction is, apart from the attractive pairing term, repulsive on the average. Hence, the interaction of a valence nucleon with nucleons of the same kind in closed shells is repulsive. We find that the effective interaction between protons and neutrons, or rather the $T = 0$ part of the nuclear interaction, is attractive. This follows also from the symmetry energy term in various formulae of binding energies. A repulsive term proportional to $(N-Z)^2 = N^2 + Z^2 - 2NZ$ indicates an average *attraction* between the Z protons and N neutrons. That attraction compensates the average repulsion between identical nucleons.

After the incursion into nuclear energies let us come back to atoms. We can now ask what gives rise to the large increase when a shell closure is approached. It is amusing to see some nuclear physicists trying to project from nuclear energies to the atomic case. In a very reputable review article it is stated: "The extra binding below the shell closure may be due to pairing energy, just as in atoms, ... electron binding energies can be seen to increase steadily as a shell is filled as a result of pairing energy..." Such explanation is obviously wrong. The Coulomb interaction between electrons is *repulsive*. Had it been the only factor, ionization energies would have

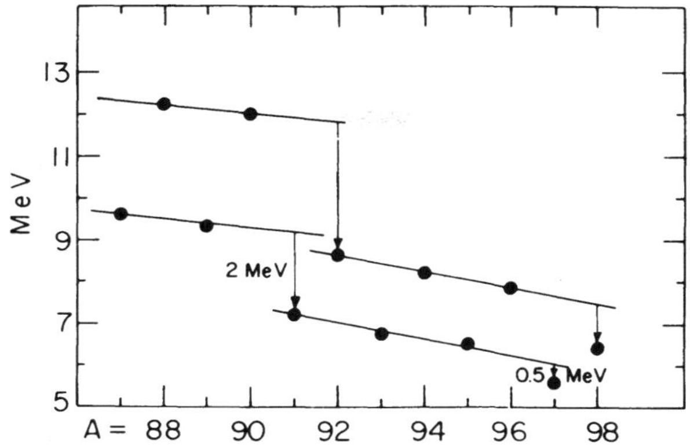

FIGURE 10.2. Neutron separation energies of Zr isotopes.

decreased as a function of electron number, as is the case for separation energies of identical nucleons. The explanation of the rise is rather simple. The sharp peaks in the atomic case are of ionization energies of *neutral atoms*. As an electron is added, the central charge is increased by one unit. The effect of the added charge in the center is much stronger than the repulsion of the extra electron. In other words, the screening of the central charge by electrons in the same orbit is not very effective. When a new shell begins, the electrons have higher kinetic energies and the screening of the central charge by electrons in closed shells is much more effective. Hence follows the sharp drop beyond shell closures.

We now have a better understanding of separation energies in nuclei and atoms. We can make a correct analogy between atoms and nuclei (apart from the strong pairing energy in the latter). See the following Table. In

in atoms	in nuclei
electrostatic repulsion between electrons	average effective repulsion between identical nucleons
strong field due to central charge	field due to attraction between protons and neutrons

order to display the average repulsion between identical nucleons we considered nuclei where the number of nucleons of the other kinds is fixed. To pursue the analogy with atoms we may add also nucleons of the other kind. This will compensate the decrease in separation energies due to the $T = 1$ repulsion but will not lead to high peaks as in atoms.

There is another way to demonstrate the analogy between nuclei and atoms. We may look at a fixed central charge and plot ionization energies as a function of the number n of added electrons. This plot is seen in Fig. 9.3 where the central charge is that of the calcium nucleus, $Z = 20$. We see

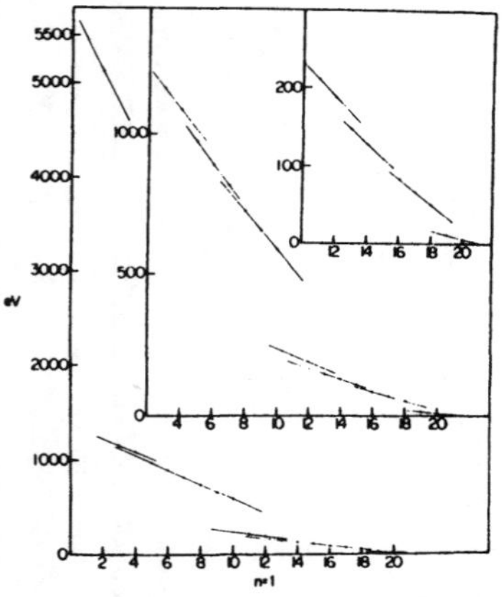

FIGURE 10.3. Electron ionization energies in the Ca atom.

indeed that ionization energies *decrease* as a function of n. Beginnings of new major shells appear as larger drops in ionization energies. The drops between s-orbits and following p-orbits are much smaller. There are also small but visible breaks in the middle of p-orbits. These do not indicate any sub-shells but are due to the Pauli principle.

In nuclei, the coupling scheme of identical nucleons is not determined by their average repulsion. The strong attractive part of the $T = 1$ effective interaction leads to nucleon pairing and the seniority scheme. This leads to the quadratic dependence of the repulsive term and the linear behavior of separation energies throughout a j-orbit or a major shell. In atoms, on the other hand, lowest electron states are those with minimum pairing, or simply those with maximum antisymmetry of the spatial part of the wave function. In the case of p-electrons there are only 3 independents states of orbital motion ($m = -1, 0, 1$) for each of them. Hence, 3 electrons may couple to a state with $L = 0$ which is fully antisymmetric in their space coordinates (and fully *symmetric* in their spins, $S = 3/2$). When a fourth electron is added the state cannot be fully space antisymmetric and has a relatively higher energy due to the Coulomb repulsion. These considerations may be made quantitative as follows. To calculate energies of p^n electron configurations we may use a method due to Racah. In the two electron p^2 configuration there are only 3 antisymmetric states allowed by the Pauli principle. These are the space antisymmetric ^3P state ($S = 1, L = 1$) and the two space symmetric states $^1S(S = 0, L = 0)$ and $^1D(S = 0, L = 2)$. Any two body (scalar) operator whose eigenvalues in these states are equal to the electron-electron interaction energies in those states [$V(^3P), V(^1S)$, and $V(^1D)$] may be used to calculate the interaction energies in all p^n electron configurations. A simple operator of this kind may be chosen as

$$V_{12} = a + 2b(\mathbf{s}_1 \cdot \mathbf{s}_2) + 2c(\boldsymbol{\ell}_1 \cdot \boldsymbol{\ell}_2), \tag{26}$$

with suitably chosen coefficients a, b, and c. The terms in (26) do not represent dynamical spin-spin or orbit-orbit interactions. The spin-spin term is due to the different space symmetry properties of the P state and S, D states. The $(\ell_1 \cdot \ell_2)$ term is due to the different energies of the $S = 0$ states with $L = 0$ and $L = 2$. In order to determine a, b, and c we make use of the general expression

$$\begin{aligned} \sum_{i<k} V_{ik} &= a\sum_{i<k} 1 + b\sum_{i<k} 2(\mathbf{s}_i \cdot \mathbf{s}_k) + c\sum_{i<k} 2(\ell_i \cdot \ell_k) \\ &= a\frac{n(n-1)}{2} + b\Big[\Big(\sum \mathbf{s}_i\Big)^2 - \sum \mathbf{s}_i^2\Big] + c\Big[\Big(\sum \ell_i\Big)^2 - \sum \ell_i^2\Big] \\ &= \frac{n(n-1)}{2}a + b\Big[S(S+1) - \frac{3}{4}n\Big] + [L(L+1) - 2n]. \end{aligned} \tag{27}$$

Applying (27) to the $n = 2$ case we obtain

$$\begin{aligned} V(^3P) &= a + \tfrac{1}{2}b - 2c, \\ V(^1S) &= a - \tfrac{3}{2}b - 4c, \\ V(^1D) &= a - \tfrac{3}{2}b + 2c. \end{aligned} \tag{28}$$

The linear equations (28) may be solved and a, b, c expressed in terms of $V(^3P)$, $V(^1S)$, and $V(^1D)$. Due to the mutual repulsion the lowest states of the p^n electron configuration are those with highest spatial antisymmetry. These are the ^3P state of the p^2 configuration and ^4S state ($S = \frac{3}{2}, L = 0$) of the p^3 configuration. For higher electron numbers there are no longer fully antisymmetric states in space coordinates. Using the correspondence between electrons and electron holes (in the closed p-orbit), we conclude that the lowest state of the p^4 configuration is ^3P as in the p^2 configuration. In the p^5 configuration there is only one antisymmetric state of a p-hole, ^2P. The p^6 configuration is that of the closed p-orbit with $S = 0$ and $L = 0$ (^1S). In the following Table energies of the lowest p^n configuration are presented as obtained from (27).

(a)	(b)	Energy	Separation energy	Difference
p	^2P	ϵ	ϵ	—
p^2	^3P	$2\epsilon + a + \frac{1}{2}b - 2c$	$\epsilon + a + \frac{1}{2}b - 2c$	$a + \frac{1}{2}b - 2c$
p^3	^4S	$3\epsilon + 3a + \frac{3}{2}b - 6c$	$\epsilon + 2a + b + 4c$	$a + \frac{1}{2}b - 2c$
p^4	^3P	$4\epsilon + 6a - b - 6c$	$\epsilon + 3a - \frac{5}{2}b$	$a + \frac{7}{2}b + 4c$
p^5	^2P	$5\epsilon + 10a - 3b - 8c$	$\epsilon + 4a - 2b - 2c$	$a + \frac{1}{2}b - 2c$
p^6	^1P	$6\epsilon + 15a - \frac{9}{2}b - 12c$	$\epsilon + 5a - \frac{3}{2}b - 4c$	$a + \frac{1}{2}b - 2c$

(a) Configuration, (b) Lowest state.

From the table follows that provided a, b and c have constant values throughout the p-shell, the ionization energies of 1, 2 and 3 electrons should

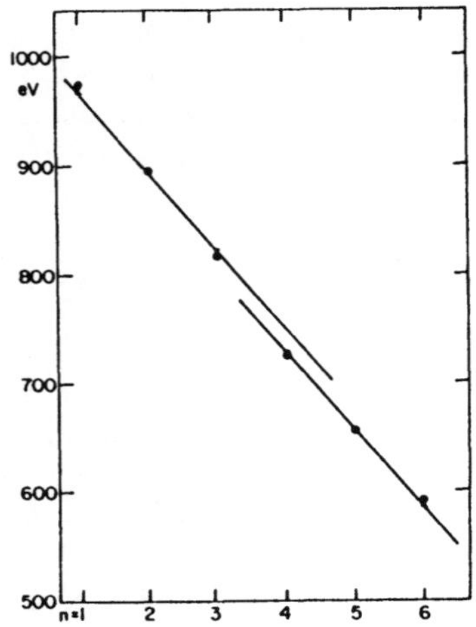

FIGURE 10.4. Experimental and calculated electron ionization energies in the first p-orbit.

lie on a straight line and those of 4, 5, and 6 electrons should also lie on a straight line parallel to the former. Between $n = 3$ and $n = 4$ there may be a jump whose size depends on the relative values of the b and c coefficients. The experimental values for the first p-orbit follow very closely this pattern as seen in Fig. 9.4. The full circles are the experimental points and the straight parallel lines go through values which were calculated as follows.

The interaction (26) with 3 independent coefficients is the most general one between identical particles in the p-orbit. Matrix elements of the Coulomb interaction between electrons, e^2/r_{12}, are linear combinations of only *two* radial integrals (F_0 and F_2). The actual effective interaction between electrons in atoms is modified by configuration mixing. The situation is like in the nuclear shell model but the modification is much less pronounced. Nevertheless, we shall use the Coulomb interaction to calculate F_0 and F_2 and express the coefficients $a, b,$ and c as linear combinations of them. Unlike the situation in nuclear physics the problem is well defined and the interaction is the well-known Coulomb interaction. The single p-electron separation energy ϵ cannot be simply calculated. The combined field of the central charge and the electrons in the first and second s-orbits is rather complicated. Hence, in Fig. 9.4 the value of ϵ was adjusted to give the best fit to the data. The values of F_0 and F_2 were calculated by taking wave functions of the lowest p-orbit in a hydrogen-like atom with $Z = 20 - 4 = 16$. Given the crude approximation, the calculated results seem to be in good agreement with experiment. We should only hope that some day it will be possible to carry out a similar calculation of nuclear energies and obtain equally good agreement with experiment.

10.1 References

[1] G. Racah, *Phys. Rev.* **63**, 367 (1943).

[2] A.K. Kerman, *Ann. Phys. (N.Y.)* **12** 300 (1961) .

[3] A. de-Shalit and I. Talmi, *Nuclear Shell Theory* (Academic Press, New York, 1963).

[4] I. Talmi, *Nucl. Phys.* **A172**, 1 (1971).

11

The Relativistic Oscillator and Mass Formulas

Arno Bohm[1] and Piotr Kielanowski[2]

The mass spectrum of hadrons is as important a problem in relativistic physics as the energy spectrum of molecules and nuclei was in non-relativistic quantum mechanics. In molecular physics the spectrum is explained in terms of single particle excitations and collective rotations and vibrations [1]. In nuclear physics the "parts" of the collective model are also rotations and vibrations [2], with the main difference between molecular and nuclear collective motions being that in molecular physics the rotational and vibrational mode are well separated from each other and from the particle excitations and in nuclear physics these "parts" are not well separated from each other and rotation-vibration particle interactions play an important role. Hadrons are relativistic extended objects and if the analogy between molecular and nuclear physics persists also for the relativistic domain we would expect the collective mode of hadrons to be relativistic rotators and relativistic oscillators.

Relativistic oscillators have been considered in the past [3,4,5] and most recently by Moshinsky *et al.* [6]. The Dirac oscillator of reference [6] is based on the idea that, as the non-relativistic oscillator Hamiltonian is quadratic in both momenta and coordinates, the Dirac equation for the relativistic oscillator should be linear not only in momenta but also in coordinates. We shall here discuss a further development of these ideas, which also used as the starting point the Dirac equation and in which the relativistic Hamiltonian is constructed in such a way that the usual quadratic Hamiltonian results in the non-relativistic limit [7].

In the same way as for the n-particle Dirac oscillator of reference [6] we separate the center of mass motion from the intrinsic motion. Like in all relativistic theories of extended objects, the center of mass motion is governed by the Poincaré group $\mathcal{P}$ with the generator P_μ (total momentum) and $J_{\mu\nu}$ (angular momentum). The P_μ, $J_{\mu\nu}$ fulfill the commutation

---

[1]The University of Texas, Austin, Texas, USA.

[2]CINVESTAV-IPN, México; also at the Institute of Theoretical Physics, Warsaw University, Poland.

relations of the Poincaré group. With them one defines

$$w^{\mu} \equiv \tfrac{1}{2}\varepsilon^{\mu\nu\rho\sigma}\hat{P}_{\nu}J_{\rho\sigma}, \quad \hat{P}_{\nu} \equiv P_{\nu}M^{-1}, \quad M \equiv (P_{\mu}P^{\mu})^{1/2}, \tag{1}$$

and the spin tensor

$$\Sigma_{\mu\nu} \equiv \varepsilon_{\mu\nu\rho\sigma}\hat{P}^{\rho}w^{\sigma}. \tag{2}$$

The internal variables, which describe the motion of the extended object relative to the center of mass, are different for the Dirac oscillator and our quantum relativistic oscillator and also for all the other relativistic theories of extended objects: The Dirac oscillator [6] uses the ordinary relative momenta and positions. The relativistic oscillating quark model [3] uses 4-dimensional intrinsic momenta π_{μ} ($\mu = 0,1,2,3$), and intrinsic positions ξ_{μ} which fulfill the relativistic canonical commutation relations. The one-mode straight line string [8] uses variables which were obtained from the conventional string variables [9] a^{μ}_{-m} and a^{μ}_{m}, $\mu = 0,1,2,3$, $m = 1,2,3,\ldots$ in a way in which quantization does not destroy (3+1) dimensional relativistic invariance. All these models use intrinsic momentum and position operators which have commuting components and which fulfill canonical commutation relations:

$$[\hat{\xi}_{\mu},\hat{\xi}_{\nu}] = 0, \quad [\hat{\pi}_{\mu},\hat{\pi}_{\nu}] = 0, \quad [\hat{\xi}_{\mu},\hat{\pi}_{\nu}] = -i\eta_{\mu\nu}, \quad \mu,\nu = 0,1,2,3. \tag{3}$$

In contrast our intrinsic variables fulfill the commutation relations

$$[\xi_{\mu},\xi_{\nu}] = -\frac{i}{(Mc)^2}\Sigma_{\mu\nu}, \tag{4}$$

$$[\pi_{\mu},\pi_{\nu}] = -\frac{i}{\alpha'(Mc)^2}\Sigma_{\mu\nu}, \tag{5}$$

$$[\xi_{\mu},\pi_{\nu}] = -\breve{g}_{\mu\nu}\frac{\hat{P}_{\rho}\Gamma^{\rho}}{\alpha'(Mc)^2}, \tag{6}$$

$$[\Sigma_{\mu\nu},\Sigma_{\rho\sigma}] = -i(\breve{g}_{\mu\rho}\Sigma_{\nu\sigma} + \breve{g}_{\nu\sigma}\Sigma_{\mu\rho} - \breve{g}_{\mu\sigma}\Sigma_{\nu\rho} - \breve{g}_{\nu\rho}\Sigma_{\mu\sigma}). \tag{7}$$

In order that ξ_{μ} and π_{μ} are in the right units of distance and momenta we had to introduce a new constant α' of dimension $(\mathrm{GeV})^{-2}$ which specifies the physical system (like the spring constant of the non-relativistic oscillator or the spring tension of the relativistic string [9]). We have also included the velocity of light c in (4–6) for use in demonstrating the nonrelativistic limit. The quantity

$$\breve{g}^{\nu}_{\mu} \equiv \eta^{\nu}_{\mu} - \hat{P}_{\mu}\hat{P}^{\nu} \tag{8}$$

is the projector onto the plane perpendicular to the center of mass momentum $P_{\nu} : \breve{g}^{\nu}_{\mu}P^{\mu} = 0$.

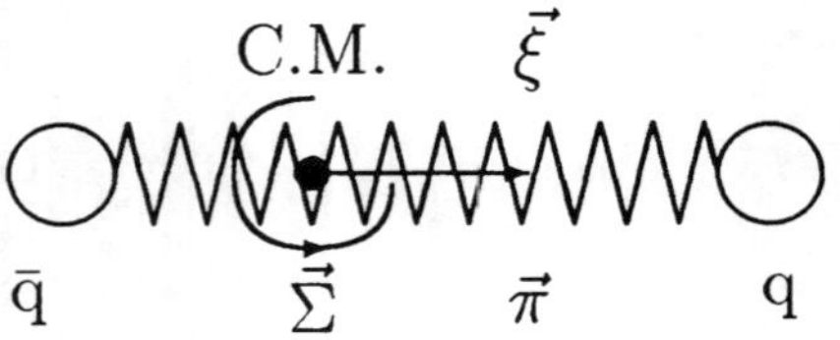

FIGURE 11.1. Oscillator part of the flux tube model with the string tension $2\pi/\alpha'$.

We are considering here the simplest possible example of a relativistic extended object, the intrinsic structure of which is described by a position 4-vector ξ_μ, a momentum 4-vector π_μ and the spin tensor (angular momentum of the extended object relative to the center of mass) $\Sigma_{\mu\nu}$. One can imagine this as a flux tube (depicted as a string in Fig. 11.1) connecting a quark and an antiquark (or a quark and a di-quark for baryons). ξ_μ is the direction and distance along the tube axis, π_μ the relative momentum of the quark-antiquark describing the vibrations of the flux spring along its axis and $\Sigma_{\mu\nu}$ describes the angular momentum of the flux spring relative to its center of mass. This angular momentum is the sum of the total quark spin which has the component $s = 1$ along the tube axis for the ρ-meson-tower because the spin of quark and antiquark line up along the tube axis ($s = \frac{1}{2}$ for baryons) and the angular momentum of the flux tube.

The commutation relations (5–7) of the observables for the intrinsic motion is the Lorentz-covariant version of the commutation relations for an underlying spectrum generating group, which for our relativistic extended object is the group $SO(3,2)_{S_{\mu\nu}\Gamma_\mu} \supset SO(3,1)_{S_{\mu\nu}}$ with the commutation relations

$$\begin{aligned}[][S_{\mu\nu}, S_{\xi,\sigma}] &= -i(\eta_{\mu\xi}S_{\nu\sigma} + \eta_{\nu\sigma}S_{\mu\xi} - \eta_{\mu\sigma}S_{\nu\xi} - \eta_{\nu\xi}S_{\mu\sigma}), \\ [S_{\mu\nu}, \Gamma_\xi] &= -i(\eta_{\mu\xi}\Gamma_\nu - \eta_{\nu\xi}\Gamma_\mu), \\ [\Gamma_\mu, \Gamma_\nu] &= -iS_{\mu\nu},\end{aligned} \tag{9}$$

where

$$\eta_{\mu\nu} = (+1, -1, -1, -1), \quad \mu, \nu = 0, 1, 2, 3. \tag{10}$$

This is a generalization of Dirac's group $SO(3,2)_{\frac{1}{2}\sigma_{\mu\nu}\frac{1}{2}\gamma_\mu}$ which is generated by $\frac{1}{2}$ of the Dirac γ-matrices and σ-matrices. It is the smallest possible simple group that qualifies as a relativistic spectrum generating group and will give therefore the simplest possible model of a relativistic extended object.

In the rest frame of the center of mass, $\hat{p}_{\mu_{\rm rest}} = (1,0,0,0)$, the intrinsic position, intrinsic momentum and the spin tensor are given by

$$\begin{aligned}
\xi_0 &\overset{\text{rest}}{=} 0, &\quad \pi_0 &\overset{\text{rest}}{=} 0, &\quad \Sigma_{0i} &\overset{\text{rest}}{=} 0, \\
\xi_i &\overset{\text{rest}}{=} S_{0i}\frac{1}{Mc}, &\quad \pi_i &\overset{\text{rest}}{=} -\Gamma_i\frac{1}{\alpha' Mc}, &\quad \Sigma_{ij} &\overset{\text{rest}}{=} S_{ij}.
\end{aligned} \tag{11}$$

The intrinsic variables ξ_μ, π_μ, $\Sigma_{\mu\nu}$ have the properties

$$P_\mu \xi^\mu = 0, \quad P_\mu \pi^\mu = 0, \quad P_\mu \Sigma^{\mu\nu} = 0. \tag{12}$$

This means that the intrinsic motion is three-dimensional space-like and there are no ghosts (time-like vibrations).

So far we have considered the two aspects of the quantum theory of a relativistic oscillator which can be described by a group. The first aspect is, that the center of mass motion is described by the symmetry group $\mathcal{P}$ which provides the definition of the mass as the eigenvalue of the operator $M = (P_\mu P^\mu)^{1/2}$. The second aspect is, that the intrinsic motion (vibration in the case of an oscillator) is described by the spectrum generating group [in our model by $SO(3,2)$] or some other algebra of observables. The third aspect of the quantum theory of a relativistic oscillator (or any other relativistic extended object) is concerned with the combination of the center of mass motion and the intrinsic motion. This is done in all the models in a similar way by postulating the relativistic hamiltonian $\mathcal{H}$ or the relativistic "wave equation". The relativistic Hamiltonian determines the Heisenberg equation of motion $d\mathcal{O}/d\tau = 1/(i\hbar)[\mathcal{O},\mathcal{H}]$ and provides a constraint relation between the mass operator and the intrinsic observables. This gives the mass formula.

As an example let us consider the relativistic mass point (which of course does not have an intrinsic structure). The relativistic Hamiltonian is $\mathcal{H} = v(P_\mu P^\mu - m_0^2) = v\Phi$ where v is the Lagrange multiplier. The equation of motion for the position is $dQ_\nu/d\tau = \dot{Q}_\nu = 1/(i\hbar)[Q_\nu,\mathcal{H}] = -2P_\mu\eta^\mu_\nu v = -2vP_\nu$. The gauge constraint $\dot{Q}_\nu\dot{Q}^\nu = 1$ determines the multiplier $v^2 = -(2M)^{-1}$. The constraint relation $\Phi = 0 = P_\mu P^\mu - m_0^2$ determines the mass spectrum $P_\mu P^\mu = m_0^2$ (which is trivial for the relativistic mass point), and the relativistic wave equation $(-\Box - m_0^2)\varphi(x) = 0$.

For a relativistic object with intrinsic structure the method is the same, but the Hamiltonian and the wave equation are more complicated. For Moshinsky's n-particle oscillator the wave equation was obtained from the n-fold product of the Dirac equation by the replacement of the intrinsic momenta $\vec{p}_s{}' \to \vec{p}_s{}' - i\vec{r}_s B$, $s = 1,\cdots n$, where $\vec{r}_s$ is the intrinsic position and B is the n-fold product of β matrices [6]. For the relativistic oscillating quark model [3] the relativistic Hamiltonian is

$$\mathcal{H} = v(P_\mu P^\mu + \pi^{(c)}_\mu \pi^{(c)\mu} + \xi^{(c)}_\mu \xi^{(c)\mu} - m_0^2), \quad \Omega = \text{system constant}, \tag{13}$$

where $\pi_\mu^{(c)}$ and $\xi_\mu^{(c)}$ are the canonical intrinsic momenta and position operators. For the unconstrained version of the conventional relativistic string [9] the relativistic Hamiltonian is:

$$\mathcal{H} = vL_0 = v(P_\mu P^\mu + \frac{1}{\alpha'}\sum_{m=1}^{\infty} a_{-m\mu}a^\mu_{+m} - \text{constant}) \tag{14}$$

where a^μ_{-m}, a^μ_{+m} are the creation and annihilation operators fulfilling the canonical commutation relations.

For the single-mode version of the relativistic string [8] the Hamiltonian is the same as in (14) only that all $a^\mu_{\mp m} = 0$ for $m \neq 1$.

For our quantum relativistic oscillator the relativistic Hamiltonian is postulated to be

$$\mathcal{H} = v(P_\mu P^\mu - \frac{1}{\alpha'}\frac{P_\mu}{M}\Gamma^\mu), \tag{15}$$

where M is the mass-operator in (1) and Γ^μ is the generator of $SO(3,2)$ which in the 4-dimensional Dirac representation of $SO(3,2)$ would be represented by $\frac{1}{2}\gamma_\mu$.

The relativistic Hamiltonian (15) is of all the above Hamiltonians the one most reminiscent of the Dirac equation for the electron, but one does not recognize anything of the non-relativistic oscillator in (15). The similarity with the Dirac electron-positron system is only superficial because we do not use the 4-dimensional representation. Instead we use the infinite dimensional representations of $SO(3,2)_{S_{\mu\nu},\Gamma_\mu}$ which are denoted in the lowest (highest) weight notation by $D(s+1,s)$ [10,11]. The connection of (15) with the oscillator is real as one can see in the following way: if one looks at ξ_i in the rest frame given by (11) one sees that it is the Lorentz generator S_{0i}, which in the Inönü-Wigner group contraction goes into the non-relativistic commuting (intrinsic) position operator. If one takes the nonrelativistic contraction limit $c \to \infty$ from the Poincaré to the Galilei group, then $SO(3,2)$ contracts into the algebra of the three-dimensional harmonic oscillator [7] and ξ_n and π_m, $(n,m = 1,2,3)$ become commuting operators that fulfill

$$[\xi_m^{(c\to\infty)}, \pi_n^{(c\to\infty)}] = -i\eta_{mn} = i\delta_{mn}. \tag{16}$$

Thus in the non-relativistic limit we obtain from (4–6) the canonical commutation relations and a careful calculation shows that the relativistic Hamiltonian (15) leads to the ordinary oscillator Hamiltonian (plus the kinetic energy of the center of mass)

$$\mathcal{H} = \frac{\vec{P}^2}{2m} + \frac{1}{4m}\vec{\pi}^{(\infty)^2} + \frac{1}{4m\hbar\alpha^2}\vec{\xi}^{(\infty)^2}. \tag{17}$$

This contraction limit can only be performed for the representation $D(s+1=1, s=0)$ (which describes a flux tube with total quark spin

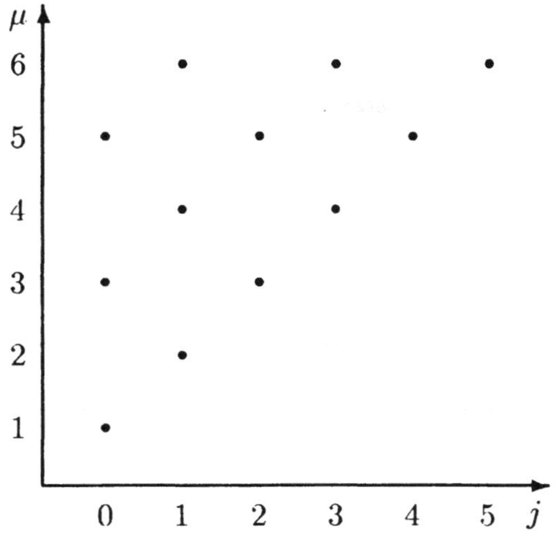

FIGURE 11.2. Weight diagram of the representation $D(1,0)$ which is identical with the energy diagram of the three-dimensional non-relativistic oscillator.

$s = 0$) and not for the representation $D(s+1=2, s=1)$ (which should be used for the ρ-meson tower) but this should be sufficient to call all models described by the $D(s+1,s)$ before the nonrelativistic limit is taken a quantum relativistic oscillator (with total quark spin s).

In order to determine the spectrum of our relativistic oscillator with $s = 1$ we consider the weight diagram of the representation $D(s+1,s)$. The weights are the values (μ, j) that characterize the irreducible representations of the maximal compact subgroup $K = SO(2)_{\Gamma_0} \times SO(3)_{S_{ij}}$ contained in $D(s+1,s)$.
Thus at rest, $\mu =$ eigenvalue Γ_0 and $j(j+1) =$ eigenvalue $(\frac{1}{2}S_{ij}S_{ij})$, which means that j is the spin of the extended object (hadron spin) and μ is another quantum number. Figure 11.2 shows the weight diagram of $D(1,0)$ for which $s = 0$ spin-free relativistic oscillator. Each dot gives the value of a pair (μ, j) that appears in $D(1,0)$. One can see that this diagram is identical with the energy diagram of the three-dimensional oscillator with vibrational quantum numbers $(\mu - 1)$. The diagram does not change in the contraction $c \to \infty$ of $SO(3,2)$ into the algebra of the 3-dimensional oscillator. Thus $(\mu - 1) =$ eigenvalue $(\hat{P}_\rho \Gamma^\rho)$ [which is the same as the eigenvalue (Γ_0) on the states at rest] should also be the vibrational quantum number of the relativistic oscillator. Figure 11.3 shows the K-type of $D(2,1)$ with $s = 1$ (this representation cannot be contracted into the simple oscillator). The dots of Fig. 11.3 correspond to the mesons with total quark spin $s = 1$. In Fig. 11.3 we show the assignment of meson mass levels.

The mass spectrum obtained from the relativistic Hamiltonian (15) for

$$\frac{1}{\alpha'} = 1.04\ (\mathrm{GeV})^2, \quad \lambda^2 = 0.02\ (\mathrm{GeV})^2.$$

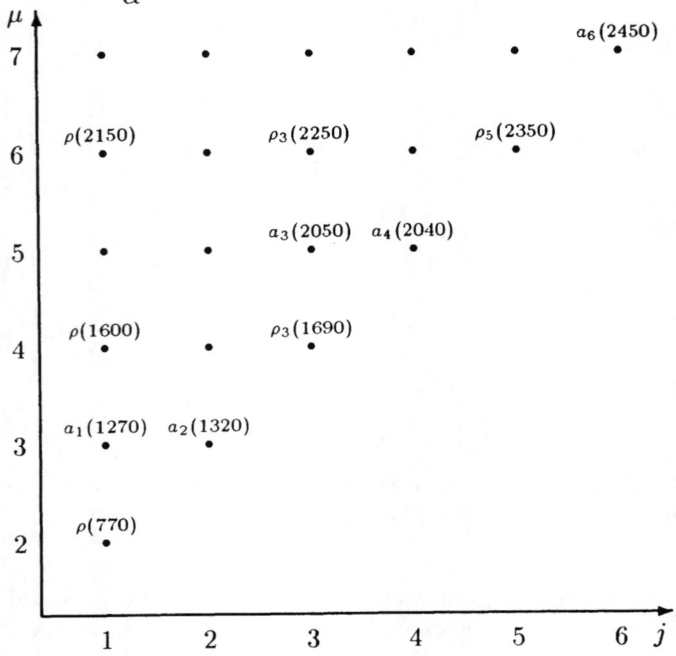

FIGURE 11.3. Weight diagram of the representation $D(2,1)$ with the assigment of mesons. Each dot represents a meson with spin j and vibrational quantum number μ. The mesons assigned to each dot are from the Particle Data Table, *Phys. Lett. B* **239** (1990).

the representation $D(2,1)$ is

$$m^2 = m_0^2 + \frac{1}{\alpha'}(\mu - 1), \quad \mu - 1 = 1, 2, 3, \cdots, \quad j = 1, 2, \cdots, \mu - 1. \qquad (18)$$

It thus predicts that the masses of resonances lie on linear Regge trajectories with slope α' and that the masses on the main trajectory $j = (\mu - 1)$ are the same as the masses of the "daughters" $j = 1, 2, \cdots, \mu - 2$ which have the same value of μ. Empirically this is not quite correct. We have therefore fitted the masses in Fig. 11.3 to the formula

$$m^2 = m_0^2 + \frac{1}{\alpha'}(\mu - 1) + \lambda^2 j(j+1), \quad \mu - 1 = 1, 2, 3, \cdots, \quad j = 1, 2, \cdots, \mu - 1. \qquad (19)$$

where λ^2 is another parameter. The values of the parameters α' and λ^2 determined by this χ^2 fit are:

$$\frac{1}{\alpha'} = 1.04\ (\mathrm{GeV})^2, \quad \lambda^2 = 0.02\ (\mathrm{GeV})^2. \qquad (20)$$

The smallness of the empirical value for λ^2 in (20) shows that the relativistic oscillator with the Hamiltonian (15) gives a very good description of the meson tower which contains those resonances which are usually associated with the ρ-meson trajectory (of slope α') and its daughters. For the baryon trajectory similar results can be obtained.

The other relativistic oscillator models *e.g.* reference [6], give slightly different predictions. And even if the experimental data may not be conclusive there seems to be very little doubt that part of the spectrum is due to relativistic vibrations. Compared to the Dirac oscillator and all the other relativistic oscillator models our quantum relativistic oscillator contains the most radical deviations from the conventional assumptions of the intrinsic variables. But except for that it is a straightforward generalization of established ideas into the relativistic domain.

11.1 References

[1] G. Herzberg, *Molecular Spectra and Molecular Structure* (D. van Nostrand Publishers, 1966).

[2] A. Bohr and B. Mottelson, *Nuclear Structure* (Benjamin, 1969) Vol. II.

[3] R.P. Feynman, M. Kislinger, and F. Ravndal, *Phys. Rev. D* **3**, 2706 (1971).

[4] T. Takabayashi, *Prog. Theor. Phys. Suppl.* **67**, 1 (1979) and references therein.

[5] See also Y.S. Kim and E. Noz, *Theory and Applications of the Poincaré group* (Reidel Publishing, 1986) for additional references.

[6] M. Moshinsky and A. Szczepaniak, *J. Phys. A* **22**, L817 (1989); M. Moshinsky, G. Loyola, A. Szczepaniak, C. Villegas, and N. Aquino, *Proceedings of the Rio de Janeiro International Workshop* (World Scientific, 1990), p. 271; M. Moshinsky, G. Loyola and A. Szczepaniak, *Anniversary Volume for J.J. Giambiaggi* (World Scientific, 1990).

[7] A. Bohm, M. Loewe, P. Magnollay, M. Tarlini, R.R. Aldinger, L.C. Biedenharn, and H. van Dam, *Phys. Rev. D* **31**, 2304 (1985); *Phys. Rev. D* **32**, 791 (1985); *Phys. Rev. D* **32**, 2828 (1985); *Phys. Rev. Lett.* **53**, 2292 (1984).

[8] V.I. Borodulin, G.P. Pron'ko, A.V. Razumov, L.D. Soloviev, and O.L. Zorin, *Proceedings of the International Conference on High Energy Physics* (Bari, Italy, July 1985); G.P. Pron'ko, *Quantum Theory of*

String in the Four-Dimensional Space-Time (Serpukhov Preprint 86–42); M.S. Plyushchay *et al., Light Meson Spectroscopy* (Serpukhov Preprint 87–24).

[9] See *e.g.* J. Scherk, *Rev. Mod. Phys.* **47**, 123 (1975); F. Rohrich, in *Recent Developments in Particle and Field Theory*, W. Dittrich, editor (Vieweg Braunschweig, 1977); M.B. Green, J.H. Schwarz, and E. Witten, *Superstrings* (Cambridge University Press, 1987).

[10] N.T. Evans, *J. Math. Phys.* **8**, 170 (1967); A.O. Barut and A. Bohm, *J. Math. Phys.* **11**, 2938 (1970); C. Fronsdal, *Phys. Rev. D* **12**, 3819 (1975); E. Angelopoulos, in *Quantum Theory, Groups, Fields and Particles*, p. 101, Ed. by A.O. Barut (Reidel, 1983) and reference 7.

[11] The representations of $SO(3,2)$ have also been disscussed by M. Moshinsky, see E. Chacón, P.O. Hess, and M. Moshinsky, *J. Math. Phys.* **28**, 2223 (1987) and references therein.

12

Relativistic Equations in External Fields

Matías Moreno[1] and Rosa María Méndez-Moreno[2]

ABSTRACT A general method to find exact non-relativistic equivalent problems for a large class of potential-interactions in Dirac-like Hamiltonians is presented. The included interactions are the supersymmetric ones and "large" projected non-supersymmetric interactions. The identification of the corresponding Lorentz structure for the Dirac equation in $3+1, 4+1$ and $2+1$ space-time dimensions is performed. A general outline of previous relevant works is presented.

12.1 Introduction

It was shown recently that for a large class of potential problems in the Dirac equation a stability of the Dirac sea can be obtained as a result of a custodial supersymmetry [1]. These potentials include the Dirac oscillator, recently revived by Marcos Moshinsky and Adam Szczepaniak [2, 3, 4, 5, 6, 7] and the Dirac particle in a magnetic field [8] . The stability of the Dirac sea was proved via the existence of a closed and exact Foldy-Wouthuysen transformation which can be explicitly constructed. Generally the supersymmetric quantum mechanics algebra requires a second-order Hamiltonian in the fermionic (differential) operators. In Ref. [1], it was shown that first-order Dirac formalism can also lead to a supersymmetric energy spectrum by extending the notion of a supersymmetric Hamiltonian. A similar result has also been obtained for a massless Dirac fermion in a uniform magnetic field case [9]. Both the massless and the massive cases are relevant for the study of the integer quantum Hall effect [10, 11] and are included in the general formalism.

In this work we will first give a revision of the Dirac oscillator. In this problem we apply the Foldy-Wouthuysen transformation (FWT) and obtain the eigenvalues and eigenfunctions of the transformed Hamiltonian. Then the general framework, of which the Dirac oscillator is a special case,

[1]Instituto de Física and Facultad de Ciencias, UNAM.
[2]Departamento de Física, Facultad de Ciencias, UNAM.

is presented [3, 1, 12]. We review here these works and develop the ideas in a way in which the generality of the methods is more transparent. In particular, we will begin here with the definition of a supersymmetric Hamiltonian and discuss in this general case the concept of stability of the Dirac sea. Then, we apply these tools to arbitrary odd potentials for a Dirac equation in $3+1$ dimensions. To show the versatility of the method, we then we consider a new example, the case of a Dirac equation in $4+1$ and $2+1$ dimensions.

We finally discuss a new method that incorporates all the supersymmetric potentials and half of the even, non-supersymmetric, potentials into an equivalent non-relativistic Schrödinger-Pauli like equation. In particular, the new framework includes a recently discovered type of oscillator potential that also has exact solutions [13, 14]. We give some examples of this procedure: the original Spanish oscillator and also the bound state spectra of a Coulomb-like interaction.

12.2 The Dirac oscillator, a study case

In their original presentation Moshinsky and Szczepaniak [2] consider an equation which is linear in coordinates and momenta. They propose a non-minimally coupled interaction through the substitution

$$\mathbf{p} \rightarrow \pi := \mathbf{p} - im\omega\beta\mathbf{r}, \tag{2.1}$$

here m is the particle rest mass and ω is the oscillator frequency. Throughout this work we will use the conventions of Bjorken and Drell book [15]. The linearity in $\mathbf{r}$ was obtained by the same procedure which yields the Dirac equation starting from the Klein-Gordon equation

$$i\partial_t\psi = [\alpha\cdot(\mathbf{p} - im\omega\beta\mathbf{r}) + \beta m]\psi, \tag{2.2}$$

with α and β, the usual 4×4 matrices. For positive energies they find exact solutions which are closely related to the non-relativistic eigenfunctions

$$\begin{aligned} \psi &= |N(l\tfrac{1}{2})jm\rangle \\ &= \sum_{\mu,\sigma}\langle l\mu, \tfrac{1}{2}\sigma|jm\rangle Y_{l\mu}(\Theta,\phi)\chi_{\frac{1}{2}\sigma}R_{Nl}(r), \end{aligned} \tag{2.3}$$

where $\langle l\mu, \frac{1}{2}\sigma|jm\rangle$ is a Clebsch-Gordan coefficient, $Y_{l\mu}(\Theta,\phi)$ are spherical harmonics, $\chi_{\frac{1}{2}\sigma}$ are half spin bispinors and $R_{Nl}(r)$ are the non-relativistic radial wavefunctions with $N=2n+l$, the total energy quantum number and $n=0,1,2,\ldots$ is the radial quantum number. The solutions give finite and infinite energy states, that have eigenvalues

$$E^{+} = \begin{cases} \sqrt{m^2 + m\hbar\omega(2(N-j)+1)}, & j = l+\frac{1}{2}, \\ \sqrt{m^2 + m\hbar\omega(2(N+j)+3)}, & j = l-\frac{1}{2}, \end{cases} \tag{2.4}$$

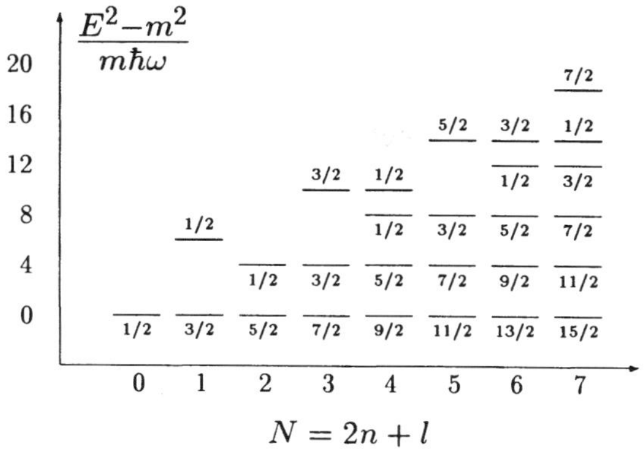

FIGTIRE 12.1. Positive energy spectrum for the Dirac oscillator. Labels for the $j = l + \frac{1}{2}$ states are bellow and those for $j = l - \frac{1}{2}$ states are above.

or

$$\frac{E^{+2} - m^2}{m\hbar\omega} = \begin{cases} 4n, & j = l + \frac{1}{2}, \\ 4(n+l) + 2, & j = l - \frac{1}{2}. \end{cases} \tag{2.5}$$

In Fig. 12.1 the energy spectrum is shown for this case.

The Dirac oscillator potential is equivalent to a frame dependent description of an anomalous magnetic coupling to an electric field [3]. It has been shown that such potential yields an exact Foldy-Wouthuysen [16] transformation (FWT) as well as an exact Cini-Touschek one [17]. With the Foldy-Wouthuysen transformation the stability of the Dirac sea is obtained, since the positive and negative energy solutions are decoupled through the FWT Ref. [3]. The FWT belong to the non-relativistic limit. The Cini-Touschek transformation, in which the Hamiltonian written in a closed form is purely odd, corresponds to the ultra-relativistic limit.

Applying the FWT, which is an unitary transformation, the Foldy-Wouthuysen Hamiltonian must be even,

$$i\partial_t - H_{FW} := e^{iS}(i\partial_t - H)e^{-iS}, \tag{2.6}$$

as usual one defines an even operator, $\mathcal{E}$, if $[\beta, \mathcal{E}] = 0$ and an odd operator, $\mathcal{O}$, if $\{\beta, \mathcal{O}\} = 0$. For an arbitrary operator A, the transformed generator A' is given by

$$A' = e^{iS} A e^{-iS} = A + [iS, A] + \frac{1}{2!}[iS, [iS, A]] + \cdots. \tag{2.7}$$

If $\{iS, A\} = 0$ then $A' = e^{2iS}A = Ae^{-2iS}$ which simplifies this form. The operator we are looking for has the form

$$iS = \beta\mathcal{O}_H\theta = \beta\,\alpha\cdot\pi\,\theta = \gamma\cdot\pi\,\theta, \tag{2.8}$$

with $\mathcal{O}_H$ the odd part of the Hamiltonian and π given in Eq. (2.1). One must require that the commutator of θ with β, $\alpha\cdot\pi$ and $\gamma\cdot\pi$, must be zero.

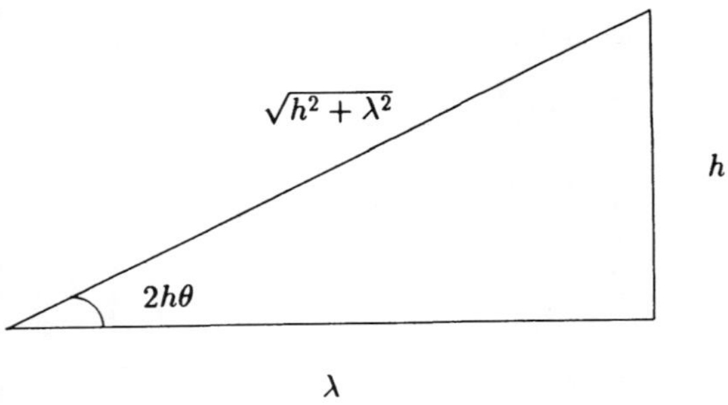

FIGURE 12.2. Triangle construction for the Foldy-Wouthuysen transformation.

Then we can write

$$e^{iS} = \cos(h\theta) + \gamma \cdot \pi \sin(h\theta) \tag{2.9}$$

where h is required to be an even root of

$$\begin{aligned} h^2 &= (\alpha \cdot \pi)^2 = -(\gamma \cdot \pi)^2 \\ &= \pi^+ \cdot \pi + i\sigma \cdot (\pi^+ \times \pi) \\ &= p^2 + m^2\omega^2 r^2 - m\omega\hbar\beta(3 + 4/\hbar\, \mathbf{L} \cdot \mathbf{S}). \end{aligned} \tag{2.10}$$

Then the transformed Hamiltonian is

$$H_{FW} = \alpha \cdot \pi[\cos(2h\theta) - \frac{m}{h}\sin(2h\theta)] + \beta[m\cos(2h\theta) + h\sin(2h\theta)], \tag{2.11}$$

and with θ selected in such a way that $\tan(2h\theta) = h/m$, we finally have

$$\begin{aligned} H_{FW} &= \beta\sqrt{m^2 + h^2} \\ &= \sqrt{m^2 + p^2 + m^2\omega^2 r^2 - m\omega\hbar\beta(3 + 4/\hbar\, \mathbf{L} \cdot \mathbf{S})}, \end{aligned} \tag{2.12}$$

this may be verified with the help of the triangle construction as in Fig. 12.2.

The eigenfunctions and the eigenvalues of this Hamiltonian can immediately be written as in Eq. (2.3) and

$$E^2_{\varepsilon Njl} - m^2 = [4n + 2l(1 - \varepsilon\delta) + 3 - \varepsilon(2 + \delta)]\ m\hbar\omega, \tag{2.13}$$

with

$$j = l + \tfrac{1}{2}\delta, \qquad \varepsilon, \delta = \pm 1; \tag{2.14}$$

ε determines the sign of the energy, δ the relation between total and orbital angular momentum and the units are such that the light velocity $c = 1$. It is worthwhile noting that the eigenfunction of the initial Hamiltonian H is $\psi' = e^{-iS}\psi$, the spectrum of the two Hamiltonians is the same.

The infinite and finite degeneracies of the non-relativistic Hamiltonian spectrum can be accounted for by the existence of a symmetry algebra $so(4) \oplus so(3,1)$ [18], the relation of this to the Susy Hamiltonian has been studied in Refs. [1, 19]. These degeneracy can be removed by the methods shown bellow.

12.3 Extended supersymmetric Hamiltonians

Next the connection with the supersymmetric quantum mechanics (SQM) will be done. A general supersymmetric (Susy) Hamiltonian, H_S satisfies

$$[Q, H_S] = 0 = [Q^\dagger, H_S] \tag{3.1}$$

with Q and $Q^\dagger$ fermionic operators, *i.e.* such that $Q^2 = 0 = Q^{\dagger 2}$. In particular a supersymmetric Hamiltonian can be of the form

$$H_S = \{Q, Q^\dagger\} \tag{3.2}$$

and

$$[Q^\dagger Q, QQ^\dagger] = 0, \tag{3.3}$$

where the $QQ^\dagger$ and $Q^\dagger Q$ commute with H_S. If the ground state is nondegenerate and normalizable, Susy is unbroken. The states $|n\rangle$, $|n_+\rangle = Q^\dagger|n\rangle$ and $|n_-\rangle = Q|n\rangle$ are degenerate, but $|n_+\rangle$ or $|n_-\rangle$ are null.

The connection between the Dirac equation and the supersymmetric quantum mechanics can be provided [1] if the Dirac Hamiltonian can be written as

$$H = Q + Q^\dagger + \lambda, \tag{3.4}$$

with λ a hermitean operator. These operators must satisfy the following anticommutation relations.

$$\{Q, \lambda\} = \{Q^\dagger, \lambda\} = 0. \tag{3.5}$$

Then the Hamiltonian commutes with $QQ^\dagger$ and $Q^\dagger Q$ and we get for the squared Hamiltonian

$$H^2 = \{Q, Q^\dagger\} + \lambda^2 =: h^2 + \lambda^2, \tag{3.6}$$

with $h^2 = \{Q, Q^\dagger\}$ and where h is required to be an even root of this operator. If we take the definition $\hat{\lambda} = \lambda/\sqrt{\lambda^2}$, the Foldy-Wouthuysen Hamiltonian is written

$$H_{FW} = \hat{\lambda}\sqrt{\{Q, Q^\dagger\} + \lambda^2}, \tag{3.7}$$

for this form to hold one needs a λ with non-null eigenvalues. A FWT is generated by

$$iS = \hat{\lambda}(Q + Q^\dagger)\theta, \tag{3.8}$$

where θ must satisfy

$$\tan(2h\theta_{FW}) = h/\sqrt{\lambda^2}. \tag{3.9}$$

In our notation θ has units of h^{-1} and from the last equation it follows that

$$\sqrt{\lambda^2}\,\theta_{FW} = f(h^2/\lambda^2), \tag{3.10}$$

which shows that θ_{FW} exists with the required properties. A Cini-Touschek transformed Hamiltonian can also be obtained with an analogous procedure, it is given by

$$H_{CT} = (Q + Q^\dagger)\sqrt{1 + \frac{\lambda^2}{\{Q, Q^\dagger\}}}. \tag{3.11}$$

One can write a matrix representation for Q and $Q^\dagger$ in the form

$$Q = \begin{pmatrix} 0 & 0 \\ A & 0 \end{pmatrix}, \qquad Q^\dagger = \begin{pmatrix} 0 & A^+ \\ 0 & 0 \end{pmatrix}. \tag{3.12}$$

Because $\hat{\lambda}$ commutes with H_{FW} and its eigenvalues are ± 1, Eq. (3.7) implies that, in general, the energy spectra has two branches; between them there exists a gap determined by the lowest eigenvalues of λ^2 and h^2. If $\hat{\lambda}$ has negative eigenvalues and H^2 is not bounded from above one gets an unstable ground state for the Susy Hamiltonian. The usual remedy for this situation is to introduce a Dirac sea. The important feature of the Susy interactions is that the definition of this sea is independent of the size of the couplings, we call this property the stability of the Dirac sea.

From the hermiticty of λ and the conditions in Eq. (3.5) one can show that the FW Hamiltonian in Eq. (3.7) anticommutes with Q and $Q^\dagger$. The standard definition of a Susy Hamiltonian is that it must commute with the fermionic operators Q and $Q^\dagger$, as in Eq. (3.1). The anticommutation condition implies that, in order to construct observables which are constants of motion in the Dirac-Susy Hamiltonian Eq. (3.4), one must form bosonic operators out of the fermionic ones. Furthermore, the specific form of the FW hamiltonian in Eq. (3.7) —the product of a commuting Susy Hamiltonian times an anticommuting square root of the unit, $\hat{\lambda}$— has an important byproduct. It implies that all the usual properties of the Susy Hamiltonians are valid except the positive definite nature of the standard Susy energy spectrum.

To conclude this section let us notice that there is an alternative way to construct the FWT. The alternative method is based on the construction of an $SU(2)$ algebra; we will not pursue this method here. The method has been presented in Ref. [12].

12.4 Dirac equation in 3 + 1 dimensions

Let us now see how the method of last section applies to the usual Dirac equation in 3+1 dimensions. If one identifies $\lambda = \beta m$, it follows that, in the standard representation for the Dirac matrices the matrix A of Eq. (3.12) is of the form

$$A = \sigma \cdot (\mathbf{p} + \mathbf{C}) + C_5, \tag{4.1}$$

where the C_i are arbitrary complex potentials and σ are Pauli matrices, *i.e.* A belongs to the quaternions. This form for A can be expressed also in terms of a non-minimal substitution

$$\pi_I = p_I + A_I + i\beta E_I \quad \text{with } I = 1, 2, 3, 5. \tag{4.2}$$

where the fifth index in 3 + 1 dimensions satisfies

$$p_5 = 0 \quad \text{and} \quad \alpha_5 = i\beta\gamma^5. \tag{4.3}$$

The eight potentials that Eq. (3.12) allows can be identified with

- A magnetostatic vector potential, $\mathbf{A}$, coupled minimally.
- An anomalous magnetic moment coupled to an electrostatic field, $\mathbf{E}$.
- A pseudoscalar potential, E_5.
- The time-like, or "electrostatic", part of an axial-vector potential, A_5.

With the Foldy-Wouthuysen Hamiltonian H_{FW}, given by

$$H_{FW} = \beta\sqrt{m^2 + h^2}, \tag{4.4}$$

and using the basic properties of the Pauli matrices or from the Dirac algebra one gets the h^2 of the form

$$h^2 = \pi^+ \cdot \pi + \pi_5^+ \pi_5 + [\pi_5^+ \pi + \pi^+ \pi_5 + i\pi^+ \times \pi] \cdot \sigma. \tag{4.5}$$

The different terms in the last equation are given by

$$\begin{aligned}
\pi^+ \cdot \pi &= (p + A)^2 + E^2 + \beta\nabla \cdot \mathbf{E} + i\beta[A^i, E^i], &(4.6)\\
\pi_5^+ \pi_5 &= A_5^2 + E_5^2 + i\beta[A_5, E_5], &(4.7)\\
\pi_5^+ \pi + \pi^+ \pi_5 &= \{(\mathbf{p} + \mathbf{A}), A_5\} + \{\mathbf{E}, E_5\} &(4.8)\\
&\quad +\beta(\nabla E_5) + i\beta[\mathbf{A}, E_5], -i\beta[A_5, \mathbf{E}], &\\
\pi^+ \times \pi &= -i(\nabla \times \mathbf{A}) + \mathbf{A} \times \mathbf{A} + \mathbf{E} \times \mathbf{E} &(4.9)\\
&\quad +i\beta[\mathbf{p} \times \mathbf{E} - \mathbf{E} \times \mathbf{p} + \mathbf{A} \times \mathbf{E} - \mathbf{E} \times \mathbf{A}]. &
\end{aligned}$$

A more explicit form for this relations can be found in Ref. [1]. It must be remarked that the last results are also valid with colored potentials. These results follow for *arbitrary* values of the coupling.

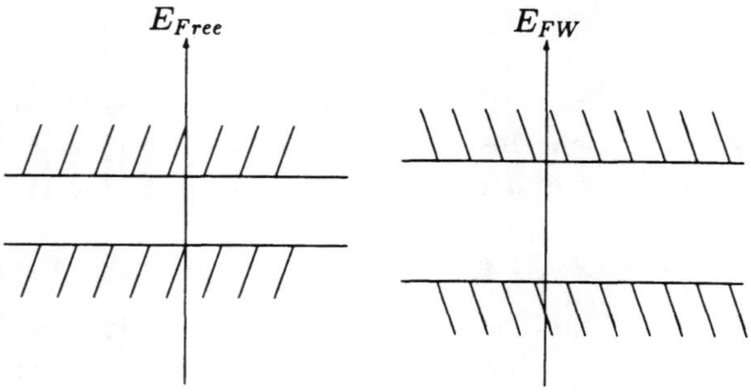

FIGURE 12.3. Schematic Dirac seas for the free case and Susy potentials.

Very important is that the positive and negative energy states do not mix, and the gap size is not decreased. The next figure pictures the Dirac sea in the free and the cases with interaction. After the second quantization is done, one gets a stable Dirac sea, see Fig. 12.3.

As a more specific example let us consider a Dirac particle minimally coupled to a constant, not necessarily uniform, magnetic field. The Hamiltonian for this case is

$$H = m\beta + \alpha \cdot (\mathbf{p} + q\mathbf{A}), \tag{4.10}$$

and the Foldy-Wouthuysen Hamiltonian reduces in this case to

$$H_{FW} = \beta\sqrt{m^2 + (\mathbf{p} + q\mathbf{A})^2 + \sigma \cdot \nabla \times \mathbf{A}}, \tag{4.11}$$

which shows that the solutions for the non-relativistic Pauli Hamiltonian

$$H_P = \frac{(\mathbf{p} + q\mathbf{A})^2}{2m} + g\frac{q}{2\hbar m}\mathbf{S} \cdot \mathbf{B} \tag{4.12}$$

with gyromagnetic ratio $g = 2$, are also solutions for the relativistic FW Hamiltonian and that the non-relativistic eigenenergies, E_{NR} are simply related to the exact relativistic one, E, by

$$E = \pm m\sqrt{1 + \frac{2}{m}E_{NR}}. \tag{4.13}$$

Even as the properties of an electron in a uniform magnetic field have been extensively studied, it is interesting to extend this analysis to the supersymmetric quantum mechanics (SQM). In the integer quantum Hall effect, studied by Laughlin, a non-relativistic electron with Pauli magnetic moment and a gyromagnetic ratio of two is an example of a supersymmetric system. This study has been extended for a system with the electric field boosted to zero and for massless electrons in the realm of the SQM [9]. Our analysis shows that the massless limitation can be overcame and that consequently the relativistic corrections to the Landau levels do not destroy the quantum Hall effect.

The Dirac oscillator is another example of Susy Hamiltonian that fits into our scheme. It corresponds to a chargeless particle with an anomalous magnetic moment, for example a neutron, in the presence of a constant electric field. The Dirac Hamiltonian is for this case

$$H = m\beta + \alpha \cdot (\mathbf{p} + i\kappa\beta\mathbf{E}). \tag{4.14}$$

In this case the Foldy-Wouthuysen Hamiltonian has the form

$$H_{FW} = \beta\sqrt{m^2 + p^2 + \beta\kappa\nabla \cdot \mathbf{E} - \beta\kappa\sigma \cdot (\mathbf{E} \times \mathbf{p} - \mathbf{p} \times \mathbf{E}) + (\kappa E)^2}, \tag{4.15}$$

as follows from Eqs. (4.5) to (4.9) To get from this case the Dirac oscillator one simply has to consider the case were

$$\mathbf{E} = \sqrt{k/2}\ \mathbf{r}, \tag{4.16}$$

from the last two equations one obtains the Hamiltonian in Eq. (2.12). A possible disadvantage of the Dirac oscillator potential in practical applications is its infinite degeneracy that stems from Eq. (2.13). The degeneracy can be lifted within the formalism described in the previous section if the condition $\lambda = \beta m$ —which has been assumed in this section— is lifted, see Ref. [20].

12.5 Susy Dirac equation in $4+1$ and $2+1$ dimensions

If the condition $p_5 = 0$ of Eq.(4.3) is not required the Dirac equation of last section applies to $4+1$ dimensions. With an important difference in the interpretation of the Lorentz structure of the potentials. In the $4+1$ case only three different Lorentz potentials arise with the identifications of Table I.

Because the underlying Susy algebra is not affected in going from $3+1$ to $4+1$ dimensions the SQM in both dimensionalities is basically the same. The Susy potentials are in the $4+1$ case: the four space-like components of the vector potential and the four components T_{i0} of the tensor potential.

Next let us study the Dirac equation in $2+1$ dimensions, both in the free particle case, as well as in the presence of external fields. The Dirac equation in the free case is given by

$$(\gamma \cdot \mathbf{p} - m)\psi = 0, \tag{5.17}$$

4 + 1	3 + 1
scalar: 1	scalar: 1
vector: 5 V_i V_4	vector: 4 + pseudoscalar: 1 V_i P
tensor: 10 T_{ij} T_{4i}	tensor: 6 + axial-vector: 4 T_{ij} A_i^5

TABLE 12.1. Lorentz structure of the potentials in 3 + 1 and 4 + 1 dimensions. The number of components is indicated in each case.

where γ has only the components $\gamma^0 = \beta$, γ^1 and γ^2. Here $\boldsymbol{\alpha} = \gamma^0\boldsymbol{\gamma}$. The α_i and β satisfy the well known anticommutator relations, *i.e.*

$$\{\alpha_i, \alpha_j\} = 2\delta_{ij}, \qquad \{\alpha_i, \beta\} = 0, \qquad \beta^2 = 1 \tag{5.18}$$

with α_i and β hermitean. Then in 2 + 1 the Pauli matrices are solutions

$$\alpha_1 \to \sigma_1, \qquad \alpha_2 \to \sigma_2, \qquad \beta = \gamma^0 = \alpha_0 \to \sigma_3. \tag{5.19}$$

Finally we have the next relations: $\gamma^1 = \sigma_3\sigma_1 = i\sigma_2$ and $\gamma^2 = \sigma_3\sigma_2 = -i\sigma_1$. With this results the Dirac equation is written as

$$[\sigma_3 i\partial_t + i\sigma_2 i\partial_x - i\sigma_1 i\partial_y - m]\psi = 0. \tag{5.20}$$

Now for the case with interactions it is easy to observe that there are four independent ones. One is the scalar and the other three are vectorial ones. Two of them are supersymmetric even in the massive case. From the Dirac equation,

$$[\gamma \cdot (p + qV) + S + m]\psi = 0. \tag{5.21}$$

The supersymmetric potentials are $V_x(r)$ and $V_y(r)$. In particular $V_x = \frac{1}{2}yB$ and $V_y = -\frac{1}{2}xB$, as it is proposed in the study of the quantum Hall effect in the symmetric gauge.

12.6 Beyond supersymmetry

Consider now the combination of the potential recently proposed in Ref. [13], we will refer to this type of projected potential as the Spanish oscillator. This relativistic equation is

$$i\partial_t\psi = \left[\alpha \cdot \mathbf{p} + \frac{1+\beta}{2}\frac{kr^2}{2} + m\beta\right]\psi, \tag{6.1}$$

or more generally,

$$i\partial_t \psi = \left[\alpha \cdot \mathbf{p} + \frac{1+\beta}{2} V(r) + m\beta\right] \psi, \tag{6.2}$$

where $V(r)$ is an arbitrary potential. The last equation also reveals this oscillator as a special combination of a scalar and the time-like component of a vector interaction. According to our previous discussion this is clearly not a Susy potential. Let us define

$$\psi_+ \quad = \quad \tfrac{1}{2}(1+\beta)\psi, \tag{6.3}$$

and

$$\psi_- \quad = \quad \tfrac{1}{2}(1-\beta)\psi. \tag{6.4}$$

Then we have the next two equations

$$i\partial_t \psi_+ \quad = \quad \alpha \cdot \mathbf{p}\psi_- + V\psi_+ + m\beta\psi_+, \tag{6.5}$$

and

$$i\partial_t \psi_- \quad = \quad \alpha \cdot \mathbf{p}\psi_+ - m\beta\psi_-. \tag{6.6}$$

If one considers energy eigenfunctions $\psi(t) = e^{-iEt}\psi$ then the last equations become

$$(E-m)\psi_+ \quad = \quad \alpha \cdot \mathbf{p}\psi_- + V\psi_+ \tag{6.7}$$

and

$$(E+m)\psi_- \quad = \quad \alpha \cdot \mathbf{p}\psi_+. \tag{6.8}$$

Therefore

$$(E-m)\psi_+ = \left[\frac{p^2}{E+m} + V\right] \psi_+, \tag{6.9}$$

which is a Schrödinger-like equation with

$$\varepsilon \quad = \quad E - m \tag{6.10}$$

and

$$2m^* \quad = \quad E + m. \tag{6.11}$$

This is a translated effective "non-relativistic" energy and an effective mass. Notice that in this case E can be greater or smaller than zero.

The Spanish Dirac oscillator is the Spanish plus supersymmetric. The substitution here is

$$\mathbf{p} \rightarrow \pi \quad = \quad \mathbf{p} + \mathbf{A}(\mathbf{r}) + i\beta \mathbf{E}(\mathbf{r}) \tag{6.12}$$

and

$$\pi^{+} \quad = \quad \mathbf{p} + \mathbf{A}(\mathbf{r}) - i\beta \mathbf{E}(\mathbf{r}) \tag{6.13}$$

with the next equation

$$\varepsilon\psi_{+} = \frac{\alpha \cdot \pi^{+} \alpha \cdot \pi}{2m^{*}} \psi_{+} + V\psi_{+}. \tag{6.14}$$

Notice that in general the Pauli-like Hamiltonian in non-hermitean.

Next we have two examples of the energy spectra of the projected potentials. First we will consider the Spanish harmonic oscillator

$$\varepsilon\psi = \left(\frac{p^2}{2m^*} + \frac{kr^2}{2} \right) \psi \tag{6.15}$$

with the harmonic oscillator frequency $\omega = \sqrt{k/m^*}$ and the eigenvalues spectrum

$$\varepsilon \quad = \quad \omega(N + \tfrac{3}{2}). \tag{6.16}$$

Then the quadratic spectrum is

$$\varepsilon^2 \quad = \quad \omega^2(N + \tfrac{3}{2})^2 \tag{6.17}$$

and then

$$(E - m)^2 \quad = \quad \frac{k}{m^*}(N + \tfrac{3}{2})^2, \tag{6.18}$$

and with the substitution for the effective mass

$$(E + m)(E - m)^2 \quad = \quad 2(N + \tfrac{3}{2})^2 k, \tag{6.19}$$

with $N = 2n + l$.

Finally we consider the "Spanish" Coulomb potential, for the bound states region. So, the Schrödinger equation is in this case

$$\varepsilon\psi \quad = \quad \left(\frac{p^2}{2m^*} - \frac{\alpha}{r} \right) \psi, \tag{6.20}$$

and the eigenvalues are

$$\varepsilon \quad = \quad -m^* \alpha^2 \frac{1}{2N^2}. \tag{6.21}$$

With the substitution of the effective mass and the eigenvalues,

$$(E - m) = -\frac{E + m}{2}\frac{\alpha^2}{2N^2}, \tag{6.22}$$

solving for the energy, we have

$$E = m\frac{1 - \alpha^2/4N^2}{1 + \alpha^2/4N^2}. \tag{6.23}$$

12.7 Conclusions

We have seen that a generalization of the standard supersymmetric quantum mechanics is needed to include the Dirac equation in a Susy framework. This framework is quite general and can in principle be applied to a large class of Hamiltonians, which include those whose square are the usual SQM. In this work we have used this framework to discuss the Dirac equation in one time and 2, 3 and 4 spatial dimensions. In the usual $3+1$ case the connection of the general framework with two important cases was remade. In particular, the supersymmetric nature of a Dirac particle in a magnetic field and the Dirac oscillator were established.

A very interesting extrapolation of the framework to non-Susy potentials was pointed out. For these potentials an equivalent exact non-relativistic problem can be defined.

12.8 References

[1] M. Moreno, R. Martínez, and A. Zentella, *Mod. Phys. Lett.* **A 5,** 949 (1990).

[2] M. Moshinsky and A. Szczepaniak, *J. Phys.* **A 22,** L817 (1989).

[3] M. Moreno and A. Zentella, *J. Phys.* **A 22** L821 (1989).

[4] D. Itô, K. Mori, and E. Carriere, *N. Cimento* **51 A,** 1119 (1967).

[5] P.A. Cook, *Lett. Nuovo Cimento* **1,** 419 (1971).

[6] W. Pauli, *Rev. Mod. Phys.* **13**, 203 (1941).

[7] E. Eriksen, *Phys. Rev.* **111**, 1011 (1958).

[8] R.J. Hughes, V. Alan Kostelecky, and M.M. Nieto, *Phys. Rev.* **D 34**, 1100 (1986).

[9] R.J. Hughes, V. Alan Kostelecky, and M.M. Nieto, *Phys. Lett.* **B 171,** 226 (1986).

[10] R.B. Laughlin, *Ann. Phys.* **191,** 163 (1989); *ib.*, *Phys. Rev.* **D 23**, 5632 (1981).

[11] M. de Crombrugghe and V. Rittenberg, *Ann. Phys.* **151,** 99 (1983).

[12] M. Moreno, R. Martínez, and A. Zentella, *Phys. Rev.* **D 43**, 2036 (1991).

[13] M. Centelles, X. Viñas, M. Barranco, and P. Schuck, *Nucl. Phys.* **A 519**, 73 (1990).

[14] V.I. Kukulin, G. Loyola, and M. Moshinsky, *Phys. Lett.* submitted for publication.

[15] J. Bjorken and S. Drell, *Relativistic quantum mechanics,* McGraw-Hill, New York, 1964.

[16] L. Foldy and S. Wouthuysen, *Phys. Rev.* **78,** 29 (1950).

[17] M. Cini and T. Touschek, *Nuovo Cimento* **7,** 422 (1950).

[18] C. Quesne and M. Moshinsky, *J.Phys.* **A 23,** 2263 (1990).

[19] C. Quesne, *Int.J.Mod.Phys.* **A 6,** 1567 (1991).

[20] O. Castaños, A. Frank, R. López, and L.F. Urrutia, *Phys. Rev.* **D 43**, 544 (1991).

13

A Parallelism Between Quantum Gravity and the IR Limit in QCD (Emergence of Hadron and Nuclear Symmetries)

Yuval Ne'eman[1]

ABSTRACT In the non-perturbative IR region, the action of QCD contains a long-range component, acting as an effective "strong gravity". It is generated by the exchange of a color neutral pair of gluons $G_{\mu\nu}(x) \sim \eta_{ab}B^a_\mu(x)B^b_\nu(x)$. The $G_{\mu\nu}$ acts formally as a Riemannian metric with $J^P = O^+, 2^+$ quanta coupled symmetrically to nuclear matter and generating the IBM paradigm, Regge trajectories and the string-like features of hadrons.

13.1 Symmetries in Nuclei: the IBM Quadrupolar Algebraics

Marcos Moshinsky has made important contributions to Physics. Specifically, his work has dealt with issues in Nuclear Physics, Statistical Physics, etc. In particular, he has greatly advanced the understanding of symmetries in nuclei. As a mathematical physicist, however, the impact of his work trascends by far the limits of the areas in which he has actually worked. It is a pleasure to dedicate to him this paper upon his 70th birthday.

The Arima-Iachello Interacting Boson Model [1] represents one of the main advances in recent years in our understanding of the spectrum of nuclear excitations in nuclei. The IBM ansatz is based on the observation that in most even-even nuclei, the lowest excitations are 0^+ and 2^+, with close energies, realized in proton or neutron pairs. A $U(6)$ symmetry is then phenomenologically postulated, with the six states in $(0^+, 2^+)$ as the *basic symmetric objects* (the quarks in particle $SU(6)$. Note that the $(0^+, 2^+)$ sextet of excitations already appears in the presence of a single nucleon pair sitting above a closed shell, so that it seems plausible to look for the

[1]Tel-Aviv University, University, and the University of Texas, Austin, USA.

originating mechanism at the level of the fundamental forces involved (*i.e.* Quantum Chromo-dynamics, "QCD"), rather than in the collective motion.

The conventional view of QCD as acting between hadrons involves meson exchange, *i.e.* the Yukawa exchange of massive quark/anti-quark pairs, with masses ranging between 0.114 to 1.3 Gev, thus generating short range forces. *Such forces do not fit an excitation spectrum that skips the* 1^- *dipoles altogether*; the study of potentials indicates that gravity-like forces do just that [2] —a somewhat surprising feature for the conventional Yukawa view with respect to nuclear forces.

Interestingly, another set of symmetries of nuclear states does involve *quadrupolar excitations* [3] directly. These symmetries involve a compact [3] $SU(3)$ (as for the harmonic oscillator) or a non- compact [4],[5] $\widehat{SL}(3,R)$ or, alternatively, the inhomogeneous [6] $T(5)(\sigma)\widehat{SO}(3)$. The distinction arises in the commutation relations between the quadrupole moments. It is possible to construct a unifying picture, explaining the selection of the dynamical groups according to various nuclear structure criteria. The entire system is contained [7] in $Sp(3,R)$.

The observation of features characteristic of gravity, dominating a strong interaction, represents yet another indication that this (IR) sector *involves a gravity-like component intrinsic to strong interactions.*

13.2 Gravity-like features in hadron dynamics

Before the rise of QCD an *ad hoc* "strong gravity" hypothesis [8] was tried, in which the f^0 meson (with $J = 2^+$ and a mass of 1270 MeV) was given a central role as the "strong graviton". It was assumed that at the fundamental level there are therefore two massless 2^+ gauge graviton fields ("f/g gravity") and that their mixing results in one massless (Einstein's) and one massive (f^0) state. However, in view of the f^0's less "special" standing within the context of the quark model or of QCD —as just one more quark-antiquark structure— its postulated gauge- field nature in f/g gravity can now at most be regarded as "phenomenological" and deriving from the accidental value of its spin, in a dispersion relations approach [9].

Several other features of hadrons and strong interactions bear a resemblance to gravity. In particular, it was this resemblance which made it possible for Yoneya and for Scherk and Schwarz [10] to reinterpret String Theory, born as a rather successful dynamical model for hadrons and strong interactions (Dual Models [11]) with a highly suggestive excitations spectrum —into a theory of gravity.

One such seemingly gravity-induced feature is the system of Regge trajectories. These exhibit $\Delta j = 2$ intervals in angular momentum, a fact which led Dothan *et al.*, [4] to identify the trajectories with unitary irreducible representations of $SL(3,R)$ and of its double-covering group $\widetilde{SL}(3,R)$ [12].

Physically, it was assumed [4] that the $sl(3, R)$ algebra is generated by the time-derivates of gravitational quadrupoles (*i.e.* shape-pulsation rates), thus involving moments of inertia. The emergence of such inertial features of gravity (related to the gravitational interaction through the Principle of Equivalence) in an essentially strong "nuclear" interaction thus provides yet another piece of this puzzle.

13.3 Flavor $SU(3)$ is generated by QCD, once the fifth is set aside

We interrupt our discussion of the unexplained algebraics of hadrons and nuclei in order to look at the other main algebraic feature of hadrons, namely the Eightfold Way $SU(3)$, and see how it has been explained in terms of QCD. Franco Iachello has often compared the IBM to the 1961 postulation of $SU(3)$ in particle physics [13] (a "flavor" global symmetry in the present dynamical picture). $SU(3)$ was indeed induced from purely phenomenological considerations; first and foremost from the identification of the observed pattern in the hadron spectrum, then from hundreds of detailed predictions [14].

Returning to these days in 1961-73, we should note that the symmetry breaking appeared to introduce paradoxical notions. As a matter of fact, the postulation of $SU(3)$ as an *approximate* symmetry of the strong interactions met with great resistance originally [15]. How could a broken symmetry propagate in a strong interaction and yet preserve its perturbative features? Unitarity allows us to cut a scattering process any number of times; each time we introduce a complete set of intermediate states, we have a $\sim \lambda^8$ spurion too, and if it were strongly coupled we should have gotten dominating contributions of that nature. And yet the symmetry-breaking did lend itself smoothly to a perturbative treatment, as demonstrated by the success [14] of the Gell-Mann-Okubo, or of the Coleman-Glashow mass formulae etc.

The answer to that riddle was first suggested in my "Fifth Interaction" paper [16]. I assumed that what had been taken to represent the strong interactions were really two different effects. The non-pertubative part (responsible for Regge trajectories etc.) is really flavor-$SU(3)$ invariant, while the $\sim \lambda^8$ symmetry-breaking represents a totally different force which I named the "Fifth". Its nature is perturbative, thus settling the paradox. Note that the name "Fifth Interaction" was later plagiarized and used to describe the supposed departure from the Principle of Equivalence in the Eötvös experiment).

This solution is presently realized in QCD. The true "strong" interaction, with its non-perturbative features, is now known to be generated by Quantum Chromodynamics (QCD), a force induced by "color" $SU(3)$. It is flavor-invariant, since $[SU(3)_{\text{color}}, SU(3)_{\text{flavor}}] = 0$.

The $SU(3)_{\text{flavor}}$-breaking force is presently inserted through the various quark masses and relegated to the "Higgs sector" —a simplified model of one new force for each mass. The true constitution of the Fifth is now part of the "Generations Puzzle", a mystery first glimpsed at with the discovery of the muon (remember I.I. Rabi's "who ordered that?"). It is the Fifth that distinguishes the s quark from the u and the μ and its neutrino from the electron set. I did in fact make the conjecture that the Fifth coupled to the μ as it does to the s in that 1965 paper. An understanding of the "fundamental" content of that Fifth will probably have to wait for the results of experiments in the TeV region [17].

In the physics of particle and fields, the emergence of flavor-$SU(3)$ symmetry in thus understood presently as resulting from the fact that color- and flavor-$SU(3)$ commute. With QCD coupling to color-$SU(3)$, this component of the strong interactions has to be flavor-$SU(3)$ invariant at the same time.

In various studies [18] we have similarly tried to present dynamical models which could explain the other hadronic main algebraic feature, namely the Regge sequences. In what follows, we present a comprehensive QCD-generated derivation of all above gravity-like algebraic features in hadrons [20]: Regge and Toller $SO(4)$ sequences and quadrupolar currents, at the same time as the string-like dynamics, confinement, and scaling. And yet, all results of asymptotic freedom (the UV limit in hadrons) remain unhindered, *i.e.* the fit to the Non Relativistic Quark Model (NRQM) stands. At the same time, our theory provides a QCD-generated derivation of both the Arima-Iachello IBM and the $Sp(3,R)$ quadrupole-related symmetries in nuclei.

13.4 "Effective" strong gravity is induced by QCD

Since the hadron lowest ground states are colorless (QCD) and in the approximation of an external QCD potential (in analogy to the treatment of the hydrogen atom in the Schrödinger equation), the hadron spectrum above these levels will be generated by color-singlet quanta, whether made of dressed 2-gluon configurations, 3-gluons,... Every possible configuration will appear. No matter what the mechanism responsible for a given flavor state, the next vibrational, rotational or pulsed excitation will correspond to the "addition" of one such collective color-singlet multigluon quantum superposition.

In the fully relativistic QCD theory, these contributions have to come from summations of appropriate Feynman diagrams, in which dressed n-gluon configurations are exchanged. We rearrange the sum by lumping together contributions from n-gluon irreducible parts, $n = 2, 3, \ldots \infty$ and with the same Lorentz quantum numbers. The simplest such system will have the quantum numbers of gluonium, *i.e.* $n = 2$. The color singlet external field can thus be constructed from the QCD gluon field as a sum (η_{ab} is the $SU(3)$ Killing-Cartan metric, d_{abc} is the totally symmetric $8 \times 8 \times 8 \to 1$ coefficient)

$$B^a{}_\mu B^b{}_\nu \eta_{ab} + B^a{}_\mu B^b{}_\nu B^c{}_\sigma d_{abc} + \cdots . \tag{4.1}$$

In the above, $B^a{}_\mu$ is the dressed gluon field. It will be useful for the applications to separate the "flat connection" $N^a{}_\mu$, *i.e.* the zero-mode of the field. Writing for the curvature or field strength, $F^a{}_{\mu\nu} = \partial_\mu B^a{}_\nu - \partial_\nu B^a{}_\mu - if^a{}_{bc} B^c{}_\nu$, we define

$$B^a{}_\mu = N^a{}_\mu + A^a{}_\mu, \tag{4.2}$$

so that $F(N) = 0$, *i.e.*, $\partial_\mu N^a{}_\nu - \partial_\nu N^a{}_\mu = if^a{}_{bc} N^b{}_\mu N^c{}_\nu$, or, in formal language,

$$dN = N \wedge N. \tag{4.3}$$

Note that (4.3) implies that $N^a{}_\mu$ is the Cartan left-invariant of $SU(3)$ in a Soft-Group-Manifold [20] version of the 8+4 dimensional $SU(3) \otimes M$, M-Minkowski space, after spontaneous fibration [21] and contraction of the holonomic $SU(3)$ indices. Equation (4.3) is therefore also the BRS equation for the ghosts C, when we replace $d \to s$ and $N \to C$.

We can now rewrite the 2–gluon configuration as

$$G_{\mu\nu} = B^a{}_\mu B^b{}_\mu \eta_{ab}, \tag{4.4}$$

which looks very much like a spacetime metric. It is an *effective* spacetime metric representing some of the geometric features induced in the spacetime base-manifold of the color-$SU(3)$ Principal Bundle. We assume that (4.4) is the dominating configuration in the excitation systematics and note that Lorentz invariance forces this metric to obey a Riemanian constraint

$$D_\sigma G_{\mu\nu} = 0, \tag{4.5}$$

where D_σ is the covariant derivative of the "effective gravity" (the connection will be given by a Christoffel symbol constructed with the metric (4.4)). The separation of the flat part of $B^a{}_\mu$ in Eq. (4.2) reproduces the separation of a tetrad $e^a{}_\mu(x) = \delta^a{}_\mu + f^a{}_\mu(x)$ into the flat background piece and the quantum gravitational contribution. As a result, $G_{\mu\nu}(x)$ itself can be separated similarly.

We note two points:

1. Out of the 10 components of $G_{\mu\nu}$ in (4.4), the 6 that survive after the 4 Riemannian constraints (4.5) have spin/parity assignments $J^P = 0^+, 2^+$. This suggests a relationship with the IBM model systematics, in which the fundamental excitation was selected with these quantum numbers, to fit the phenomenology. We noted in Section 1 that the absence of dipolar excitations is in itself an indication that a gravity-like force is involved.

2. An effective Riemannian metric induces Einsteinian dynamics. However, *our correspondence is between low-energy (IR) QCD, with its strong coupling, and the high-energy (UV) strong coupling region of our effective gravity* (and not with the weak coupling Newtonian limit). *Gravity becomes a strong force in the quantum regime* (two Planck mass particles attract each other gravitationally $10^{19} \times 10^{19} \sim 10^{38}$ times stronger than two nucleons).

The quantum gravity Lagrangian includes curvature-quadratic counterterms generated by the renormalization procedure and corresponds to the (effective) invariant action,

$$I_{\text{inv}} = -\int d^4x\sqrt{-G}(\alpha R_{\mu\nu}R^{\mu\nu} + \beta R^2 - \gamma\kappa^{-2}R) \tag{4.6}$$

This is the Stelle [22] action, when used for true gravity. It was shown by Stelle to be renormalizable, a feature befitting our present application, since QCD is renormalizable and any piece of it should preserve the finiteness feature; but (4.6) is not unitary, which makes it unutilizable for true gravity —but which befits the present application: a "piece" of QCD should not be unitary, QCD being an irreducible theory.

Stelle's main result, however, was to show that renormalizability is caused by $1/p^4$ propagators. *But $1/p^4$ propagators are dynamical equivalent to confinement!* [23].

Such propagators were generally introduced in QCD in an *ad hoc* manner. Here, they stem from our basic premise. Note that although we only assumed that the lowest states are color-singlets, the $1/p^4$ propagators will cause any colored state to be bound and confined; adding a quark or gluon to a color-singlet hadron will polarize the vacuum, creating pairs until the configuration becomes color-neutral.

In recent years, quadratic Lagrangians like (4.6) with, in addition, torsion- squared terms, have been investigated classically in the context of the Poincaré Gauge Theory of Gravity [24]. The exact solutions display, aside from the Newtonian potential M/r, a component behaving $\sim r^2$, dominating the strong-field limit and originating in the curvature-squared terms as in (4.6).

There is one more feature that relates to Eq. (4.6). It has been shown [25] that the String, as a gravitational theory, is equivalent to an action such as (4.6), *i.e.* with quadratic counterterms. From a different viewpoint, the embedding of the String in curved "target" spacetimes has been interpreted [26] as a series of constraints on the manifold's geometry due to the necessity of preserving the cancellation of the conformal anomaly. Such constraints are regarded in String Theory as replacing Einstein's equation in fixing the geometry of the target space —and their lowest terms are also those of (4.6). As a result, our *ansatz* explains the good fit of String Theory in its original hadron version, in reproducing the IR region features: color confinement, string flux-tubes etc.

One more comment relates to "f/g gravity" [8] and the f^0. In that hypothesis, it was necessary to assign the mass of the f^0 to some unknown Higgs-type effect, in order to forego writing a geometric Einsteinian equation for a massive graviton, a doubtful procedure. In the present "effective" picture, *the* $G_{\mu\nu}$ *"effective" QCD-induced metric field is massless* in the *ansatz* approximation, because of the gauge invariance generated by te Covariance Group (the diffeomorphisms) of the "effective" Einsteinian theory.

At this point, it is important to note that the exchange of two gluons according to Eq. (4.1) *does not correspond to the Yukawa-type exchange of "gluonioum"*, a possible bound state of two gluons whose mass is estimated (from lattice calculations) to be of the order of 1.5 GeV. The correspondence between a pole in the s-channel and an exchange in the t-channel is based on the Klein-Gordon nature of the usual $1/p^2$ propagator. But this does not hold for a $1/p^4$ propagator! At most, such a propagator would describe the difference between two such exchanges!

Note that our "effective" gravity, representing the coherent action of two gluons resembles very much the model in which gravity itself is not a fundamental force [27], as suggested by Sakharov and developed by Adler and by Zee.

13.5 The algebraics of hadrons and nuclei (classical and quantum)

The external field $G_{\mu\nu}(x)$ transforms under Lorentz tranformation as a (reducible) second rank symmetric tensor field, with Abelian components: $[G_{\mu\nu}, G_{\rho\sigma}] = 0$. Algebraically, $G_{\mu\nu}$ and the Lorentz generators form the algebra of $T_{10}(\sigma)SO(1,3)$, an inhomogeneous Lorentz group with tensor "translations". This is a classical relativistic algebra.

For the quantum case, when the gluon field is expanded in creation and annihilation operators, we can write,

$$G_{\mu\nu} = T_{\mu\nu} + U_{\mu\nu},$$

where the quadrupolar excitation-rate is given by

$$T_{\mu\nu} = \eta_{ab} \int d\tilde{k} \left[\alpha^{a+}{}_{\mu}(k)\alpha^{b+}{}_{\nu}(k)e^{2ikx} + \alpha^{a}{}_{\mu}(k)\alpha^{b}{}_{\nu}(k)e^{-2ikx}\right] \qquad (5.1)$$

for (infinite) $gl(3,R)$ non-compact excitation bands, whereas

$$U_{\mu\nu} = \eta_{ab} \int d\tilde{k} \left[\alpha^{a+}{}_{\mu^+}(k)\alpha^{b}{}_{\nu}(k) + \alpha^{a}{}_{\mu}(k)\alpha^{b+}{}_{\nu}(k)\right]. \qquad (5.2)$$

generates finite $u(3)$ spectral multiplets. We have use of the canonical transformation:

$$[\alpha^a{}_\mu(k) + \tfrac{1}{2}N^a{}_\mu e^{ikx}] \quad \rightarrow \quad \alpha^a{}_\mu(k), \tag{5.3a}$$

$$[\alpha^{a+}{}_\mu(k) + \tfrac{1}{2}N^a{}_\mu e^{-ikx}] \quad \rightarrow \quad \alpha^a_{\mu^+}(k). \tag{5.3b}$$

Using $[\alpha^a{}_\mu(k), \alpha^{b+}{}_\nu(k')] = \partial^{ab}\partial_{\mu\nu}\partial(k-k')$, one verifies that the operators $T_{\mu\nu}$ and $U_{\mu\nu}$, together with the operators

$$S_{\mu\nu} = \eta_{ab}\int d\tilde{k}\,[\alpha^{a+}{}_{\mu^+}(k)\alpha^b{}_\nu(k) - \alpha^a{}_\mu(k)\alpha^{b+}{}_\nu(k)] \tag{5.4}$$

close respectively on the $gl(4;R)$ and $u(1,3)$ algebras. Note that the largest (linearly realized) algebra with generators quadratic in the α^+_μ, α_μ operators is $sp(1,3;R)$, where the notation "1,3" implies a definition over Minkowski space,

$$sp(1,3;R) \rightarrow \left| \begin{array}{llllll} \rightarrow & u(1,3) & \rightarrow & u(1,3) & \rightarrow \\ \rightarrow & gl(4;R) & \rightarrow & sl(4;R) & \rightarrow \\ \rightarrow & t_{10}(\sigma)so(1,3) & \rightarrow & t_g(\sigma)so(1,3) & \rightarrow \end{array} \right| \rightarrow so(1,3). \tag{5.5}$$

The $gl(4;R)$ algebra represents a Spectrum-Generating Algebra for the set of states of a given flavor [18].

The now return to the expansion in Eq. (4.1). The $sl(4;R)$ is generated by the 2–gluon configuration. What about 3-gluon and n-nucleon exchanges? The corresponding algebras do not close and generate the full Ogievetsky algebra [28] of the diffeomorphisms in Minkowski space, the 4-dimensional analog of the Virasoro algebra (the algebra of diffeomorphism on the circle).

Had we considered the entire (infinite) sequence when writing Eq. (4.1), we would have generated this diff$(4;R)$. The maximal linear subalgebra of diff$(4;R)$ is $gl(4;R)$. The remaining generators [*i.e.* the quotient diff$(4;R)/gl(4,R)$] can be explicity realized in terms of the $gl(4,R)$ generators [29], both for tensors and for spinors. In our case, this would involve functions as matrix elements of the representation of our generators $T_{\mu\nu}$ and $S_{\mu\nu}$ in Eq. (5.1) and (5.3a). As in General Relativity, the entire "$G_{\mu\nu}$-covariance" can be realized in terms of the invariant action given in Eq. (4.6).

Now diff$(3,1;R)$ can also be represented linearly. It will then involve infinite, ever more massive, repetitions of the representations of $sl(4;R)$. In either way, we find that using $sl(4;R)$ takes care of the entire sequence in Eq. (4.1).

The inhomogeneous versions of the algebras in Eq. (5.5), *i.e.* their semi-direct product with the translations t_4, are relevant to the Hilbert space spectrum of states. In the case of $u(1,3)$ in Eq. (5.5), when selecting a time-like vector (for massive states), the stability subgroup is the compact $u(3)$ with finite representations —as against non-compact $gl(3;R)$ for $sl(4;R)$. This fits with the situation in nuclei, as well shall see in the coming sections.

13.6 Hadron systematics

Dynamically, we have discussed the role of 2-gluon excitations in generating the transition to a hadronic excited level, from any given hadron state. To this we can now add scaling symmetry: the general linear group $GL(4,R)$ decomposes into its unimodular $SL(4,R)$ subgroup, the R^+ of scale transformations and Parity Π : $GL(4,r) = [\Pi(\sigma)SL(4,R)] \otimes R^+$. This scaling symmetry corresponds to the observations in deep-inelastic photon-nucleon scattering experiments. Color confinement too manifests itself algebraically at several levels. The $sl(4;R)$ subalgebra preserves the 4-dimensional measure, a geometric realization of confinement as a dynamical 4-volume-preserving rotation-deformation-vibration pulsation mechanism.

When dealing with the hadron Hilbert space states, momenta come in and the translations thus have to be algebraically adjoined. Thus $sa(4;R) = t_4(\sigma)sl(4;R)$. The massive states of the hadron spectrum are then classified according to the stability subalgebra, here

$$sa'(3;R) = t'_3(\sigma)sl(3;R).$$

The t'_3 quantum numbers are trivialized, as is done with the formal translations t'_2 of the Euclidean 2-dimensional stability subalgebra for the massless states in the Poincaré group representations.

With J. Pecina-Cruz and J. Lemke in Austin, I have recently investigated the Casimir operator of $sa(n,R)$ —in particular $sa(3,R)$ and $sa(4,R)$ [30]. The invariant of $sa(2,R)$ was calculated in the seventies. Note that $gl(n,R)$ has no invariant [31]. The $sl(n,R)$ have one invariant for each n. It is of order $n(n+1)/2$ and has the general form

$$C(n) = F^{\mu\nu\ldots\sigma}{}_{\alpha}{}^{\beta}{}_{\gamma}{}^{\delta\ldots}{}_{\zeta}{}^{\xi} P_\mu P_\nu \cdots P_\sigma L^\alpha{}_\beta L^\gamma{}_\delta \cdots\cdots L^\zeta{}_\xi. \qquad (6.1)$$

It is a homogeneous polynominal of degree n in the translations P and of degree $n(n-1)/1$ in the $sl(n,R)$ generators L (including both the T^μ_ν and S^u_ν above). The decomposition displays the $sa'(n-1,R)$ subgroups with their own Casimir invariant. The study of these invariants will assist in the classification of hadron states and in the understanding of the mass spectrum.

Meanwhile, hadron states are characterized by the $sl(3;R)$ subalgebra, whose infinite representations correspond to Regge trajectories [4]. These preserve the measure in 3-space: a Regge trajectory described by such a representation corresponds to a given "bag".

Spinors span the infinite representations of the double-covering groups $\widetilde{SA}(4;R) = T_4(\sigma)\widetilde{SL}(4;R)$.

We collected all hadronic field configurations obtained by successive application of the quantum 2-gluon field's $sl(4;R)$ dynamical algebra, into an infinite-component field (manifield). These manifields are subject to the following constraints:

1. Owing to Eq. (4.5), the wave-equations have to be Lorentz covariant.
2. The lowest manifield components should fit the basic quark system field configuration in its Lorentz content.
3. They should transform according to non-unitary representations of the $\widetilde{SO}(1,3) \subset \widetilde{SL}(4;R)$ in order to meet the experimental fact that a boosted particle keeps its spin quantum number

These natural requirements of our 2-gluon picture determine uniquely the selection of manifields and their equations [19] while the "non-unitary" condition on $\widetilde{SO}(1,3)$ representations is achieved by making use of "A-deunitarized" $\widetilde{SL}(4;R)$ unitary irreducible representations [32].

For mesons, take a manifield $\Phi, \Box + M^2)\Phi = 0$, transforming according to the ladder representation $D(1/2,1/2)^A$. For the 3-quark octet configuration we use the manifield Ψ, $(i\chi^\mu \partial_\mu - M)\Psi = 0$ transforming according to the spinorial multiplicity-free $SL(4;R)$ representation $[D(1/2,0) \oplus D(0,1/2)]^A$. For the decuplet 3-quark configuration we take a manifield Ψ_ρ fulfilling $(i\chi^\mu \partial_\mu - M)\Psi_\rho = 0$. It transforms according to the spinorial multiplicity-free representation $[D(1/2,0) \oplus D(0,1/2)]^A \otimes (1/2,1/2)$.

We find a good fit with the experimental data. Note that this classification was originally suggested and realized phenomenologically [18] and formulated using field theory and QCD as an *ad hoc ansatz.* The present work purports to supply solid dynamical foundations deriving from what appears to be an extremely versatile approximation for QCD in the IR region [19].

13.7 The interacting boson model in nuclei

In nuclei, we look again at the *ansatz* defining the IBM model $U(6)$ symmetry described in section 1: a fundamental 0^+ and 2^+ quasi-degenerate set of excitations. Such a set would never arise from the exchange of the lighter quark-antiquark pairs (*i.e.* mesons with spins 0 and 1). This would not generate quadrupole excitations. Skipping the 1^-dipole is generally intimately connected with the (tensor) gravitational potential [2]. The 2^+ mesons such as the $f^0(1270)$ are represented by tensor fields and could thus do it in the "old" strong gravity model, had their range not been too short. Our QCD-induced effective gravity as described by Eq. (1–6) appears to supply the correct answer, as we shall now see [33].

Out of the 10 components of $G_{\mu\nu}$ the 6 that survive the 4 Riemannian constraints (4.5) have spin/parity assignments $J^P = 0^+, 2^+$. It is mainly this feature, suggesting a relationship with the IBM systematics, that we apply here.

Algebraically, $G_{\mu\nu}(x)$ carries the 10-dimensional (non-unitary) irreducible representation of $GL(4,R)$. In true gravity, this is a geometric group, the linear subgroup of the Covariance Group Diff$(4,R)$. Here it is a dynamical construct, except for the geometric Lorentz subroup. The nonrelativistic subgroup of $SL(4,R)$ (the traceless piece, including shears, aside from Lorentz transformation) is $SL(3,R)$. Under this group, the 0^+ and 2^+ states span together one irreducible 6-dimensional representation. The couplings to the "effective" gravity are given by the $SL(4,R)$ group; they will thus be $SL(3,R)$ invariant.

There is thus full justification, in this picture, for the IBM postulate of a $U(6)$ symmetry between the defining states! Note also that this $SL(3,R)$, obtained from the basic QCD fields, takes on a geometric interpretation, once we use $G_{\mu\nu}$ as a formal metric field. In that picture, $SL(3,R)$ predicts the conservation of 3-volume, *i.e. incompressibility*, for nuclei and hadrons (where it justifies the "bag" model as an approximation of hadron dynamics).

When applying "effective gravity" to nuclei, it is natural to assume that closed shells assume the role of "vacua", as rigid structures. "Graviton" excitations should then be searched for in the valence nucleon systematics. In this sector of even-even nuclei the $G_{\mu\nu}$ quanta can indeed excite nucleon pairs; the overwhelming preponderance of proton-proton and neutron-neutron over proton-neutron pairs can be fully explained in terms of the Clebsch-Gordan coefficients in the direct channel. Dynamically for one pair, we assume that the pairing force itself is due to the exchange of a "strong graviton" between the two nucleons. The paired system then displays further excited states with the absorption of additional such quanta. The picture now is of an external field supplying these quanta, perhaps like the role of the electromagnetic field in the hydrogen atom in the Schrödinger equation treatment. It is thus natural that proton pairs and neutron pairs should have the same energy difference between 0^+ and 2^+, since these are due to the same flavor-independent component of QCD —precisely for the same reason that the Eightfold Way (flavor) $SU(3)$ invariance is due to the flavor- independence of QCD.

In estimating [19] the amplitude for such pair excitations, we note that the 2-gluon exchange here is dominated by a pole in the direct channel at $k^2 = 4m^2$, m the nucleon mass. Remembering that "effective strong gravitation" excitations are also seen in hadrons, where they cause $\Delta J = 2$ sequences of resonances along Regge trajectories, we get a value for the effective $1/\kappa^2$ in Eq. (4.6). It is given by the slope of the Regge trajectories, roughly $1/\kappa^2 \sim 1\ (\mathrm{GeV})^2$.

We now return to the Hamiltonian corresponding to Eq. (4.6), as translated into our effective dynamics. The curvature R corresponds in true gravity to terms $\partial\Gamma + \Gamma \cdot \Gamma$; the Christoffel formula gives

$$\Gamma \sim G^{-1}\partial G \text{ and } R \sim \partial(G^{-1}\partial G) + (G^{-1}\partial G)(G^{-1}\partial G),$$

where G stands for $G_{\mu\nu}$. When $\underline{b}$ and $\underline{b}^+$ represent the destruction and creation of a 6-dimensional "strong graviton" quantum,

$$\begin{aligned} H &= \frac{1}{M^3}\int dk\,[C_1(k^2/\kappa^2)(\underline{b}^+\cdot\underline{b}) + C_2(k^2/\kappa^2)(\underline{b}^+\cdot\underline{b})(\underline{b}^+\cdot\underline{b}) \\ &\quad + A_1k^4(\underline{b}^+\cdot\underline{b})(\underline{b}^+\cdot\underline{b}) + A_2k^4(\underline{b}^+\cdot\underline{b})(\underline{b}^+\cdot\underline{b})(\underline{b}^+\cdot\underline{b}) \\ &\quad + A_3k^4(\underline{b}^+\cdot\underline{b})(\underline{b}^+\cdot\underline{b})(\underline{b}^+\cdot\underline{b})(\underline{b}^+\cdot\underline{b})]. \end{aligned} \tag{7.1}$$

The coefficients C_i and A_i respectively contain dynamical information relating to the γ and α, β terms in (4.6), an approximation for the non-linear effect of the $\sqrt{-G}$, the reduced matrix element for the coupling to the nucleon pair, etc. M is a mass parameter that takes care of the dimensionality.

We select M to be of the order of the impacted system, *i.e.* the valence nucleons, $M \sim 20$ GeV. For k^2 we use the dispersion relation result mentioned above, *i.e.* $\langle k^2\rangle \sim 4\ (\mathrm{GeV})^2$. Using our string-Regge result $(1/\kappa^2) \sim 1\ (\mathrm{GeV})^2$, and assuming the C coefficients to be of the order of unity, we get for the first terms $\langle \Delta E\rangle \sim 0.5$ MeV. This is of course by far no more than an order of magnitude check, but it seems the values are roughly in the right ballpark. The values will decrease for larger M. Our Hamiltonian (7.1) is of course equivalent to the IBM Hamiltonian with higher order terms.

13.8 Quadrupolar symmetries in nuclei

In nuclei the relevant symmetries in Eq. (5.5) correspond to the non-relativistic subgroups, *i.e.* to the $Sp(3,R)$ and the related groups $T_5(\sigma)SO(3)$, $SL(3,R)$ and $SU(3)$. Moreover, averaging over the color $SU(3)$ degrees of freedom, which are summed over anyhow, in Eqs. (7–10) and integrating over the $d\hat{k}$, we get the $Sp(3,R)$ generators of Rowe [7], as expressed in terms of shell-model harmonic oscillators.

It is interesting that this picture in which the system of creation and annihilation operators of the Arima-Iachello $U(6)$ and those of the $Sp(3,R)$ of deformed nuclei are constructed as, respectively, quadratic and linear expressions involving the same dynamical elements validates a "guess" expressed by Lipkin some time ago [34].

13.9 References

[1] A. Arima and F. Iachello, *Phys. Rev. Lett.* **35** (1975) 1069; *ib., Ann. Phys.* **99** (1976) 253; **111** (1978) 201; **123** (1979) 468; F. Iachello and I. Talmi, *Rev. Mod. Phys.* **54** 339 (1987).

[2] See for example C.W. Misner, K.S. Thorne, and J.A. Wheeler, *Gravitation* (W.H. Freeman and Co. Publ., San Francisco, 1973) Sect. 36.1.

[3] J.P. Elliott, *Proc. Roy. Soc.* **A245** (1958) 128, 562; J.P. Elliott and M. Harvey, *Proc. Roy. Soc.* **A272** (1963) 557.

[4] Y. Dothan, M. Gell-Mann, and Y. Ne'eman, *Phys. Lett.* **17** (1965) 148.

[5] L. Weaver and L.C. Biedenharn, *Phys. Lett.* **32B** (1970) 326; *Nucl. Phys.* **A185** (1972) 1.

[6] H. Ui, *Prog. Theor, Phys.* **44** (1970) 153; L. Weaver, L.C. Biedenharn, and R.Y. Cusson, *Ann. Phys.* **77** (1973) 250.

[7] S. Goshen and H.J. Lipkin, *Ann. Phys.* **6** (1959) 301; D.J. Rowe, in *Dynamical Groups and Spectrum Generating Algebras*, A. Bohm, Y. Ne'eman, and A.O. Barut editors, World Scientific, Singapore (1989), p. 287; G. Rosensteel and D.J. Rowe, *Phys. Rev. Lett.* **47** (1981) 223; J.P. Draayer and K.J. Weeks, *Phys. Rev. Lett.* **51** (1983) 1422.

[8] C.J. Isham, A. Salam, and J. Strathdee, *Phys. Rev.* **D8** (1973) 2600; *ibid.* **D9** (1974) 1702; *ibid. Lett. N. Cimento* **5** (1972) 969.

[9] M. Gell-Mann, Phys. Rev. **125** (1962) 1067, footnote on p. 38; P.G.O. Freund, *Phys. Lett.* **2** (1962) 136.

[10] T. Yoneya, *Prog. Theor. Phys.* **51** (1974) 1907; J. Scherk and J.H. Schwarz, *Nucl. Phys.* **B81** (1974) 118.

[11] G. Veneziano, *Lett. N. Cimento* **A57** (1968) 190.

[12] D.W. Joseph, University of Nebraska preprint, unpublished (1969); Y. Ne'eman, *Ann. Inst. H. Poincaré* **28** (1978) 639; Y. Ne'eman and Dj. Šijački, *Int. J. Mod. Phys.* **A2** (1987) 1655.

[13] Y. Ne'eman, *Nucl. Phys.* **26**, (1961) 222; M. Gell-Mann, Report CTSL-20, unpublished.

[14] M. Gell-Mann and Y. Ne'eman, *The Eightfold Way* (W.A. Benjamin, NY, 1964).

[15] C.N. Yang and R.J. Oakes, *Phys. Rev. Lett.* **125** (1963) 1067.

[16] Y. Ne'eman, *Phys. Rev.* **134** (1965) B1355.

[17] Y. Ne'eman, *Nucl. Phys. (Proc. Supp. Sec.)* **B13** (1990) 582.

[18] Y. Ne'eman and Dj. Šijački, *Phys. Lett.* **157B**, 267 (1985); *ib. Phys. Rev.* **D37**, 3267 (1988).

[19] Dj. Šijački, and Y. Ne'eman, *Phys. Lett.* **B247** (1990) 571.

[20] Y. Ne'eman and T. Regge, *Riv. N. Cimento Ser. 3,* **1** (1978) 5.

[21] J. Thierry-Mieg and Y. Ne'eman, *Ann. Phys. (NY)* **123** (1979) 247.

[22] K.S. Stelle, *Phys. Rev.* **D16** (1977) 953.

[23] See for example J. Kiskis, *Phys. Rev.* **D11** (1975) 2178; G.B. West, *Phys. Lett.* **115B** (1982) 468.

[24] F.W. Hehl, Y. Ne'eman, J. Nisch, and P. v.d. Heyde, *Phys. Lett.* **B78** (1978) 102; P. Baeckler and F.W. Hehl, in *From SU(3) to Gravity*, E. Gotsman and G. Tauber editors (Cambridge University Press, Cambridge, 1985), p. 341.

[25] S. Deser and A.N. Redlich, *Phys. Lett.* **176B** (1986) 350.

[26] E.S. Fradkin and A.A. Tseytlin, *Phys. Lett.* **B158** (1985) 316.

[27] A.D. Sakharov, *Dokl. Akad, Nauk SSSR* **171** (1967) 70; S.L. Adler, *Rev. Mod. Phys.,* **54** (1982) 729; A. Zee, *Phys. Rev.* **D23** (1981) 858.

[28] V.I. Ogievetsky, *Lett. N. Cimento* **8** (1973) 988.

[29] F.W. Hehl and Y. Ne'eman, "Spacetime as a Continuum with Microstructure and Metric-Affine Gravity", to be published in Ivananko Festschrift, World Scientific Publ.

[30] J. Pecina-Cruz, J. Lemke, and Y. Ne'eman, to be published.

[31] M. Perroud, *J. Math. Phys.* **24** (1983) 1381.

[32] Dj. Šijački, and Y. Ne'eman, *J. Math. Phys.* **16** (1985) 2457.

[33] Dj. Šijački, and Y. Ne'eman, *Phys. Lett.* **B250** (1990) 1.

[34] H.J. Lipkin, *Nucl. Phys.* **A350** (1980) 16.

14

On Rainich-Misner-Wheeler Conditions in Nonlinear Electrodynamics

Jerzy F. Plebański[1] and Maciej Przanowski[2]

ABSTRACT It is my pleasure to participate in this meeting in honor of the anniversary of 70+50 years of Dr. Marcos Moshinsky. From my limited scope, the material I have chosen is this on Rainich-Misner-Wheeler conditions in non-linear electrodynamics, devised by M. Przanowski and myself. It deals with an important complementary remark to the papers by H. Salazar, A. García Díaz and myself on the solutions of the Taub-NUT generalized metrics [1], [2].

We start with a useful Lemma:

Lemma 1. Let the energy-momentum tensor $T_{\mu\nu}$ be of the type $[2S\text{-}2T]_2$ and let $T_{\mu\nu}$ satisfy the Einstein equations with cosmological constant Λ,

$$R_{\mu\nu} - \tfrac{1}{2}Rg_{\mu\nu} = -8\Pi T_{\mu\nu} + \Lambda g_{\mu\nu}. \tag{1}$$

Then, the energy conditions

$$T_{\mu\nu}t^\mu t^\nu > 0, \tag{2a}$$

$$T_{\mu\nu}t^\nu \qquad \text{is non-spacelike} \tag{2b}$$

for every timelike vector t^μ (*i.e.* $t^\mu t_\mu < 0$),

$$C_{\mu\nu}t^\mu t^\nu < 0 \qquad \text{for every timelike } t^\mu, \tag{3a}$$

$$R + 4\Lambda \leq 0. \tag{3b}$$

Conversely, let

$$C_{\mu\nu}n^\mu n^\nu < 0 \qquad \text{for some non-spacelike vector} \tag{4a}$$

$$R + 4\Lambda \leq 0. \tag{4b}$$

then the energy conditions $(2a)$ and $(2b)$ are satisfied.

---

[1] Departamento de Física, CINVESTAV, México.

[2] Institute of Physics, Technical University of Łódź, Poland.

As it is well known, the energy-momentum tensor of a *general nonlinear electromagnetic field* is of the type $[2S\text{-}2T]_2$, so our Lemma holds in this case.

In Ref. [1] it has been shown that Einstein equations in the presence of the general nonlinear electromagnetic field admitting the *duality rotation* read

$$C_{12} = -C_{34} = -(DE + BH), \tag{5}$$
$$R = -4\Lambda - 8M^+ + 4(DE + BH), \tag{6}$$

where

$$M^+ = M^+\left(\tfrac{1}{2}(D^2 + B^2)\right), \quad E = \frac{\partial M^+}{\partial D}, \quad H = \frac{\partial M^+}{\partial B}. \tag{7}$$

Thus we have:

$$C_{12} = -C_{34} = -2x\frac{dM^+}{dx}, \tag{8}$$
$$R = -4\Lambda - 8M^+ + 8x\frac{dM^+}{dx}, \tag{9}$$

where

$$x := \tfrac{1}{2}(D^2 + B^2). \tag{10}$$

From (8), (9), and Lemma 1, one finds that the energy conditions give

$$\frac{M^+}{x} \geq \frac{dM^+}{dx} > 0. \tag{11}$$

We then assume that (compare Ref. [1])

$$\frac{\partial^2 M^+}{\partial D^2} \neq 0. \tag{12}$$

Consequently

$$\frac{dM^+}{dx} + D^2\frac{d^2M^+}{dx^2} \neq 0 \quad \text{and} \quad \frac{dM^+}{dx^2} + B^2\frac{d^2M^+}{dx^2} \neq 0. \tag{13}$$

Therefore, since $0 \leq D^2 \leq 2x, 0 \leq B^2 \leq 2x$ one infers that

$$\frac{dM^+}{dx} + x\frac{d^2M^+}{dx^2} \equiv \frac{d}{dx}\left(x\frac{dM^+}{dx}\right) \neq 0. \tag{14}$$

Now we introduce the invariant

$$I := \sqrt{C^{\mu\nu}C_{\mu\nu}} = -2C_{12} = 2C_{34}. \tag{15}$$

Substituting (15) into (8) and (9) one has

$$I = 4x\frac{dM^+}{dx}, \tag{16a}$$
$$R = -4\Lambda - 8M^+ + 2I. \tag{16b}$$

From (14) and (16a) we obtain

$$\frac{dI}{dx} \neq 0, \tag{17}$$

and this means that (16a) gives x as a function of I:

$$x = x(I). \tag{18}$$

Substituting (18) into (16b) one gets R *as a function of* I

$$R = R(I) = -4\Lambda - 8M^+(x(I)) + 2I. \tag{19}$$

Differentiating (19) with respect to I we find

$$\frac{dM^+}{dx} \cdot \frac{dx}{dI} = -\frac{1}{8}\left(\frac{dR}{dI} - 2\right) \neq 0. \tag{20}$$

Then (16a) and (20) yield

$$\frac{dx}{x} = \frac{dI}{I} - \frac{1}{2I}\frac{dR}{dI}dI \tag{21}$$

or

$$x = aI \exp\left(-\frac{1}{2}\int \frac{1}{I}\frac{dR}{dI}\,dI\right), \tag{22}$$

where a is a positive constant.

Example (i). Let $R =$ constant. Then by (22)

$$x = aI, \qquad M^+ = -\tfrac{1}{2} - \tfrac{1}{8}R + x/4a. \tag{23}$$

From the correspondence to the linear theory we assume

$$\lim_{x\to 0}\frac{dM^+}{dx} = I \to 1/4a = 1, \quad a = \tfrac{1}{4}. \tag{24}$$

Finally

$$x = \tfrac{1}{4}I, \qquad M^+ = -\tfrac{1}{2}\Lambda - \tfrac{1}{8}R + x. \tag{25}$$

This is, of course, the *linear electrodynamics*.

Example (ii). Let R and I appear to be some functions of a coordinate, say r, *i.e.* $R = (r)$, $I = I(r)$. Then, (22) give:

$$x = a\, I(r)\, \exp\left(-\frac{1}{2}\int \frac{\dot{R}(r)}{I(r)} dr\right), \tag{26}$$

where $\dot{R} \equiv dR/dr$. This is the case of Ref. [1]. Indeed, one has (compare Eq. (3.11) of Ref. [1]):

$$\begin{aligned} R &= -4\Lambda + \frac{1}{r^2+\ell^2} C, \qquad C = C(r), \\ I &= \frac{1}{2}\left(\frac{1}{r^2+\ell^2}\ddot{C} + 4\frac{C - r\dot{C}}{(r^2+\ell^2)^2}\right). \end{aligned} \tag{27}$$

Substituting (27) into (26) we obtain

$$x = \frac{a}{2}\frac{1}{(r^2+\ell^2)^2}$$

or taking

$$a =: \frac{e^2 + \breve{g}^2}{b^2} \tag{28}$$

we get

$$x = \frac{e^2 + \breve{g}^2}{2b^2(r^2+\ell^2)^2}, \tag{29a}$$

i.e. Eq. (3.18) of Ref. [1]. Then (compare (3.19) of Ref. [1]):

$$M^+ = -b^2(e^2 + \breve{g}^2)^{-1} x r^2 \left(\frac{1}{r}C\right)^{\cdot}. \tag{29b}$$

Now one can easily observe that we are able to follow backwards. Indeed, let

$$R = R(I) \qquad \text{and} \qquad \frac{dR}{dI} - 2 \neq 0. \tag{30}$$

Therefore, from the formula

$$M^+ = -\tfrac{1}{2}\Lambda - \tfrac{1}{8}(R - 2I) \tag{31}$$

we find I as a function of M^+, $I = I(M^+)$.

Then from (16a) we find x as a function of M^+ by

$$x = a \exp\left(4\int \frac{dM^+}{I}\right) = a\, I \exp\left(-\frac{1}{2}\int \frac{1}{I}\frac{dR}{dI}\, dI\right), \quad a > 0. \tag{32}$$

From (32) we have:

$$\frac{dx}{dI} = \frac{x}{I}\left(1 - \frac{1}{2}\frac{dR}{dI}\right) \neq 0. \tag{33}$$

Then (32) gives I as a function of x and by (31) we obtain M^+ as a function of x

$$M^+ = M^+(x). \tag{34}$$

It is easy to show that

$$\frac{dM^+}{dx} = \frac{1}{4}\frac{I}{x} \Rightarrow I = 4x\frac{dM^+}{dx}. \tag{35}$$

Concluding, we see that assumption (30) enables us to reconstruct the Einstein equations (16*a*), (16*b*) with ***nonlinear electromagnetic sources admitting the duality rotation.***

It remains only to assure that the energy-momentum tensor is of the type $[2S{-}2T]_2$. The condition which assures this fact can be stated in the form

$$C^{\mu\nu}C_{\rho\nu} = \tfrac{1}{4}(C^{\omega\tau}C_{\omega\tau})\delta^\mu_\rho, \qquad C^{\omega\tau}C_{\omega\tau} \neq 0. \tag{36}$$

Gathering, we arrive at the fundamental theorem.

Theorem 1. Let a space-time geometry satisfy (locally) the following conditions

(i) $C^{\mu\nu}C_{\rho\nu} = \frac{1}{4}(C^{\omega\tau}C_{\omega\tau})\delta^\mu_\rho$,

$$C^{\mu\nu}C_{\mu\nu} \neq 0, \tag{37}$$

(ii) $C_{\mu\nu}n^\mu n^\nu < 0$ for some non-spacelike vector, and

$$R + 4\Lambda \leq 0, \tag{38}$$

(iii) R is a function of $I := \sqrt{C^{\mu\nu}C_{\mu\nu}}$, $R = R(I)$, such that

$$\frac{dR}{dI} - 2 \neq 0, \tag{39}$$

Then there exist $x \neq 0$ and a function $M^+ = M^+(x)$ such that Einstein equations (8) and (9) are satisfied. Moreover, the energy conditions (11) are also satisfied. The variable x is defined by (22) with accuracy to the transformation $x \to ax$, $a = \text{constant} > 0$.

Example (iii). Born-Infeld nonlinear electrodynamics. Assume that

$$R + 4\Lambda = -2\left(\sqrt{I^2 + c} - \sqrt{c}\right) \tag{40}$$

where $c \geq 0$ is a constant parameter. Then by (22) one has

$$\begin{aligned} x &= aI\exp\left(-\frac{1}{2}\int\frac{-2\cdot 2I}{2\sqrt{I^2+c}}dI\right) \\ &= aI\exp\left(\int\frac{dI}{\sqrt{I^2+c}}\right) = aI\left(I + \sqrt{I^2+c}\right). \end{aligned} \tag{41}$$

By the "limiting condition"

$$\lim_{x\to 0+} \frac{dM^+}{dx} = \Lambda \Rightarrow \lim_{x\to 0+} \frac{I}{4x} = 1, \tag{42}$$

one finds that $c > 0$ and

$$4a\sqrt{c} = 1 \Rightarrow a = 1/4\sqrt{c}. \tag{43}$$

Then from (41) with (43) we get

$$I = \frac{4x}{\sqrt{1 + 9x/\sqrt{c}}}. \tag{44}$$

Consequently, (31), (40), and (44) give:

$$M^+ = \frac{\sqrt{c}}{4}\left(\sqrt{1 + \frac{8}{\sqrt{c}}x} - 1\right). \tag{45}$$

Finally, substituting

$$c = 16b^4 \tag{46}$$

one has

$$M^+ = b^2\left(\sqrt{1 + 2x/b^2} - 1\right). \tag{47}$$

This is of course the *Born-Infeld electrodynamics.*

It is interesting to point out that for $c = 0$ we have

$$R + 4\Lambda = -2I, \tag{48}$$

$$x = aI^2, \tag{49}$$

$$M^+ = \tfrac{1}{2}I = \sqrt{x}/2\sqrt{a}. \tag{50}$$

Then

$$\frac{dM^+}{dx} = \frac{1}{4\sqrt{ax}}, \qquad \lim_{x\to 0+} \frac{dm^+}{dx} = +\infty. \tag{51}$$

Therefore we *cannot* assume the "*limit to the linear electrodynamics*". Now we consider the Maxwell equations.

It is well known that in terms of local null tetrad (e^1, e^2, e^3, e^4) one has

$$C_{ab}e^a \times e^b = I(e^{(3} \times e^{4)} - e^{(1} \times e^{2)}). \tag{52}$$

Then, the following forms

$$\xi = \tfrac{1}{2}\xi_{\mu\nu}dx^\mu \wedge dx^\nu := e^1 \wedge e^2, \quad \xi_{\mu\nu} = -\xi_{\nu\mu}, \tag{53a}$$

$$\eta = \tfrac{1}{2}\eta_{\mu\nu}dx^\mu \wedge dx^\nu := e^3 \wedge e^4, \quad \eta_{\mu\nu} = -\eta_{\nu\mu}, \tag{53b}$$

are defined (locally) by $C_{\mu\nu}$ *with precision to sing.* Moreover if a local orientation of a space-time is defined then assuming that "Hodge $*$-operation" is appropriately defined we can put

$$\eta = *\xi, \qquad \eta_{\mu\nu} = *\eta_{\mu\nu}. \tag{54}$$

Given x and $M^+(x)$ one defines (compare Ref. [1])

$$\begin{aligned} D + iB &:= \sqrt{2x}\, e^{i\Phi} &(55)\\ E + IH &:= (D + iB)\frac{dM^+}{dx} = \sqrt{2x}\, e^{i\Phi}\frac{dM^+}{dx} &(56)\end{aligned}$$

where Φ is a *phase.*

The electromagnetic 2-form ω is defined by

$$\begin{aligned} \omega &= (D + iB)\, e^1 \wedge e^2 + (E + iH)\, e^3 \wedge e^4 \\ &= \sqrt{2x}\, e^{i\Phi}\xi + \sqrt{2x}\, e^{i\Phi}\frac{dM^+}{dx}(*\xi). \end{aligned} \tag{57}$$

The Maxwell equations read

$$d\omega = 0 \tag{58}$$

and, modulo the special Bianchi identities, they lead to

$$\begin{aligned} &\frac{dM^+}{dx} \left[(\Gamma_{312} - \Gamma_{321})\, e^3 + (\Gamma_{421} - \Gamma_{412})\, e^4\right] \\ +&\left(\frac{dM^+}{dx}\right)^{-1} \left[(\Gamma_{134} - \Gamma_{143})\, e^1 + (\Gamma_{243} - \Gamma_{234})\, e^2\right] = i\, d\Phi. \end{aligned} \tag{59}$$

Simple calculations show that Eq. (59) can be written as follows:

$$\left[\frac{dM^+}{dx}\xi^{\mu\nu}\xi_{\rho\mu;\nu} + \left(\frac{dM^+}{dx}\right)^{-1} \xi^{\mu\nu}\xi_{\rho\mu;\nu}\right] dx^\rho = i\, d\Phi. \tag{60}$$

By (16a), and (32) we have

$$\frac{dM^+}{dx} = \frac{I}{4x} = \frac{1}{4a}\exp\left(\frac{1}{2}\int \frac{1}{I}\frac{dR}{dI}dI\right). \tag{61}$$

Now, as

$$\xi^{\mu\nu}\xi_{\rho\nu} = 0, \tag{62}$$

one obtains

$$\begin{aligned} &\left[\frac{1}{4a}\exp\left(\frac{1}{2}\int \frac{1}{I}\frac{dR}{dI}dI\right)\xi^{\mu\nu}_{;\nu}\xi_{\mu\rho} \right.\\ +&\left. \quad 4a\exp\left(-\frac{1}{2}\int \frac{1}{2}\frac{dR}{dI}dI\right)\xi^{\mu\nu}_{;\nu}\xi_{\mu\rho}\right] dx^\rho = i\, d\Phi. \end{aligned} \tag{63}$$

Especially in the case of linear electrodynamics, we have

$$(\xi^{\mu\nu}_{;\nu}\xi_{\mu\rho} + \xi^{\mu\nu}_{;\nu}\xi_{\mu\rho})dx^{\rho} = i\,d\Phi \tag{64}$$

(compare with D. Kramer *et. al.*). It is easy to show that (64) can be written in another from (*e.g.* Rainich, Misner, and Wheeler, D. Kramer *et al.*)

$$i\sqrt{-g}\,\varepsilon_{\mu\nu\rho\sigma}\frac{C^{\nu}_{\tau}C^{\tau\rho;\sigma}}{C^{\alpha\beta}C_{\alpha\beta}}dx^{\mu} = i\,d\Phi. \tag{65}$$

As we see from Eq. (63), in nonlinear electrodynamics one deals with more fine objects that $C_{\mu\nu}$, *i.e.* with bivectors $\xi_{\mu\nu}$ (or, in another words, with 2-dimensional distributions defined by $C_{\mu\nu}$, Π_{1-2}, Π_{3-4}). Concluding, we get the following theorem.

Theorem 2. Let the assumption of Theorem 1 be satisfied and let there exist a constant $a > 0$ and a function Φ such that Eq. (63) holds. Then the space-time admits (locally) the existence of a nonlinear electromagnetic field such that the Einstein-Maxwell equations are satisfied. The electromagnetic 2-form ω is defined by

$$\omega = e^{i\Phi_0}\left[\sqrt{2x}\,e^{i\Phi}\xi + \sqrt{2x}\frac{dM^+}{dx}e^{i\Phi}(*\xi)\right], \tag{66}$$

where the objects x and $M^+(x)$ are defined as before *i.e.*,

$$\begin{aligned} x &= aI\exp\left(-\frac{1}{2}\int\frac{1}{I}\frac{dR}{dI}dI\right), \\ M^+(x) &= -\tfrac{1}{2}\Lambda - \tfrac{1}{8}\left[R\,I(x) - 2I(x)\right], \end{aligned} \tag{67}$$

and Φ_0 is any real constant ("complexion").

Theorems 1 and 2 give the Rainich-Misner-Wheeler "already unified field theory" for nonlinear electrodynamics with duality rotation. We see that such electrodynamics can be formulated in terms of geometrical objects characterizing space-time.

Finally we make some important remarks. From the physical point of view it is natural to impose the following conditions

$$(i) \qquad \lim_{x\to 0+}\frac{dM^+}{dx} = 1, \tag{68}$$

$$(ii) \qquad \lim_{x\to 0+} M^+(x) = 0. \tag{69}$$

By (16*a*) and (32), the condition (68) is equivalent to

$$(i') \qquad \lim_{I\to 0+}\exp\left(\frac{1}{2}\int\frac{1}{I}\frac{dR}{dI}dI\right) = 1. \tag{70}$$

Consequently, (70) defines the constant $a > 0$ uniquely!

Then the condition *(ii)* under the assumption *(i)* appears equivalent to

$$(ii') \qquad \lim_{I \to 0+} R(I) = -4\Lambda. \tag{71}$$

Therefore it is natural to impose the conditions (70) and (71). In this case the constant $a > 0$ in Theorem 2, is fixed by equation (70)! Of course the condition (63) means locally that the left hand side of (63) is a closed 1-form (Poincaré Lemma).

14.1 References

[1] H. Salazar, A. García Díaz, and J. Plebański, *J. Math. Phys.* **28**, 2171 (1987).

[2] H. Salazar, A. García Díaz, and J. Plebański, *J. Math. Phys.* **30**, 2689 (1989).

15

Hamiltonian Formulation of a Simple Covariant Harmonic Oscillator for Bosons and Fermions

Adam Szczepaniak[1]

ABSTRACT We study three covariant two-body equations (one for bosons and two for fermions) which admit light cone Hamiltonian description. The general form of covariant and Hamiltonian solutions together with the dynamical operators of the Hamiltonian formalism is studied, and an explicit wave function for the $J^P = 0^-$ channel is given. The relevance to a realistic quark model calculation is discussed.

15.1 Introduction

A great success of the constituent quark model [10] over the past twenty years has been of a constant source of stimulation for the study of relativistic wave functions for light hadrons involving only a few degrees of freedom. A relatively well established picture is that at low energies hadrons behave much like bound systems of relativistic particles. Although relativity and quantum mechanics call for a quantum field description, the fact that at low energies an infinite number of degrees of freedom seems to be frozen out, gives hope that there may exist a suitable bassis of relativistic few body wave functions relevant to a description of hadronic structure. Within a Hamiltonian formalism there exists a natural framework for the description of relativistic systems of fixed number of particles —the light cone formulation [1], [2]. This is due to the fact that a field theory quantized on the surface of a light cone leads to a particularly simple vacuum structure which in turn allows for a sensible expansion of a hadronic state in a Fock series of states, each with fixed number of constituents [3], [5]. In this article we discuss examples of covariant, two body wave equations which admit a light cone Hamiltonian description and lead to a harmonic oscillator type of spectrum. A basis of solutions to such equations can be

[1]Florida State University, Tallahassee, USA.

used as a simple generalization of the nonrelativistic oscillator models of confinement.

In relativistic quantum mechanics all single particle physical states are classified according to covariant representations of the Poincaré group. Covariance means that under the action of the Poincaré group, generators states transform as a finite irreducible representation. Irreducibility means that the states have defined values of the two Casimir operators $P^2 = M^2$ and $W^2 = M^2 J(J+1)$, which specify the mass and internal spin of the particle, respectively. For example, the covariant transformation of a spin $S = J = 1/2$ state due to Lorentz rotation Λ is of the form

$$\Psi_\alpha^c(x) \rightarrow S_{\alpha\beta}(\Lambda)\Psi_\beta^c(\Lambda^{-1}x),$$

where Ψ is a covariant wave function defined as

$$\Psi_\alpha^c(x) = {}^c\langle x, \alpha|\Psi\rangle, \tag{1.1}$$

and $|x,\alpha\rangle^c$ form a covariant coordinate space basis of a single spin-1/2 particle state. We shall refer to such systems as elementary systems. For composite systems built up out of N elementary systems, one constructs the algebra of generators of the Poincaré transformations of the composite system as a direct sum of generators of N elementary generators; for a composite system to represent a physical states requires that it transforms in an covariant way under the action of the composite algebra. Thus in a relativistic quantum mechanics dynamical problem one looks for a set of covariant equations for the composite wave function. Solutions of these, which transform covariantly, are thus identified with physical states. Covariant wave equations by themselves do not represent the full solution to a dynamical quantum problem. One has to be able to interpret the wave functions as probability amplitudes, *i.e.* define a Hilbert space structure in a space of covariant solutions. In this respect a covariant Hamiltonian formulation is of particular interest since it makes a connection between special relativity and quantum mechanics. In a Hamiltonian approach one looks for a realization of the 10 Poincaré group generators in terms of some set of dynamical variables associated with a given hypersurface Σ in Minkowski space [1]. The generators which have orbits inside Σ form the stability group and are usually referred to as kinematical ones; the remaining are referred to as dynamical or Hamiltonian, since they describe the evolution of the system in time, defined as a direction not contained in Σ. A solution to the dynamics is given as a set of eigenstates of a maximal number of commuting generators restricted to the surface Σ. These form a representation of the Poincaré group which, however, need not be covariant. Restriction of the Poincaré generators to some hypersurface amounts to a loss of the information contained in the original covariant commutation relations, and can be restored as a set of constraints between projected generators. Such constraints strongly restrict the possible forms of interactions entering the Hamiltonians. Furthermore, for a covariant set of equations to admit a Hamiltonian description it is necessary that they admit Σ to be an initial value surface in a Cauchy problem. The most natural choice for Σ is to take it to be a flat spacelike surface. This is the "instant" formulation where the evolution takes place in the ordinary time variable. In a nonrelativistic Hamiltonian formalism, where instead of the Poincaré group one deals with a Galilean group, this is a unique way in which a Cauchy problem can be formulated. However, for relativistic systems the choice of Σ is rather arbitrary; any hypersurface Σ which does not contain timelike directions is admissible. From a practical point of view, however, one looks for a surface with maximal symmetry to reduce the number of Hamiltonians. In this respect the most suitable choice for Σ is a surface of the light cone; the stability group contains 7 operators and only 3 operators are dynamical, as compared to the "instant" formulation with 6 and 4 operators respectively. In the following we shall briefly review the main features of the light cone Hamiltonian formulation of relativistic many body dynamics following closely the article by Leutwyler and Stern [2].

15.2 Light cone Hamiltonian formalism

We shall use the following notation: for an arbitrary four vector A, we denote the Minkowski coordinates by $A = A^\mu = (A^0, \mathbf{A})$ and we introduce light cone coordinates by a transformation

$$A^+ \equiv A^0 + A^3, \quad A^- \equiv \tfrac{1}{2}(A^0 - A^3), \quad A^s \equiv (A^1, A^2), \quad s = 1,2,$$

$A^\mu = (A^+, A^-, A^s)$. The scalar product of two vectors in light cone coordinates is

$$A \cdot B = g_{\mu\nu} A^\mu B^\nu = A^+ B^- + A^- B^+ - A^s B^s .$$

The hypersurface Σ of the light cone Hamiltonian formulation is defined as

$$\Sigma = \{x : x^- = 0\}.$$

On the basis of the 10 generators of the Poincaré group $\hat{P}^\mu, \hat{M}^{\mu\nu}$ we define

$$\begin{gathered}\{\hat{P}^+,\, \hat{P}^s,\, \hat{E}^s \equiv \hat{M}^{-s},\, \hat{K} \equiv \hat{M}^{+-}\},\\ \{\hat{J}^\xi \equiv \hat{M}^{12} + \tfrac{1}{\hat{P}^-}(\hat{E}^1 \hat{P}^2 - \hat{E}^2 \hat{P}^1)\},\\ \{\hat{M}^2 \equiv 2\hat{P}^- \hat{P}^+ - \hat{P}^s \hat{P}^s,\, \hat{M}\hat{J}^s \equiv \hat{W}^s - \tfrac{\hat{P}^s}{\hat{P}^-}\hat{W}^-\},\end{gathered} \tag{2.1}$$

with $\hat{W}_\mu = \epsilon_{\mu\nu\rho\sigma} \hat{P}^\nu \hat{M}^{\rho\sigma}$ being the Pauli-Lubanski spin operator. The above grouping of the operators has been done according to the way they act on Σ. The first six generators in Eq. (2.1) map Σ onto itself generating a stability group and forming the kinematical algebra where the only nonvanishing commutators are

$$[\hat{K}, \hat{E}^s] = -i\hat{E}^s, \quad [\hat{K}, \hat{P}^-] = -i\hat{P}^-, \quad [\hat{E}^s, \hat{P}^r] = -i\delta_{sr}\hat{P}^- . \tag{2.2}$$

An operator $\hat{J}^\xi$ also leaves Σ invariant, *i.e.* it belongs to the stability group of Σ. However, it also commutes with the generators of the kinematical group and thus it will simply label the irreducible representations of the kinematical algebra, *i.e.*, it acts in the representation space in a way independent of the action of the kinematical generators. Furthermore, the last three generators together with $\hat{J}^\xi$ close to form an $u(2)$ algebra $\hat{J}^i \equiv (\hat{J}^\xi, \hat{J}^s)$

$$[\hat{J}^i, \hat{J}^j] = i\epsilon_{ijk}\hat{J}^k, \qquad [\hat{M}, \hat{J}^i] = 0, \tag{2.3}$$

and as $\hat{J}^\xi$ they commute with the kinematical operators. We shall refer to the two operators $\hat{M}^2$, $\hat{M}\hat{J}^s$ as Hamiltonians. Since they do not commute with the projection on Σ, they govern the evolution of a system and are in general interaction dependent. As compared to the "instant" classification of the Poincaré algebra, where the interaction dependence appears in the mass operator and three boost operators, in a light cone classification the dynamics enter through the mass operator and two operators related to the total spin [*cf.* Eq. (2.1)].

We shall now study the representations of the algebra constructed above. It has a product structure of

$$\mathcal{K} \oplus \mathcal{D}. \tag{2.4}$$

It is useful to start with representations of the kinematical generators since by definition they do not involve interactions and their action on a basis of states should be simple. In particular we choose a basis in which the light cone momentum operator $\hat{\tilde{P}} = (\hat{P}^-, \hat{P}^s)$ is diagonal,

$$\hat{\tilde{P}}|\tilde{P}, \lambda\rangle^h = \tilde{P}|\tilde{P}, \lambda\rangle^h. \tag{2.5}$$

Here λ stands for additional labels which may be necessary to completely specify the representation, and the superscript h indicates that this is a basis in a Hamiltonian formalism. In a basis defined as above, a representation for the operators $\hat{E}^s$, $\hat{K}$ can be easily found from the commutation relations given in Eq. (2.2) by

$$\hat{E}_s|\tilde{P}, \lambda\rangle^h = iP^-\frac{\partial}{\partial P^s}|\tilde{P}, \lambda\rangle^h, \quad \hat{K}|\tilde{P}, \lambda\rangle^h = iP^-\frac{\partial}{\partial P^-}|\tilde{P}, \lambda\rangle^h.$$

It is worthwhile noting that the action of the operator $\hat{K}$ is transitive on the half plane $P^+ > 0$. Thus any massive physical state with arbitrary momentum P^+ can be obtained by a simple kinematical transformation from a state at rest. This is one of the major advantages of the light cone hamiltonian formulation; the boosts are kinematical, thus any process which involves matrix elements between states of different three-momenta can be easily evaluated within the light cone formalism, in contrast to the "instant" Hamiltonian approach. As we have already noticed, the operator $\hat{J}^\xi$ acts on basis states [Eq. (2.5)] in a $\tilde{P}$-independent way and in order for the representation to be irreducible, Schur's lemma implies that $\hat{J}^\xi$ is proportional to the identity operator. We thus choose

$$\hat{J}^\xi|\tilde{P},\lambda\rangle^h = \lambda|\tilde{P},\lambda\rangle^h.$$

In the case of elementary systems we call the corresponding representation an elementary one. Since the mass operator is the Casimir operator of the product algebra $\mathcal{K}\otimes\mathcal{D}$ for the elementary representation, we simply have

$$\hat{M}^2|\tilde{P},\lambda\rangle^h = M^2|\tilde{P},\lambda\rangle^h.$$

The remaining two dynamical operators $\hat{M}\hat{J}^1$, $\hat{M}\hat{J}^2$ contain a part which acts on a space labeled by an index λ; thus vectors $|\tilde{P},\lambda\rangle^h$ with $-S\cdots<\lambda<\cdots S$ span a basis of an u(2) matrix representation of dimension S for operators $\hat{J}^i$. The fact that the main part of the relativistic invariance relies on the operators forming an u(2) algebra is a general property of the Poincaré group, and is the same as in the original "instant" construction of unitary irreducible representations due to Wigner [4] in which the dynamical operators act on a space of inner variables in a **P**-independent way. However, as we have already pointed out, a particular feature of the light cone representation is that the kinematical variables act on the $\tilde{P}$ space in a way independent of internal degrees of freedom. In other words, out of the Poincaré algebra we have constructed an algebra with a product structure of Eq. (2.4). Thus in general we shall be able to write

$$|\tilde{P},n\rangle^h = |\tilde{P}\rangle^h\otimes|n\rangle^h, \tag{2.6}$$

with n labeling the dynamical variables specifying the representation of the dynamical algebra (for an elementary representation, $n=\lambda$).

For a composite system of N-particles we define

$$\Sigma_N = \{x_a : x_a^- = 0,\ a = 1,\ldots,N\}, \tag{2.7}$$

and we shall require that a composite representations transform under the action of a stability group of Σ_N as a product of N elementary representations. In the following we shall discuss a general case of two interacting particles. A stability algebra of Σ is spanned by

$$\hat{\tilde{P}} = \hat{\tilde{P}}_1 + \hat{\tilde{P}},\quad \hat{E}^s = \hat{E}_1^s + \hat{E}^s,\quad \hat{K} = \hat{K}_1 + \hat{K}_2.$$

A basis for a product representation of the composite kinematical algebra is spanned by vectors $|\tilde{p}_1,\lambda_1,\tilde{p}_2,\lambda_2\rangle^h$ and is given by Eq. (2.5) with $\tilde{p}=\tilde{p}_1+\tilde{p}_2$. In order to bring the states $|\tilde{p}_1,\lambda_1;\tilde{p}_2,\lambda_2\rangle^h$ of a composite representation into a canonical form, Eq (2.6), we transform a set of variables $\tilde{p}_1,\tilde{p}_2$ into

$$\tilde{P}\equiv\tilde{p}_1+\tilde{p}_2,\quad \xi\equiv\frac{p_1^- - p_2^-}{2P^-},\quad k^s\equiv\frac{p_2^-p_1^s - p_1^-p_2^s}{2P^+}.$$

The relative momentum variable $\tilde{k} \equiv (\xi, k^s)$ is invariant under the action of the composite stability group. Thus the desired transformation is

$$|\tilde{p}_1, \lambda_1; \tilde{p}_2, \lambda_2\rangle^h = |\tilde{P}\rangle^h \otimes |\tilde{k}, \lambda_1, \lambda_2\rangle^h. \tag{2.8}$$

The representation of a kinematical group is unitary in the norm

$$\begin{aligned} &{}^h\langle\tilde{P}'| \otimes {}^h\langle\tilde{k}', \lambda_1', \lambda_2'|\tilde{k}', \lambda_1, \lambda_2\rangle^h \otimes |\tilde{P}\rangle \\ &\qquad = 16\pi^3 P^- \delta^3(\tilde{P}' - \tilde{P})^h\langle\tilde{k}', \lambda_1', \lambda_2'|\tilde{k}, \lambda_1, \lambda_2\rangle^h. \end{aligned} \tag{2.9}$$

The factor $16\pi^3$ comes from the single particle phase space volume element, which in the light come coordinates takes the form

$$\frac{d^4p}{(2\pi)^4} 2\pi\delta(p^2 - m^2)\theta(p^0) = \frac{dp^- d^2p^s}{16\pi^3 p^-}.$$

The norm in the inner space [last term in Eq. (2.9)] cannot be specified unless one requires that some of the generators of the individual elementary representations be hermitian (correspond to a physical observable) in a given norm. For example, if one requires that two elementary representations are independently unitary then

$${}^h\langle\tilde{k}', \lambda_1', \lambda_2'|\tilde{k}, \lambda_1, \lambda_2\rangle^h = 16\pi^3(\tfrac{1}{4} - \xi^2)\delta(\xi' - \xi)\delta^2(k'^s - k^s)\delta_{\lambda_1'\lambda_1}\delta_{\lambda_2'\lambda_2},$$

is just a product of single particle norms. The Hamiltonians $\hat{M}^2$, $\hat{M}\hat{J}^s$ act only on a space spanned by the vectors $|\tilde{k}, \lambda_1, \lambda_2\rangle^h$, and for the operator $\hat{J}^\xi$ we have

$$\hat{J}^\xi |\tilde{k}, \lambda_1, \lambda_2\rangle^h = (\lambda_1 + \lambda_2)|\tilde{k}, \lambda_1, \lambda_2\rangle^h.$$

The last requirement, to guarantee that the space of two particle states [Eq. (2.8)] form a basis of a unitary irreducible representation of the Poincaré group, is to request that in the general case of interacting particles the potentials entering the Hamiltonians depend on the internal variables $k_\perp$, ξ and λ_i, that they satisfy the commutation relations of Eq. (2.3), and that they are self-adjoint in the norm of Eq. (2.9). A light cone wave function in momentum space is defined through $(-S_1 \ldots < \lambda_1 < \ldots S_1, -S_2 \ldots < \lambda_2 \ldots S_2)$

$$\Psi^h(\tilde{P}, \tilde{k})_{\lambda_1, \lambda_2} \equiv {}^h\langle\tilde{P}| \otimes \langle\tilde{k}, \lambda_1, \lambda_2|\Psi\rangle.$$

As we mentioned in the Introduction, it is not sufficient to satisfy the Poincaré invariance in order for a basis states to transform in a covariant way. The requirement of covariance leads to a unique relation between the covariant (c) and Hamiltonian (h) states. For example, for a system of two spin-0 particles, covariance implies

$$\begin{aligned} |\tilde{X}, \tilde{x}\rangle^h &\equiv \left|X \equiv \tfrac{1}{2}(x_1 + x_2), x \equiv x_1 - x_2\right\rangle^c\Big|_{X^- = x^+ = 0}, \\ &= \int \frac{dP^- d^2p^s}{16\pi^3 P^-} e^{i\tilde{P}\cdot\tilde{X}} \int \frac{d\xi\, d^2k^s}{16\pi^3(\frac{1}{4} - \xi^2)} e^{i\tilde{k}\cdot\tilde{x}} |\tilde{P}\rangle^h \otimes |\tilde{k}\rangle^h, \end{aligned} \tag{2.10}$$

with

$$\begin{aligned} \tilde{X} &\equiv (X_-, X_s), & \tilde{P}\cdot\tilde{X} &= X_-P^- + X_sP^s \\ \tilde{x} &\equiv (x_\xi \equiv x_-P^- + x_sP^s, x_s), & \tilde{k}\cdot\tilde{x} &= \xi x_\xi + k^s x_s. \end{aligned} \tag{2.11}$$

This relation is nontrivial since it introduces a relation between the p^- and x_- variables on a light cone which is not defined if one only assumes that the vectors $|\tilde{p}\rangle^h$ have the right transformation properties under a stability group transformation. For two spin-$\frac{1}{2}$ particles, the analogous relation has the form

$$\begin{aligned} |\tilde{X},\tilde{x},\lambda_1,\lambda_2\rangle^h &\equiv \left|X,x,\alpha_1=(\gamma_1,\lambda_1),\alpha_2=(\gamma_2,\lambda_2)\right\rangle^c\Big|_{X^-=x^+=0,\gamma_1=\gamma_2=+} \\ &= \int \frac{dP^-d^2p^s}{16\pi^3P^-}e^{i\tilde{P}\cdot\tilde{X}} \\ &\quad\times\int\frac{d\xi\, d^2k^s}{16\pi^3(\frac{1}{4}-\xi^2)^{\frac{1}{2}}}e^{i\tilde{k}\cdot\tilde{x}}|\tilde{P}\rangle^h\otimes|\tilde{k},\lambda_1,\lambda_2\rangle^h, \end{aligned}$$

where the 4-valued Dirac index of a covariant vector α [*cf.* Eq. (1.1)] has been written in terms of two 2-valued indexes

$$\alpha = (\gamma = (+,-),\ \lambda = (-\tfrac{1}{2},+\tfrac{1}{2})),$$

with γ labeling the two (upper *vs.* lower) components of the positive energy solutions to the free Dirac equation in a light cone canonical basis for the γ-matrices (see Appendix A), and λ labeling the helicity of the upper components. A light cone wave function in coordinate space satisfies then

$$\begin{aligned} \Psi^h(\tilde{X},\tilde{x})_{\lambda_1,\lambda_2} &= {}^h\langle\tilde{X},\tilde{x},\lambda_1,\lambda_2|\Psi\rangle \\ &= {}^c\big\langle X,x,(\gamma_1,\lambda_1,),(\gamma_2,\lambda_2)|\Psi\big\rangle\big|_{X^-=x^-=0,\gamma_1=\gamma_2=+}. \end{aligned} \tag{2.12}$$

Covariant generators of the Poincaré group, which after projection on a surface Σ define the Hamiltonians [*cf.* Eq. (2.1)] do not leave Σ invariant. Thus a projection of those generators in general will not respect covariance. It can be shown that to restore covariance the Hamiltonians must obey certain nonlinear equations, called the angular conditions. For the two spin-0 composite system the angular condition has the form

$$x_\xi \hat{J}^\xi + x_s\hat{M}\hat{J}^s = 0.$$

15.3 Covariant harmonic oscillator model for two spin-0 constituents

The requirement of covariance strongly constrains the dynamics in a Hamiltonian formulation. Furthermore, a set of covariant wave equations for a composite system of N particles admits a Hamiltonian description if it is complete, *i.e.*, if it fixes the behaviour of the wave functions in all individual time variables. The immediate consequence of the last requirement is that for a system of N particles a set of N covariant equations is needed for a system to have a Hamiltonian description. For an $N = 2$ particle system we shall write those equations in a particular form,

$$\hat{C}_a \Phi^c(x_a) = 0, \qquad a = 1, 2, \tag{3.1}$$

with the $\hat{C}_i$'s being local covariant operators acting on a covariant wave function Φ^c. In order for a system of Eq. (3.1) to have a nontrivial solution the operators $\hat{C}_i$ have to satisfy the weak constraint in a Dirac sense [6],

$$[\hat{C}_1, \hat{C}_2] \sim 0. \tag{3.2}$$

The covariant equation for two spin-0 particles was proposed a long time ago by Feynman, Kislinger, and Ravndal [7], and is usually referred to as the FKR model. One of the two covariant equations is a generalization of a Klein-Gordon mass shell constraint, and for two equal-mass particles it has the form

$$2\hat{C}_1 \Phi^c = [\hat{p}_1^2 - m^2 + \hat{p}_2^2 - m^2 + \hat{V}(x_1, x_2)]\Phi^c = 0. \tag{3.3}$$

We shall refer to such type of constraints as leading dynamical ones. The second equation fixes the relative time behaviour. It can have a rather arbitrary form; here we shall use a particulary simple one, namely

$$\hat{C}_2 \Phi^c = (\hat{p}_1 - \hat{p}_2)(\hat{p}_1 + \hat{p}_2)\Phi^c = 0.$$

For states that belong to a given representation of the Poincaré group it is useful to define a projection operator $\hat{L}$ on a total momentum $P = p_1 + p_2$. Thus for an arbitrary four-vector A we define

$$A = \hat{L}A + (1 - \hat{L})A \equiv A_L + A_T = \frac{(A \cdot P)P}{P^2} + \left(A - \frac{(A \cdot P)P}{P^2}\right). \tag{3.4}$$

Furthermore, total and relative four momentum and four position vectors are

$$\begin{aligned} P &= p_1 + p_2, & X &= \tfrac{1}{2}(x_1 + x_2), \\ q &= \tfrac{1}{2}(p_1 - p_2), & x &= x_1 - x_2. \end{aligned}$$

The compatibility condition of Eq. (3.2) now takes the form

$$[\hat{q}_L, \hat{V}(x_T^2, x_L^2)] = 0.$$

The constraint is satisfied provided

$$\hat{V}(x_T^2, x_L^2) = \hat{V}(x_T^2).$$

A harmonic-oscillator-type interaction is obtained by setting

$$\hat{V}(x_T^2) = 2\hat{x}_T^2,$$

where the factor of 2 is chosen by convenience, and where we choose units such that the oscillator strength is $\omega = 1$. The set of two constraints commutes with the Casimir operators of the Poincaré group

$$\begin{aligned} \hat{M}^2 &= \hat{P}^2, \\ \hat{W}^2 &= \hat{W}_\mu \hat{W}^\mu, \\ \hat{W}_\mu &= \epsilon_{\mu\nu\rho\sigma} \hat{P}^\nu (\hat{x}^\rho \hat{q}^\sigma - \hat{x}^\sigma \hat{q}^\rho). \end{aligned} \tag{3.5}$$

Thus we shall look for solutions which correspond to given eigenvalues, M^2 and $M^2 L(L+1)$, of the two Casimirs respectively; these solutions we identify as the covariant wave function of the bound state of mass M and spin S. Since both $\hat{C}_1$ and $\hat{C}_2$ commute with the 4-vector operator $\hat{P}$, it is convenient to look for solutions in the center of momentum frame (c.m.), *i.e.*,

$$\begin{aligned} \hat{P}\Phi^c(X,x) &= (P^0 = M, \mathbf{0})\Phi^c(X,x), \\ \hat{x}_T \Phi^c(x,x) &= (0, \hat{\mathbf{x}})\Phi^c(X,x), \\ \hat{q}_T \Phi^c(X,x) &= (0, \hat{\mathbf{q}})\Phi^c(X,x). \end{aligned} \tag{3.6}$$

Thus in c.m. frame

$$\Phi^c(X,x) = \Phi^c(X^0, \mathbf{x})$$

and

$$\hat{W}\Phi^c(X^0,\mathbf{x}) = (0, \hat{\mathbf{W}})\Phi^c(X^0,\mathbf{x}) = (0, \hat{\mathbf{x}} \times \hat{\mathbf{q}})\Phi^c(X^0,\mathbf{x}).$$

The solutions to the pair of constraints in the c.m. frame have the form

$$\begin{aligned} \Phi^c(X^0, \mathbf{x}; n, L, \lambda) &= e^{-iMX^0} R_{nL}(\mathbf{x}^2) Y_L^\lambda\left(\frac{\mathbf{x}}{|\mathbf{x}|}\right) \\ &= e^{-iP\cdot X_L} R_{nL}(-x_T^2) Y_L^\lambda\left(\frac{x_T}{-x_T^2}\right), \end{aligned}$$

with a spectrum given by

$$M^2 = 4m^2 + 8(2n + L + 2), \tag{3.7}$$

and

$$W^2 = M^2 L(L+1).$$

A solution in an arbitrary frame is obtained from a covariant transformation law

$$\Phi^c(X,x) = \Phi^c(\Lambda X, \Lambda x),$$

and has the general form

$$\Phi^c(X,x;n,L,\lambda,P) = e^{-iP\cdot X} x_T^{\mu_1} \dots x_T^{\mu_L} \epsilon^{\lambda}_{\mu_1\dots\mu_L}(P) R_{nL}(-x_T^2),$$

with $\epsilon^m_{\mu_1\dots\mu_l}(P)$ being a standard polarization tensor.

The above set of covariant equations admit a light cone Hamiltonian description. To get an expression for the light cone Hamiltonians, $\hat{M}, \hat{M}\hat{J}^s$ we have to restrict the leading dynamical constraint, Eq. (3.3), to a light cone surface Σ_2 [*cf.* Eq. (2.7)]. However, the constraint $\hat{C}_1$ explicitly involves a derivative with respect to x^- (terms proportional to $\hat{q}^+$):

$$\begin{aligned} \hat{C}_1 &= \tfrac{1}{4}\hat{P}^2 + 2\hat{q}^+\hat{q}^- - \hat{q}^s\hat{q}^s - m^2 \\ &\quad + 2\hat{x}^+\hat{x}^- - \hat{x}_s\hat{x}_s - \frac{(\hat{x}^+\hat{P}^- + \hat{x}^-\hat{P}^+ + \hat{x}_s\hat{P}^s)^2}{\hat{P}^2}. \end{aligned} \tag{3.8}$$

Those terms can be eliminated with the use of the relative time fixing constraint, giving

$$\hat{P}^-\hat{q}^+ = \hat{P}^s\hat{q}^s - \frac{\hat{M}^2 + \hat{P}^s\hat{P}^s}{2}\frac{\hat{q}^-}{\hat{P}^-}. \tag{3.9}$$

Equations (3.8) and (3.9), when restricted to the light cone surface, give a nonlinear operator equation for the mass eigenvalues

$$\hat{M}^2\Phi^h(\tilde{X},\tilde{x}) = M^2\Phi^h(\tilde{X},\tilde{x}),$$

with

$$M^2(\tfrac{1}{4} - \hat{\xi}^2) - \hat{x}_\xi^2/M^2 - (\hat{k}^s\hat{k}^s + \hat{x}_s\hat{x}_s) - m^2 = 0, \tag{3.10}$$

and $[k^s, x_r] = -i\delta^s_r$, $[\xi, x_\xi] = -i$, with the light cone relative momenta and conjugated position vectors defined as in Eq (2.11). Similarly the two dynamical angular momentum operators [*cf.* Eq. (2.1)] can be found from Eq. (3.5) together with Eq. (3.9). It is not necessary however to solve the complicated eigenvalue equations for the light cone Hamiltonians to obtain the light cone wave functions. These can be easily found by projecting the covariant eigensolutions to the light cone surface [*cf.* Eq. (2.12)],

$$\begin{aligned} \Phi^h(\tilde{X},\tilde{x};n,L,\lambda,\tilde{P}) &= {}^c\langle X,x|\Phi^c;n,L,\lambda,\tilde{P}\rangle\big|_{X^+=x^+=0} \\ &= e^{-i\tilde{P}\cdot\tilde{X}}|\tilde{x}_T|^L Y_L^\lambda\left(\frac{\tilde{x}_T}{|\tilde{x}_T|}\right) R_{nL}(\tilde{x}_T^2), \end{aligned}$$

with $\tilde{x}_T \equiv (x_\xi/M, x_s)$. We note that in an arbitrary frame the part of the light cone wave function which describes the internal dynamics (which depends on $\tilde{x}_T$) does not depend on the total momenta P; this has been advertised already as a particular feature of the light cone formulation. The wave function in momentum space is given by a Fourier transform with the kernel as in Eq. (2.10) leading to

$$\Phi^h(\tilde{P},\tilde{k};n,L,\lambda;\tilde{P}') = 8\pi^3\delta^3(\tilde{P}-\tilde{P}')|\tilde{k}_T|^L Y_L^\lambda\left(\frac{\tilde{k}_T}{|\tilde{k}_T|}\right) R_{nL}(\tilde{k}_T^2),$$

with $\tilde{k}_T = (M\xi, k^s)$. Finally we note that the Hamiltonians in momentum space are self adjoint in the product norm of Eq. (2.9).

Other alternatives to the FKR light cone models [2, 8] have been proposed. They usually differ by choosing a different relative time-fixing constraint. Leutwyler and Stern have shown that if one insists on having purely local expressions for the light cone Hamiltonians [as opposed to the case discussed here, *cf.* Eq. (3.10)], the freedom in choosing a potential $\hat{V}(x)$ and a constraint $\hat{C}_2$ is very limited. Different forms of a FKR oscillator lead also to different norms in a space of light cone solutions, with the simplest one corresponding to the case presented above.

15.4 Covariant harmonic oscillator models for two spin-1/2 constituents

In this Section we shall study two examples of covariant Hamiltonian dynamics for a fermion-antifermion system which lead to a harmonic oscillator spectrum. There have been many attempts to generalize Dirac's ideas [6] to formulate a covariant description of many-body fermion systems (for the most recent results see Ref. [9]). We shall proceed here with the simplest idea, which is to generalize the Klein-Gordon constraint to describe two Dirac particles. We start with two constraints of the form

$$\begin{aligned}\hat{D}_1''\Psi^c &= [\hat{p}_1\cdot\gamma_1 - m + \hat{V}_1]\Psi^c = 0,\\ \hat{D}_2''\Psi^c &= \Psi^c[\hat{p}_2\cdot\gamma_2 + m + \hat{V}_2] = 0.\end{aligned} \tag{4.1}$$

Here, Ψ^c is a 4×4 matrix in Dirac indices which transforms under the Lorentz group as a product $\Psi_\alpha\overline{\Psi}_\beta$ of two Dirac spinors, *i.e.*

$$\Psi^c(x_1,x_2) = S(\Lambda)\Psi^c(\Lambda^{-1}x_1,\Lambda^{-1}x_2)S^{-1}(\Lambda),$$

and the matrices γ_a, $a=1,2$, act on a first and second Dirac index of Ψ^c from left and right respectively. To obtain an analogous set of covariant equations as for the boson case, *i.e.*, the first one representing the sum of two mass-shell constraints and the second fixing the relative time behaviour, we first multiply the two equations in Eq. (4.1) by $\hat{p}_1\cdot\gamma_1 + m$ and $\hat{p}_2\cdot\gamma_2 - m$ respectively, to get

$$\begin{aligned}\hat{D}_1'\Psi^c &= [\hat{p}_1^2 - m^2 - (\hat{p}_1\cdot\gamma_1 + m)\hat{V}_1]\Psi^c = 0,\\ \hat{D}_2'\Psi^c &= \Psi^c[\hat{p}_2^2 - m^2 - \hat{V}_2(\hat{p}_2\cdot\gamma_2 - m)] = 0.\end{aligned}$$

Finally the two equations are

$$\begin{aligned}
\hat{D}_1\Psi^c &= \tfrac{1}{2}(\hat{D}'_1 + \hat{D}'_2)\Psi^c \\
&= (\tfrac{1}{4}\hat{P}^2 + \hat{P}\cdot\hat{q} + \hat{q}^2 - m^2)\Psi^c - (\hat{p}_1\cdot\gamma_1 + m)\tfrac{1}{2}\hat{V}_1\Psi^c \\
&\quad - \Psi^c\tfrac{1}{2}\hat{V}_2(\hat{p}_2\cdot\gamma_2 - m) = 0, \\
\hat{D}_2\Psi^c &= (\hat{D}'_1 - \hat{D}'_2)\Psi^c \\
&= \hat{P}\cdot\hat{q}\Psi^c - (\hat{p}_1\cdot\gamma_1 + m)\hat{V}_1\Psi^c \\
&\quad + \Psi^c\hat{V}_2(\hat{p}_2\cdot\gamma_2 - m) = 0.
\end{aligned}$$

From a general class of potentials we choose those which satisfy

$$\begin{gathered}
(\hat{p}_1\cdot\gamma_1 + m)\hat{V}_1 = \hat{V}_2(\hat{p}_2\cdot\gamma_2 - m) = \hat{V}, \\
\hat{V}_1 = \frac{1}{(\hat{p}_1\cdot\gamma_1 + m)}\hat{V}, \qquad \hat{V}_2 = \hat{V}\frac{1}{(\hat{p}_2\cdot\gamma_2 - m)},
\end{gathered}$$

in order to bring the constraint $\hat{D}_2$ to the form of $\hat{C}_2$. The compatibility condition $[\hat{D}_1, \hat{D}_2] \sim 0$ requires $\hat{V} = \hat{V}(\hat{x}_T^2)$ and the two equations finally have the form of

$$\begin{aligned}
\hat{D}_1\Psi^c &= (\tfrac{1}{4}\hat{P}^2 + \hat{q}_T^2 + \tfrac{1}{2}\hat{V}(\hat{x}_T^2) - m^2)\Psi^c = 0, \\
\hat{D}_2\Psi^c &= \hat{q}_L\Psi^c = 0.
\end{aligned}$$

Thus, for a two fermion system, starting from the Dirac-type constraints with nonlocal potentials we have been able to arrive at the fermion version of an FKR model. The spectrum is of course identical to the one in the scalar case, Eq. (3.7). However, the difference in solutions relies on the fact that the total spin operator

$$\begin{aligned}
\hat{W}_\mu &= \epsilon_{\mu\nu\rho\sigma}\hat{P}^\nu(\hat{x}^\rho\hat{q}^\sigma - \hat{x}^\sigma\hat{q}^\rho + \sigma_1^{\rho\sigma} - \sigma_2^{\rho\sigma}), \\
\hat{W}^2\Psi^c &= M^2 J(J+1)\Psi^c, \qquad (4.2)
\end{aligned}$$

contains extra terms acting on individual spins. A basis for covariant solutions to the two constraints with given M^2 an S is spanned by functions of the form

$$\Psi^c(X,x) = e^{-iP\cdot X}\left\{S(x), \frac{P}{\sqrt{P^2}}\cdot\gamma V_P(x), x\cdot\gamma V_x(X,x), \frac{P_\mu}{\sqrt{P^2}}x_\nu\sigma^{\mu\nu}T(x)\right\} \qquad (4.3)$$

for states with normal parity with S, V_P, V_x, and T being scalar functions of the form $x^{\mu_1}\ldots x^{\mu_L}\epsilon^{\lambda_L}_{\mu_1\ldots\mu_L}R_{nL}(-x_T^2)$, $L = J \pm 1$, and

$$\begin{aligned}
\Psi^c(X,x) &= e^{-iP\cdot X}\left\{P(x), \frac{P}{\sqrt{P^2}}\cdot\gamma\gamma_5 A_P(x),\right. \\
&\qquad \left. x\cdot\gamma\gamma_5 A_x(X,x), \epsilon_{\mu\nu\rho\sigma}\frac{P^\mu}{\sqrt{P^2}}x^\nu\sigma^{\rho\sigma}T(x)\right\} \qquad (4.4)
\end{aligned}$$

for states with abnormal parity. Since the constraints commute separately with the orbital and spin part of the covariant spin operator $\hat{W}^{\mu}$, the eigensolutions are linear combinations of the basis vectors of the same orbital angular momentum in a rest frame with arbitrary numerical coeficients. For example, a set of covariant wave functions describing the state with $S = 0$, and odd parity (a basis to study a "pion" in a relativistic constituent quark model) is

$$\Psi^c_\pi(X, x; n, 0, P) = \left(A + B\frac{P}{\sqrt{P^2}} \cdot \gamma\gamma_5\right) R_{n0}(-x_T^2)$$

with arbitrary coeficients A and B. The light cone Hamiltonian description of this covariant model is very similar to the scalar case of the FKR model. The mass operator $\hat{M}^2$ satisfies the Eq. (3.10). The expressions for the angular momentum operators $\hat{M}\hat{J}^s$ can be found from Eqs. (2.1) and (4.2) for spin-1/2 constituents

$$\hat{M}\hat{J}^s = \hat{W}^s - \frac{\hat{P}^s}{\hat{P}^-}\hat{W}_+\big|_{x_1^+ = x_2^+ = 0, \gamma_1 = \gamma_2 = +},$$

with the use of Eq. (3.9) to project the $\hat{q}^+$ operator, and a representation of the gamma matrices given in Appendix A to relate the redundant "bad" components ($\gamma_a \neq +$, $a = 1, 2$) to the "good" ones [2]. The general expression is rather complicated and is of little interest since the basis for the light cone wave functions, as in the scalar case, can be obtained by projecting the covariant solutions onto the light cone surface [*cf.* Eq. (2.12)] without having to solve the eigenstate problem for the light cone Hamiltonians. For example, a set of light cone wave function for a "pion" in this model is given by

$$\Psi^h_\pi(\tilde{X}, \tilde{x}; n, \tilde{P})_{\lambda_1, \lambda_2} = e^{-i\tilde{P}\cdot\tilde{X}} 2\frac{P^-}{\sqrt{P^2}}[\sigma_2]_{\lambda_1\lambda_2} R_{n0}(\tilde{x}_T^2), \tag{4.5}$$

and in momentum space by

$$\Psi^h_\pi(\tilde{P}, \tilde{k}; n, \tilde{P}')_{\lambda_1, \lambda_2} = 16\pi^3 \frac{P^-}{\sqrt{P^2}} \delta^3(\tilde{P} - \tilde{P}')[\sigma_2]_{\lambda_1\lambda_2} R_{n0}(\tilde{k}_T^2). \tag{4.6}$$

We shall now discuss another set of covariant equations leading to a harmonic oscillator spectrum, which is a generalization of the Dirac oscillator model for two particles [11]. A leading constraint for the fermion-antifermion Dirac oscillator equation has the covariant form of

$$\begin{aligned} \hat{H}_1\Psi^c \; = \; & (\hat{u}\cdot\gamma_1)(\hat{p}_1\cdot\gamma_1 - m)\Psi^c + \Psi^c(\hat{p}_2\cdot\gamma_2 + m)(\hat{u}\cdot\gamma_2) \\ & - i(\hat{u}\cdot\gamma_1)(\hat{x}_T\cdot\gamma_1)(\hat{u}\cdot\gamma_1)\Psi^c(\hat{u}\cdot\gamma_2) \\ & - (\hat{u}\gamma_1)\Psi^c(\hat{u}\cdot\gamma_2)i(\hat{x}_T\cdot\gamma_2)(\hat{u}\cdot\gamma_2) = 0, \end{aligned} \tag{4.7}$$

with $\hat{u}$ being an operator with eigenvalues u satisfying $u \cdot u = 1$, and $x_T = (1 - \hat{L})(x_1 - x_2)$ [*cf.* Eq. (3.4)]. Since the constraint $\hat{H}_1$ commutes with the total momentum operator $\hat{P}$ we shall use

$$\hat{u}\Psi^c(P) = \frac{\hat{P}}{\hat{P}^2}\Psi^c(P) = \frac{P}{P^2}\Psi^c(P).$$

If we neglect the two last terms in Eq. (4.7) corresponding to the interaction between the fermion pair, we get the two free two-fermion Dirac constraint in the form used by Barut [12]. As before, the relative time fixing constraint is taken in the simple form

$$\hat{H}_2\Psi^c = \hat{q}_L\Psi^c = 0. \tag{4.8}$$

Equation (4.7) can be rewritten in the form

$$\begin{aligned}\hat{H}_1\Psi^c \quad &= \quad \sqrt{P^2}\Psi^c + (\hat{u}\cdot\gamma_{L1})(\hat{p}_1\cdot\gamma_{T1} - m)\Psi^c + \Psi^c(\hat{p}_2\cdot\gamma_{T2} + m)(\hat{u}\cdot\gamma_{L2}) \\ &\quad - i(\hat{u}\cdot\gamma_{L1})(\hat{x}_T\cdot\gamma_{T1})(\hat{u}\cdot\gamma_{L1})\Psi^c(\hat{u}\cdot\gamma_{L2}) \\ &\quad + i(\hat{u}\cdot\gamma_{L1})\Psi^c(\hat{u}\cdot\gamma_{L2})(\hat{x}_T\cdot\gamma_{L2})(\hat{u}\cdot\gamma_{L2}),\end{aligned} \tag{4.9}$$

with

$$\gamma_{La} = \frac{(\gamma_a\cdot P)P}{P^2}, \quad \gamma_{Ta} = \gamma_a - \gamma_{La}, \quad a = 1,2.$$

It is straightforward to verify that $[\hat{H}_1, \hat{H}_2] \sim 0$. The two constraints have a particularly simple form in a c.m. frame where in addition to the relations given in Eq. (3.6) we have

$$\gamma_{La} = (\beta_a, \mathbf{0}), \ \gamma_{Ta} = (0, \alpha_a),$$

and the two constraints, Eqs. (4.9) and (4.8), reduce to

$$\begin{aligned}\sqrt{P^2}\Psi^c = M\Psi^c \quad &= \quad [\alpha\cdot\hat{\mathbf{q}}_T + m\beta]\Psi^c + \Psi^c[\alpha\cdot\hat{\mathbf{q}}_T - m\beta] \\ &\quad - i\alpha\cdot\hat{\mathbf{x}}_T\beta\Psi^c - i\beta\Psi^c\beta\alpha\cdot\hat{\mathbf{x}}_T\hat{q}_L\Psi^c = 0.\end{aligned} \tag{4.10}$$

A potential that is linear in the relative position variable in a Dirac equation leads to harmonic oscillator type spectra. Solutions to Eqs. (4.10) have been been found in Ref. [11] in terms of a noncovariant basis; a set of solutions in an arbitrary frame has been obtained by Moshinsky *et al.* [13] by boosting the noncovariant basis of solutions in a rest frame to a frame with arbitrary $\mathbf{P}$. Here however, we are interested in covariant solutions which are eigenstates of both $\hat{M}^2$ and $\hat{W}^2$. Covariant solutions for arbitrary S are of the form given by Eqs. (4.3) and (4.4). However, since neither the spinor nor the orbital part of the covariant spin operator commute with the constraints, the eigensolutions will couple solutions of different Dirac and orbital (S, V_P, V_x etc.) structures in a nontrivial way. Again as an example we look for solutions which would correspond to a basis for a "pion" state in the constituent quark model. The most general form for the state with $J^P = 0^-$ in a Dirac oscillator model has the form

$$\begin{aligned}\Psi^c_\pi(X,x;P) &= e^{-iP\cdot X}[\gamma_5 P(x) + P\cdot\gamma\gamma_5 A_P(x) + x\cdot\gamma\gamma_5 A_x(x) \\ &\quad + \epsilon_{\mu\nu\rho\sigma}P^\mu x^\nu \sigma^{\rho\sigma}T(x)].\end{aligned}$$

Substituting this general form of solution into Eq. (4.9), we get (for $x_T^2 < 0$)

$$\begin{aligned}\tfrac{1}{4}M^2 P(-x_T^2) &= (-\hat{q}_T^2 - \hat{x}_T^2 + m^2 + 3)P(-x_T^2), \\ \tfrac{1}{2}MA_P(-x_T^2) &= mP(-x_T^2), \\ \tfrac{1}{2}Mx_T T(-x_T^2) &= (\hat{q}_T + i\hat{x}_T)P(-x_T^2), \\ A_x(-x_T^2) &= 0.\end{aligned}$$

The solution of the above set is given by

$$\begin{aligned}\Psi^c_\pi(X,x;n,P) &= e^{-iP\cdot X}\left[\gamma_5 + \frac{P}{\sqrt{P^2}}\cdot\gamma\gamma_5 \right. \\ &\quad \left. + \frac{2i}{\sqrt{P^2}}\epsilon_{\mu\nu\rho\sigma}P^\mu x_T^\nu \sigma^{\rho\sigma}\right] R_{n0}(-x_T^2),\end{aligned}$$

and the masses in our model are given by

$$M^2 = 4m^2 + 16n.$$

The light cone Hamiltonian solutions are obtained in a usual way and they satisfy the same properties as the ones found for the $\hat{D}_a$-constraint model. For the pion we get a Hamiltonian basis spanned by a set of functions

$$\begin{aligned}\Psi^h_\pi(\tilde{X},\tilde{x};n,\tilde{P})_{\lambda_1,\lambda_2} &= 2\frac{P^-}{\sqrt{P^2}}e^{-i\tilde{P}\cdot\tilde{X}}R_{n0}(\tilde{x}_T^2) \\ &\quad \times\left[[\sigma_2]_{\lambda_1\lambda_2} + 4(\tilde{x}_T^1[I]_{\lambda_1\lambda_2} - i\tilde{x}_T^2[\sigma_3]_{\lambda_1\lambda_2})\right], \quad (4.11)\end{aligned}$$

and

$$\begin{aligned}\Psi^h_\pi(\tilde{P},\tilde{k};n,\tilde{P}')_{\lambda_1,\lambda_2} &= 16\pi^3\frac{P^-}{\sqrt{P^2}}\delta^3(\tilde{P}-\tilde{P}')R_{n0}(\tilde{k}_T^2) \\ &\quad \times\left[[\sigma_2]_{\lambda_1\lambda_2} + 4(\tilde{k}_T^1[I]_{\lambda_1\lambda_2} - i\tilde{k}_T^2[\sigma_3]_{\lambda_1\lambda_2})\right], \quad (4.12)\end{aligned}$$

in momentum space.

15.5 Summary and conclusions

We have given a recipe for constructing covariant states and a whole basis (complete and orthogonal in a given norm) of light cone Hamiltonian wave functions for different models of harmonic oscillator confinement. In particular Eqs. (4.5), (4.6), (4.11) and (4.12) give explicit forms for such a basis describing the quark model pion state. The light cone wave functions are extremely useful to study model dependence of the so-called static properties of the hadronic states, magnetic moments, charge radii, form factors, etc. [5], [14]. For example, an exact QCD result for the pion form factor is expressed in terms of the light cone partonic amplitudes, and can be symbolically written as [5]

$$F_\pi(Q^2) = \sum_n \int [d\tilde{k}] \sum_\lambda \psi^\dagger_{n/\pi}(\tilde{k}, \lambda)\psi_{n/\pi}(\tilde{k}, Q, \lambda),$$

where the amplitudes $\psi_{n/\pi}(\tilde{k}, \lambda)$ are the light cone wave function of the n-partonic contribution to the pion. The calculations of the variety of static properties of the pion is thus straightforward with the use of the basis constructed above. The existence of complete bases is of particular interest, since it allows for dynamical calculations, mass spectra, transition rates, in addition to simple static calculations. For example, with QCD-inspired effective potentials in a valence Fock sector (as the Richardson potential for heavy quarkonia [15]) the above covariant bases can be used to obtain realistic wave functions in a fully relativistic constituent quark model.

15.6 Appendix A

The relation between the on-shell spin-$\frac{1}{2}$ light cone wave function,

$$\Psi^h(\tilde{p})_\lambda = {}^h\langle\tilde{p}, \lambda|\Psi\rangle,$$

and a covariant 4-component one $(\alpha = (\gamma = (+,-), \lambda = (-\frac{1}{2}, +\frac{1}{2})))$,

$$\Phi^c(x)_\alpha = (\Phi^c(x)_{+,\lambda}, \Phi^c(x)_{-,\lambda}),$$

is given by [2]

$$\begin{aligned}\Phi^c(x)_{+,\lambda} &= \int \frac{dp^- d^2p^s}{16\pi^3(p^-)^{1/2}} \\ &\quad \times \exp\left[-i\left(p^- x^+ + \frac{m^2 + p^s p^s}{2p^-} x^- + p^s x_s\right)\right] \Phi^h(\tilde{p})_\lambda, \\ \Phi^c(x)_{-,\lambda} &= 2(\partial_+)^{-1}(m - \epsilon_{rs}\sigma_r\partial_s)\Phi^c(x)_{+,\lambda}.\end{aligned}$$

The covariant wave function Φ^c is a positive energy solution of the Dirac equation

$$(-i\partial \cdot \gamma - m)\Phi^c = 0,$$

with

$$\gamma^0 = \begin{pmatrix} 0 & -i \\ i & 0 \end{pmatrix}, \quad \gamma^1 = \begin{pmatrix} i\sigma_2 & 0 \\ 0 & -i\sigma_2 \end{pmatrix},$$
$$\gamma^2 = \begin{pmatrix} -i\sigma_1 & 0 \\ 0 & i\sigma_1 \end{pmatrix}, \quad \gamma^3 = \begin{pmatrix} 0 & -i \\ -i & 0 \end{pmatrix}.$$

15.7 REFERENCES

[1] P.A.M. Dirac, *Rev. Mod. Phys.* **21**, 392 (1949).

[2] H. Leutwyler and J. Stern, *Ann. Phys.* **112**, 94 (1978).

[3] S. Weinberg, *Phys. Rev. B* **133**, 1318 (1964).

[4] E. Wigner, *Ann. Math.* **40**, 149 (1939).

[5] S.J. Brodsky, T. Huang, and G.P. Lepage, in *Quarks and Nuclear Forces*, Ed. by D. Frios and B. Zeitnitz, Springer Tracts on Modern Physics Vol. 100 (Springer, New York, 1982).

[6] P.A.M. Dirac, *Lecture on Quantum Mechanics*, Yeshiva University, Belfer Graduate School of Science (New York, 1964).

[7] R.P. Feynman, M. Kislinger, and F. Ravndal, *Phys. Rev. D* **3**, 2706 (1971).

[8] M.G. Fuda, *Phys. Rev. D* **42**, 2898 (1990).

[9] H.W. Crater and P. Van Alstine, *Phys. Rev. D* **36**, 3007 (1987); H. Sazdjian, *Phys. Rev. D* **33**, 3401 (1986); S.J. Wallace, in *Proceedings of the Relativistic Nuclear Many-Body Physics Workshop.* Ohio State University, June 1988 (World Scientific, Singapore, 1989).

[10] S. Godfrey and N. Isgur, *Phys. Rev. D* **32**, 189 (1985), and references therein.

[11] M. Moshinsky and A. Szczepaniak, *J. Phys. A* **22**, (1989); M. Moshinsky, G. Loyola, and A. Szczepaniak, The two body Dirac oscillator. In *Anniversary Volume in Honor of J.J. Giambiagi* (World Scientific, Singapore, 1989).

[12] A.O. Barut and S. Komy, *Fortsch. Phys.* **33**, 309 (1985); A.O. Barut and G.L. Srobel, *Few-Body Sys.* **1**, 167 (1985).

[13] M. Moshinsky, G. Loyola and C. Villegas, *J. Math. Phys.* **32**, 372 (1991).

[14] Z. Dziembowski, *Phys. Rev. D* **37**, 706 (1988), and references therein.

[15] J. Richardson, *Phys. Lett. B* **82**, 272 (1979).

16

Doorway States in Classical Physics

J. Flores[1], T.H. Seligman, J.M. Alvarez-Tostado, O. Novaro, J.L. Mateos[2]

ABSTRACT The study of two coupled systems having discrete eigenvalues leads naturally to the existence of doorway states, whether the coupled systems have a classical or quantal nature. Doorway states in classical phenomena are here exemplified by a membrane coupled to an acoustic cavity. These ideas are useful also to understand seismic data collected in Mexico City for recent earthquakes, which are naturally interpreted along the lines of the acoustic model presented here.

16.1 Introduction

In 1971 Marcos Moshinsky developed a mechanical model for doorway states consisting of an array of springs and strings which behave according to classical physics [1]. In our recent scientific excursion into the study of seismic response of sedimentary basins [2], [3], we have found evidence that classical doorway states might be relevant in this field also. Due to the classical nature of the seismic problem and true to Marcos' idea of using simple models, we shall here develop an acoustical model which is useful to clarify the doorway state concept and which is almost exactly soluble in the sense that for simple shapes it requires at most one numerical step.

The seismological problem at hand is the seismic response of the Valley of Mexico. Here the seismic waves, generated at the source travel through bedrocks and enter a basin consisting of two coupled geological formations: the consolidated alluvial sediments which form the bulk of the basin material and the very soft lacustrine clays that partially cover these sediments in several regions of the basin's surface [2]. The elastic waves enter through the sediments and reach the clays over which a sizeable part of the downtown area lies and where almost all seismic damage is historically located

[1]Centro Universitario de Comunicación de la Ciencia, UNAM.
[2]Instituto de Física, UNAM.

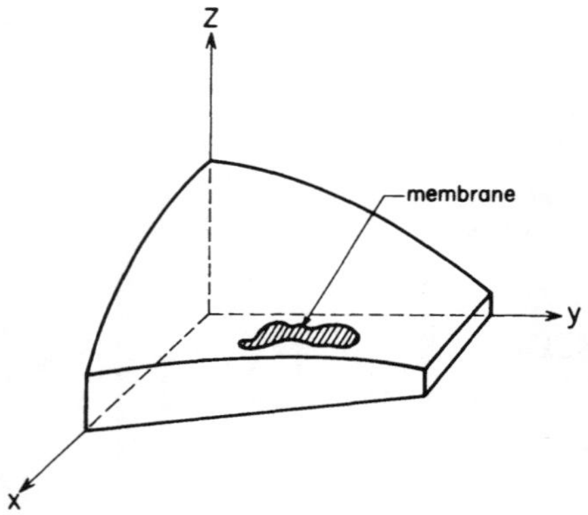

FIGURE 16.1. The coupled system formed by an acoustic cavity with an elastic membrane. Both cavity and membrane can be of arbitrary shape.

in Mexico City earthquakes. In particular, the strong motion at the clay region produced the tragedy caused by the September 1985 earthquakes.

We have shown [3] that, due to this geological structure, a resonant two-step process takes place which causes the tremendous amplification and long duration at certain frequencies of the seismic motion at the soft-clay zones. If we consider the sediments and soft-clays as two coupled systems, it is reasonable to assume that we are facing a typical doorway-state phenomenon. Indeed, the purpose of this paper is to show that this phenomenon exists in elastic and acoustic systems and, furthermore, that it is actually easier to understand the doorway state concept for these systems than it is in nuclear physics [4]. Here enters the model.

We consider an acoustic resonator with stiff walls except for an elastic membrane which forms part of a flat portion of the resonator (See Fig. 16.1). This problem has been dealt with before (see, for example Ref. [5]) but we shall treat it here from the point of view of doorway states. The membrane provides us with a perfect set of "doorway" states, while the resonator contains the next generation of "compound" states. Since any exterior wave has to pass through the membrane before reaching the cavity, the concept of two generations of states is in our case quite literal.

In Section 2 we define the acoustic model and its parameters. We then outline the technique used to solve the problem in Section 3. In Section 4 we present numerical results for stationary states, analyzing how the strength of the coupling between membrane and cavity affects our results; we study also the effects of the symmetries of the problem. Finally, some general conclusions for the concept of doorway state in classical physics are drawn and we show how they apply to interpret the seismic data obtained in Mexico City.

16.2 The acoustical model

We consider the system shown in Fig. 16.1, formed by a cavity of arbitrary shape bounded in some flat region by a plane elastic membrane, also of arbitrary shape. Except for the membrane, the cavity walls are perfectly rigid. The interior and exterior regions of the cavity are filled with two different fluids. We see that, in distinction to quantum–mechanical problems, we might switch off the external coupling by choosing the outside to be the vacuum. We should study the scattering of sound waves from the membrane, as the rest of the system is stiff and therefore only susceptible of shape–elastic (geometric) scattering. As is well known from scattering theory [6], resonant effects are dominated by the internal states of the system as long as the external coupling is not so strong as to hide the very effect of resonant scattering.

For the above model we have a large number of free parameters that allow us to play with all kinds of relations between cavity and membrane. The three relevant characteristics from the physical point of view are: the cavity level density, that of the membrane, and the membrane–cavity coupling. Each of these characteristics depends on several of the free parameters. The strength of the coupling depends on the shape and size of the membrane and on the mass density of both membrane and fluid inside the cavity. The distribution of the coupling between different states depends sensitively on the geometry, both through exact and approximate symmetries and through the node patterns. The latter may arise more or less accidentally and play an important role, particularly if the wavelength in the cavity is large compared to the size of the membrane. The cavity level density, in turn, depends on the fluid mass density although can always be adjusted independently by changing the size of the cavity. The level density of the membrane depends on its size, its surface mass density σ and tension τ.

In any case, we see that we have a sufficient number of free parameters to change the three relevant physical characteristics. In particular, we can alter the membrane level density, even if we insist, as we shall do in the next section, on large tensions in order to assure small displacements of the membrane. For our purposes in this paper, we require a low membrane level density and a large one for the cavity; the coupling should be effective over a range of five to twenty cavity states.

The concept of doorway state arises from our model in the most natural way, since any wave from the exterior will have to pass through the membrane before exciting the cavity states. We therefore hope that the classical doorway states are more easily visualizable than their quantum–mechanical counterparts. In a similar fashion, other concepts of complex scattering and radiation dynamics —such as the spreading width and the escape width—, become equally simple to understand, and their properties can be easily displayed by analytical, numerical or even experimental evidence.

16.3 The mathematical setting

We found that a reasonably simple solution of the acoustical problem could be obtained assuming small displacements of the membrane [7]. This is easy to fulfill for the regime we are interested in, as a large tension implies a low membrane level density and also small membrane displacements, even for large sound intensity in the cavity. We can now follow Morse and Feshbach and obtain the equations for the velocity potential $\bar{\psi}$ in the cavity and for the membrane displacement $\bar{\phi}$.

Both $\bar{\phi}$ and $\bar{\psi}$ obey wave equations:

$$\nabla^2\bar{\phi} - \frac{1}{c_1^2}\frac{\partial^2\bar{\phi}}{\partial t^2} = -\frac{\bar{p}}{\tau}, \tag{3.1}$$

$$\nabla^2\bar{\psi} - \frac{1}{c_2^2}\frac{\partial^2\bar{\psi}}{\partial t^2} = 0, \tag{3.2}$$

where $c_1 = \sqrt{\tau/\sigma}$ and c_2 are the membrane and fluid wave velocities respectively. The acoustic pressure $\bar{p}$ produces the force that drives the membrane, and is related to $\bar{\psi}$ by

$$\bar{p} = -\rho\frac{\partial\bar{\psi}}{\partial t}. \tag{3.3}$$

where ρ is the fluid mass density.

The functions $\bar{\phi}$ and $\bar{\psi}$ are coupled also through the boundary conditions. The membrane displacement $\bar{\phi}$ should be zero along the rim of the membrane, since it is fixed to the rigid walls of the cavity. On the other hand, $\bar{\psi}$ should obey Neumann boundary conditions

$$\frac{\partial\bar{\psi}}{\partial n} = 0 \tag{3.4}$$

at all points of the cavity walls except at the membrane, where both the velocity of the fluid and that of the membrane should coincide:

$$\frac{\partial\bar{\psi}}{\partial n} = \frac{\partial\bar{\phi}}{\partial t}. \tag{3.5}$$

Since we are interested in obtaining the normal modes of the coupled system, we assume harmonic motion with frequency ω and call $\phi(x,y)$ and $\psi(x,y,z)$ the time independent part of $\bar{\phi}$ and $\bar{\psi}$, respectively. We then use the Green's function method, and write

$$\begin{aligned}\phi(x,y) &= -\int_S G_1(x,y\,|\,x_0,y_0;\omega)\frac{p(x_0,y_0,0)}{\tau}dS \\ &= -\omega\frac{\rho}{\tau}\int_S G_1(x,y\,|\,x_0,y_0;\omega)\,\psi(x_0,y_0,0)\,dS,\end{aligned} \tag{3.6}$$

where G_1 is the membrane Green's function, given in terms of the eigenmodes ϕ_n and eigenvalues Ω_n by

$$G_1(x, y|x_0, y_0; \omega) = c_1^2 \sum_n \frac{\phi_n(x, y)\phi_n(x_0, y_0)}{(\omega^2 - \Omega_n^2)} . \tag{3.7}$$

In a similar fashion, we have

$$\begin{aligned} \psi(\mathbf{r}) &= -\int_S G_2(x, y, z \mid x_0, y_0, 0; \omega) \frac{\partial \psi(x_0, y_0, 0)}{\partial n} dS \\ &= -\omega \int_S G_2(x, y, z \mid x_0, y_0, 0; \omega)\, \phi(x_0, y_0)\, dS, \end{aligned} \tag{3.8}$$

where (3.5) was taken into account. Here G_2 is the fluid Green's function, given by

$$G_2(x, y, z \mid x_0, y_0, 0; \omega) = c_2^2 \sum_n \frac{\psi_n(x, y, z)\psi_n(x_0, y_0, 0)}{(\omega^2 - \omega_n^2)} , \tag{3.9}$$

where ψ_n and ω_n are the acoustic eigenfunctions and eigenvalues of the cavity, respectively. In Eqs. (3.6) and (3.8), the integration is over the surfaces occupied by the membrane in the plane $z = 0$.

If we now expand ψ and ϕ in terms of the corresponding eigenstates,

$$\psi = \sum_{n=1}^{N} A_n \psi_n \qquad \phi = \sum_{m=1}^{M} B_m \phi_m , \tag{3.10}$$

and substitute them in Eqs. (3.6) and (3.8), we obtain, after taking advantage of the orthogonal properties of ϕ_n and ψ_n, the following set of coupled equations:

$$\begin{aligned} (\omega^2 - \omega_l^2) A_l + \omega c_2^2 \sum_{m=1}^{M} B_m I_{ml} &= 0, \\ (\omega^2 - \Omega_k^2) B_k + \omega \frac{\rho}{\tau} c_1^2 \sum_{n=1}^{N} A_n I_{kn} &= 0, \end{aligned} \tag{3.11}$$

where the integrals

$$I_{ij} = \int_S \phi_i(x, y) \psi_j(x, y, 0) dS \tag{3.12}$$

depend on the shapes of both membrane and cavity.

The system of equations (3.11) can be put in matrix form as follows:

$$(\omega^2 \mathcal{R} + \omega \mathcal{K} - \mathcal{S}) \mathbf{X} = 0, \tag{3.13}$$

where

$$\mathcal{R} = \begin{pmatrix} \mathcal{I}_N & 0 \\ 0 & \Lambda \mathcal{I}_M \end{pmatrix}, \tag{3.14a}$$

$$\mathcal{K} = \begin{pmatrix} & & & c_0^2 I_{11} & \dots & c_0^2 I_{M1} \\ & 0 & & \vdots & \ddots & \vdots \\ & & & c_0^2 I_{1N} & \dots & c_0^2 I_{MN} \\ c_0^2 I_{11} & \dots & c_0^2 I_{1N} & & & \\ \vdots & \ddots & \vdots & & 0 & \\ c_0^2 I_{M1} & \dots & c_0^2 I_{MN} & & & \end{pmatrix}, \tag{3.14b}$$

and

$$\mathcal{S} = \begin{pmatrix} \omega_1^2 & & & & & \\ & \ddots & & & 0 & \\ & & \omega_N^2 & & & \\ & & & \Lambda\Omega_1^2 & & \\ & 0 & & & \ddots & \\ & & & & & \Lambda\Omega_M^2 \end{pmatrix}. \tag{3.14c}$$

Here $\mathcal{I}_j$ is the $j \times j$ unit matrix and $\Lambda = \sigma c_2^2/\rho$. The matrix $\mathcal{K}$ is real and symmetric. In Eq. (3.13) the vector $\mathbf{X}$ is an $(N + M)$-dimensional vector

$$\mathbf{X} = \begin{pmatrix} A_1 \\ \vdots \\ A_N \\ B_1 \\ \vdots \\ B_M \end{pmatrix} \tag{3.15}$$

formed by the expansion coefficients of Eq. (3.10).

The system of equations (3.13) can be converted into an eigenvalue problem by doubling the space dimensionality, using a $2(N + M)$ dimensional vector

$$\chi = \begin{pmatrix} \mathbf{X} \\ \mathbf{Y} \end{pmatrix}, \qquad \text{with} \qquad \mathbf{Y} = \omega \mathbf{X}. \tag{3.16}$$

It becomes

$$\begin{pmatrix} 0_{M+N} & \mathcal{I}_{M+N} \\ -\mathcal{R}^{-1}\mathcal{S} & \mathcal{R}^{-1}\mathcal{K} \end{pmatrix} \chi = \omega \chi, \tag{3.17}$$

which can be solved by the usual methods of diagonalization. The matrix 0_j is the null $j \times j$ matrix.

We have thus reduced the equations of motion to matrix form and can truncate the basis for different values of N and M. All we need to know are the eigenstates of cavity and membrane in the range of frequencies we are

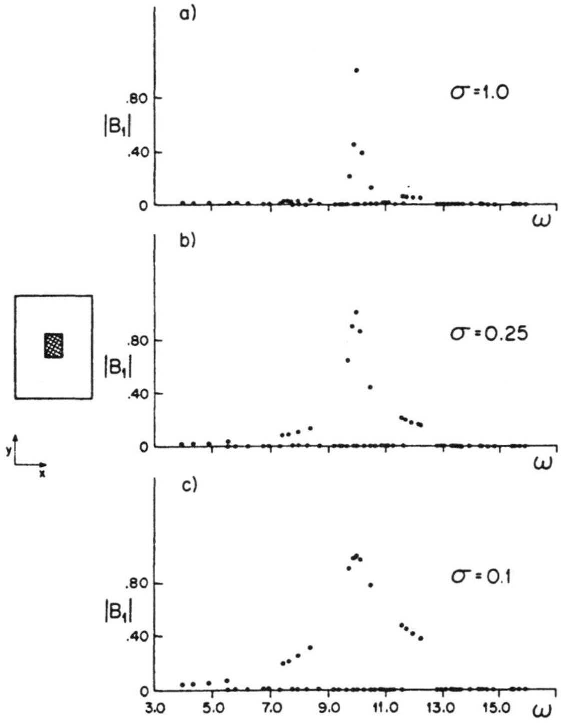

FIGURE 16.2. The coefficient $|B_1|$ depicted as a function of the eigenvalues ω of the coupled system. As shown in the inset, the rectangular membrane is located at the center of the smallest wall of the rectangular cavity. We show three cases a, b and c, which correspond to increasing coupling strength. The latter grows inversely with the surface mass density σ. A spreading width $\Gamma^{\downarrow}$ cannot be defined, due to the strong fluctuations of $|B_1|$ as a function of ω.

interested in. In this work we shall use simple shapes: rectangular cavities and membranes. For these, eigenstates and eigenfrequencies are available analytically. The localization of the membrane is arbitrary, thus permitting a variable coupling. By changing the membrane's position we can destroy or preserve symmetry. If we wish to use arbitrary shapes these could be considered by numerical methods.

16.4 Numerical results

We now show a set of figures which illustrate the effect of the coupling strength between membrane and cavity states. We emphasize the significance of the location of the membrane in the cavity wall, as well as the influence on the results of the coupling strength and distribution.

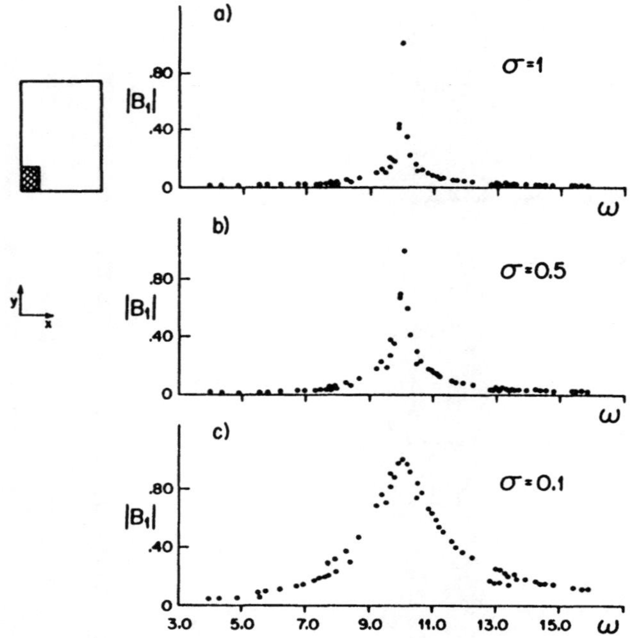

FIGURE 16.3. The coefficient $|B_1|$ as a function of the ω for an asymmetric position of the membrane. The coupling strength increases from figures *a* to *c* and so do the number of states affected. Due to the smoothness of the graphs, specially for case *c*, a spreading width $\Gamma^{\downarrow}$, which increases with the coupling strength, is now well defined.

The cavity was chosen to be a rectangular box with incommensurable edges $L_x = 1$, $L_y = \sqrt{2}$ and $L_z = \pi$. The membrane, placed on the smallest face of the cavity, was chosen in turn as a rectangle similar to this face and with its sides reduced by a factor of ten. The position of the membrane was taken either in the center (Fig. 16.2) or in a corner (Fig. 16.3). We consider these cases for different coupling strengths between membrane and cavity. The change of the coupling strength was achieved varying simultaneously the density σ and the tension τ of the membrane in such a way that the eigenfrequencies of the membrane are not affected. The ratio of level densities of membrane to cavity was chosen of the order of 1/100. This large difference of densities allows us to focus in only one membrane state, making the analysis of its role as a doorway state simpler.

What we are really interested in is the admixture of this membrane state with the eigenstates of the coupled system. This admixture is crucial to calculate the probability with which an external wave can populate a given resonant state [4]. As we have seen in the previous section, in particular in Eqs. (3.10) and (3.15), this admixture is characterized by the absolute value $|B_i|$ of the overlap of the i-th membrane state with the coupled state

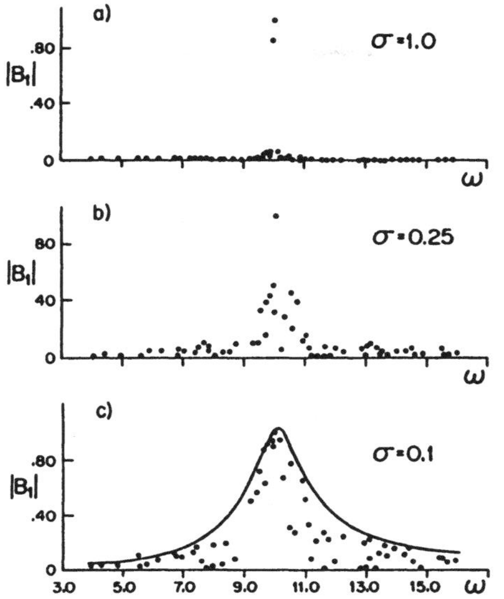

FIGURE 16.4. The coefficient $|B_1|$ as a function of the ω for an asymmetric position of the membrane. The coupling strength increases from figures a to c and so do the number of states affected. Due to the smoothness of the graphs, specially for case c, a spreading width $\Gamma^{\downarrow}$, which increases with the coupling strength, is now well defined.

under consideration. Since our choice of the ratio of level densities allows us to concentrate on the lowest membrane state, $|B_1|$ will be the quantity of interest here. In the following set of figures, we therefore present $|B_1|$ as a function of the eigenfrequencies of the coupled system, and see under what conditions can ϕ_1 be considered a doorway state.

Let us first locate the membrane in a symmetrical position, as indicated in the inset of Fig. 16.2. In this case, the symmetry of box and membrane are closely related to each other and exact and approximate selection rules for the coupling between states exist. Indeed, certain coupled states are not excited at all. For example, the membrane ground-state wave function can not excite any cavity function having a nodal plane which divides the membrane into two equal parts. As a consequence of the symmetry $|B_1|$ collapses to zero for several frequencies, and its distribution becomes erratic. It does not have a well defined envelope, as can be seen in Fig. 16.2. The very concept of spreading width, which would be the width of the distribution of $|B_1|$ around Ω_1 and called $\Gamma^{\downarrow}$ in the nuclear physics literature, is not well defined. The same effect has been found for certain quantum systems [8]. This conclusion is independent of the coupling strength, which grows as the surface mass density σ decreases, as will be seen by comparing Figs. 16.2a, 16.2b and 16.2c.

We now place the membrane asymmetrically at one corner of the cavity, as indicated in the inset of Fig. 16.3. Now the selectivity of states disappears and $|B_1|$ as a function of ω follows a smooth curve. We see that the spread of membrane amplitudes over the frequency spectrum is in good approximation a Lorentzian, independently of the coupling strength. The width $\Gamma^{\downarrow}$ of these distributions is now well defined and increases with the coupling strength, as can be seen by comparing Figs. 16.3*a*, 16.3*b*, and 16.3*c*.

The calculations for complicated shapes of cavity and membrane are basically the same as above. The eigenfunctions ψ_n and ϕ_n of Eq. (3.10), however, must be obtained numerically. For the eigenfunction of the membrane, *i.e.* for a two-dimensional problem with Dirichlet boundary condition this is quite simple. For the cavity, on the other hand, the problem has one more dimension and because of the Neumann boundary condition its solutions become more involved. Yet it is for the latter that we need a large number of states. In view of these difficulties a complete calculation for such a situation was postponed and we decided to simulate a chaotic coupling by choosing the integrals I_{ij} of Eq. (3.12) as independent Gaussian random variables. This assumption is plausible for these integrals if a least one of the two systems is chaotic and we consider highly-excited states. According to an argument by Berry [9] the wave function should tend to Gaussian white noise at high excitations, whereas scars [10] –the major known deviation from Gaussian noise– have integrals that vanish for increasing excitation. Figure 16.4. It shows the result of such a calculation. The distribution of amplitudes now fluctuates but not as much as in the more symmetrical case of Fig. 16.2. An envelope can still be adjusted as done in Fig. 16.4*c*, where the Lorentzian curve taken from the corresponding case of Fig. 16.3*c* is shown for comparison. The results are therefore intermediate between the symmetrical and asymmetrical positions of the membrane. Note that these spectra (Figs. 16.3 and 16.4) with different degrees of fluctuations evoke the image of intermediate structures, as known in Nuclear Physics [11].

16.5 Conclusions

If we wish to interpret the above results in terms of scattering theory, the width observed in the Figs. 16.3 and 16.4 would correspond to the spreading width $\Gamma^{\downarrow}$. This quantity $\Gamma^{\downarrow}$ determines the transfer to the cavity except in cases where the coupling to the exterior region could lead to an escape width $\Gamma^{\uparrow}$ which would be much larger than $\Gamma^{\downarrow}$. We may therefore predict that for well broken symmetries, where $\Gamma^{\downarrow}$ may be defined rigorously, the concept of doorway state is well founded. In other words, the spreading width $\Gamma^{\downarrow}$ will correspond to a frequency range centered around the eigenmodes of the membrane. These are then the doorway states which allow the incoming

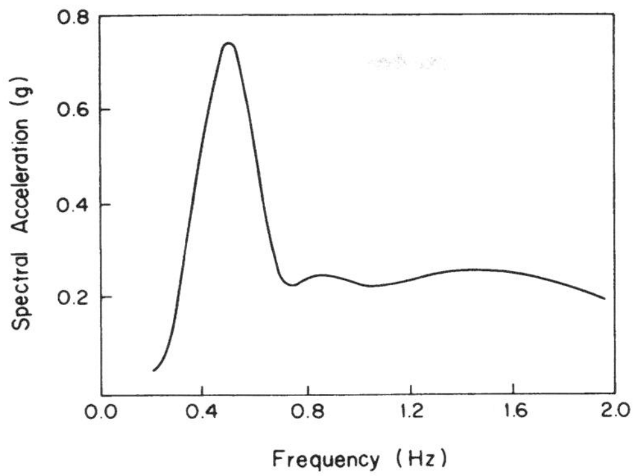

FIGURE 16.5. Spectral accelerations at a clay site during the September 19, 1985 earthquake (magnitude $M = 8.1$) given in units of the gravity constant g. Notice that the largest accelerations ($\geq g/2$) are all for frequencies clustered around the doorway state frequency of the sediments, ~ 0.45Hz.

waves to penetrate and excite the eigenstates of the cavity that fall precisely within those frequency ranges.

Assuming poor resolution, dips corresponding to the membrane frequencies in the absorption spectrum and also peaks in the excitation spectrum of the cavity would be found. The spectra would thus show a typical intermediate structure and the very fact that this structure is associated to the membrane states, through which all waves have to pass before reaching the interior, justifies the name of doorway state. Under well-defined circumstances those intermediate structures become giant resonances as discussed in [12].

We have thus presented an in principle exactly solvable model of two coupled systems having discrete elastic or acoustic eigenstates and which presents the general characteristics of doorway states and intermediate structures. That classical systems show this behaviour –generally associated to quantum phenomena– may or may not be considered surprising. In any case this situation is not at all limited to the simple example of the cavity-membrane coupled system discussed above. In fact it could be expected that, in general, elastic waves entering coupled systems might present a similar behaviour.

Let us now return to the problem of the seismic response of the Valley of Mexico and view it in this light. It seems natural to assume that the eigenstates of the sediment may act as doorway states for the clays and it is easy to verify that the spectral density of the sediment is very low as compared to the one of the clays, due to the large differences in the velocities of the seismic waves involved. Thus only those frequencies that are clustered around the valley sediment's eigenmodes (known to have their lowest frequency around 0.45 Hz [13]) should be observed in the accelerograms located in downtown Mexico City. This is indeed the case as shown by the Fourier response spectrum of the large earthquake of September 19, 1985. Very large amplifications in the soft-clay are centered around the sediment's lowest eigenmode which acts as a doorway state. The form of the Fourier spectra is in fact almost a textbook image of a giant resonance as shown in Fig. 16.5.

In order to have a giant resonance we know that we need doorway states plus a mechanism that concentrates most of the strength in a single state; such mechanism is not contained in our acoustic model but is discussed for the situation of the Valley of Mexico in Ref. [14]. In this particular case, the giant resonance may be the culprit of the tragedy that downtown Mexico City experienced in 1985 (we must remember that most of the collapses occurred in buildings which had or reached precisely the frequency of 0.45 Hz in their structural oscillations, *i.e.*, buildings between 6 and 12 stories high), and therefore a thorough understanding of the coupling between the sedimentary valley and the ancient lake bed clays is a matter of life and death.

We see then that exactly solvable models of classical doorway states, intermediate structures and giant resonances help us to understand more precisely the seismic response of sedimentary basins with very soft clays, a situation commonly present in several urban centers in Mexico, California, Japan, and other parts of the world.

16.6 References

[1] Mello, P.A. and Moshinsky, M., Mechanical models for reactions involving isobaric analog and doorway states, *Am. J. Phys.* **39**, 54–65 (1971).

[2] Flores, J., Novaro, O., and Seligman, T.H., Possible resonance effect in the distribution of earthquake damage in Mexico City, *Nature,* **326**, 783–785 (1987).

[3] Seligman, T.H., Alvarez-Tostado, J.M., Mateos, J.L., Flores, J., and Novaro O., Resonant response models for the Valley of Mexico-I. The elastic inclusion approach, *Geophys. J. Int.* **99**, 789–799 (1989).

[4] Feshbach, H., Kerman, A.K. and Lemmer, R.H., Intermediate structure and doorway states in nuclear reactions, *Ann. Phys.* (New York) **41**, 230–286 (1967).

[5] Baker, W.P., Kriegsmann, G.A., and Reiss, E.L., Acoustic scattering by baffled cavity-backed membranes, *J. Acoust. Soc. Am.* **83**, 423–432 (1988).

[6] Newton, R.G., *Scattering theory of waves and particles* (McGraw-Hill, New York, 1966).

[7] Morse P.M., and Feshbach, H., *Methods of Theoretical Physics* (McGraw-Hill, New York, 1953).

[8] Weidemüller, H.A. and Seligman, T.H., work in progress (1991).

[9] Berry, M.V., In: *Chaotic Behaviour of Deterministic Systems*, Ed. by G. Iooss, R. Hellman, and R. Stora, Les Houches Summer School Proceedings No. 36 (North-Holland, New York, 1981). Chap 4.

[10] Heller, E.J., Bound-state eigenfunctions of classically chaotic Hamiltonian systems: Scars of periodic orbits, *Phys. Rev. Lett.* **53**, 1515 (1984).

[11] Temmer, G.M., Isospin and its consequences in nuclear physics, In: *Fundamentals in Nuclear Theory*, Chap. III, Ed. by A. de-Shalit and C. Villi (IAEA Vienna, 1967).

[12] Bromley, D.A., Trends in nuclear physics: The challenge of precision, In: *Proceedings of the International Conference on Nuclear Physics II*, Ed. by J. de Boer and H. J. Mang, (North-Holland, Amsterdam, 1973).

[13] Herrera, I.E., Rosenbluth, E. and Rascón, O.A., In: *Proceedings of the Third World Conference on Earthquake Engineering*, Ed. by Van Roekel, J.H. (Wellington, New Zealand, 1965) p. I61-I74.

[14] Mateos, J.L., Flores, J., Novaro O., Alvarez-Tostado, J. M., and Seligman, T.H., Generation of evanescent P-waves and site effects in the Mexico City basin *Tectonophysics* (submitted, 1991).

17

Resonant States and the Decay Process

Gastón García-Calderón[1]

ABSTRACT It is shown that the time evolution of the nondecay amplitude, for an arbitrary initial state and an interaction that vanishes beyond a distance, may be written as an infinite sum of terms involving the full set of resonant states associated to the complex energy eigenvalues of the problem. Each term of the sum consists of the product of coefficients formed by the overlap of the initial state with the respective resonant state multiplied by a Moshinsky function that depends on the corresponding complex energy eigenvalue and the time. The above representation is used to obtain an exact one-level formula for the decay amplitude and to discuss the nonexponential contributions to the decay process.

17.1 Introduction

The time evolution of quantum decay is a subject as old as quantum theory. Around two years after Schrödinger published his papers on wave mechanics, Gamow [1] and Gurney and Condon [2] considered the problem of α-decay in atomic nuclei and found that the desintegration rate of α particles with time follows the exponential decay law $\exp(-t/\tau)$ where τ stands for the lifetime. Subsequent theoretical work, however, pointed out to an approximate validity of the exponential decay. The pionnering work in 1957 by Khalfin [3] and since then by numerous authors [4]-[16] has led to the conclusion that deviations from purely exponential behaviour are to be expected at very short and very long times compared with the lifetime of the system. It is of interest to mention that most experiments supporting the validity of the exponential decay law [17]-[19] refer to the long time region covering a span of time that goes from a few up to around 60 lifetimes. In fact the first experiment reporting a measurement of the short time region was published in 1988 [19]. So far no experiment has provided evidence for deviations from purely exponential decay.

Most studies on the time evolution of decay refer to spherically symmetric systems. Regarding the long time region $t \gg \tau$, there seems to be a

[1]Instituto de Física, UNAM.

consensus that the nondecay amplitude should behave, for the case of zero angular momentum, as $1/t^3$. However for short times $t \ll \tau$ one finds no agreement among different approaches. They may even lead to results that seem paradoxical [20].

The description of nonexponential terms of decay seems to depart from the simplicity of the description of the purely exponential part as obtained by Gamow [1], where two quantities seem to suffice to describe the nondecay amplitude: ϵ, that represents the energy of the decaying fragment, and the width Γ that stands for the decay rate and in fact provides an estimate of the lifetime of the systen through the well known relationship $\tau \approx \hbar/\Gamma$.

In order to describe the decay process Gamow [1] considered solutions of the Schrödinger equation which at large distances consist of purely outgoing waves. This leads to complex energy eigenvalues and correspondingly to a definition of resonant states for which the density at each point of space decays exponentially in time, *i.e.*, $|u_n(r,t)|^2 \propto \exp[-\Gamma_n t/\hbar]$. However the amplitude of these wave functions increases exponentially in space, representing the presence, at large distances, of particles emitted earlier from a stronger source. Due to this, the usual rules for normalization and completeness of eigenfunctions are not valid, and consequently, it has been generally believed that resonant or Gamow states are not a useful notion for calculations. However during the last two decades or so studies on the properties of resonant states have led to a consistent normalization rule for these states and also to learn how to use them in eigenfunction expansions [21]. In particular resonant states may be regarded as part of a complete set of states, containing also bound states and a continuum of wave functions [22].

The aim of this work is to show that the nondecay amplitude for the time evolution of decay may be written as a infinite sum of terms involving resonant states and the corresponding complex energy eigenvalues. Each term of the sum consists of coefficients formed by the overlap of the initial wave function with the respective resonant state multiplied by a Moshinsky function that depends on the complex energy eigenvalue and the time. As a result it is obtained that in addition to an exponential decay term each resonant state and its corresponding complex energy eigenvalue yields also a contribution that decays nonexponentially.

17.2 The nondecay amplitude

The relevant quantity for the time evolution of decay is the nondecay amplitude $A(t)$,

$$A(t) = \int \phi^*(r,0)\phi(r,t)\,dr. \tag{2.1}$$

The above expression provides the probability amplitude that at time t a system characterized by a Hamiltonian H remains in the initial state $\phi(r, t = 0)$. The corresponding nondecay probability is therefore given by $|A(t)|^2$. For reasons that will become clear below it is assumed that the interaction $V(r)$ is of finite range. For the sake of simplicity the discussion will refer to the case of a particle of zero angular momentum. The Hamiltonian of the problem is

$$H = -d^2/dr^2 + V(r). \tag{2.2}$$

Eq. (2.1) may also be written as

$$A(t) = \int_0^R dr \int_0^R dr'\, \phi^*(r,0) g(r,r';t) \phi(r',0), \tag{2.3}$$

where $\phi(r,0)$ and $\phi(r',0)$ stand for the initial arbitrary states prepared at time $t = 0$ confined initially within the interaction region $r < R$ and $g(r,r';t)$ is the retarded time dependent Green function that is responsible for the time evolution of the problem since

$$\phi(r,t) = \int_0^R g(r,r';t) \phi(r',0)\, dr'. \tag{2.4}$$

17.3 Time-dependent Green function and resonant states

The retarded time-dependent Green function may be written, using Laplace transform techniques, as [22],

$$g(r,r';t) = \frac{i}{2\pi} \int_{ic_o-\infty}^{ic_o+\infty} G^+(r,r';k) e^{-ik^2t}\, 2k\, dk, \tag{3.1}$$

where $G^+(r,r';k)$ corresponds to the outgoing Green function associated to the Schrödinger equation of the problem and the integration contour c_o in the $k-$plane is shown in Fig. 17.1. The units utilized are $\hbar = m = 1$ and consequently the energy E and the wave number k are related by $E = k^2$. The outgoing Green function or propagator is an appropriate quantity to establish a relationship with the resonant states of the problem. The connection between decay and the complex poles of the propagator goes back to some remarks by Peierls [23]. In particular the idea that the residues at the complex poles of the propagator correspond to the resonant states of the problem has been considered by a number of authors [22], [24]-[27]. It is well known that for a finite range interaction the outgoing Green function $G^+(r,r';k)$, as a function of k, can be extended analytically to the whole complex $k-$plane where it possess an infinite number of poles

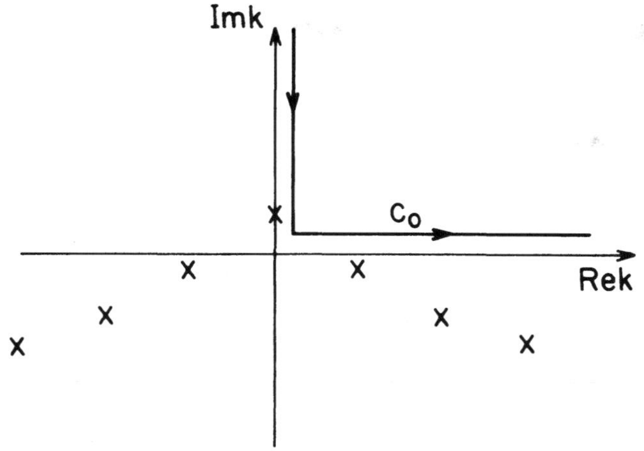

FIGURE 17.1. Integration contour C_o and location of poles in k-plane.

distributed in a well known form [21], [28]. Purely imaginary poles situated on the upper half $k-$plane correspond to bound states of the problem whereas purely imaginary poles seated on the lower half $k-$plane are related to antibound states. On the other hand complex poles are only found on the lower half $k-$plane and are associated with resonant states. To each complex pole at $k_n = \alpha_n - i\beta_n$ with $\alpha_n, \beta_n > 0$ there corresponds, from time reversal invariance, a complex pole k_{-n} situated symmetrically with respect to the imaginary axis, *i.e.*, $k_{-n} = -k_n^*$. A few poles are exhibited schematically in Fig. 17.1.

Resonant states are defined as solutions of the Schrödinger equation [1], [22]

$$\frac{d^2u_n(r)}{dr^2} + \left[k_n^2 - V(r)\right] u_n(r) = 0, \tag{3.2}$$

where the potential $V(r)$ vanishes beyond a distance, *i.e.*, $V(r) = 0, r > R$. The function $u_n(r)$ obeys at the origin the boundary condition

$$u_n(r = 0) = 0, \tag{3.3}$$

and at distances $r \geq R$ it describes a situation where there are no incident particles, namely $u_n(r) \propto \exp(ik_n r)$, and hence it fulfills the outgoing boundary condition

$$\left[\frac{du_n(r)}{dr}\right]_{r=R_-} = ik_n u_n(R), \qquad r \geq R. \tag{3.4}$$

Gamow [1] showed that the above boundary conditions lead to complex energy eigenvalues $k_n^2 = E_n = \epsilon_n - i\Gamma_n/2$. The above boundary conditions also provide the bound and antibound solutions corresponding, respectively, to imaginary positive, $k_n = i\gamma_m$, and imaginary negative, $k_n = -i\delta_n$, values of k_n. It is shown in the appendix that the residues $C_n(r,r')$ at the complex poles $\{k_n\}$ of the outgoing Green function correspond precisely to the resonant states of the problem, namely, [see (A.15)],

$$C_n(r,r') = \frac{u_n(r)u_n(r')}{2k_n}, \tag{3.5}$$

provided they are normalized according to (A.16). A similar result applies, of course, for bound and antibound poles [22].

It is now possible to write a representation of the time dependent Green function involving resonant states plus an integral contribution [22]. This may be done by deforming appropriately the integration contour C_0 in the k−plane . Since the variation of $G^+(r,r';k)$ with k is at most exponential [28], the behaviour with k of the integrand in (3.1) is dominated by $\exp(-ik^2t)$. Hence a useful choice that avoids the fact that the factor $\exp(-ik^2t)$ diverges for large values of k in the upper right and lower left quadrants, while it decreases rapidly, for $t > 0$, in the other two quadrants is shown in Fig. 17.2. By deforming the contour one passes over some poles of the outgoing Green function. These include those corresponding to bound states and the subset of complex poles such that Re $E_n >$ Im E_n associated with the so called proper resonant states. By extending the contour C_0' up to infinity one readily obtains that the semi-circles C_R yield a vanishing contribution so that one is left with a infinite sum of terms plus an integral contribution arising from the straight line C_L at 45°. Therefore the time-dependent Green function may be written as [22],

$$\begin{aligned} g(r,r';t) &= \sum_b^M u_b(r)u_b(r')\exp(-ik_b^2t) + \sum_r^\infty u_r(r)u_r(r')\exp(-ik_r^2t) \\ &\quad + \frac{i}{2\pi}\int_{C_L} G^+(r,r';k)\mathrm{e}^{-ik^2t}\,2k\,dk, \end{aligned} \tag{3.6}$$

where the sums run over the bound (b) and proper resonant poles (r) of the outgoing Green function.

Inserting (3.6) into (2.3) gives the nondecay amplitude

$$A(t) = \sum_{p=b,r}^{\infty} C_p\bar{C}_p\,\mathrm{e}^{-ik_p^2t} + N(t), \tag{3.7}$$

where p denotes both bound and proper resonant poles and the coefficients C_p and $\bar{C}_p$ are given, respectively, by

$$C_p = \int_0^R \phi(r,0)u_p(r)\,dr, \qquad \bar{C}_p = \int_0^R \phi^*(r,0)u_p(r)\,dr, \tag{3.8}$$

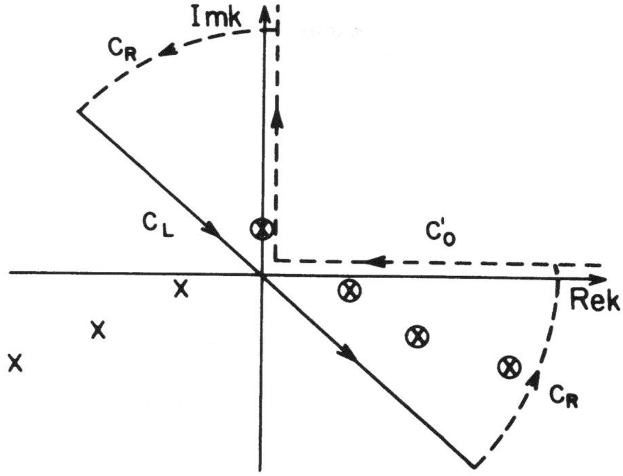

FIGURE 17.2. Contour leading to Eq. (3.6).

and the integral term $N(t)$ reads,

$$N(t) = \frac{i}{2\pi} \int_0^R dr\, \phi(r,0) \int_0^R dr'\, \phi(r',0) \int_{C_L} G^+(r,r';k)\, e^{-ik^2t}\, 2k\, dk. \tag{3.9}$$

Equation (3.7) for the nondecay amplitude $A(t)$ is interesting because it exhibits explicitly the contribution from the exponentially decaying terms, each "weighted" by a coefficient $C_p\bar{C}_p$, plus the integral contribution $N(t)$ that presumably is responsible for the nonexponential contribution to decay. A representation resembling Eq. (3.7) has been considered also by Sudarshan [13].

The evaluation of $N(t)$, as it stands, seems a formidable task. The problem is how to calculate the integral along the line C_L involving the Green function (see Fig. 17.2). Peierls [29] has shown that for long times this integral yields the expected $t^{-3/2}$ behaviour. He discussed the time evolution of a similar quantity, the wave function given by (2.4), using also the representation for $g(r,r';t)$ given by (3.6). The integral may be evaluated for long times using the steepest descent method. The straight line through the origin in the k−plane (see Fig. 17.2) is the line of steepest descent of the factor $\exp -ik^2t$ and $k=0$ is the saddle point. Hence one may expand the Green function by a Taylor series around $k=0$. The first term in the expansion vanishes because the integrand becomes an odd function of k and hence one must go to the second term in the Taylor expansion. The resulting expression for $N(t)$ reads

$$N(t) \approx \frac{2\gamma}{\pi} \left\{ \frac{\partial}{\partial k} G^+ \right\}_{k=0} \int_0^\infty e^{-z^2t} z^2 dz = \frac{\gamma}{2\pi^{1/2}} \left\{ \frac{\partial}{\partial k} G^+ \right\}_{k=0} \frac{1}{t^{3/2}}, \tag{3.10}$$

where G^+ denotes the outgoing Green function and the change of variable $k = \gamma z$ with $\gamma = (-i)^{1/2}$ has been performed to evaluate the integral. Substitution of (3.10) into (3.7) yields for the nondecay amplitude

$$A(t) \approx \sum_{p=b,r}^{\infty} C_p\bar{C}_p e^{-ik_p^2 t} + \frac{\gamma}{2\pi^{1/2}} \int_0^R dr\, \phi(r,0) \int_0^R dr'\, \phi(r',0) \left\{ \frac{\partial}{\partial k} G^+ \right\}_{k=0} \frac{1}{t^{3/2}} \qquad (3.11)$$

The nonexponential character of $A(t)$ after a long time is explicitly exhibited by the above equation. As pointed out above the notion of long time is to be referred to the time scale set up by the lifetime of the system $\tau = \hbar/\Gamma$. Here Γ corresponds to the smallest of the set of decay widths $\{\Gamma_p\}$ of the problem.

Regarding the small time behaviour of $A(t)$ one may be tempted to expand the exponential factor in (3.9). However it may be proved that such procedure leads to misleading results [20].

One may conclude at this point that although the expression for the nondecay amplitude, Eq. (3.7), includes a resonant expansion that exhibits explicitly the time evolution of the exponentially decaying terms, the nonexponential term, (3.9), seems unappropriate to study the corresponding time evolution, particularly if one is interested in a comparison between exponential and nonexponential contributions to decay. In the next section it is shown how to describe the nonexponential contribution using resonant states.

17.4 Full discrete expansion of $\mathbf{g(r,r';t)}$ and $\mathbf{A(t)}$

Let us denote by $I(t)$ the integral term that appears in (3.6),

$$I(t) = \frac{i}{2\pi} \int_{C_L} G^+(r,r';k) e^{-ik^2 t}\, 2k\, dk \qquad (4.1)$$

that is responsible, as discussed in the previous section, for the nonexponential contributions to the time evolution of decay. It is interesting to see that by considering some properties od the outgoing Green function it is also possible to obtain a resonant expansion for $I(t)$. This is achieved by expanding $G^+(r,r';k)$ in terms of all its singularities on the complex k-plane using Cauchy integral theorem. Let us therefore consider the contour integral

$$J = \frac{1}{2\pi i} \int_C \frac{G^+(r,r';k)}{k'-k}\, dk', \qquad (4.2)$$

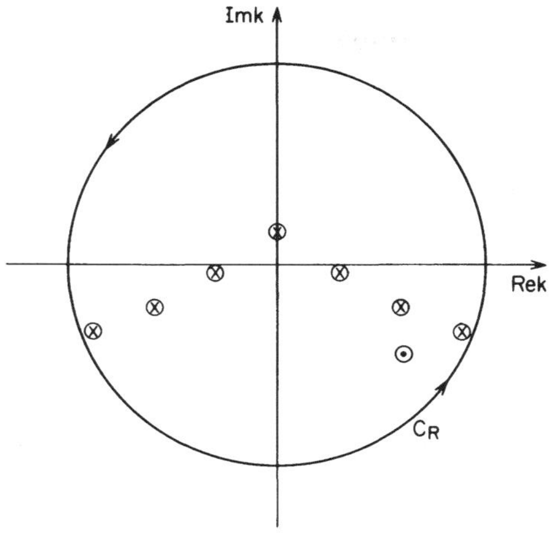

FIGURE 17.3. Integration contour for Cauchy expansion of $G^+(r, r'; k)$.

where C represents a large closed contour of radius L in the k–plane about the origin in the *clockwise* direction which excludes all the poles, k_n and the value at $k' = k$. Using Cauchy theorem it follows that $J = 0$ and hence one may be write

$$J = \frac{1}{2\pi i}\left[-\int_{C_R}\frac{G^+(k')}{k'-k}dk' + \sum_n \int_{C_n}\frac{G^+(k')}{k'-k}dk' + \int_{C_k}\frac{G^+(k')}{k'-k}dk'\right] = 0 \tag{4.3}$$

where for simplicity of notation $G^+(r, r'; k)$ is denoted by $G^+(k')$ and C_R represents a large circle centered at the origin; the contours C_n encircle each of the poles k_n that are enclosed by C_R, and the contour C_k encloses the value $k' = k$. All these contours are in the counterclockwise direction as shown in Fig. 17.3.

Using the theorem of residues and (3.5) one may write (4.3) as

$$G^+(r, r'; k) = \sum_n^N \frac{u_n(r)u_n(r')}{2k_n(k-k_n)} + \frac{1}{2\pi i}\int_{C_R}\frac{G^+(r, r'; k)}{k'-k}dk'. \tag{4.4}$$

The above equation is defined for all $r, r' \geq 0$. However we need to know $G^+(r, r'; k)$ only along the internal region of the interaction, *i.e.*, $(r, r') < R$. In this case it can be proved that the outgoing Green function goes exponentially to zero along all directions in the complex k–plane [30], *i.e.*,

$$G^+(r, r'; k) \to 0 \text{ as } |k| \to \infty, \qquad (r, r') < R, \tag{4.5}$$

and therefore by extending the radius L of the contour C [see (4.2)] up to infinity one then obtains that the integral term along C_R in (4.4) vanishes exactly. One is left therefore with an infinite sum over the full set of resonant terms [27], [31]-[33]

$$G^+(r,r';k) = \sum_n^{\infty} \frac{u_n(r)u_n(r')}{2k_n(k-k_n)}, \qquad (r,\, r') < R. \tag{4.6}$$

The above expression remains convergent if one of the variables, r or r', remains inside the interaction radius R while the other is evaluated at the boundary value R [34]. However at $r = r' = R$ the expansion of $G^+(r,r';k)$ requires at least two subtractions [34]. Substitution of the expansion given by (4.6) into (A.1) followed by the addition and subtraction of the quantity $k_n^2 u_n(r)$, using (3.2) yields

$$\sum_n^{\infty} \frac{k+k_n}{k_n} u_n(r)u_n(r') = \delta(r-r'). \tag{4.7}$$

The above equation implies that the following relationships must be fulfilled [32]

$$\sum_n^{\infty} \frac{u_n(r)u_n(r')}{k_n} = 0, \qquad (r,r') < R, \tag{4.8}$$

and

$$\frac{1}{2}\sum_n^{\infty} u_n(r)u_n(r') = \delta(r-r'), \qquad (r,r') < R. \tag{4.9}$$

Now since

$$\frac{1}{2k_n(k-k_n)} \equiv \frac{1}{2k}\left[\frac{1}{k-k_n} + \frac{1}{k_n}\right], \tag{4.10}$$

one may substitute (4.10) into (4.6), using (4.8), to write

$$G^+(r,r';k) = \frac{1}{2k}\sum_n^{\infty} \frac{u_n(r)u_n(r')}{(k-k_n)}. \tag{4.11}$$

Let us now substitute (4.11) into (4.1), to obtain

$$I(t) = \sum_n^{\infty} u_n(r)u_n(r')\left\{-\frac{i}{2\pi}\int_{C_L} \frac{\mathrm{e}^{-ik^2t}}{k_n-k}\,dk\right\}, \qquad (r,r') < R. \tag{4.12}$$

The term within brackets in (4.12) corresponds to a known function [35]. In particular making a suitable change of variable it may be written in terms of the Moshinsky function $M(k_n,t)$ [20], [35],[2]

$$M(k_n,t) = -\frac{i}{2\pi}\int_{-\infty}^{\infty} \frac{\mathrm{e}^{-u^2}}{(iy_n-u)}\,du = \frac{1}{2}\mathrm{e}^{y_n^2}\,\mathrm{erfc}\,(y_n) \tag{4.13}$$

[2]The author's notation for the Moshinsky function $M(k,t)$ corresponds to $M(0,k,t)$ in Ref. 36.

with $u = -\exp(i\pi/4)kt^{1/2}$ and $y_n = -\exp(-i\pi/4)k_n t^{1/2}$. Therefore (4.12) may be written as

$$
\begin{aligned}
I(t) \;=\; & \sum_b^M u_b(r)u_b(r')M(-k_b,t) + \sum_a^N u_a(r)u_a(r')M(k_a,t) \\
& + \sum_r^\infty u_r(r)u_r(r')M(-k_r,t), \qquad (r,r') < R, \qquad (4.14)
\end{aligned}
$$

where the subscripts b, a, and r refer, respectively to bound, antibound and resonant terms. The sum over r includes the contribution from the poles $k_{-r} = -k_r^*$, situated on the third quadrant of the k-plane, and corresponding eigenfunctions $u_{-r} = u_r^*$. Notice that $M(-k_r,t)$ correponds to $y_n = \exp(-i\pi/4)k_n t^{1/2}$ and similarly for the Moshinsky function for k_b. Substitution of (4.14), [recalling (4.1)], into (3.6) yields the following representation for the time dependent Green function,

$$
\begin{aligned}
g(r,r';t) \;=\; & \sum_b^M u_b(r)u_b(r')\left[e^{-ik_b^2 t} - M(-k_b,t)\right] \\
& + \sum_a^N u_a(r)u_a(r')M(k_a,t) \qquad (r,r') < R \\
& + \sum_r^\infty u_r(r)u_r(r')\left[e^{-ik_r^2 t} - M(-k_r,t)\right]. \qquad (4.15)
\end{aligned}
$$

Using the symmetry relationship obeyed by the Moshinsky function [20], [35]

$$M(k_n,t) = e^{-ik_n^2 t} - M(-k_n,t) \qquad (4.16)$$

allows to write (4.15) in the compact form

$$g(r,r';t) = \sum_{n,all}^\infty u_n(r)u_n(r')M(k_n,t), \qquad (r,r') < R, \qquad (4.17)$$

where the sum runs over the full set of poles of the problem. Substitution of (4.17) into (2.3) gives the full discrete expansion of the nondecay amplitude,

$$A(t) = \sum_{n,all}^\infty C_n\bar{C}_n M(k_n,t), \qquad (4.18)$$

where the expansion coefficients are given by (3.8). Using the relationships obeyed by the set of states $\{u_n(r)\}$, *i.e.*, Eqs. (4.7) and (4.8), one readily obtains that the coefficients C_n and $\bar{C}_n$ fulfill the sum rules,

$$\frac{1}{2}\sum_{n,all}^\infty C_n\bar{C}_n = 1 \qquad (4.19)$$

and

$$\sum_{n,all}^{\infty} \frac{C_n \bar{C}_n}{k_n} = 0 \tag{4.20}$$

Since bound and antibound states are stationary functions, in what follows for the sake of simplicity, the discussion is restricted to the resonant states of the problem. Hence one may write $A(t)$ as

$$A(t) = \sum_{r=1}^{\infty} \left\{ C_r \bar{C}_r e^{-ik_r^2 t} - C_r \bar{C}_r M(-k_r, t) + (C_r \bar{C}_r)^* M(-k_r^*, t) \right\}, \tag{4.21}$$

where the contribution from the poles seated on the third quadrant of the k–plane is exhibited explicitly. Similarly (4.19) and (4.20) may be written as

$$\sum_{r=1}^{\infty} \operatorname{Re} C_r \bar{C}_r = 1 \tag{4.22}$$

and

$$\sum_{r=1}^{\infty} \operatorname{Im} \frac{C_r \bar{C}_r}{k_r} = 0. \tag{4.23}$$

17.4.1 Short and long time behaviour

The function $M(-k_r, t)$ may be expanded as [35]

$$M(-k_r, t) = \frac{1}{2} \sum_{q}^{\infty} \frac{(iy_n)^q}{\Gamma[(q/2) + 1]}, \tag{4.24}$$

where $y_n = \exp(-i\pi/4) k_n t^{1/2}$. Obviously a similar expansion holds for $M(-k_r^*, t)$. From (4.24) it follows that for $t = 0$, $M(k_r, 0) = 1/2$, then (4.21) at $t = 0$, using (4.22), yields $A(0) = 1$ as it should be.

For very short times one may in principle expand in (4.21) the exponential and the Moshinky functions and interchange the order of summation over q and r to extract the leading dependence on time. This requires that terms of the form

$$\sum_{r}^{\infty} C_r \bar{C}_r k_n^q \tag{4.25}$$

converge. However in the case of a solvable problem it is obtained that (4.25) diverges [20]. Therefore it does not seem in general possible to write an analytic dependence of $A(t)$ for short times.

The case of very long times may be obtained by using the asymptotic expansion for $\exp(z^2)\mathrm{erfc}(z)$ [20], [35] to write,

$$\begin{aligned} A(t) &\approx \sum_r^{\infty} C_r\bar{C}_r e^{-ik_r^2 t} \\ &\quad - \frac{1}{2(\pi i)^{1/2}} \sum_r^{\infty} \left\{ \left[\frac{C_r\bar{C}_r}{k_r} - \left(\frac{C_r\bar{C}_r}{k_r} \right)^* \right] \frac{1}{t^{1/2}} \right\} \\ &\quad + \frac{i}{4(\pi i)^{1/2}} \sum_r^{\infty} \left\{ \left[\frac{C_r\bar{C}_r}{k_r^3} - \left(\frac{C_r\bar{C}_r}{k_r^3} \right)^* \right] \frac{1}{t^{3/2}} + \cdots \right\} \end{aligned} \tag{4.26}$$

Notice that the term within brackets proportional to $t^{-1/2}$, in view of (4.22), cancels exactly. Therefore

$$\begin{aligned} A(t) &\approx \sum_r^{\infty} C_r\bar{C}_r e^{-ik_r^2 t} \\ &\quad + \frac{i}{4(\pi i)^{1/2}} \sum_r^{\infty} \left\{ \left[\frac{C_r\bar{C}_r}{k_r^3} - \left(\frac{C_r\bar{C}_r}{k_r^3} \right)^* \right] \frac{1}{t^{3/2}} + \cdots \right\}. \end{aligned} \tag{4.27}$$

17.5 Example

In this section the formalism developed in the preceeding sections is applied to the solution of the same exactly solvable problem as discussed in ref. 20, namely, the decay of a s-wave particle with initial wave function at $t = 0$,

$$\phi(r,0) = \sqrt{\frac{2}{a}} \sin Kr, \qquad K = \frac{m\pi}{a}, \tag{5.1}$$

and m is an integer. The potential is a δ-function of strength λ at the point $r = a$, namely

$$V(r) = \lambda\delta(r-a). \tag{5.2}$$

The solutions of (3.2), using the interaction (5.2), may be written as

$$\begin{aligned} u_n(r) &= B_n \sin k_n r, \qquad r \le a, \\ u_n(r) &= D_n e^{ik_n r}, \qquad r \ge a. \end{aligned} \tag{5.3}$$

Here $A(t)$ depends only on the solution along the internal region of the potential. Consequently the normalization coefficient B_n may be obtained using (A.16), namely

$$B_n = \sqrt{\frac{2(\lambda a - 2ik_n a)}{a(1 + \lambda a - 2ik_n a)}}. \tag{5.4}$$

The coefficients C_n and $\bar{C}_n$, defined by (3.8), may be obtained using (5.1), (5.2), and (5.3). Since $\phi(r,0)$ is real, hence $C_n = \bar{C}_n$, and in addition the coeficients appear as a product in (4.8), it is convenient to write

$$C_n\bar{C}_n = C_n^2 = \frac{1}{\lambda a}\left[\frac{1}{1+\lambda a - 2ik_n a}\right]\frac{4K^2k_n^2}{(K^2+k_n^2)^2}. \tag{5.5}$$

The k_n may be obtained from the conditions of continuity of the resonant states at the boundary value $r = a$. They correspond to the complex roots of

$$2ik_n + \lambda a(\mathrm{e}^{2ik_n a} - 1) = 0. \tag{5.6}$$

The units utilized are $\hbar = 2m = 1$. They are slightly different from those of Ref. [20] that read $\hbar = m = a = 1$. Hence the corresponding condition of equivalence is $\lambda = 2b$ where b stands for the strength of the δ interaction in Ref. [20].

Using (5.6) in(4.18) gives

$$A(t) = \sum_{n,all}^{\infty}\frac{1}{\lambda a}\left[\frac{1}{1+\lambda a - 2ik_n a}\right]\frac{4K^2k_n^2}{(K^2+k_n^2)^2}M(k_n,t). \tag{5.7}$$

The above expression corresponds exactly to the equation (4.35) of Ref. [20]. In that work the behaviour of (5.7) is discussed for short and long times and therefore it will not be repeated here. Regarding the expansion coefficients $C_n\bar{C}_n$ in (4.18) is worth mentioning the following. For an initial wave function (5.1) with $j = 1$, the corresponding momentum is $K = \pi$ and the resonant poles, for a strength of the potential $\lambda = 100.0$, nearer to K correspond to $n = 1, 2, 3$. These poles are at $k_1 = 3.110527 - i0.000956$, $k_2 = 6.221286 - i0.003803$, and, $k_3 = 9.332499 - i0.00847933$.

There are no bound or antibound poles for this problem. Now it is interesing to see that the real part of the expansion coefficients $C_n\bar{C}_n = C_n^2$ reads: $\mathrm{Re}\,C_1^2 = 0.9996789$, $\mathrm{Re}\,C_2^2 = 0.000179$, and $\mathrm{Re}\,C_3^2 = 0.000051$. One sees that the main contribution to the sum rule (4.19), or (4.22), arises from the first term, namely, that corresponding to a resonant state whose momentum is very close to the momentum of the initial wave function. Incidentally, the first 25 terms yield $\sum_{r=1}^{25}\mathrm{Re}\,C_n^2 = 0.999998$ already very close to unity. Similarly (4.20), or (4.23), yields $\sum_n^{25}\mathrm{Im}C_n^2/k_n = 7.61\,10^{-9}$ that is very close to zero.

17.6 Exact one-level decay formula

The usual formula of the nondecay amplitude for purely exponential decay reads [37]

$$A(t) = \mathrm{e}^{-i\epsilon_n}\mathrm{e}^{-\Gamma_n t/2}. \tag{6.1}$$

Here, of course, it is assumed that the overlap between the decaying state and the initial arbitrary state at $t = 0$ is unity.

One may derive an exact one-level decay formula from (4.18) and the sum rules (4.19) and (4.20). Indeed let us assume that in the above expressions only one term differs from zero, say, the mth. Then using the symmetry relation (4.16), one may write (4.18) as

$$A(t) = C_m \bar{C}_m e^{-ik_m^2 t} - C_m \bar{C}_m M(-k_m, t) + (C_m \bar{C}_m)^* M(-k_m^*, t). \quad (6.2)$$

Then from (4.19) and (4.20) one obtains, respectively,

$$\tfrac{1}{2}\left[C_m \bar{C}_m + (C_m \bar{C}_m)^*\right] = 1 \quad (6.3)$$

and

$$\frac{C_n \bar{C}_m}{k_m} - \frac{(C_n \bar{C}_m)^*}{k_m^*} = 0. \quad (6.4)$$

From (6.3) and (6.4) it follows immediately that

$$C_m \bar{C}_m = \frac{2k_m}{k_m + k_m^*} = 1 - i\frac{\beta_m}{\alpha_m}, \quad (6.5)$$

where $k_m = \alpha_m - i\beta_m$. Therefore, using (6.4), one may write (6.2) as

$$\begin{aligned} A(t) \;=\; & \left(1 - i\frac{\beta_m}{\alpha_m}\right) e^{-ik_m^2 t} \\ & - \left(1 - i\frac{\beta_m}{\alpha_m}\right) M(-k_m, t) \\ & + \left(1 + i\frac{\beta_m}{\alpha_m}\right) M(-k_m^*, t). \end{aligned} \quad (6.6)$$

An expression similar to (6.6) has been obtained by Moshinsky [20], [38] for the nondecay amplitude of a compound particle in the framework of a schematic theory of nuclear reactions. It is of interest to note that the approach used by Moshinsky differs completely from the one presented here. For example in his approach the interaction region is reduced to the point $r = 0$.

17.6.1 Very Short times

The short time behaviour of the nondecay amplitude (6.6) may be obtained by expanding the exponential function and the Moshinsky functions (see (4.22)). The leading term in t reads

$$A(t) \approx 1 - i\left(\frac{i}{\pi}\right)^{1/2} \left[C_m \bar{C}_m k_m - (C_m \bar{C}_m)^* k_m^*\right] t^{1/2} = 1 - \left(\frac{i}{\pi}\right)^{1/2} \frac{\Gamma_m}{\alpha_m} t^{1/2} \quad (6.7)$$

and therefore the corresponding nondecay probability $P(t) = |A(t)|^2$ is

$$P(t) \approx 1 - \sqrt{\frac{2}{\pi}} \frac{\Gamma_m}{\alpha_m} t^{1/2}. \tag{6.8}$$

17.6.2 Very Long Times

For long times one may use (4.25) to write

$$A(t) \approx \frac{i}{4(\pi i)^{1/2}} \left[\frac{C_m \bar{C}_m}{k_m^3} - \left(\frac{C_m \bar{C}_m}{k_m^3} \right)^* \right] \frac{1}{t^{3/2}} + \cdots . \tag{6.9}$$

Using (6.5) one may then write

$$A(t) \approx \frac{1}{4(\pi i)^{1/2}} \frac{1}{\alpha_m} \frac{\Gamma_m}{(\epsilon_m^2 + \Gamma_m^2)} \frac{1}{t^{3/2}}. \tag{6.10}$$

From the above expression it follows that $P(t) \propto \sin t^{-3}$.

It is worth mentioning that the expression for the nondecay formula obtained by Moshinsky [38] in 1951, represents to my knowledge the first derivation of a decay formula valid in the full interval $0 \leq t \leq \infty$ and therefore the first formula to include nonexponential contributions to decay. However Moshinsky did not discuss the nonexponential character of that formula and therefore his work passed unnoticed in the literature on this subject. Figs. 17.4 and 17.5 illustrate, repectively, a calculation for the short time region, and the exponential and long time time regions, of the nondecay probability (6.6) as a function of the lifetime $\tau = 1/\Gamma$. The only input is the ratio of the resonance energy to the resonance width, namely, $R = \epsilon/\Gamma = 100$.

17.7 Conclusion

The main results of this work are given by both by formula (4.18), that provides the time evolution of the nondecay amplitude as a purely discrete infinite sum of terms involving the resonant states of the problem, and by formula (6.6), that yields an exact one-level expression of the nondecay amplitude. In both cases a discussion of the exponential and nonexponential contributions to the time evolution of decay is given. One may see the one-level nondecay formula as the result of a condition on the initial state, namely, that initial state whose overlap with all resonant states except one, gives a vanishing contribution. This is in contrast with the general case where each resonant state contributes a certain amount to the decay amplitude.

For the long time region it is found that both formulas yield the well known $t^{-3/2}$ behaviour. For the short time limit the situation is different.

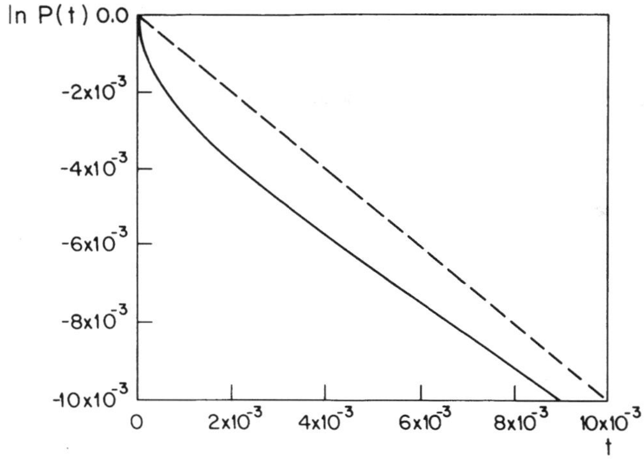

FIGURE 17.4. Plot of the nondecay probability ln $P(t)$ *vs.* the lifetime t for the very short time region using the exact one-level expression of the nondecay amplitude [Eq. (6.6)] (continuous line). For comparison a plot involving the purely exponential decay amplitude [Eq.(6.1)] (broken line) is also presented. The required input is the ratio of the resonance energy to the decay width, chosen as $R = \epsilon/\Gamma = 100$. I thank Verónica Riquer for the computing calculations needed to draw this figure.

In the case of the full resonant expansion it does not seem possible in general to obtain an analytic expression for the small time behaviour. However for the δ potential numerical results [20] indicate that the first term after unity in the nondecay amplitude goes as t^{α}, with $\alpha < 1$. On the other hand for the one-level nondecay amplitude it is obtained analytically that the first term after unity behaves as $t^{1/2}$.

17.A Appendix: Determination of the residue at the pole of the outgoing Green function

It is well known that for a finite range interaction the outgoing Green function $G^+(r,r';k)$, as a function of k, can be extended analytically to the whole complex k−plane where it possess an infinite number of poles distributed in a well known form [21], [28]. In what follows it is shown that the residue of $G^+(r,r';k)$ near one of its complex poles is proportional to the functions $u_n(r)$ and $u_n(r')$. In doing so I follow my own work [22] adapted to a description in the k−plane . The outgoing Green function associated to the Hamiltonian given by Eq. (2.2) satisfies the equation

$$\frac{\partial^2}{\partial r^2}G^+(r,r';k) + \left[k^2 - V(r)\right] G^+(r,r';k) = \delta(r-r'), \qquad (A.1)$$

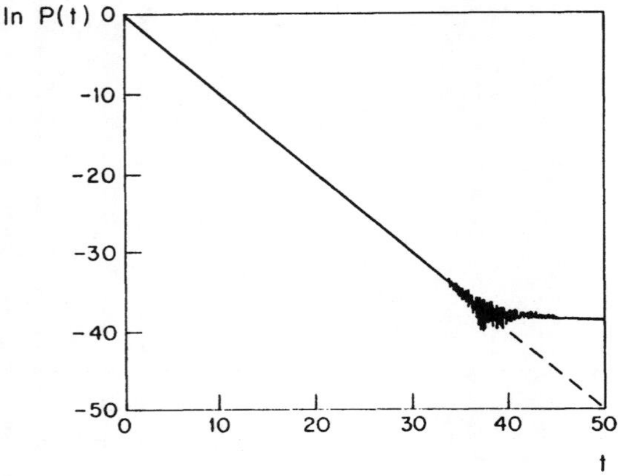

FIGURE 17.5. Plot of the nondecay probability ln $P(t)$ *vs.* the lifetime t for the exponential and very long time regions. The continuous line refers to the exact calculation [Eq. (6.6)]. The broken line corresponds to a calculation for pure exponential decay [Eq. (6.1)]. The input is the same as in Fig. 17.4, *i.e.*, $R = 100$. I thank Verónica Riquer for the computing calculations needed to draw this figure.

with the boundary conditions at $r = 0$ and $r = R$ given respectively by

$$G(0, r'; k) = 0 \tag{A.2}$$

and

$$\left[\frac{\partial}{\partial r} G^+(r, r'; k)\right]_{r=R_-} = ikG^+(r = R, r'; k). \tag{A.3}$$

Near a complex pole k_n one then may write

$$G^+(r, r'; k) = \frac{C_n(r, r')}{k - k_n} + \chi(r, r'; k), \tag{A.4}$$

where $C_n(r, r'; k)$ is the residue and $\chi(r, r'; k)$ is regular at the pole. Substitution of Eq. (A.4) into Eq. (A.1) leads after some simple algebra to the result

$$\frac{1}{k - k_n}\left\{\frac{d^2 C_n(r, r')}{dr^2} + \left[k^2 - V(r)\right] C_n(r, r')\right\}$$
$$+\left\{\frac{\partial^2 \chi(r, r'; k)}{\partial r^2} + \left[k^2 - V(r)\right] \chi(r, r'; k)\right\} - \delta(r - r') = 0. \tag{A.5}$$

Addition and substraction of $k_n^2 C_n(r, r')/(k - k_n)$ to the previous equation and taking the limit $k \to k_n$ leads to the expressions

$$\frac{\partial^2 C_n(r, r')}{\partial r^2} + \left[k_n^2 - V(r)\right] C_n(r, r') = 0, \tag{A.6}$$

and

$$\frac{\partial^2 \chi(r,r';k_n)}{\partial r^2} + \left[k_n^2 - V(r)\right] \chi(r,r';k_n) + C_n(r,r') = \delta(r-r'). \qquad (A.7)$$

Now, substitution of Eq. (A.4) into the boundary conditions given by Eqs. (A.2) and (A.3), adding and subtracting $k_n C_n(r,r')/(k-k_n)$ and taking the limit $k \rightarrow k_n$, yields

$$C_n(0,r') = 0, \qquad (A.8)$$

$$\chi(0,r';k_n) = 0, \qquad (A.9)$$

and also

$$\left[\frac{\partial}{\partial r} C_n(r,r')\right]_{r=R_-} = ikC_n(R,r'), \qquad (A.10)$$

$$\left[\frac{\partial}{\partial r} \chi(r,r';k_n)\right]_{r=R_-} = ik\chi(R,r';k_n) + \frac{i}{2k_n} C_n(R,r'). \qquad (A.11)$$

One sees that Eq. (A.6) for $C_n(r,r')$ and its boundary conditions, Eqs. (A.8) and (A.10), are identical with Eq. (3.2) for $u_n(r)$ and its boundary conditions, Eqs. (3.3) and (3.4). Consequently these functions are proportional,

$$C_n(r,r') = u_n(r)P(r'). \qquad (A.12)$$

An explicit expression for $P(r')$ may be obtained as follows. Multiply Eq. (3.2) by $G^+(r,r';k)$ and Eq. (A.1) by $u_n(r)$, subtract one from the other and integrate from $r=0$ to $r=R$. The result may be written as

$$\left[u_n(r)\frac{\partial}{\partial r}\chi(r,r';k_n) - \chi(r,r';k_n)\frac{d}{dr}u_n(r)\right]_0^R$$

$$+ \int_0^R u_n(r)C_n(r,r')dr = \int_0^R u_n(r)\delta(r-r')dr, \qquad (A.13)$$

where the prime denotes derivative with respect to r. It then follows, using Eqs. (3.3), (3.4), (A.2), (A.3), and (A.12) that

$$P(r') = \frac{u_n(r')}{2k_n\left\{\int_0^R u_n^2(r)dr + iu_n^2(R)/2k_n\right\}}. \qquad (A.14)$$

Hence the residue of the outgoing Green function at the pole k_n may be written as,

$$C_n(r,r') = \frac{u_n(r)u_n(r')}{2k_n}, \qquad (A.15)$$

provided the resonant states are normalized according to the condition

$$\int_0^R u_n^2(r)dr + \frac{i}{2k_n} u_n^2(R) = 1. \tag{A.16}$$

It is worth noting that the residue of the Green function at the pole k_n in the complex k−plane differs from the corresponding pole k_n^2 in the k^2 − *plane* by a the factor $1/2k_n$ [22].

17.B References

[1] G. Gamow, Zur Quantentheorie des Atomkernes, *Z. Phys.* **51**, 204–212 (1928).

[2] R.W. Gurney and E.U. Condon, Quantum mechanics and radioactive disintegration, *Phys. Rev.* **33**, 127–140 (1929).

[3] L.A. Khalfin, On the theory of the decay of a quasi-stationary state, *Sov. Phys. Doklady* **2**, 340–344 (1957); L.A. Khalfin, Contribution to the decay theory of a quasi-stationary state, *Sov. Phys. JETP* **6**, 1053–1063 (1958).

[4] P.T. Matthews and A. Salam, Relativistic field theory of unstable particles, *Phys. Rev.* **112**, 283–287 (1958).

[5] M. Lévy, On the validity of the exponential decay law for the decay of an unstable particle, *Nuovo Cimento* **14**, 612–624 (1959).

[6] R.G. Winter, Evolution of a quasi-stationary state, *Phys. Rev.* **123**, 1503–1507 (1961).

[7] R.G. Newton, The exponential decay law of unstable systems, *Ann. Phys.* (N. Y.), **14**, 333–345 (1961).

[8] R. Jacob and R.G. Sachs, Mass and lifetime of unstable particles, *Phys. Rev.* **121**, 350–356 (1961).

[9] M.L. Goldberger and K. M. Watson, Lifetime and the decay of unstable particles in S-matrix theory, *Phys. Rev.* **136**, B1472–B1480 (1964).

[10] M.V. Terent'ev, On the exponential decay law of nonstable particle, *Ann. Phys.* (N.Y.) **74**, 1–42 (1972).

[11] M. Razavy, Time evolution of quasi-stationary states, *Int. J. Theor. Phys.* **13**, 237–257 (1975).

[12] C.B. Chiu and E.C.G. Sudarshan, Time evolution of unstable quantum systems and a resolution of Zeno's paradox, *Phys. Rev.* **D16**, 520–529 (1977).

[13] E.C.G. Sudarshan, C.B. Chiu, and V. Gorini, Decaying states as complex energy eigenvectors in generalized quantum mechanics, *Phys. Rev.* **D18**, 2914–2929 (1978).

[14] A. Peres, Nonexponential decay law, *Ann. Phys.* (N.Y.) **129**, 33–46 (1980).

[15] J. Berger and S.G. Eckstein, Criterion for nonexponential decay, *Phys. Rev.* **A26**, 1226–1227 (1982).

[16] E.J. Robinson, Nonobservability of early-time departures from Fermi's "golden rule", *Phys. Rev. Lett.* **52**, 2309–2312 (1984).

[17] N.N. Nikolaev, Verification of the exponential decay law, *Sov. Phys. Usp.* **11**, 522 (1986)

[18] D.K. Butt and A.R. Wilson, A study of the radioactive decay law, *J. Phys.* **A5**, 1248 (1972).

[19] E.B. Norman, S.B. Gazes, S.G. Crane, and D.A. Bennett, Tests of the exponential decay law at short and long times, *Phys. Rev. Lett.* **60**, 2246–2249 (1988).

[20] G. García-Calderón, G. Loyola, and M. Moshinsky, The decay process: an exactly soluble problem and its implications. In these Proceedings.

[21] V.I. Kukulin, V.M. Krasnopol'sky, and J. Horáček, *Theory of Resonances, Principles and Applications* (Kluwer, Dordrecht 1989).

[22] G. García-Calderón and R.E. Peierls, Resonant states and their uses, *Nucl. Phys.* **A265**, 443–460 (1976).

[23] R.E. Peierls, Interpretation and properties of propagators. In: *The Proccedings of the 1954 Glasgow Conference on Nuclear and Meson Physics*, edited by E. H. Bellamy and R. G. Moorhouse (Pergamon Press, London and New York, 1955), p. 296–299.

[24] W.J. Romo, Inner product for resonant states and shell-model applications, *Nucl. Phys.* **A116**, 618–636 (1968).

[25] T. Berggren, On the use of resonant states and eigenfunction expansions of scattering and reaction amplitudes, *Nucl. Phys.* **A109**, 265–287 (1968).

[26] T.A. Minelli and F. Zardi, A dynamical approach to the S-matrix theory of nuclear reactions, *N. Cimento* **6A**, 655 (1971).

[27] R.M. More, Theory of decaying states, *Phys. Rev.* **A4**, 1782–1790 (1973).

[28] R.G. Newton, *Scattering Theory of Waves and Particles* (Second edition, Springer-Verlag, New York, 1982) Chapter 12.

[29] R.E. Peierls, Decaying states (unpublished, 1975).

[30] G. García-Calderón and M. Berrondo, Note on the asymptotic energy behaviour of the Green function for a cut-off potential, *Lett. N. Cimento* **26**, 562–566 (1979).

[31] F.A. Gareev, M.H. Gitzatkulov, and S.A. Goncharov, Expansion of continuum functions on resonance wave functions and amplitudes, *Nucl. Phys.* **309**, 381–421 (1978).

[32] G. García-Calderón, On the overcompletness of the set of bound, antibound and resonant states, *Lett. N. Cimento,* **33**, 253–257 (1982).

[33] G. García-Calderón and A. Rubio, An expansion of continuum wave functions in terms of resonant states II. Solvable models, *Nucl. Phys.* **A458**, 560–572 (1976).

[34] G. García-Calderón, An expansion of continuum wave in terms of resonant states, *Nucl. Phys.* **A261**, 130–140 (1976).

[35] M. Abramowitz and I.A. Stegun, *Handbook of Mathematical Functions* (Dover Publications Inc., New York, 1972), p. 297–298.

[36] H.M. Nussenzveig, Moshinsky functions, resonances and tunneling. *In these Proceedings.*

[37] L.D. Landau and E.M. Lifshitz, *Quantum Mechanics, Non Relativistic Theory* (Pergamon Press, Oxford, and Addison Wesley, Reading, 1965), p.512.

[38] M. Moshinsky, Boundary conditions and time dependent states, *Phys. Rev.* **84**, 525–533 (1951).

18

The Decay Process: An Exactly Soluble Example and its Implications

Gaston García-Calderón, Gerardo Loyola[1] and Marcos Moshinsky[2]

ABSTRACT There is agreement on the exponential decay law for unstable states in a certain range of times but not on what happens for very long or very short times. In particular for the latter a paradox appears from the contradictory results of two very straightforward calculations. To understand the matter better we consider s-states limited by a δ function radial potential of infinite height at a distance $r = a$. At time $t = 0$ we lower the height of the δ to a finite value b and obtain in an explicit analytic form the decay amplitude as a function of time in the full interval $0 \leq t \leq \infty$. Our results support one of the calculations for very short times based on a schematic theory of nuclear reactions.

Preface

In view of the fact that this conference intended to celebrate the anniversaries of Marcos Moshinsky, it seemed proper that his contribution be based on an old idea but with contemporary implications. Fortunately a couple of months before the conference the authors came across a curious paradox related to the decay process and in trying to understand it this paper was born.

18.1 Introduction

Since the advent of quantum mechanics the problem of decaying states has been a recurrent one. Probably the first papers on the subject, with the emission of a particle, were the ones on α-decay by Gamow [1] and by Gurney and Condon [2] in 1928. Since then the matter has been treated by many researchers with varying results. Everybody agrees that there is a

---

[1]Instituto de Física, UNAM, México.

[2]*id.* and member of El Colegio Nacional, México.

range of times t in which the decay follows the exponential rule $\exp(-t/\tau)$ with τ being the lifetime. There is no agreement though on what happens for very sort times $t \ll \tau$, some saying [3], [4] that the first term after the unity goes as $t^{1/2}$ while others argue [5] that it behaves as t^2. There is ambiguity, in the older literature, even in the conclusions when $t \gg \tau$.

Due to all these ambiguities, we thought that an example in which the solution can be obtained exactly, and the behavior at different times clearly illustrated, could be of interest.

Based on this observation we discuss the problem, stated in the standard language of elementary quantum mechanics. But before, it we proceed to discuss a paradox that motivated this paper.

18.2 A paradox

In a schematic theory of nuclear reactions developed [3] by one of the authors (M.M.) in 1951, the state of the system, in the frame of reference where the center of mass is at rest, is represented by the two component wave function

$$\Psi = \begin{bmatrix} \psi_1(r,t) \\ \psi_2(t) \end{bmatrix} . \tag{2.1}$$

In (2.1) $\psi_1(r,t)$ describes the relative motion of the two colliding particles which for $r \neq 0$ satisfy a free particle time dependent Schrödinger equation, while $\psi_2(t)$ describes the compound particle formed by the collision. The conservation of probability then gave the boundary condition [3] relating $\psi_1(r,t)$, and $[\partial\psi_1(r,t)/\partial r]$ at $r = 0$ with $\psi_2(t)$ and $(\partial\psi_2/\partial t)$.

For stationary states, *i.e.*, when the time dependence is given by $\exp(-iEt)$, the schematic theory of nuclear reactions gives just the standard one-level Breit-Wigner formula [6]. There is though the possibility of discussing the reaction process as a function of time with arbitrary initial conditions. If at $t = 0$ we impose the initial value

$$\Psi = \begin{bmatrix} 0 \\ 1 \end{bmatrix} , \tag{2.2}$$

i.e., that the state is only in the form of a compound particle, then $\psi_2(t)$ gives us the amplitude of decay [3]. For short times we showed that [3], [7]

$$\psi_2(t) = 1 - e^{i\pi/4} a\, t^{1/2} + \cdots , \tag{2.3}$$

where a is some real constant. Thus the term in the decay amplitude after the initial value 1 goes as the square root of the time.

We proceed now to discuss a widely used argument [5] that says that the change from 1 goes as t^2 instead of $t^{1/2}$ thus giving rise to the paradox we mentioned.

Let us consider first an arbitrary Hamiltonian H but with purely discrete spectra. We shall designate its eigenstates by the angular ket $|n\rangle$ corresponding to the eigenvalue E_n where the index n can be enumerated as $n = 1, 2, 3, \cdots$. We proceed then to designate by the round ket $|K)$ an arbitrary initial state *i.e.* determined at $t = 0$. At a later time t the state $|K)$ obviously becomes [5]

$$\exp(-iHt) \mid K), \tag{2.4}$$

so the amplitude for the decay at time t is given by

$$A(K,t) = (K \mid \exp(-iHt) \mid K), \tag{2.5}$$

as $|A(K,t)|^2$ is the probability that the initial state $|K)$ remains at t in its original form. We assumed in (2.4), (2.5) that $\hbar = 1$.

Making use of the eigenstates $|n\rangle$ and eigenvalues E_n of H that were mentioned above we can also write the amplitude of decay as

$$A(K,t) = \sum_{n=1}^{\infty} (K|n\rangle \exp(-iE_n t)\langle n|K), \tag{2.6}$$

and we proceed to analyze its value for very short times.

As we can write

$$\exp(-iEt) = 1 - iEt - \tfrac{1}{2}E^2t^2 + \cdots, \tag{2.7}$$

if we *invert the order of summation* we obtain

$$\begin{aligned} A(K,t) &= (K \mid \exp(-iHt) \mid K) \\ &= 1 - it\sum_{n=1}^{\infty} E_n \mid \langle n \mid K) \mid^2 \\ &\quad - (t^2/2)\sum_{n=1}^{\infty} E_n^2 \mid \langle n \mid K) \mid^2 + \cdots, \end{aligned} \tag{2.8}$$

where we used the hermiticity property

$$(K \mid n\rangle = \langle n \mid K)^*, \tag{2.9}$$

and the normalization condition

$$\sum_{n=1}^{\infty} (K \mid n\rangle\langle n \mid K) = 1. \tag{2.10}$$

The probability of decay is then given by

$$\begin{aligned} |A(K,t)|^2 &= 1 - \left\{ \sum_{n=1}^{\infty} E_n^2 \mid \langle n \mid K) \mid^2 \right. \\ &\quad \left. - \left[\sum_{n} E_n \mid \langle n \mid K) \mid^2 \right]^2 \right\} t^2 + \cdots \end{aligned} \tag{2.11}$$

and thus we see that the term after 1 is proportional to t^2, while from (2.3) we obtain that

$$|\psi_2(t)|^2 = 1 - \sqrt{2}\, a\, t^{1/2} + \cdots, \tag{2.12}$$

so that the term after 1 goes as $t^{1/2}$.

We thus get a contradiction between the results given for short times by the schematic theory of nuclear reactions *i.e.* (2.12), as well as by other procedures, [4] and the one obtained in the reasoning leading to (2.11). To understand this paradox we discuss in the full interval $0 \leq t \leq \infty$, the simple decay process indicated in the next section.

18.3 The problem

We shall discuss the decay through a potential barrier, but take a very simple form for the latter so as to be able to solve the problem exactly.

We restrict ourselves to s-waves in the radial coordinate r for which besides the kinetic energy we have a δ function potential at the point $r = a$. Besides we assume that we begin with an infinite height for this δ function so that we have normalized bound states that vanish for $r \geq a$ and for $0 \leq r \leq a$ take the form

$$\sqrt{2/a} \sin Kr, \qquad K = m\pi/a, \tag{3.1}$$

where from now on K will be a shorthand notation for $m\pi/a$, where m is an integer.

At some time, say $t = 0$, this infinite δ barrier is lowered to a height b and so what we try to determine is the function $\psi(r, K, t)$ which satisfies the time dependent equation

$$-\frac{1}{2}\frac{\partial^2 \psi}{\partial r^2} + b\; \delta(r - a)\psi = i\,\frac{\partial \psi}{\partial t}, \tag{3.2}$$

with the initial value

$$\psi(r, K, 0) = \sqrt{2/a} \sin Kr, \qquad 0 \leq r \leq a, \tag{3.3}$$

and it vanishes for $r > a$, where we use units in which $\hbar$ and the mass of the particle are taken as 1.

Besides, once we determine $\psi(r, K, t)$ we will be interested in its scalar product with the initial state *i.e.*,

$$A(K, t) = \int_0^a \sqrt{2/a} \sin(Kr)\psi(r, K, t)\, dr, \tag{3.4}$$

as $|A(K, t)|^2$ will give us the probability that at time t the state still remains in its original form. In the next section we determine $A(K, t)$ explicitly.

18.4 The solution

To solve our problem the obvious procedure is to use the Laplace transform

$$\bar{\psi}(r, K, s) = \int_0^\infty \psi(r, K, t)e^{-st}dt, \tag{4.1}$$

and applying it to both sides of Eq. (3.2) we obtain

$$\left[-\frac{1}{2}\frac{d^2}{dr^2} + b\delta(r-a)\right]\bar{\psi}(r, K, s)$$
$$= \begin{cases} -i\sqrt{2/a}\sin Kr + is\bar{\psi}(r, K, s), & 0 \leq r \leq a, \\ is\bar{\psi}(r, K, s), & r > a. \end{cases} \tag{4.2}$$

Once $\bar{\psi}(r, K, s)$ is determined we can get $\psi(r, K, t)$ by the inverse Laplace transform over the Bromwhich contour

$$\psi(r, K, t) = \frac{1}{2\pi i}\int_{c-i\infty}^{c+i\infty} \bar{\psi}(r, K, s)e^{st}ds, \tag{4.3}$$

where c is chosen so that all the singularities of $\bar{\psi}(r, K, s)$ are to the left of this line.

In view of the fact that our final interest is the $A(K, t)$ of (3.4), we can invert the order of integration to get

$$A(K, t) = \frac{1}{2\pi i}\int_{c-i\infty}^{c+i\infty} \bar{A}(K, s)e^{st}ds, \tag{4.4}$$

with

$$\bar{A}(K, s) \equiv \int_0^a \sqrt{2/a}\sin(Kr)\bar{\psi}(r, K, s)\, dr. \tag{4.5}$$

Our analysis will start with the solution of the ordinary inhomogeneous differential equation (4.2), then determine $\bar{A}(K, s)$ defined by the integral (4.5) and finally apply the inverse Laplace transform (4.4).

18.4.1 Determination of $\bar{\psi}(r, K, k)$

We start by defining the variable k as

$$k \equiv \sqrt{2is} \quad \text{or} \quad s = -ik^2/2. \tag{4.6a,b}$$

Substituting (4.6*b*) in (4.2) we immediately see that the solution of the latter can be written as

$$\bar{\psi}(r, K, k) = D\exp(ikr), \quad r > a \tag{4.7a}$$
$$\bar{\psi}(r, K, k) = B\sin kr + C\sin Kr, \quad 0 \leq r \leq a, \tag{4.7b}$$

where from now on we substitute the s in $\bar{\psi}(r, K, s)$ and $\bar{A}(K, s)$ by the k of (4.6).

We note that we take only the outgoing wave in $r > a$ as in terms of a real s it gives a bounded solution at ∞. Furthermore we take the sine in $0 \leq r \leq a$, as we require that the solution vanishes at $r = 0$.

From the equation (4.2) $C \sin K\, r$ will be a solution of (4.2) in $0 \leq r \leq a$ only if

$$C = \frac{2i\sqrt{2/a}}{k^2 - K^2}, \tag{4.8}$$

furthermore $\bar{\psi}(r, K, k)$ must be continuous at $r = a$ and, due to the potential $\delta(r - a)$, when we integrate equation (4.2) with respect to r in an interval around a, we get

$$-\frac{1}{2}\left[\frac{\partial\psi(r, K, k)}{\partial r}\right]_{a-0}^{a+0} + b\psi(a, K, k) = 0. \tag{4.9}$$

Thus we obtain the two linear equations

$$D\exp(ika) - B\sin ka = 0, \tag{4.10a}$$

$$D\left[b - (ik/2)\right]\exp(ika) + B(k/2)\cos ka$$
$$= -i\sqrt{2/a}K(k^2 - K^2)^{-1}(-1)^m, \tag{4.10b}$$

where we used the fact that as $Ka = m\pi$ we have $\sin(Ka) = 0$, $\cos(Ka) = (-1)^m$. From (4.10) we see then that

$$B = \frac{(-2iK)\sqrt{2a}(-1)^m \exp(ika)}{(k^2 - K^2)\{ka + iba[1 - \exp(2ika)]\}}, \tag{4.11}$$

and thus the solution of (4.2) for $0 \leq r \leq a$, given by (4.7b), is completely defined.

18.4.2 Determination of $\bar{A}(K, k)$

Using (4.7b) the evaluation of the integral (4.5) is immediate and we obtain

$$\bar{A}(K, k) = 2i(k^2 - K^2)^{-1} + U(K, k)V(k), \tag{4.12}$$

where

$$U(K, k) = \frac{4iK^2 ka}{(k^2 - K^2)^2}, \tag{4.13}$$

$$V(k) = \frac{1}{2ika}\frac{[1 - \exp(i2ka)]}{\{ka + iba[1 - \exp(i2ka)]\}}, \tag{4.14}$$

and again we made use of the fact that $Ka = m\pi$.

Our next step is to understand the analytic behavior of (4.14), for which first we must discuss the nature of its poles *i.e.*, the values of k of which the curly bracket in the denominator of (4.14) vanishes. These turn out to be given by the same equation as that appearing when we deal with the purely outgoing solution of the stationary problem corresponding to (3.2).

18.4.3 THE RESONANT STATES AND THE ANALYTIC BEHAVIOR OF $V(k)$

If in (3.2) we take a time dependent behavior of the form $\exp(-ik^2t/2)$, the solution for purely outgoing states will be

$$D\exp(ikr) \quad \text{for} \quad r > a, \tag{4.15a}$$

$$B\sin kr \quad \text{for} \quad 0 \leq r \leq a. \tag{4.15b}$$

The continuity of the wave function at $r = a$ and the boundary condition (4.9) lead now to the equations (4.10) but in which the right hand of (4.10*b*) also is zero. Thus we have homogeneous linear equations and the determinant must vanish to have a solution of the problem, thus giving the equation

$$ka + iba[1 - \exp(i2ka)] = 0, \tag{4.16}$$

which is precisely the one determining the poles of the $V(k)$ of (4.14).

To explore the characteristics of the complex k satisfying (4.16) we introduce the variable $z = ka$ for which the equation becomes

$$z + iba[1 - \exp(2iz)] = 0. \tag{4.17}$$

If $ba \gg 1$ this equation indicates that the roots tend to $n\pi$. For $ba \gg 1$ we can then write

$$z_n = n\pi + \xi_n, \tag{4.18}$$

where $|\xi_n|$ is close to 0 for all $n\pi \ll ba$. In general we have that absolute value of the real part of ξ_n is always smaller than π, *i.e.*,

$$|\mathrm{Re}\,\xi_n| < \pi. \tag{4.19}$$

For $ba \gg 1$ and using (4.18) and

$$\exp(2i\xi_n) \simeq 1 + (2i\xi_n) + (1/2)(2i\xi_n)^2, \tag{4.20}$$

we obtain a second degree equation for ξ_n, so that the z_n become

$$z_n = [(n\pi) - (n\pi/2ba)] - i(n\pi/2ba)^2, \tag{4.21}$$

where we disregard 1 as compared to $2ba$. We note that the real values $n\pi$ for z_n, valid when $ba = \infty$, are modified slightly by the term $-(n\pi/2ba)$, but the imaginary parts, given by the square of this term, are still smaller.

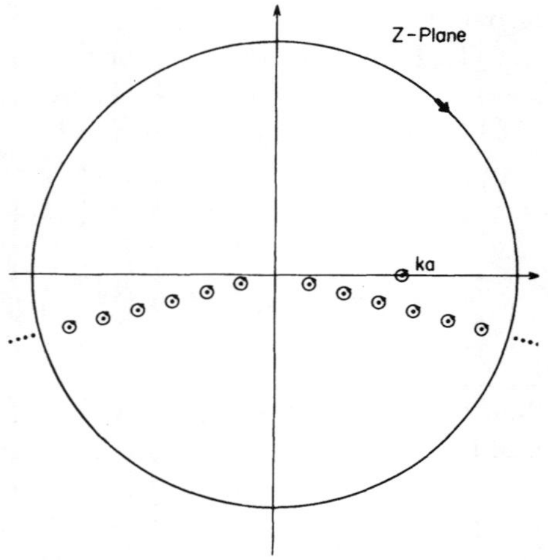

FIGURE 18.1. The contour of the integral (4.22). As the poles $z = ka$ or $z = k_n a$, $n = 1, -1, 2, -2, \cdots$, are surrounded by circles, in the inside of the contour the integrand is analytic and thus the integral (4.22) vanishes as is indicated. Furthermore the large circle of the contour contributes nothing if its radius goes to ∞, as the absolute value of the integrand goes to zero as $|z|^{-2}$ if $|z| \to 0$. Thus we obtain the relation (4.23) from the residues of all the poles including $z = ka$.

Thus for $ba \gg 1$ we have states slightly displaced from $n\pi$ with very narrow widths.

The values $k_n = z_n/a$ are then poles of $V(k)$ of (4.14) in the lower right quadrant of the complex plane k and close to the real axis. Besides if z_n is a zero of (4.17) then $-z_n^*$ is also a zero as can be seen by conjugation of (4.17). Thus we have poles on the lower left quadrant that are a reflection on the imaginary axis of those on the lower right quadrant. We shall enumerate the poles on the right as k_n, $n = 1, 2, 3 \ldots$ in order of increasing real part and those on the left s k_n, $n = -1, -2, -3, \cdots$.

We note now that the function $V(k)$ of (4.14) has no poles on the upper half of the k plane and it goes there as $|k|^{-2}$ when $|k| \to \infty$. On the other hand in the lower half plane it goes as $|k|^{-1}$ when $|k| \to \infty$. Thus all over the k plane when $|k| \to \infty$, $V(k) \to 0$ at least as $|k|^{-1}$.

To expand $V(k)$ in terms of its poles we consider an integral in a complex plane z of the form

$$\frac{1}{2\pi i} \int_C \frac{V(z)}{z - ka} dz = 0, \tag{4.22}$$

where $V(z)$ is given by (4.14) when we replace ka by z. The contour in Fig. 18.1 includes a large circle that goes to ∞ and small circles that include the poles $z_n = k_n a$, $n = 1, -1, 2, -2, \cdots$ of (4.17), as well as ka. Thus the integrand of (4.22) is analytic in the region, which is the reason that integral vanishes as indicated in (4.22). Furthermore the large circle contributes nothing as the radius goes to ∞ as the integrand goes to 0 at least as $|z|^{-2}$. Thus we obtain by the calculus of residues the relation

$$V(k) = \sum_{n=-\infty}^{\infty} d_n(k - k_n)^{-1}, \tag{4.23}$$

where k_n are the poles of $V(k)$ given by (4.16) and the residues are determined by

$$\begin{aligned} d_n &= \lim_{k \to k_n} [(k - k_n)V(k)] \\ &= i(2a^2)^{-1} [2ba\ k_n + ib(1 + 2ba)]^{-1}. \end{aligned} \tag{4.24}$$

This is shown in the Appendix, where we also derive the sum rules

$$\sum_{n=-\infty}^{\infty} \left[d_n(K \pm k_n)^{-1}\right] = 0, \tag{4.25}$$

$$\sum_{n=-\infty}^{\infty} \left[d_n(K \pm k_n)^{-2}\right] = (K^2 a)^{-1}, \tag{4.26}$$

with $Ka = m\pi$.

18.4.4 Determination of $A(K,t)$

From (4.4) we see that

$$\begin{aligned} A(K,t) &= \frac{1}{2\pi i}\int_{c-i\infty}^{c+i\infty} \bar{A}(K,s)\exp(st)ds \\ &= \frac{1}{2\pi}\int_{-\infty+ic}^{\infty+ic} \bar{A}(K,k)\exp(-iEt)dE \\ &= \frac{1}{2\pi}\int_C k\bar{A}(K,k)\exp(-ik^2t/2)dk, \end{aligned} \tag{4.27}$$

where $\bar{A}(K,k)$ is given by (4.12). In (4.27) we passed from the variable s to $E = is$ with the corresponding change in the limits of integration. We then wrote $E = k^2/2$ and the contour is given by the curve C' in Fig. 2c. We can complete it with the dotted *arc* of circle as on it the $\exp(-ik^2t/2) \to 0$ if $|k| \to \infty, (\pi/2) \leq \arg k \leq \pi$. As $\bar{A}(K,k)$ is analytic in the upper half plane we can deform C' to the contour C in which small circles bypass the poles $k = \pm K$ on the real axis.

Using (4.12)-(4.14) and (4.23) we can write

$$A(K,t) = \frac{1}{2\pi}\int_C (2ik)(k^2 - K^2)^{-1}\exp(-ik^2t/2)dk$$

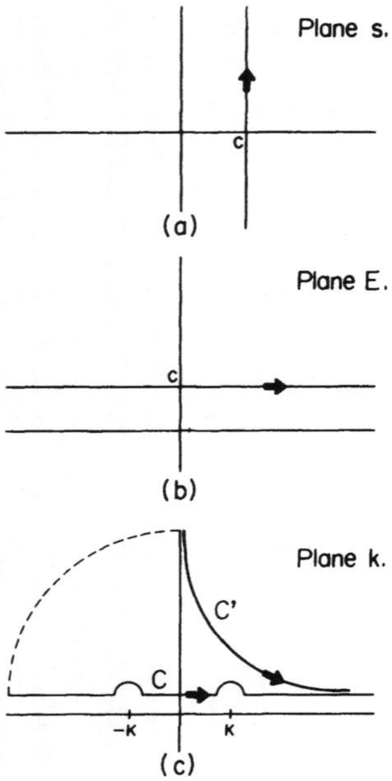

FIGURE 18.2. We passed from the variable s in Fig. 18.2a to $E = is$ in Fig. 18.2b with the corresponding change in the limits of integration. We then wrote $E = (k^2/2)$ and the contour is given by the curve C' in Fig. 18.2c. We can complete it with the dotted arc of circle as on it the $\exp(-ik^2t/2) \to 0$ if $|k| \to \infty, (\pi/2) \leq \arg k \leq \pi$. As $A(K,k)$ is analytic in the upper half plane we can deform C' to the contour C in which small circles bypass the poles $k = \pm K$ on the real axis.

$$+\frac{ia}{2\pi}\int_C \left\{ \sum_{n=-\infty}^{\infty} d_n \frac{4K^2k^2}{(k^2-K^2)^2} \frac{1}{(k-k_n)} \right\} \times \exp(-ik^2t/2)dk. \tag{4.28}$$

From the theory residues we see that

$$\left\{ \sum_{n=-\infty}^{\infty} \frac{d_n(4K^2k^2)}{(k^2-K^2)^2(k-k_n)} \right\}$$
$$= \left\{ \frac{K^2}{(k-K)^2} \sum_{n=-\infty}^{\infty} \frac{d_n}{(K-k_n)} - \frac{K^2}{(k+K)^2} \sum_{n=-\infty}^{\infty} \frac{d_n}{(K+k_n)} \right.$$
$$- \frac{K}{(k-K)} \sum_{n=-\infty}^{\infty} \frac{d_nk_n}{(K-k_n)^2} + \frac{K}{(k+K)} \sum_{n=-\infty}^{\infty} \frac{d_nk_n}{(K+k_n)^2}$$
$$\left. + \sum_{n=-\infty}^{\infty} \frac{4K^2k_n^2d_n}{(K^2-k_n^2)^2(k-k_n)} \right\}, \tag{4.29}$$

which can also be checked directly by taking as common denominator the one of the left hand side of (4.29) and evaluating the numerator.

From the sum rule (4.25) the first two terms on the right hand side of (4.29) vanish. We see then that in the second term of (4.28) we just have to evaluate the integral [3], [8], [9]

$$\begin{aligned} M(\lambda,t) &= \frac{i}{2\pi}\int_C (k-\lambda)^{-1}\exp(-ik^2t/2)dk \\ &= \tfrac{1}{2}\exp y^2 \operatorname{erfc}(y), \end{aligned} \tag{4.30}$$

where

$$y = -e^{-i\pi/4}\lambda\sqrt{t/2}, \tag{4.31}$$

and λ takes the value $\lambda = k_n, n = 1,-1,2,-2,\cdots$, and $\lambda = \pm K$. For the convenience of the reader the derivation of the explicit form (4.30) of $M(\lambda,t)$ is also discussed in the Appendix.

We note now that

$$\begin{aligned} \sum_{n=-\infty}^{\infty}\frac{(\pm k_n)d_n}{(K\pm k_n)^2} &= \sum_{n=-\infty}^{\infty}\frac{d_n}{(K\pm k_n)} \\ &\quad - K\sum_{n=-\infty}^{\infty}\frac{d_n}{(K\pm k_n)^2} = -\frac{1}{Ka}, \end{aligned} \tag{4.32}$$

where we used the summation rules (4.25) and (4.26). Thus the contribution of the third and fourth term of (4.29) to the second integral in (4.28) is

$$-\left[M(K,t)+M(-K,t)\right] = -\exp(-iK^2t/2), \tag{4.33}$$

as shown also in the Appendix.

The first integral in (4.28) can be written as

$$\begin{aligned} &\frac{1}{2\pi}\int_C (2ik)(k^2-K^2)^{-1}\exp(-ik^2t/2)\,dk \\ &= -\frac{1}{2\pi i}\int_{-\infty+ic}^{\infty+ic}(E-\varepsilon)^{-1}\exp(-iEt)\,dE = \exp(-i\varepsilon t) \end{aligned} \tag{4.34}$$

where $E = k^2/2$, $\varepsilon = K^2/2$. Thus (4.34) exactly cancels (4.33), and from the last term in (4.29) we obtain for $A(K,t)$ of (4.28) the expression

$$A(K,t) = a\sum_{n=-\infty}^{\infty}\left[\frac{4K^2k_n^2d_n}{(K^2-k_n^2)^2}M(k_n,t)\right], \tag{4.35}$$

with $M(k_n,t)$ given by (4.30) when $\lambda = k_n$, and the k_n being the complex roots of Eq. (4.16). In the Appendix we show, as expected, that $A(K,0) = 1$.

It is interesting to note that a similar formula holds for the decay amplitude $\psi_2(t)$, with the initial condition (2.2), in the schematic theory of nuclear reactions, where there are only two poles of the S matrix which we designate by k_+, k_-. The $\psi_2(t)$ is then given by [3], [7]

$$\psi_2(t) = 2(k_+ - k_-)^{-1}\left[k_+ M(k_+,t) - k_- M(k_-,t)\right], \tag{4.36}$$

and thus there is no summation but the same functions $M(k_\pm, t)$ of (4.30) are involved as in the $A(K,t)$ of (4.35).

We thus have determined the amplitude of decay $A(K,t)$ for all times *i.e.* $0 \le t \le \infty$. As (4.35) involves an infinite sum we have also to discuss its convergence. This implies finding an upper bound to the absolute value in the summation (4.35) starting with some arbitrary N of n. We saw though in subsection 4c that the real part of k_n goes as $n\pi$. Thus we could select our N such that for a given t we have

$$|y_N| = |-e^{-i\pi/4} k_N \sqrt{t/2}| \gg 1 \tag{4.37}$$

and use the asymptotic form of $\exp(y_n^2)\operatorname{erfc}(y_n)$ when $|y_n| \gg 1$ to discuss the convergence. As we have to do this anyway for large times we defer the discussion of convergence to the next section.

18.5 The behavior of $A(K,t)$ for large times

We first define large times as being such that

$$|y_n| = |-e^{-i\pi/4} k_n (t/2)^{1/2}| \gg 1, \tag{5.1}$$

for all $n = 1, -1, 2, -2, \cdots$. In this case we can use the asymptotic behavior $\exp(y^2)\operatorname{erfc} y$ for large $|y|$ given by

$$\begin{aligned} e^{y^2}\operatorname{erfc} y &= \pi^{-1/2}y^{-1} - (1/2)\pi^{-1/2}y^{-3} + \cdots, \\ &\quad \text{if } |y| \to \infty, -\pi/2 \le \arg y \le \pi/2, &\text{(5.2a)} \\ e^{y^2}\operatorname{erfc} y &= 2e^{y^2} + \pi^{-1/2}y^{-1} - (1/2)\pi^{-1/2}y^{-3} + \cdots, \\ &\quad \text{if } |y| \to \infty,\ \pi/2 \le \arg y \le 3\pi/2, &\text{(5.2b)} \end{aligned}$$

to show that if (5.1) is satisfied then we can write

$$\begin{aligned} A(K,t) &= a\sum_{n=1}^{\infty} \frac{8K^2k_n^2 d_n}{(K^2-k_n^2)^2}\exp(-ik_n^2 t/2) \\ &\quad + a\pi^{-1/2}\sum_{n=-\infty}^{\infty} \frac{2K^2k_n^{-1}d_n}{(K^2-k_n^2)^2} e^{-i3\pi/4}(t/2)^{-3/2} + \cdots \end{aligned} \tag{5.3}$$

where we show in the Appendix that the term corresponding y^{-1}, *i.e.* $t^{-1/2}$, vanishes.

We can use first (5.3) to see that the series in (4.35) converges for which, from the arguments at the end of the last section, it is sufficient to show that the asymptotic form (5.3) of $A(K,t)$ is convergent.

We note that k_n, $n = 1,2,3,\cdots$ has positive real part and has negative imaginary part and thus

$$|\exp(-ik_n^2 t/2)| < 1 \tag{5.4}$$

Furthermore we have from (4.24) that $|d_n|$ is proportional to $|k_n|^{-1}$ for large n while

$$|k_n| \geq n\pi. \tag{5.5}$$

Thus the two series in (5.3) can be majorized by converging series and both (5.3) and the general result (4.35) for $A(K,t)$ are convergent.

Let us now look at the first series in (5.3) which obviously indicates an exponential decay associated with each resonant state but with different amplitudes. Because of the denominator $(K^2 - k_n^2)^{-2}$ where $K = m\pi/a$ and k_n is very close to $(n\pi/a)$ if $ba \gg 1$ as shown in (4.21), we see that if the height of the barrier is very large the main contribution will come in the summation when $n = m$ and the lifetime of the state will then be

$$\tau = \left[\operatorname{Im} k_m^2/2\right]^{-1} \simeq (2ba/m\pi)^2. \tag{5.6}$$

Thus we have mainly the exponential decay of the original bound state (3.1) due to the lowering of the infinite δ barrier at $t = 0$ to the height b.

For times t much larger than τ, the preponderant term in the decay amplitude will be the one associated with $t^{-3/2}$ which is also a well known result [11].

Thus we have analyzed the behavior of the decay amplitude for large times and we turn now our attention to very short times and the paradox with which we initiated this article.

18.6 The behavior of $A(K,t)$ for very short times

Since y of (4.31) is proportional to $t^{1/2}$, if we want to discuss the behavior of $A(K,t)$ for very short times it is convenient to consider the expansion

$$\exp y^2 \operatorname{erfc} y = \sum_{r=0}^{\infty} a_r\, y^r, \tag{6.1}$$

where [9]

$$a_{2r} = (r)^{-1}, \quad a_{2r+1} = -\frac{2}{\sqrt{\pi}} \sum_{k=0}^{r} \frac{(-1)^k}{(r-k)k(2k+1)}. \tag{6.2a,b}$$

We note from (4.31)

$$y_n^r = (-1)^r e^{-i\pi r/4} k_n^r (t/2)^{r/2}. \tag{6.3}$$

Thus if in (4.35) we interchanged the order summation we would have terms of the form

$$\sum_{n=-\infty}^{\infty} \left[4K^2 k_n^{r+2} d_n (K^2 - k_n^2)^{-2}\right] (-1)^r e^{-i\pi r/4} (t/2)^{r/2}. \tag{6.4}$$

As for large n, $|d_n|$ is proportional to the reciprocal of $|k_n|$ and the k_n is close to $n\pi/a$ it is clear that the series (6.4) diverges for $r \geq 2$. Thus the interchange of the summation on r in (6.1) with the summation on n in (4.35) *is not an allowed operation.* This suggests in turn that we can not exchange the summation for the series defining $\exp(-iE_n t)$ with respect to the one on n in (2.6) and therefore the argument leading to the decay probability (2.11) for short times is not valid. Thus our paradox disappears.

Furthermore one can rewrite the probability amplitude of (4.35) by use of (4.30) in the form

$$\begin{aligned} A(K,t) &= \frac{a}{2} \sum_{n=-\infty}^{\infty} \{4K^2 k_n^2 d_n (K^2 - k_n^2)^{-2} \\ &\quad \times \exp(-ik_n^2 t/2)\, \mathrm{erfc}[-e^{-i\pi/4} k_n \sqrt{t/2}]\}, \end{aligned} \tag{6.5}$$

and then calculate it numerically for values of the parameters b, a given in the caption of Fig. 18.3, where also we indicate the number of terms k_n that we consider.

The curve in Fig. 18.3 for $|A(K,t)|^2$ as function of t has derivative $-\infty$ at $t = 0$ indicating that the behavior of $|A(K,t)|^2$ at very small times should go as t^α with α smaller than 1 and not as t^2 as suggested by the argument of (2.11) which we showed to be inaccurate. In the latter case the derivative of $|A(K,t)|^2$ at $t = 0$ would be 0 as (2.11) shows.

18.7 Conclusion

We obtained for the full time interval $0 \leq t \leq \infty$ the decay amplitude $A(K,t)$ for the problem of an s-state with a δ function barrier of height b at distance a. We discussed the behavior of $A(K,t)$ in the region in which it is exponential as well as in the cases of very large and very short times where it is not.

Our results are valid not only for the case of a δ interaction, but in more general situations as shown, from another view point, in the paper of G. García-Calderón in this book [4].

Had we taken an arbitrary potential but cut it at $r = a$ and enclosed it in an infinite potential barrier at this distance, we would have a discrete

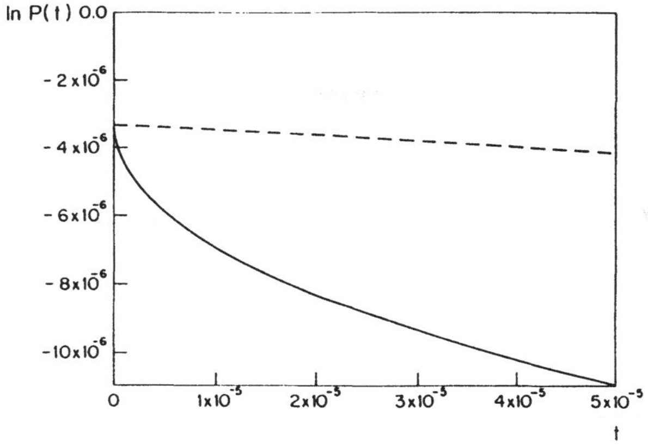

FIGURE 18.3. The above figure shows a plot of $\ln P(t)$ where $P(t) \equiv |A(t)|^2$ versus time t at very short times for both the exact expression for $A(t)$ given by Eq. (4.35) (continuous line) and for a purely exponential decay (broken line) in the case of the delta potential involving the first 25 resonances. The parameters of the delta potential, in which $\hbar$ and the mass of the particle are unity, are $b = 50.0$ and $a = 1.0$ and we choose for the initial state $m = 1$ so that $K = \pi$. The time scale is fixed by the lifetime associated to the first pole, *i.e.*, $\tau = [2(\text{Re}\, k_1)(\text{Im}\, k_1)]^{-1} = 42.02$. Notice that due to the finite number of terms in Eq. (4.35) the non-decay probability $P(t)$ does not start exactly at 1. Notice also that $dP/dt = -\infty$ at $t = 0$. The figure clearly shows that the exact non-decay probability deviates from the purely exponential contribution at very short times as discussed in the text. We acknowledge the help of Verónica Riquer for the required calculations and elaboration of this figure.

set of states. Choosing one of them we could analyze its decay when the infinite barrier at $r \geq a$, but *not* the potential, is removed at $t = 0$.

The only change that occurs now is that $V(k)$ of (4.14) is replaced by

$$kV(k) = (2ika)^{-1}\left[1 - S(k)\exp(i2ka)\right], \tag{7.1}$$

where $S(k)$ is the S matrix of the problem, while $U(K, k)$ of (4.13) remains essentially the same.

The analytic properties of $kV(k)$ for cut potentials have been discussed [12] and thus expansions of the form (4.23) could be carried out and from there the rest of the analysis follows. Thus the expression (6.5) for $A(K,t)$ remains valid with only slight modifications, for any cut potential with the initial conditions indicated above. The $k_n, n = 1, -1, 2, -2, \cdots$ are now the poles of the S matrix for the cut potential in question.

A detailed discussion of the general cut potential problem will be given elsewhere.

18.8 Acknowledgements

We are indebted to Prof. H.M. Nussenzveig not only for his notation for the function $M(\lambda, t)$, but also for many discussions along the years, as well as for the fact that his contribution to this book deals with problems analyzed with similar techniques.

18.A Appendix

18.A.1 Determination of the Residue d_n at the Pole k_n

In this section we will discuss some of the pending points mentioned in the text.

We indicated that the $V(k)$ of (4.14) can be expanded in poles as in (4.23). The residue d_n is then given by

$$\begin{aligned} d_n &= \lim_{k \to k_n} [(k - k_n)V(k)] \\ &= \frac{1 - \exp(2ik_n a)}{2ik_n a} \left[\frac{d}{dk} \{ka + iba[1 - \exp(i2ka)]\} \right]^{-1}_{k=k_n} \\ &= i(2a^2)^{-1} [2k_n ba + ib(1 + 2ba)]^{-1}, \end{aligned} \tag{A.1}$$

where we used the relation

$$k_n a + iba\,[1 - \exp(i2k_n a)] = 0, \tag{A.2}$$

to eliminate the $\exp(i2k_n a)$ appearing in (A.1).

We note that when $|k_n| \gg b$ as well as $n \gg 1$, we can write d_n as proportional to k_n^{-1}, a property that is used in the discussion of convergence.

18.A.2 The Sum Rules

We can write

$$V(k) = [1 - \exp(i2ka)]\, W(k) = \sum_{n=-\infty}^{\infty} \frac{d_n}{(k - k_n)}, \tag{A.3}$$

where

$$W(k) = (i2ka)^{-1} \{ka + iba\,[1 - \exp(i2ka)]\}^{-1}. \tag{A.4}$$

Clearly then if $k = \pm K = \pm(m\pi/a)$ the left hand side of (A.3) vanishes and we have the sum rule

$$\sum_{n=-\infty}^{\infty} \frac{d_n}{K \pm k_n} = 0. \tag{A.5}$$

If we now take the derivative of both sides of (A.3) with respect to k and evaluate it at $k = K$ we obtain

$$-i2a\exp(i2Ka)W(K) = -\sum_{n=-\infty}^{\infty} \frac{d_n}{(K-k_n)^2}. \tag{A.6}$$

From (A.4) we see that $W(K) = (2iK^2a^2)^{-1}$ so we can finally write

$$\sum_{n=-\infty}^{\infty} \frac{d_n}{(K \pm k_n)^2} = \frac{1}{K^2 a}, \tag{A.7}$$

as the argument could also be repeated for $k = -K$.

18.A.3 Derivation of $M(k_n, t)$

Let us introduce the variable

$$u = -e^{i\pi/4} k\sqrt{t/2}, \tag{A.8}$$

in which case we could write

$$M(k_n, t) = \frac{i}{2\pi}\int_{-\infty}^{\infty} (k-k_n)^{-1}\exp(-u^2)dk, \tag{A.9}$$

as the poles of $V(k)$ are all below the real axis so the contour C can be reduced to the abscissa.

From the differentiation of (A.8) in which t is a parameter, we have

$$dk = -e^{-i\pi/4}(t/2)^{-1/2}du, \tag{A.10}$$

so we can write

$$M(k_n, t) = \frac{i}{2\pi}\int_{C'} (iy_n - u)^{-1}\exp(-u^2)du, \tag{A.11}$$

where the contour C' is now a straight line forming a 45° degree angle with the abscissa and y_n is given by

$$y_n = -e^{-i\pi/4} k_n\sqrt{t/2}. \tag{A.12}$$

The line forming the contour C' in the u-plane can be joined by arcs of circle with the abscissa. As $\exp(-u^2)$ on these arcs vanishes when their radius tend to ∞ we can change the contour C' to the real axis thus getting

$$\begin{aligned} M(k_n, t) &= \frac{i}{2\pi}\int_{-\infty}^{\infty} (iy_n - u)^{-1}\exp(-u^2)\,du \\ &= \tfrac{1}{2}\exp(y_n^2)\operatorname{erfc} y_n, \end{aligned} \tag{A.13}$$

according to a result of Reference [13].

18.A.4 Relations between $M(K,t)$ and $M(-K,t)$

We have from (4.30),

$$M(K,t) = \tfrac{1}{2}\exp(y^2)\,[1 - \operatorname{erfc} y]\,, \tag{A.14}$$

with y being given by (4.31) with $\lambda = K$. As erfc y is an odd function of y we can write

$$\begin{aligned} M(K,t) &= \tfrac{1}{2}\exp(y^2)\,[1 + \operatorname{erfc}(-y)] \\ &= \tfrac{1}{2}\exp(y^2)\,[2 - \operatorname{erfc}(-y)] \\ &= \exp(y^2) - M(-K,t). \end{aligned} \tag{A.15}$$

Thus we obtain the relation

$$M(K,t) + M(-K,t) = \exp(-iK^2t/2), \tag{A.16}$$

mentioned in (4.33).

18.A.5 Proof that for Large Times the Coefficient of $t^{-1/2}$ Vanishes

We see from (4.35), when using the expressions (4.30) for $M(k_n,t)$ and the expansions (5.2) for large y, that the term containing $t^{-1/2}$ has the coefficient

$$\begin{aligned} &\sum_{n=-\infty}^{\infty} \frac{4K^2 k_n d_n}{(K^2 - k_n^2)^{-2}} \\ &= \sum_{n=-\infty}^{\infty} \left\{ (4K^2 k_n d_n) \left[\frac{1}{2K} \left(\frac{1}{K - k_n} + \frac{1}{K + k_n} \right) \right]^2 \right\} \end{aligned} \tag{A.17}$$

$$\begin{aligned} &= \sum_{n=-\infty}^{\infty} \frac{k_n d_n}{(K - k_n)^2} + \sum_{n=-\infty}^{\infty} \left[\frac{d_n}{K - k_n} - \frac{d_n}{K + k_n} \right] \\ &\quad + \sum_{n=-\infty}^{\infty} \frac{k_n d_n}{(K + k_n)^2} \\ &= K \sum_{n=-\infty}^{\infty} \frac{d_n}{(K - k_n)^2} - K \sum_{n=-\infty}^{\infty} \frac{d_n}{(K + k_n)^2} = 0, \end{aligned} \tag{A.18}$$

where we used the summation rules (4.25), (4.26).

18.A.6 Proof that $A(K,0) = 1$

The decay amplitude $A(K,t)$ at $t = 0$ becomes the scalar product of the normalized initial state with itself *i.e.* $A(K,0) = 1$.

We proceed to show that it also follows from (4.35) when $t = 0$. For this we first note that when $t = 0$, $y = 0$ and $M(k_n, 0) = (1/2)$ and thus

$$\begin{aligned}
A(K,0) &= \frac{a}{2}\sum_{n=-\infty}^{\infty}\frac{4K^2k_n^2d_n}{(K^2-k_n^2)^2} \\
&= -\frac{a}{2}\sum_{n=-\infty}^{\infty}\frac{4K^2d_n}{K^2-k_n^2} + 2K^4a\sum_{n=-\infty}^{\infty}\frac{d_n}{(K^2-k_n^2)^2} \\
&= -2K^2a\sum_{n=-\infty}^{\infty}\frac{d_n}{2K}\left[\frac{1}{K-k_n}+\frac{1}{K+k_n}\right] \\
&\quad +\frac{aK^2}{2}\sum_{n=-\infty}^{\infty}d_n\left[\frac{1}{K-k_n}+\frac{1}{K+k_n}\right]^2 \\
&= \frac{aK^2}{2}\left\{\sum_{n=-\infty}^{\infty}\frac{d_n}{(K-k_n)^2}+\sum_{n=-\infty}^{\infty}\frac{d_n}{(K+k_n)^2}\right\} \\
&\quad +\frac{1}{K}\sum_{n=-\infty}^{\infty}d_n\left[\frac{1}{K-k_n}+\frac{1}{K+k_n}\right] \qquad \text{(A.19)} \\
&= (aK^2/2)(2/aK^2) = 1, \qquad \text{(A.20)}
\end{aligned}$$

where again we made use of the sum rules (4.25) (4.26).

18.B References

[1] G. Gamow, *Z. Phys.* **51**, 204 (1928).

[2] W. Gurney and E.U. Condon, *Phys. Rev.* **33**, 127 (1929).

[3] M. Moshinsky, *Phys. Rev.* **84**, 525 (1951).

[4] G. García-Calderón, *in these Proceedings.*

[5] G. Breit and E.P. Wigner, *Phys. Rev.* **49**, 519, 642 (1936).

[6] A. Peres, *Ann. Phys. (N.Y.)* **129**, 33 (1980); K. Grotza and H.V. Klapdor, *Phys. Rev.* **C 30**, 2098 (1984).

[7] M. Moshinsky, G. Loyola, and J.L. Mateos, "Time dependent model for heavy ion collision with a single resonance", Proceedings of the XIV Symposium on Nuclear Physics, Oaxtepec, México (World Scientific 1991, Singapore).

[8] G. Beck and H.M. Nussenzveig, *N. Cimento* **16**, 441 (1960). See also the contribution of H.M. Nussenzveig *in these Proceedings.*

[9] M. Moshinsky, G. Loyola, and C. Villegas, *Rev. Mex. Fís.* **37**, 369 (1991).

[10] I.S. Gradshteyn and I.M. Ryzhik, *Tables of Integrals, Series and Products* (Academic Press, New York 1980) p. 931, Formula 8.254.

[11] M.L. Goldberger and K.M. Watson, *Collision Theory* (John Wiley & Sons, New York, 1964) Chapter VIII and in particular p. 450.

[12] G. García-Calderón and M. Berrondo, *Lett. N. Cimento, Serie 2,* **26**, 562 (1979).

[13] M. Abramowitz and I.A. Stegun, *Handbook of Mathematical Functions* (Dover Publications Inc., New York, 1972) p. 297, formula 7.1.4.

19

Moshinsky Functions, Resonances and Tunneling

Herch M. Nussenzveig[1]

ABSTRACT A brief review of the chief properties and applications of Moshinsky functions is given. Applications surveyed include diffraction in time, time-dependent perturbation theory, one-dimensional tunneling, the decay of a localized state, and resonance scattering.

19.1 Introduction

In 1951, in connection with a time-dependent scattering problem, Moshinsky [1] introduced a function that he later interpreted [2], [3] in terms of a new transient effect, *diffraction in time*. This function plays an important role in describing the time dependence of many process in nonrelativistic quantum mechanics, particularly in connection with tunneling and resonance scattering.

We survey here the main features and applications of Moshinsky functions. For convenience, their definition and most useful mathematical properties are collected in Section 2. The original application to diffraction in time, as well as applications to time-dependent perturbation theory, are discussed in Section 3. Section 4 describes recent applications to one-dimensional tunneling. In time-dependent scattering theory, the main applications are to the decay of a localized state (Section 5) and to the treatment of resonance scattering (Section 6).

It is a pleasure to dedicate this work to Professor Marcos Moshinsky on his 70th birthday, as a token of his manifold contributions to physics and to scientific collaboration in Latin America.

19.2 The Moshinsky function

Consider a beam of nonrelativistic particles travelling in the x direction, with momentum k (we employ units such that $\hbar = m = 1$) and confined

[1]Depto. de Física, Pontifícia Universidade Católica, Rio de Janeiro, Brazil.

to the left half-space by a perfectly absorbing shutter at $x = 0$. If the shutter is suddenly removed at $t = 0$, the wave function for $t > 0$ is the Moshinsky function $M(x, k, t)$. It is the solution of the free-particle Schrödinger equation

$$i\partial_t \psi + \tfrac{1}{2}\psi'' = 0 \tag{2.1}$$

(where primes denote derivatives with respect to x) that satisfies the initial condition

$$\psi(x, 0) \equiv M(x, k, 0) = \theta(-x)\exp(ikx), \tag{2.2}$$

in which θ is the Heaviside step function.

The free-particle Schrödinger propagator

$$K_0(x, t \mid x', t') \equiv U(x - x', t - t'), \tag{2.3}$$

is the solution of (2.1) that reduces to $\delta(x - x')$ for $t = t'$. The function U is given by [4]

$$U(x, t) = \frac{\exp(ix'/2t)}{\sqrt{2\pi i t}}, \tag{2.4}$$

where

$$\sqrt{i} \equiv \exp(i\pi/4). \tag{2.5}$$

By (2.2), we have

$$M(x, k, t) = \int_{-\infty}^{0} \exp(ikx')U(x - x', t)\, dx'. \tag{2.6}$$

Substituting U by (2.4) and completing the square in the exponent of the integrand, we find

$$M(x, k, t) = \tfrac{1}{2}v(x, k, t)\,\mathrm{erfc}\,(e^{-i\pi/4}w) \tag{2.7}$$

where

$$v(x, k, t) \equiv \exp\left[i(kx - Et)\right], \qquad E = \tfrac{1}{2}k^2, \tag{2.8}$$

$$w \equiv (x - kt)/\sqrt{2t}, \tag{2.9}$$

and

$$\mathrm{erfc}\, z = \frac{2}{\sqrt{\pi}} \int_z^{\infty} \exp(-\zeta^2) d\zeta \tag{2.10}$$

is the complementary error function.

Although we have assumed that k is real so far, all of the above expressions remain valid for $\mathrm{Im}\, k < 0$, thus extending the definition of the Moshinsky function to values of k in the lower half of the complex plane. For such points we find the alternative representation

$$M(x, k_0, t) = \frac{i}{2\pi} \int_{-\infty}^{\infty} \frac{v(x, k, t)}{k - k_0} dk, \quad \mathrm{Im}\, k_0 < 0, \tag{2.11}$$

which follows from the fact that both sides are solutions of (2.1) satisfying the initial condition (2.2) with $k = k_0$.

For the solution of initial-value problems, it is useful to determine the Laplace transforms of U and M. Defining

$$\tilde{f}(s) \equiv \int_0^\infty \exp(-st)f(t)dt, \tag{2.12}$$

the Laplace transform of U, by (2.1), satisfies

$$\tilde{U}'' + 2is\tilde{U} = 2i\delta(x), \tag{2.13}$$

and that of M, by (2.2), satisfies

$$\tilde{M}'' + 2is\tilde{M} = 2i\theta(-x)\exp(ikx). \tag{2.14}$$

The solution of (2.13) is

$$\tilde{U}(x,s) = \exp(iz|x|)/z \tag{2.15}$$

where

$$z = \sqrt{2is} \tag{2.16}$$

and we employ: $|x|'' = 2\delta(x)$. Similarly, by solving (2.14) separately for $x < 0$ and for $x > 0$ and matching the results at $x = 0$, we find [5]

$$\tilde{M}(x,k,s) = \frac{i\exp(izx)}{z(z-k)}, \quad x > 0, \tag{2.17}$$

with a corresponding expression for $x < 0$.

We can now employ these results to obtain the Schrödinger propagator for a one-dimensional delta-function potential [6], [7], *i.e.*, the solution $K_\xi(x,t|x',0)$ of

$$i\partial_t\psi + \tfrac{1}{2}\psi'' = b\delta(x-\xi)\psi \tag{2.18}$$

that reduces to $\delta(x - x')$ for $t = 0$. The Laplace-transformed equation is

$$\tilde{K}_\xi'' + 2is\tilde{K}_\xi - 2b\delta(x-\xi)\tilde{K}_\xi = 2i\delta(x-x'). \tag{2.19}$$

Setting

$$\tilde{K}_\xi(x,s \mid x',0) \equiv \tilde{U}(x-x',s) + \tilde{\mathcal{K}}_\xi(x,s \mid x',0), \tag{2.20}$$

and taking into account (2.13), we find

$$\tilde{\mathcal{K}}_\xi'' + 2is\tilde{\mathcal{K}}_\xi = 2b[\tilde{U}(x-\xi,s) + \tilde{\mathcal{K}}_\xi(\xi,s \mid x',0)]\delta(x-\xi). \tag{2.21}$$

By comparison with (2.13), this suggests the *Ansatz*

$$\tilde{\mathcal{K}}_\xi(x,s \mid x',0) = c\tilde{U}(x-\xi,s)\tilde{U}(x'-\xi,s)$$

which satisfies the symmetry property of the propagator. Inserting into (2.21) and employing (2.13), we obtain

$$c = -ib/[1 + ib\tilde{U}(0, s)],$$

so that (2.20) yields

$$\tilde{K}_\xi(x, s \mid x', 0) = \tilde{U}(x - x', s) - \frac{ib}{1 + ib\tilde{U}(0, s)}\tilde{U}(x - \xi, s)\tilde{U}(\xi - x', s). \quad (2.22)$$

Thus, in this representation, the propagator for the one-dimensional delta-function potential represents processes of free propagation and of propagation through the potential with a single point interaction, as might be expected.

Substituting (2.15) into (2.22), we get

$$\tilde{K}_\xi(x, s \mid x', 0) = \frac{\exp(iz|x - x'|)}{z} - \frac{ib}{z(z + ib)}\exp[iz(|x - \xi| + |\xi - x'|)]$$

and, according to (2.15) and (2.17), this finally yields

$$K_\xi(x, t \mid x', 0) = U(x - x', t) - bM(|x - \xi| + |\xi - x'|, -ib, t). \quad (2.23)$$

Thus, the exact propagator for the one-dimensional delta-function potential is also expressed in terms of the Moshinsky function.

It is readily verified [7] that the partial derivatives of $M(x, k, t)$ with respect to x, k or t can all be expressed in terms of $U(x, t)$ and $M(x, k, t)$. Another very useful result is the indefinite integral

$$\int \exp(ikx)M(ax + b, c, t)dx$$
$$= \frac{\exp(ikx)}{i(k + ac)}\left[M(ax + b, c, t) - M(ax + b, -k/a, t)\right], \quad (2.24)$$

which can be obtained by partial integration; for $k \to -ac$, the l'Hôpital's rule must be applied.

The asymptotic behavior of the Moshinsky function (2.7), with k real or complex, when $|w| \gg 1$ follows from that of the complementary error function [8]. It depends on whether w is located above (A) or below (B) the second bisector of the complex w plane, *i.e.*,

$$w \in A : -\frac{\pi}{4} < \arg w < \frac{3\pi}{4}; \quad w \in B : \frac{3\pi}{4} < \arg w < \frac{7\pi}{4}. \quad (2.25)$$

The results are [9], for $|w| \gg 1$,

$$M(x, k, t) \equiv M_A(x, k, t) = \frac{it}{x - xt}U(x, t) \qquad w \in A,$$
$$\times \left[1 - \frac{i}{2w^2} + \cdots + \frac{(2n - 1)!!}{(2iw^2)^n} + R_n(w)\right] \quad (2.26)$$

$$M(x, k, t) \equiv M_B(x, k, t)$$
$$= M_A(x, k, t) + v(x, k, t) \qquad w \in B, \quad (2.27)$$

where

$$|R(w)| \leq \sqrt{\pi}(2n+1)!!/\left(2^{n+1}|w|^{2n+1}\right), \quad n = 0, 1, 2, \cdots. \tag{2.28}$$

19.3 Applications: transient effects

19.3.1 Diffraction in Time

(i) Unbounded wave train: For the Moshinsky shutter problem (2.1)-(2.2), the probability current $j(x,t)$ follows immediately from the solution (2.7). For distances from the shutter much greater than the de Broglie wavelength, $kx \gg 1$, it is given by

$$j(x,t) \approx k|M(x,k,t)|^2 = \frac{k}{2}\left\{\left[\frac{1}{2} - C\left(w\sqrt{\frac{2}{\pi}}\right)\right]^2 + \left[\frac{1}{2} - S\left(w\sqrt{\frac{2}{\pi}}\right)\right]^2\right\} \tag{3.1}$$

where w is given by (2.9) and

$$F(z) = \int_0^z \exp(i\frac{\pi}{2}\tau^2)d\tau = C(z) + iS(z) \tag{3.2}$$

is the Fresnel integral [8], which corresponds to the error function in (2.7).

The result (3.1) has the same form as the Fresnel diffraction pattern of a straight edge in optics [10], justifying the expression "diffraction in time". In classical mechanics, the shutter casts a sharp shadow in time, with

$$j_{cl}(x,t) = k\theta(t - t_{cl}), \quad t_{cl} = x/k, \tag{3.3}$$

where t_{cl} is the classical arrival time. The wave nature of matter leads to an oscillatory behavior of the current around this time, with a typical width Δt of the main Fresnel peak given by $|\Delta w| = O(1)$, *i.e.*,

$$\Delta t \sim \sqrt{t_{cl}/E} \sim \sqrt{t_{cl}T}, \tag{3.4}$$

where T is the period associated with the energy E. The proportionality to $\sqrt{t_{cl}}$ is characteristic of a diffusion process, resulting from the "wavelike diffusive" character of the Schrödinger equation.

(ii) Bounded wave packet: What happens if we replace the unbounded semi-infinite incident beam by a more realistic bounded wave packet? One possibility is to cut off the initial wave function (2.2) at $x = -d$, which leads [11] to the Fresnel diffraction pattern of a slit, with Fresnel number

$$\mathcal{F} \sim d/\sqrt{t}. \tag{3.5}$$

The shape of the initial wave packet remains recognizable (with Fresnel fringes around the edges) for large Fresnel number, but this is no longer so in the "Fraunhofer region" $\mathcal{F} \ll 1$, *i.e.*, for $t \gg t_s \equiv d^2$.

This result has a very simple interpretation. The initial position uncertainty of the wave packet, $\Delta x(0) \sim d$, generates a corresponding momentum uncertainty $\Delta k(0) \sim d^{-1}$ which, after free propagation for a time t, increases the position uncertainty by $\Delta k(0)\, t$. This is the well-known quantum spreading effect, which becomes dominant after a *spreading time* t_s such that the additional width is comparable to the original one and memory of the initial shape gets lost. Thus, the spreading time indeed marks the transition from Fresnel to Fraunhofer diffraction in time, and it is given by

$$t_s \sim 1/[\Delta x(0)]^2. \tag{3.6}$$

Another interesting model that represents an initial wave packet of finite width is obtained by taking a complex value for k in (2.2):

$$k \equiv k' - i\kappa, \quad \kappa > 0, \tag{3.7}$$

where the negative imaginary part converts the initial wave packet into a damped oscillation, vanishing exponentially as $x \to -\infty$.

To find the shape of the wave packet for $x > 0$ after removal of the shutter in this case, we exclude the neighborhodd of the wavefront $|w| \leq 1$, where Fresnel-like oscillations may occur and consider only the behavior of the solution (2.7) for $|w| \gg 1$, where, by (2.9), we now have

$$w = [(x - k't) + i\kappa t]/\sqrt{2t} \tag{3.8}$$

In order to apply the asymptotic expansions (2.26)–(2.27), we need to determine the location of w relative to the second bisector of the complex plane. Since $x > 0, t > 0$, the only way to have $w \in B$ is if

$$0 < x < (k' - \kappa)t, \tag{3.9}$$

which requires that $\kappa < k'$, *i.e.*, in (3.7), $k \in A$. Under these conditions, (2.27) applies; in all other cases, the behavior is given by (2.26). By (2.8), we have

$$|v(x,k,t)| = \exp[\kappa(x - k't)] \tag{3.10}$$

so that the last term of (2.27) corresponds to the propagation of the initial exponential wave packet without change of shape, with velocity k'. The first term $M_a(x,k,t)$, by (2.26), differs from the free particle propagator by a factor whose squared modulus is

$$t^2/[(x - k'^t)^2 + (\kappa t)^2],$$

which is peaked around the wavefront with a width κt. Since

$$\Delta x(0) \sim 1/\kappa \sim 1/\Delta k(0),$$

this term is associated with the spreading of the wave packet, with a spreading time

$$t_s \sim \kappa^{-2}. \tag{3.11}$$

At a given point x, (3.10) yields for probability density an exponential decay in time, with lifetime

$$\tau + 1/2k'\kappa. \tag{3.12}$$

Within the domain (3.9), this process competes with the spreading of the wave packet. As k approaches the second bisector from above, the two times (3.11) and (3.12) become of the same order: the initial wave packet is so sharp that the spreading begins to dominate before there is time for exponential decay. Finally, when k goes below the second bisector, no trace of exponential decay remains: the wave packet just spreads as it propagates.

19.3.2 Time-dependent Perturbation Theory

One can employ the exact propagator found for the one-dimensional delta-function potential to test standard approximations employed in time-dependent perturbation theory [5] [take $\xi = 0$ in (2.18)].

For this purpose, one can choose as initial unperturbed wave function the free-space momentum eingenfunction.

$$\psi_k(x,0) \equiv \psi_k(x) = (2\pi)^{-1/2}\exp(ikx), \qquad -\infty < x < \infty, \tag{3.13}$$

and ask for the transition probability after a time t to another momentum eigenfunction with momentum k':

$$P_{k\to k'}(t) = |\,(\psi_{k'}(x), \psi_k(x,t))\,|^2. \tag{3.14}$$

The exact propagator (2.23) allows us to obtain the exact solution $\psi_k(x,t)$ with the help of (2.24). Since the result involves only exponential and Moshinsky functions, the scalar product in (3.14) can again be evaluated exactly, by another application of (2.24) [5]. The applicability of perturbation theory requieres that the strength b of the interaction be much smaller than k and k'. For times t large as compared with the inverse energies involved (associated with k, k', and b), it is found that

$$P_{k\to k'}(t) = \frac{b^2}{(2\pi)^2}\frac{\sin^2[\frac{1}{2}(E-E')t]}{[\frac{1}{2}(E-E')]^2} \tag{3.15}$$

with $E = \frac{1}{2}k^2, E' = \frac{1}{2}k'^2$, which coincides with the well-known result of time-dependent perturbation theory. Furthermore, the corrections to this result are also obtained.

Similar results are found in the shutter problem when one leaves the shutter open during a time t and asks for the probability to observe the particle with a given energy beyond the shutter. The shift in energy is related to the time t by the time-energy uncertainty relation [3].

19.4 Applications: one-dimensional tunneling

19.4.1 Constant Delta-Function Barrier

We again take the Moshinsky shutter problem, but now with a delta-function barrier ($b > 0$) right behind the shutter, so that one must solve (2.18), with $\xi = 0$, given the initial condition (2.2).

The problem is very similar to that just discussed; the solution again follows from (2.23) and (2.24). It is given by [7]

$$\psi(x,t) = M(x,k,t) + \frac{b}{b - ik}[M(|x|, -ib, t) - M(|x|, k, t)], \tag{4.1}$$

where the first represents the free-particle solution.

By plotting the corresponding probability current at $x = 0^+$, one finds [7] an oscillatory pattern rather similar to the Fresnel free-particle result, but attenuated by transmission through the barrier and showing a *time delay* with respect to the free-particle result.

The delta-function potential may be regarded as the limit of a square barrier of height $V_0 \to \infty$ and width $a \to 0$ in such a way that $V_0 a = b$ remains constant. For finite V_0 and a, a long-standing problem, still not satisfactorily settled, is: how long does it take for a particle to tunnel through the barrier? In spite of many different proposals to define a "tunneling time", it is still not clear whether such a concept is well-defined [12].

For the thin-barrier limit represented by a delta-function potential, one would expect the traversal time to be negligible. The time delay found right behind the barrier represents a "waiting time" interference effect involving incident and reflected waves. Such a delay is not found for wave packets that are more sharply localized in x —and are consequently delocalized in momentum space. Indeed, the transmissivity of the delta-function barrier for incident momentum k is readily computed to be given by

$$T(k,b) = 1/[1 + (b/k)^2], \tag{4.2}$$

which approaches unity for high-momentum components. This gives rise to a basic difficulty in the definition of barrier-traversal times, connected with the uncertainty relation: sharper localization of an incident front tends to introduce momentum components going above the barrier top.

A path-integral evaluation of the distribution of times spent within a square barrier [13] leads to oscillatory effects related with an interference among paths that spend different amounts of time inside, which might appear in the transient current for a Moshinsky-type shutter observation.

19.4.2 Time-Dependent Barrier

What happens for a non-instantaneous removal of the Moshinsky shutter? It has been shown by Scheitler and Kleber [14] that there is still a very simple analytical solution for $t > 0$ if the coupling strength of the delta-function in (2.18) (where we take $\xi = 0$) varies according to the law

$$b \rightarrow b(t) = \Lambda/t, \tag{4.3}$$

where $\Lambda > 0$ is a constant with the dimension of length (in our units). This represents an infinite barrier at $t = 0^+$, which gradually shrinks to vanishing strength as $t \rightarrow \infty$.

The exact propagator turns out to be

$$K(x,t, \mid x',0) = U(x - x',t) - \frac{i\Lambda}{|x'| + i\Lambda} U(x + x',t). \tag{4.4}$$

We may assume that the barrier is impenetrable for $t < 0$ and take as initial wave function, accordingly,

$$\psi(x,0) = \theta(-x)\sin(kx). \tag{4.5}$$

The nature of the tunneling process is determined by the dimensionless parameter

$$\alpha = \Lambda k \tag{4.6}$$

which plays a role analogous to that of the Massey parameter [15] as a "degree of adiabaticity". If $\alpha \gg 1$, the process is adiabatic: the tunneling current at each t approaches that through a delta-function potential with the instantenous strength (4.3), namely, $kT(k, b(t))$, where the transmissivity T is given by (4.2).

On the other hand, if $\alpha \ll 1$, the barrier switch-off may be regarded as sudden, and the current behaves like that of the Moshinsky shutter problem, showing the characteristic diffraction in time oscillations, modified by the "waiting time" delay.

19.4.3 Delta-Function Model of Strong-Field Ionization

An often employed one-dimensional model for ionization in the presence of an external electric field $F(t)$ is that of a particle (charge $e = 1$) initially bound in an attractive ($b = -B < 0$) delta-function potential,

$$i\partial_t\psi + \tfrac{1}{2}\psi'' + [B\delta(x) - xF(t)]\psi = 0 \tag{4.7}$$

with the initial bound-state wave function

$$\psi(x,0) = \sqrt{b}\exp(-B|x|) \tag{4.8}$$

We do not wish to discuss this model (see Elberfeld and Kleber [16] and other references therein), but only to point out that, in the strong-field limit, the (Volkov-) propagation of (4.8) under the action of the external field alone plays an important role. This propagation can be described in terms of the Moshinsky functions $M(\pm x, -iB, t)$.

19.5 Applications: decay

19.5.1 The Delta-Shell Model

We now go over to three-dimensional tunneling out of a localized state, *i.e.*, to a model for the decay of a metastable state. This is the "delta-shell" model [17] associated with the central potential

$$V(r) = (A/2a)\,\delta(r-a), \tag{5.1}$$

where the dimensionless parameter $A > 0$ measures the opacity of the shell.

It is convenient to denote by indices 1 and 2 the solutions inside and outside of the shell, respectively. We consider only s-states, described by the radial wave functions

$$\varphi_j(r,t) \equiv r\psi_j(r,t), \quad j = 1,2. \tag{5.2}$$

The general solution of the initial-value problem is then expressed in terms of propagators as

$$\begin{aligned}\varphi_j(r,t) &= \int_0^a K_{j1}(r,r' \mid t,0)\varphi_1(r',0)\,dr' \\ &\quad + \int_a^\infty K_{j2}(r,r' \mid t,0)\varphi_2(r',0)\,dr', \quad j=1,2, \end{aligned}\tag{5.3}$$

defining four propagators K_{jk} that connect the internal and external regions.

The S-function for this problem is given by

$$S(k) = -\frac{Q(-2ika)}{Q(2ika)}, \qquad Q(z) \equiv \frac{1}{A + z - A\exp(-z)}. \tag{5.4}$$

By applying the Laplace transformation, one finds that K_{11} is given by [17]

$$K_{11}(r,r' \mid t,0) = G_1(\mid r-r'\mid, t) - G(|r+r'|,t) \tag{5.5}$$

with

$$G_1(r,t) = U(r,t) + \int_0^\infty [U(2a+r+\rho,t) + U(2a-r+\rho,t)]R_{11}(\rho)\,d\rho, \tag{5.6}$$

$$R_{11}(\rho) = A\mathcal{L}^{-1}[Q(2as)], \tag{5.7}$$

where $\mathcal{L}^{-1}$ denotes the inverse Laplace transform.

19.5.2 Totally Reflecting Cavity Limit

In the limit $A \to \infty$, K_{11} describes propagation within a spherical cavity with impenetrable walls, and (5.7) becomes

$$R_{11}(\rho) = \tfrac{1}{2}\mathcal{L}^{-1}\left[\frac{\exp(as)}{\sinh(as)}\right] = \frac{1}{2}\delta(\rho) + \frac{1}{2a}\left[1 + \sum_{n}{}' \exp(-ik_n\rho)\right] \quad (5.8)$$

where

$$k_n = n\pi/a, \quad n = \pm 1, \pm 2, \cdots \quad (5.9)$$

and the prime in the summation sign means that the sum is taken in the order of increasing $|k_n|$, so that n and $-n$ are grouped together. The points $s_n = -ik_n$ are poles of the Laplace transform in the s plane, and (5.8) may be obtained by a Cauchy partial-fraction expansion in terms of this inifinite set of poles [18].

Substituting (5.8) into (5.6) and employing (2.6), we find

$$\begin{aligned} G_1(r,t) &= U(r,t) + \tfrac{1}{2}[U(2a+r,t) + U(2a-r,t)] \\ &\quad + \frac{1}{2a}\left[M(2a+r,0,t) + M(2a-r,0,t)\right] \\ &\quad + \frac{1}{2a}\sum_{n}{}' \left[M(2a+r,k_n,t) + M(2a-r,k_n,t)\right], \end{aligned} \quad (5.10)$$

which [when substituted into (5.5)] leads to the *transient mode expansion* of the propagator in terms of the poles k_n.

On the other hand, according to the Poisson sum formula, (5.8) is also a representation of the periodic delta function [19]:

$$R_{11}(\rho) = \sum_{n=-\infty}^{\infty} \delta(\rho - 2na), \quad (5.11)$$

so that (5.6), in this limiting case, can also be written as

$$G_1(r,t) = U(r,t) + \sum_{n=1}^{\infty}[U(2na+r,t) + U(2na-r,t)], \quad (5.12)$$

which can be expressed in terms of Jacobi's theta function [20]. By Jacobi's transformation formula, it becomes

$$G_1(r,t) = \frac{1}{2a} + \frac{1}{a}\sum_{n=1}^{\infty}\cos(k_n r)\exp(-\tfrac{1}{2}ik_n^2 t). \quad (5.13)$$

The results (5.10), (5.12), and (5.13) yield three different representations for the same propagator. The interpretation of (5.12) is immediate: it corresponds to the solution by the method of images, with the point source between the "mirrors" at $r = 0$ and $r = a$ giving rise to an infinite series of images. Accordingly, this representation converges well for short times, $t \ll T$, where $T = 4a^2/\pi$ is the period of motion in the ground state.

On the other hand, (5.13) corresponds to an expansion in terms of the stationary states within the cavity, which converges well for long times, $t \gg T$, when (5.12) is poorly convergent, since many reflections between the "mirrors" are required to build up the stationary states. Conversely, many stationary states are required to represent the initial reflections, so that (5.13) converges poorly for short times. The Jacobi theta-function transformation plays a similar convergence-improvement role from short to long times in the theory of heat conduction [20].

According to (2.26) and (2.27), the transient-mode expansion (5.10) converges well both for $t \ll T$, when it is dominated by the free-particle propagator, and for $t \gg T$, when it behaves like the stationary-state expansion. Thus, the representation in terms of Moshinsky functions is the most advantageous one. In contrast with the other two, it can be immediately extended to finite A.

19.5.3 The Decay Law

For finite A, again by Cauchy expansion of the Q function in (5.7) in terms of its poles, one finds [17], [21] that (5.10) is extended to

$$\begin{aligned} G_1(r,t) &= U(r,t) + b_0[M(2a+r,0,t) + M(2a-r,0,t)] \\ & \sum_n{}' b_n[M(2a+r,k_n,t) + M(2a-r,k_n,t)], \end{aligned} \tag{5.14}$$

where

$$b_0 = \frac{A}{2a(A+1)}, \quad b_n = \frac{A}{2a(A+1-2ik_n a)}, \tag{5.15}$$

and k_n are the poles of the S-function (5.4). For a strongly-reflecting barrier, $A \gg 1$, poles with $|k_n| \ll A$ are located close to their limiting positions on the real axis for $A \to \infty$; they are given by

$$k_n a = n\pi\left(1 - \frac{1}{A+1}\right) - i\left(\frac{n\pi}{A}\right)^2 + O\left[\left(\frac{n\pi}{A}\right)^3\right], \quad |n\pi| \ll A. \tag{5.16}$$

One can also obtain the corresponding expansion for the propagators linking the internal and external regions in (5.3):

$$K_{12}(r,r' \mid t,0) = G(|r-r'|,t) - G(r+r',t) = K_{21}(r',r \mid t,0) \tag{5.17}$$

with

$$G(r,t) = \tfrac{1}{2}U(r,t) + \sum_n{}' c_n M(r,k_n,t), \qquad c_n = \frac{2ib_n k_n a}{A}. \tag{5.18}$$

The "complex energy" E_n associated with a complex pole $k_n = k'_n - i\kappa_n$ $(\kappa_n > 0)$ is given by

$$E_n = \tfrac{1}{2}k_n^2 = \tfrac{1}{2}(k_n'^2 - \kappa_n^2) - ik'_n\kappa_n \equiv E'_n - \tfrac{1}{2}i\Gamma_n, \tag{5.19}$$

and the corresponding "complex energy eigenfunction" [*cf.* (2.8)]

$$\begin{aligned} v(r,k_n,t) &= \exp\left[i(k_n r - E_n t)\right] = \exp\left[i(k'_n r - E'_n t)\right] \\ &\quad \times \exp(\kappa_n r)\exp\left(-t/2\tau_n\right), \quad \tau_n \equiv 1/\Gamma_n \end{aligned} \tag{5.20}$$

yields a probability density decaying exponentially in time with lifetime τ_n.

For a pole close to the real axis, such as (5.16), we find

$$\tau_n = \frac{2a}{k'_n\Theta_n}, \qquad \Theta_n \approx 4\left(\frac{k'_n a}{A}\right)^2, \tag{5.21}$$

where Θ_n is the transmissivity of the delta-shell barrier for $n\pi \ll A$. This expresses the lifetime [22] as the period $2a/k'_n$ of the internal motion multiplied by the number of attempts to escape $1/\Theta_n$.

Wave functions of the type (5.20) were first introduced by Gamow [23] in the theory of alpha decay. Unfortunately, the exponential decay in time goes together with an exponential increase in space [factor $\exp(\kappa_n r)$ in (5.20)], that renders (5.20) unnormalizable. This "exponential catastrophe", which reflects the fact that particles at large distances were emitted long ago, when the source strength was correspondingly higher, is obviously due to the unphysical assumption that (5.20) remains valid for arbitrarily remote times in the past, whereas a full description of a decaying system must take into account the excitation process.

This can be done by formulating the decay problem as an initial value problem [9]. For the delta-shell potential, a decay process is described by taking an initial wave function that vanishes outside [$\varphi_2(r,0) = 0$ in (5.3)]. The corresponding propagator inside follows from (5.5) and (5.14).

For $t \ll 2a/k'_1 = 4a^2/\pi$, the period of motion within the shell, the free-particle term in (5.14) is dominant. After many internal reflections, $t \gg 4a^2/\pi$, the asymptotic expansion (2.26)–(2.27) of the Moshinsky function may be applied, and we find

$$\begin{aligned} K_{11}(r,t \mid r',0) &\approx \frac{2}{a}\sum_{k_n \in A}\left[1 - \frac{1}{A+1-2ik_n a}\right]\sin(k_n r') \\ &\quad \times \sin(k_n r)\exp(-iE_n t) \\ &\quad - \sqrt{\frac{2}{\pi}}\frac{e^{i\pi/4}r'r}{(A+1)^2 t^{3/2}} + O(t^{-5/2}), \end{aligned} \tag{5.22}$$

where the summation is extended over the poles of the S-function located above the second bisector [*cf.* (2.25)].

Each term in the sum decays exponentially with the lifetime τ_n associated with the pole k_n in (5.20). However, for sufficiently large times, the exponential decay law is superseded by a decay law in t^{-3} for the probability density. This is just the asymptotic decay law for a free-particle wave packet. Indeed, for times much greater than the spreading time (3.6), the linear dimensions of an initially localized free-particle wave packet grow linearly with t, so that the normalization volume grows like t^3. For a finite-range potential, the asymptotic decay is determined by "slow" components, for which the de Broglie wavelength is much greater than the range, so that these components are almost unaffected by the potential, leading to free-particle-like decay.

Thus, the exponential decay law is not valid either for very short times (after the preparation of the decaying state) or for very long times. However, for a long-lived decaying state, it is readily seen that deviations from exponential decay would appear only after many lifetimes, so that it would be very difficult to detect them.

By (5.3), the propagator in the external region for the decay problem is K_{21}. According to (5.17), (5.18) and (2.2), this propagator corresponds to the solution of the one-dimensional free-particle Schrödinger equation that reduces, for $t \downarrow 0$, to

$$\begin{aligned} K_{21}(r, r' \mid 0, 0) \;=\; & \tfrac{1}{2}[\delta(r' - r) - \delta(r' + r)] \\ & + \sum_n{}' c_n \{\theta(r' - r) \exp[ik_n(r - r')] \\ & - \theta(-r' - r) \exp[ik_n(r + r')]\}, \qquad (5.23) \end{aligned}$$

where $0 \leq r' \leq a$ and we extend r to the whole line, introducing an image space. The free evolution of (5.23) for $t > 0$ gives, in the region $r > a$, the desired propagator.

Apart from reflection at the origin, we see that each pole k_n of the S-function gives rise in the propagator to a cutoff Gamow exponential wave function of the type (5.20), averting the exponential catastrophe because of the cutoff introduced by the excitation at $t = 0$. The free time evolution of such a wave packet was already discussed following (3.7). For $k_n \in A$, it gives rise to a Gamow wave function (5.20) with a diffuse wavefront at $r \approx (k_n' - k_n)t$ as well as to terms in $t^{-3/2}$ associated with the spreading effect, that become dominant for large times. For $k_n \in B$, spreading immediately predominates over exponential decay, which does not appear at all.

We conclude, therefore, that *the Moshinsky function is the basic propagator for a Schrödinger transient mode associated with a complex pole of the S-function.*

Transient modes have properties intermediate between those of stationary states and those of free-particle wave packets. For long-lived transient modes, within the domain of validity of the exponential decay law, stationary-state-like properties hold, with approximate orthogonality of different modes and a corresponding interpretation of expansion coefficients in terms of excitation amplitudes. However, for large times or short-lived modes, free-particle features dominate and these approximations break down.

19.6 Applications: resonance scattering

19.6.1 Delta-Shell Model

A pure scattering process within the model of Section 5 corresponds to the initial condition $\varphi_1(r,0) = 0$ in (5.3). Confining our attention to the external region, the relevant propagator K_{22} is found to be given by [17]

$$\begin{aligned} K_{22}(r,r' \mid t,0) &= U(r-r',t) - \tfrac{1}{2}[U(r+r'-2a,t) + U(r+r',t)] \\ &\quad + {\textstyle\sum_n'} c_n[M(r+r'-2a,k_n,t) - M(r+r',k_n,t)] \quad (6.1) \end{aligned}$$

where c_n is defined by (5.18).

To describe resonance scattering, it is convenient to choose an initial exponential wave packet

$$\varphi_2(r,0) = \exp[-ik(r-a)], \quad k = k' - i\kappa, \ \kappa > 0, \ r \geq a, \qquad (6.2)$$

where the mean momentum k' and the width κ are adjustable. We assume proximity to a sharp resonance, taken to be the lowest one,

$$A \gg 1, \quad |k - k_1| \ll |k_1|. \qquad (6.3)$$

With the help of (2.24), the exact external solution can be expressed in terms of Moshinsky functions. Centering the excitation at the resonance momentum, $k' = k_1$, and considering the behavior of the outgoing wave packet for times much greater than the lifetime of the resonant level but much smaller than the spreading time, one finds that this behavior, sufficiently far behind the diffuse wavefront so that diffraction in time effects can be neglected, is given by [17]

$$\begin{aligned} \varphi_{2,\text{out}}(r,t) &\approx \left\{ \frac{\exp(-\kappa\rho) + 2\kappa_1}{(\kappa - \kappa_1)\left[\exp(-\kappa\rho) - \exp(-\kappa_1\rho)\right]} \right\} \\ &\quad \times \exp[i(k_1' r - \tfrac{1}{2}k_1'^2 t)], \quad \rho \equiv k_1' t - r \gg \sqrt{2t}. \quad (6.4) \end{aligned}$$

The first term within the curly brackets represents hard-sphere scattering and the second one corresponds to resonance scattering in the one-pole approximation for the S-function,

$$S(k) \approx \exp(-2ika)\left(\frac{k - k_1^*}{k - k_1}\right). \qquad (6.5)$$

Its shape can be described in terms of a rise time and a decay time, for excitation by a broad line ($\kappa \gg \kappa_1$) or by a narrow line ($\kappa \ll \kappa_1$). It is found that the rise time is the faster one between the temporal width of the excitation and the lifetime of the resonance, while the decay time is the slower one of this pair.

One can also detect the effect of the resonance through the average time delay in the internal region, given by [24]–[26]

$$\langle \Delta t \rangle = \left\langle 2\frac{d\eta}{dE} \right\rangle_{\text{in}} \tag{6.6}$$

where η is the scattering phase shift and the angular brackets denote the expectation value over the incoming wave packet. One finds that the effective retardation is

$$\langle \Delta r \rangle = k_1' \langle \Delta t \rangle \approx -2a + 2/(\kappa_1 + \kappa), \tag{6.7}$$

where the first term corresponds to the time advance due to surface reflection (negligible under the assumed conditions) and the second one is the delay due to the resonance lifetime and to the temporal width of the excitation.

19.6.2 General Cutoff Potential

The above results can be extended to a general cutoff potential and applied to a time-dependent theory of nuclear reactions [21], [27]. The assumption of a cutoff potential with "channel radius" a leads to an explicit transient-mode expansion for the propagator K_{22} in the external region, though explicit representations in the internal region are of course no longer possible.

To obtain this expansion for s-wave scattering, still employing the notations (5.3), one notes that

$$\begin{aligned} K_{22}(r, r' \mid t, 0) &= U(r - r', t) \\ &\quad - \frac{1}{2\pi} \int_{-\infty}^{\infty} S(k) \exp\left[ik(r + r') - iEt\right] dk, \end{aligned} \tag{6.8}$$

and one substitutes for the S-function $S(k)$ a partial-fraction expansion in terms of its poles k_n. The corresponding contribution for each pole, according to (2.11), is proportional to a Moshinsky function. This agrees with the general interpretation of the Moshinsky function given in Section 5, as the basic propagator for a Schrödinger transient mode.

There is some ambiguity in the choice of a partial-fraction representation for $S(k)$, related with the ambiguity in the definition of "potential scattering" as a background to resonance scattering. One can get a well-defined representation, based on the choice of hard-sphere scattering for this purpose, by applying the Cauchy expansion [21], [28] to $\exp(2ika)S(k)$, or one can employ a Mittag-Leffler expansion of $S(k)$, where the contribution from potential scattering is not found explicitly [27].

19.6.3 Higher Angular Momenta

While we restricted the discussion to s-wave scattering in Secs. 5 and 6, the transient-mode expansion can be extended to angular momenta $l \neq 0$. One can reduce the problem essentially to that with $l \neq 0$ by employing the multipole differential operator [9], [27], $r^l(r^{-1}\partial/\partial r)^l$.

An essentially different situation occurs in semiclassical scattering, where a very large number of partial waves contribute. Under such conditions, it is advantageous to employ a complex angular momentum representation rather than a complex wave number one. The dynamical effects of the centrifugal barrier must be taken into account; tunneling through this barrier plays an important role, and resonance effects are described in terms of Regge trajectories [29].

19.7 Acknowledgements

This work was partially supported by the Brazilian agencies CNPq, CAPES and Finep. The author's participation in the Symposium was made possible by a grant from the International Centre for Theoretical Physics (Trieste).

19.8 References

[1] M. Moshinsky, *Phys. Rev.* **84**, 521 (1951).

[2] M. Moshinsky, *Phys. Rev.* **88**, 625 (1952).

[3] M. Moshinsky, *Am. J. Phys.* **44**, 1037 (1976).

[4] W. Pauli, in *Handbuch der Physik*, Ed. by S. Flügge, Vol. V/I (Springer, Berlin, 1958).

[5] M. Moshinsky, G. Loyola, and C. Villegas, *Rev. Mex. Fís.* **37**, 369 (1991).

[6] B. Gaveau and L.S. Schulman, *J. Phys. A* **19**, 1833 (1986).

[7] W. Elberfeld and M. Kleber, *Am. J. Phys.* **56**, 154 (1988).

[8] M. Abramowitz and I.A. Stegun, *Handbook of Mathematical Functions* (Dover, New York, 1970).

[9] G. Beck and H.M. Nussenzveig, *Nuovo Cimento* **16**, 416 (1960).

[10] A. Sommerfeld, *Optics* (Academic Press, New York, 1954).

[11] J. Goldemberg and H.M. Nussenzveig, Rev. Mex. Fís. **6**, 105 (1957).

[12] E.H. Hauge and J.A. Støvneng, *Rev. Mod. Phys.* **61**, 917 (1989).

[13] H.A. Fertig, Phys. Rev. Lett. **65**, 2321 (1990).

[14] G. Scheitler and M. Kleber, *Z. Phys. D* **9**, 267 (1988).

[15] H.S.W. Massey, *Rep. Prog. Phys.* **12**, 248 (1949).

[16] W. Elberfeld and M. Kleber, *Z. Phys. B* **73**, 23 (1988).

[17] H.M. Nussenzveig, *Nuovo Cimento* **20**, 694 (1961).

[18] E.C. Titchmarsh, *The Theory of Functions*, 2nd ed. (Oxford University Press, 1958).

[19] L. Schwart, *Mathematics for the Physical Sciences* (Addison-Wesley, Reading, 1966).

[20] A. Sommerfeld, *Partial Differential Equations in Physics* (Academic Press, New York, 1949).

[21] H.M. Nussenzveig, *Causality and Dispersion Relations* (Academic Press, New York, 1972).

[22] J.M. Blatt and V.F. Weisskopf, *Theoretical Nuclear Physics* (Wiley, New York, 1952).

[23] G. Gamow, *Z. Phys.* **51**, 204; *ib.* **52**, 510 (1928).

[24] E.P. Wigner, *Phys. Rev.* **98**, 145 (1955).

[25] F.T. Smith, *Phys. Rev.* **118**, 349 (1960).

[26] H.M. Nussenzveig, *Phys. Rev. D* **6**, 1534 (1972).

[27] L. Rosenfeld, *Nucl. Phys.* **70**, 1 (1965).

[28] R.E. Peierls, *Proc. Roy. Soc. A* **253**, 16 (1959).

[29] H.M. Nussenzveig, *Diffraction Effects in Semiclassical Scattering* (Cambridge University Press, 1991).

20

Nonstationary Oscillator in Quantum Mechanics

Vladimir I. Man'ko[1]

ABSTRACT We give the theory of the quantum oscillator with time-dependent parameters. Time-dependent integrals of motion are shown to play an important role in the approach. We give the application to the nonstationary Casimir effect as a basis for constructing the generator of squeezed light (resonator with moving walls) or squeezed current (parametric Josephson junction).

20.1 Introduction

The harmonic oscillator plays the main role in the study of many physical processes. The quantum harmonic oscillator is the basis for describing the properties of atoms, molecules, and nuclei. The theory of the quantum harmonic oscillator with constant parameters and its applications in physics has been given in a very clear and useful book [1] by M. Moshinsky, where all the aspects of this problem have been considered. On the other hand, for many physical processes the parametric oscillator may describe the motion of a charge in nonstationary electromagnetic fields [2] or vibronic spectra of polyatomic molecules [3].

The aim of this article is to present the theory of the quantum parametric oscillator by finding time-dependent integrals of motion which are linear forms in position and momentum operators with time-dependent coefficients. Such integrals of motion have been found for the parametric forced oscillator in Ref. [4]. The solutions of the Schrödinger equation for the parametric oscillator generalizing the Gaussian packets and states wave functions have been found in Ref. [5]. The quadratic integral of motion in momentum and position operators for the parametric oscillator has been found in Ref. [6].

The idea of applying the integral of motion method consists in finding such integrals of motion for which the commutation relations are identical with the commutation relations of the creation and annihilation operators for the usual stationary harmonic oscillator. Thus, the dynamical algebra

[1] P.N. Lebedev Institute of Physics, Moscow, USSR.

of the nonstationary harmonic oscillator turns out to be the same as the dynamical algebra of the stationary oscillator. Due to this it is possible to use the method of constructing the solutions of the stationary oscillator and apply it without any essential change in constructing the solutions of the parametric oscillator. Thus, the most essential step to construct the parametric oscillator wave functions is to find the integrals of motion with the boson creation and annihilation operator commutation relations. After that, the usual scheme of the oscillator theory may be applied. In this article we will follow this scheme.

20.2 Linear integrals of motion

Let us consider the one-dimensional parametric oscillator with time-dependent frequency described by the Hamiltonian

$$\widehat{H}(t) = \frac{\widehat{p}^2}{2} + \frac{\Omega^2(t)\widehat{q}^2}{2}. \tag{2.1}$$

For simplicity we have taken mass and the Planck constant to be equal to unity. Let us also to take the frequency at the initial instant $t = 0$ to be equal to unity also (*i.e.* $\Omega(0) = 1$). Then in coordinate representation the Schrödinger equation for the wave function of the oscillator $\Psi(q,t)$ is

$$i\frac{\partial \Psi}{\partial t} = -\frac{1}{2}\frac{\partial^2}{\partial q^2}\Psi(q,t) + \frac{\Omega^2(t)}{2}q^2\Psi(q,t). \tag{2.2}$$

The parametric oscillator has the following linear integrals of motion [4]

$$\widehat{\mathbf{A}} = \frac{i}{\sqrt{2}}\left(\epsilon(t)\,\widehat{p} - \dot{\epsilon}(t)\,\widehat{q}\right), \qquad \widehat{\mathbf{A}}^+ = \frac{-i}{\sqrt{2}}(\epsilon^*(t)\,\widehat{p} - \dot{\epsilon}^*(t)\,\widehat{q}). \tag{2.3}$$

Here the complex function $\epsilon(t)$ obeys the classical equation of motion

$$\ddot{\epsilon} + \Omega^2(t)\epsilon = 0. \tag{2.4}$$

If we take for Wronskian, which is a constant of motion for equation (2.4), the initial value

$$\dot{\epsilon}\,\epsilon^* - \dot{\epsilon}^*\epsilon = 2i, \tag{2.5}$$

the integrals of motion $\widehat{\mathbf{A}}(t)$ and $\widehat{\mathbf{A}}^+(t)$ in (2.3) will have commutation relations

$$\left[\,\widehat{\mathbf{A}}(t)\,,\,\widehat{\mathbf{A}}^+(t)\,\right] = 1. \tag{2.6}$$

Thus, we have constructed integrals of motion $\widehat{\mathbf{A}}(t)$ and $\widehat{\mathbf{A}}^+(t)$ which have the same commutation relations as the usual creation and annihilation operators for the stationary harmonic oscillator. The initial conditions for the function $\epsilon(t)$ may be taken in the form

$$\epsilon(0) = 1 \,,\ \dot{\epsilon}(0) = i. \tag{2.7}$$

Then the integral of motion $\widehat{\mathbf{A}}(t)$ at the initial instant $t = 0$ coincides with the usual annihilation operator of the stationary harmonic oscillator ($\hbar = m = \Omega(0) = 1$)

$$\widehat{\mathbf{A}}(0) = \frac{1}{\sqrt{2}}(\widehat{q} + i\widehat{p}). \tag{2.8}$$

If the frequency of the oscillator $\Omega(t)$ is constant and equal to 1, the function $\epsilon(t)$ is simply the following exponent

$$\epsilon(t) = e^{it}. \tag{2.9}$$

Thus, for the stationary harmonic oscillator the linear integral of motion in the Schrödinger representation has the form

$$\widehat{\mathbf{a}}(t) = \frac{e^{it}}{\sqrt{2}}(\widehat{q} + i\widehat{p}). \tag{2.10}$$

The integrals of motion have by definition the property that their average does not vary with time:

$$\frac{d}{dt}\langle\, \Psi(t)\,|\,\widehat{\mathbf{A}}(t)\,|\,\Psi(t)\,\rangle = 0. \tag{2.11}$$

Here $|\,\Psi(t)\,\rangle$ is a state vector in the Schrödinger representation. The integrals of motion $\widehat{\mathbf{A}}(t)$, $\widehat{\mathbf{A}}^{+}(t)$ (2.3) do not commute with the Hamiltonian.

If we introduce the evolution operator $\widehat{\mathbf{U}}(t)$ such that

$$\widehat{\mathbf{U}}(t)\,|\,\Psi(0)\,\rangle = |\,\Psi(t)\,\rangle\,, \tag{2.12}$$

the integrals of motion may be constructed in the Schrödinger representation using this operator

$$\widehat{\mathbf{I}}(t) = \widehat{\mathbf{U}}(t)\,\widehat{\mathbf{I}}(0)\,\mathbf{U}^{-1}(t). \tag{2.13}$$

We consider the unitary evolution operators corresponding to hermitian Hamiltonians. For example the operators of the form

$$|\,m,t\,\rangle\,\langle\, n,t\,| = \widehat{\mathbf{U}}(t)\,|\,m\,\rangle\,\langle\, n\,|\,\widehat{\mathbf{U}}^{-1}(t) \tag{2.14}$$

are integrals of motion for any state vectors $|\,m\,\rangle$ and $|\,n\,\rangle$. The projectors

$$|\,n,t\,\rangle\,\langle\, n,t\,| = \widehat{\mathbf{U}}(t)\,|\,n\,\rangle\,\langle\, n\,|\,\widehat{\mathbf{U}}(t) \tag{2.15}$$

are integrals of motion, too. Here $|\,n\,\rangle$ is the state vector for n^{th} excited state of the usual time-independent oscillator and the evolution operator

$\widehat{\mathbf{U}}(t)$ satisfies the Schrödinger equation with the Hamiltonian (2.1). According to the theorem about quantum integrals of motion [7], any function of the integrals of motion is an integral of motion, too. Thus, the operator

$$\widehat{\mathbf{N}}(t) = \widehat{\mathbf{A}}^{+}(t)\widehat{\mathbf{A}}\,(t) \tag{2.16}$$

is an integral of motion. The commutation relations of this operator with the integrals of motion $\widehat{\mathbf{A}}(t)$ and $\widehat{\mathbf{A}}^{+}(t)$ are the following:

$$\begin{aligned}\left[\widehat{\mathbf{A}}(t)\,,\,\widehat{\mathbf{N}}(t)\right] &= \widehat{\mathbf{A}}(t),\\ \left[\widehat{\mathbf{A}}^{+}(t)\,,\,\mathbf{N}(t)\right] &= -\widehat{\mathbf{A}}^{+}(t).\end{aligned} \tag{2.17}$$

Thus, the algebra of the operators $\widehat{\mathbf{A}}(t)$, $\widehat{\mathbf{A}}^{+}(t)$ and $\widehat{\mathbf{A}}^{+}(t)\,\widehat{\mathbf{A}}(t)$ is the same as the algebra of the operators $\widehat{\mathbf{a}}(t)\,\widehat{\mathbf{a}}^{+}(t)$ and $\widehat{\mathbf{a}}^{+}(t)\,\widehat{\mathbf{a}}(t)$ for the usual stationary quantum harmonic oscillator. Using these properties we will construct the solutions to the Schrödinger equation (2.1) in the next section.

20.3 "Ground" state and coherent states of the parametric oscillator

Following the usual scheme, let us find the state $\Psi_0(q,t)$ which is annihilated by the operator $\widehat{\mathbf{A}}(t)$, *i.e.*

$$\widehat{\mathbf{A}}(t)\,\Psi_0(q,t) = 0. \tag{3.1}$$

This gives the linear differential equation

$$i\epsilon\,(t)\,\frac{\partial\Psi_0(q,t)}{\partial q} + \dot{\epsilon}\,q\Psi_0(q,t) = 0. \tag{3.2}$$

This equation has a Gaussian normalized solution

$$\Psi_0(q,t) = \pi^{-1/4}[\epsilon\,(t)]^{-1/2}e^{i\dot{\epsilon}q^2/2\epsilon}. \tag{3.3}$$

The time-dependent phase of the solution Ψ_0 is chosen to take into account the requirement that this function must satisfy the Schrödinger equation (2.2). For the stationary oscillator we can construct the coherent states as the eigenstates of the annihilation operator. For the parametric oscillator we can find the solution for the eigenvalue problem formulated by means of the integral of motion $\widehat{\mathbf{A}}(t)$ in (2.3),

$$\widehat{\mathbf{A}}(t)\Psi_\alpha(q,t) = \alpha\Psi_\alpha(q,t). \tag{3.4}$$

Here the number α is a constant complex number. According to the properties of the integrals of motion [7], the eigenvalues of the invariants do not depend on time. The equation (3.4) has the explicit form

$$i\epsilon\,(t)\frac{\partial\Psi_\alpha(q,t)}{\partial q}+\dot{\epsilon}\,q\Psi_0(q,t)=i\sqrt{2}\,\alpha\Psi_\alpha(q,t). \tag{3.5}$$

This equation may be easily solved and we have the normalized "coherent" state wave function

$$\Psi_\alpha(q,t)=\Psi_0(q,t)\exp\left\{-\frac{|\alpha|^2}{2}+\frac{\sqrt{2}\alpha q}{\epsilon\,(t)}-\frac{\epsilon^*}{2\epsilon}\alpha^2\right\}. \tag{3.6}$$

The wave functions $\Psi_\alpha(q,t)$ have the overlap integral

$$\int_{-\infty}^{\infty}\Psi_\alpha^*(q,t)\,\Psi_\beta(q,t)\,dq=\exp\left[-\frac{|\alpha|^2}{2}-\frac{|\beta|^2}{2}+\alpha^*\beta\right]. \tag{3.7}$$

For time-independent frequency these states coincide with the quantum harmonic oscillator used in Ref. [8]. If we calculate the variances $(\delta q)^2$ and $(\delta p)^2$ for any coherent state Ψ_α we will have

$$(\delta q)^2=\tfrac{1}{2}|\epsilon\,(t)|^2 \tag{3.8}$$

and

$$(\delta p)^2=\tfrac{1}{2}|\dot{\epsilon}(t)|^2. \tag{3.9}$$

The correlation coefficient of position and momentum r, which for any state is

$$r=\frac{\frac{1}{2}\langle\widehat{q}\widehat{p}+\widehat{p}\widehat{q}\rangle-\langle\widehat{q}\rangle\langle\widehat{p}\rangle}{\delta q\,\delta p}, \tag{3.10}$$

may be calculated for the states (3.6); it is

$$r=|\epsilon\,\dot{\epsilon}|^{-1}\left(|\epsilon\,\dot{\epsilon}|^2-1\right)^{1/2}. \tag{3.11}$$

We see that the "coherent" states of the parametric oscillator (3.6) minimize the Schrödinger uncertainty relation [9]

$$\delta q\,\delta p\geq\frac{\hbar}{2}\frac{1}{\sqrt{1-r^2}}. \tag{3.12}$$

(In this equation we show the Planck constant explicitly.) For coherent states Ψ_α we have, instead of the inequality (3.12), the equality

$$\delta q\,\delta p=\tfrac{1}{2}\frac{1}{\sqrt{1-r^2}}. \tag{3.13}$$

(We set again $\hbar=1$.)

The correlated coherent state Ψ_α was constructed in Ref. [10] by requiring it to minimize the Schrödinger uncertainty relation. In fact, this state is a squeezed and correlated state. The squeezing parameters,

$$\frac{\delta q}{1/\sqrt{2}} = |\epsilon(t)| \tag{3.14}$$

and

$$\frac{\delta p}{1/\sqrt{2}} = |\dot{\epsilon}(t)|, \tag{3.15}$$

are completely determined by the complex classical trajectory $\epsilon(t)$. For the usual stationary harmonic oscillator ($\epsilon = e^{it}$) there is no correlation ($r = 0$) and no squeezing ($|\epsilon| = |\dot{\epsilon}| = 1$) for its coherent states $\Psi_\alpha(q,t)$ in (3.6). The squeezing and correlation phenomena appear due to the time-dependence of the oscillator frequency.

20.4 Fock states and transition probabilities of the time-dependent oscillator

Since the coherent states (3.6) are generating functions for the Fock states $\Psi_n(q,t)$ of the nonstationary oscillator, due to

$$\Psi_\alpha(q,t) = e^{-|\alpha|^2/2} \sum_{n=0}^{\infty} \frac{\alpha^n}{\sqrt{n!}} \Psi_n(q,t), \tag{4.1}$$

we can find the explicit form of the functions $\Psi_n(q,t)$ by comparing the expression (4.1) with the generating function for Hermite polynomials of one variable,

$$e^{-x^2+2tx} = \sum_{n=0}^{\infty} \frac{x^n}{n!} H_n(t). \tag{4.2}$$

We have

$$\Psi_n(q,t) = (n!)^{-1/2} \left(\frac{\epsilon^*}{2\epsilon}\right)^{n/2} \pi^{-1/4} \epsilon^{-1/2} H_n\left(\frac{q}{|\epsilon|}\right) \exp\left(\frac{i\dot{\epsilon}}{2\epsilon} q^2\right). \tag{4.3}$$

These functions are eigenfunctions of the integral of motion (2.16),

$$\begin{aligned} \widehat{\mathbf{N}} &= \widehat{\mathbf{A}}^+(t)\,\widehat{\mathbf{A}}(t) \\ \widehat{N}\Psi_n(q,t) &= n\,\Psi_n(q,t)\,, \quad n = 0,1,2,\cdots. \end{aligned}$$

Expanding these functions in series with respect to the functions with time-independent frequencies, we can find the transition amplitudes between the Fock states of the oscillator in the case when, after some variation, the frequency takes the initial value.

Let us introduce the SU(1,1) parameters of the linear canonical transform

$$\widehat{\mathbf{A}}(t) = u(t)\,\widehat{a} + v(t)\,\widehat{a}^{+} \,, \tag{4.4}$$

where

$$u(t) = \tfrac{1}{2}\left(\epsilon\,(t) - i\,\dot{\epsilon}\,(t)\,\right), \tag{4.5}$$

$$v(t) = -\tfrac{1}{2}\left(\epsilon(t) + i\dot{\epsilon}(t)\right). \tag{4.6}$$

Then the transition amplitude for $m \geq n$ has the form

$$\begin{aligned} c_{nm}(t) &= \left[\frac{2}{\epsilon - i\dot{\epsilon}}\right]^{1/2} \left(\frac{n!}{m!}\right)^{1/2} \\ &\times \exp\left\{\frac{i}{2}(n-m)(\varphi_v + t) - \frac{i}{2}(n+m)(\varphi_u - t) + \frac{it}{2}\right\} \\ &\times P^{(n-m)/2}_{(m+n)/2}\left(|u|^{-1}\right), \end{aligned} \tag{4.7}$$

where P^N_M is the associated Legendre polynomial, and φ_u and φ_v are the arguments of the complex numbers u and v. The transition probabilities between the excited states of the parametric oscillator may be thus expressed as follows:

$$\begin{aligned} W_{nm} &= |c_{nm}|^2 \\ &= \left(E + \tfrac{1}{2}\right)^{-1/2} \frac{n!}{m!} \left[P^{(m-n)/2}_{(m+n)/2}\left(\left(E + \tfrac{1}{2}\right)^{-1/2}\right)\right]^2, \quad m \geq n \end{aligned} \tag{4.8}$$

$$E = \tfrac{1}{4}\left(|\epsilon|^2 + |\dot{\epsilon}|^2\right). \tag{4.9}$$

Thus we see that all the properties of the oscillator depend on the complex classical solution $\epsilon\,(t)$ only. The generating function for the transition probabilities $\chi(z_1, z_2)$ is

$$\chi\,(z_1, z_2) = \sum_{n=0}^{\infty}\sum_{m=0}^{\infty} z_1^n\, z_2^m\, W_{nm}, \tag{4.10}$$

and has the following explicit form

$$\begin{aligned} \chi\,(z_1, z_2) &= \sqrt{2}\left\{(1 - z_1 z_2)^2 + (z_1 - z_2)^2 \right. \\ &\left. + \left(|u|^2 + |v|^2\right)\left[(1 - z_1 z_2)^2 - (z_1 - z_2)^2\right]\right\}^{-1/2} \end{aligned} \tag{4.11}$$

20.5 Invariants and propagator

From the formula (2.13) follow equations for the matrix elements of the evolution operator. In the coordinate representation, the matrix element is the Green function or propagator of the system. If we will take the operator

$I(0)$ in (2.13) to be either the operator $\widehat{p}$ or the operator $\widehat{q}$, we will obtain as the integrals of motion $\widehat{I}(t)$ the operators $\widehat{p}_0(t)$ and $\widehat{q}_0(t)$. The propagator $G(q,q',t)$ obeys the following system of equations [7]

$$\widehat{q}_0(t)\,G(q,q',t) = q'\,G(q,q',t), \tag{5.1}$$

$$\widehat{p}_0(t)\,G(q,q',t) = i\frac{\partial}{\partial q'}G(q,q',t), \tag{5.2}$$

and

$$\left(i\frac{\partial}{\partial t} - \widehat{H}\right)G(q,q',t) = i\,\delta(t)\,\delta(q-q'). \tag{5.3}$$

The integrals of motion $\widehat{q}_0(t)$ and $\widehat{p}_0(t)$ are connected with the invariants $\widehat{\mathbf{A}}(t)$, $\widehat{\mathbf{A}}^+(t)$ in (2.3) by the relations

$$\widehat{q}_0(t) = \frac{\widehat{\mathbf{A}}(t) + \widehat{\mathbf{A}}^+(t)}{\sqrt{2}} \tag{5.4}$$

and

$$\widehat{p}_0(t) = \frac{\widehat{\mathbf{A}}(t) - \widehat{\mathbf{A}}^+(t)}{i\sqrt{2}}. \tag{5.5}$$

Solving the system of equations (5.1)–(5.2) we obtain the explicit form of the propagator of the oscillator with time varying frequency

$$G(q,q',t) = (-2\pi i\lambda_3)^{-1/2} \times \exp\left\{-\tfrac{1}{2}i\left(q^2\lambda_3^{-1}\lambda_4 - 2qq'\lambda_3^{-1} + \lambda_1\lambda_3^{-1}(q')^2\right)\right\} \tag{5.6}$$

where the numbers λ_1, λ_3, and λ_4 are the time-dependent coefficients in the expressions for the integrals of motion (5.4) and (5.5),

$$\widehat{p}_0(t) = \lambda_1(t)\widehat{p} + \lambda_2(t)\widehat{q}, \tag{5.7}$$

$$\widehat{q}_0(t) = \lambda_3(t)\widehat{p} + \lambda_4(t)\widehat{q}. \tag{5.8}$$

These coefficients may be expressed in terms of the solution $\epsilon(t)$ to the classical equation of motion (2.4)

$$\lambda_1 = \tfrac{1}{2}(\epsilon + \epsilon^*), \tag{5.9}$$

$$\lambda_2 = -\tfrac{1}{2}(\dot{\epsilon} + \dot{\epsilon}^*), \tag{5.10}$$

$$\lambda_3 = \tfrac{1}{2}i(\epsilon - \epsilon^*), \tag{5.11}$$

and

$$\lambda_4 = \tfrac{1}{2}i(\dot{\epsilon}^* - \dot{\epsilon}). \tag{5.12}$$

We introduce the propagator in the Wigner representation as the Weyl symbol of the operator

$$\widehat{W}_{\alpha\beta} = \widehat{\mathbf{U}}(t)\,|\alpha\rangle\,\langle\beta|\,\widehat{\mathbf{U}}^{-1}(t), \tag{5.13}$$

which is a the function $W_{\alpha\beta}(q,p,t)$ of the variables p and q. If we introduce the following variable

$$z(t) = \frac{i\,\epsilon\,(t)}{\sqrt{2}}\,p - \frac{i\,\dot{\epsilon}\,(t)}{\sqrt{2}}\,q, \tag{5.14}$$

the function $W_{\alpha\beta}$ will have the expression

$$W_{\alpha\beta}(p,q,t) = 2\exp\left(-\tfrac{1}{2}|\alpha|^2 - \tfrac{1}{2}|\beta|^2 - \alpha\beta^* - 2zz^* + 2\alpha z^* + 2\beta^* z\right). \tag{5.15}$$

We can also calculate the Weyl symbol of the operator

$$\widehat{\mathbf{W}}_{mn} = \widehat{\mathbf{U}}(t)\ |\ m\rangle\ \langle n\ |\ \widehat{\mathbf{U}}^{-1}(t). \tag{5.16}$$

We have

$$W_{mn}(p,q,t) = 2^{n-m+1}(-1)^m\left(\frac{m!}{n!}\right)\, z^{n-m}\, e^{-2zz^*}\, L_m^{n-m}(4zz^*), \tag{5.17}$$

where L_m^{n-m} is a Laguerre polynomial.

20.6 "Damped" oscillator

Let us consider the system with the Hamiltonian [11]

$$\widehat{\mathbf{H}}(t) = \tfrac{1}{2}[\widehat{p}^2 e^{-2\Gamma(t)} + \Omega^2(t)\, e^{2\Gamma(t)}\,\widehat{q}^2]. \tag{6.1}$$

This system may be considered as an oscillator with damping (or gain) since the Hamiltonian (6.1) leads to the following equation of motion

$$\ddot{\widehat{q}} + 2\,\dot{\Gamma}(t)\,\dot{\widehat{q}} + \Omega^2(t)\,\widehat{q} = 0. \tag{6.2}$$

The linear integrals of motion $\widehat{A}(t)$ for this system have the form

$$\widehat{A}(t) = \frac{i}{\sqrt{2}}\left[\epsilon\,(t)\,\widehat{p} - \dot{\epsilon}\, e^{2\Gamma(t)}\widehat{q}\right]. \tag{6.3}$$

They satisfy

$$\left[\widehat{\mathbf{A}}(t)\,,\widehat{\mathbf{A}}^{+}(t)\right] = 1$$

if

$$e^{2\Gamma(t)}(\dot{\epsilon}\epsilon^* - \dot{\epsilon}^*\epsilon) = 2, \tag{6.4}$$

and the function $\epsilon\,(t)$ obeys the equation

$$\ddot{\epsilon} + 2\,\dot{\Gamma}\,\dot{\epsilon} + \Omega^2\,\epsilon = 0. \tag{6.5}$$

The coherent state $\Psi_\alpha(q,t)$ is the eigenfunction of the invariant (6.3); and this function has the form

$$\begin{aligned}\Psi_\alpha(q,t) &= \pi^{-1/4}\,\epsilon^{-1/2}\\ &\quad\times\exp\left\{\frac{i\dot\epsilon}{2\epsilon}\,e^{2\Gamma(t)}\,q^2+\frac{\sqrt{2}\alpha q}{\epsilon}-\frac{\epsilon^*}{2\epsilon}\,\alpha^2-\frac{|\alpha|^2}{2}\right\}.\end{aligned} \tag{6.6}$$

The average values of the coordinate and momentum operators in these states are

$$\langle\alpha\mid\widehat{q}\mid\alpha\rangle = \sqrt{2}\,\mathrm{Re}\,(\alpha\,\epsilon^*), \tag{6.7}$$

$$\langle\alpha\mid\widehat{p}\mid\alpha\rangle = \sqrt{2}\,e^{2\Gamma(t)}\,\mathrm{Re}\,(\alpha\,\dot\epsilon^*). \tag{6.8}$$

The quadratic integral of motion

$$\begin{aligned}\widehat{\mathbf{K}}(t) &= \tfrac{1}{2}(\widehat{\mathbf{A}}^+\widehat{\mathbf{A}}+\widehat{\mathbf{A}}\widehat{\mathbf{A}}^+)\\ &= \tfrac{1}{2}\left[|\epsilon|^2\widehat{p}^2-|\dot\epsilon|^2\,e^{4\Gamma}\,\widehat{q}^2-e^{2\Gamma}\,\mathrm{Re}\,(\dot\epsilon\,\epsilon^*)\,(\widehat{q}\,\widehat{p}+\widehat{p}\,\widehat{q})\right]\end{aligned} \tag{6.9}$$

may be used to find the Fock states Ψ_n for the damped oscillator,

$$\widehat{\mathbf{K}}\,\Psi_n = (n+\tfrac{1}{2})\,\Psi_n, \qquad n=0,1,2,\cdots. \tag{6.10}$$

We have

$$\Psi_n(q,t) = (\pi\,\epsilon^2)^{-1/4}\,\exp\left(\frac{i\,\dot\epsilon}{2\epsilon}\,e^{2\Gamma}\,q^2\right)(n!)^{-1/2}\left(\frac{\epsilon^*}{2\epsilon}\right)^{n/2} H_n\left(\frac{q}{|\epsilon|}\right). \tag{6.11}$$

If the frequency is $\Omega=1$ and the damping is $\Gamma=\gamma t$ with γ constant, the Hamiltonian (6.1) has the special symmetry

$$\widehat{\mathbf{H}}(t+\pi/\gamma) = \widehat{\mathbf{H}}(t). \tag{6.12}$$

Due to this the integral of motion $\widehat{\mathbf{A}}(t)$ in (6.3) may be taken in the form

$$\widehat{\mathbf{A}}(t) = \frac{i}{\sqrt{2}}e^{it}\left[e^{-\gamma t}\widehat{p}+(\gamma-i)\,e^{\gamma t}\,\widehat{p}\right]. \tag{6.13}$$

This integral of motion posesses the following property

$$\widehat{\mathbf{A}}\,(t+i\pi/\gamma) = -\exp\,(-\pi/\gamma)\,\widehat{\mathbf{A}}(t), \tag{6.14}$$

$$\widehat{\mathbf{A}}^+\,(\gamma+i\pi/\gamma) = -\exp\,(-\pi/\gamma)\,\widehat{\mathbf{A}}^+(t). \tag{6.15}$$

The integral of motion $\widehat{\mathbf{K}}(t)$ is invariant with respect to the imaginary time shift operation

$$\widehat{\mathbf{K}}\,(t+i\pi/\gamma) = \widehat{\mathbf{K}}(t). \tag{6.16}$$

The properties of the system with symmetry (6.12) are analogous to the quasi-energy states with Hamiltonians which are periodic with the real period

$$\widehat{\mathbf{H}}\,(t+T) = \widehat{\mathbf{H}}(t). \tag{6.17}$$

It is very interesting to consider all possible symmetry properties of "crystal structures" in space and time if taking into account the lattice in real or imaginary time for Hamiltonians obeying either (6.12) or (6.17). They may be some space-time periodic (pulsating) structures with the crystal-like properties of nonstationary quantum systems.

20.7 Casimir effect and parametric oscillator

Squeezed and correlated light has become now an object of experimental and theoretical investigation. One of the possible methods to generate squeezed light is to use the nonstationary Casimir effect when moving resonator walls produce a continuous time dependent reconstruction of the electromagnetic vacuum state. The work against the Casimir forces produces two effects. The first is the effect of photon generation from the vacuum. Thus, a resonator with mechanically trembling walls is a quantum vacuum generator of electromagnetic radiation. The second effect is the squeezing of quantized modes in the resonator due to the parametric change of vacuum energy. Both effects exist simultaneously.

Let us discuss a general scheme for the arising of the Casimir effect. For many physical systems, the Hamiltonian may be written as the sum of energies of one-dimensional oscillator modes of small linear vibrations,

$$\mathbf{H} = \sum_n \left[\hbar\omega_n(\mathbf{L})\, a_n^+ a_n + \tfrac{1}{2}\hbar\omega_n(\mathbf{L})\right].$$

The sum over the index n is the sum over all the modes in the system under consideration. The frequencies $\omega_n(\mathbf{L})$ depend on the parameters of the classical system which are described by formal vectors $\mathbf{L} = (L_1, L_2, \cdots, L_m)$. The components of these vectors are various physical parameters, for example the geometrical parameters (such as length or radius of a cylindrical resonator) magnetic and electric field components for the electron in these fields, etc. The number of these parameters is the dimension of the vector $\mathbf{L}$. If the index n ranges in an infinite interval, the energy of the ground or vacuum state

$$E_{vac} = \sum_n \frac{\hbar\,\omega_n(\mathbf{L})}{2}$$

is infinite. But the energy may be considered as the term which is infinite and does not depend on the parameters $\mathbf{L}$ and the term $E(\mathbf{L})$, depending on the parameters $\mathbf{L}$, which is the reason for Casimir forces. The reason for these forces is formally connected with the dependence of the system modes frequencies $\omega_n(\mathbf{L})$ on the external parameters $\mathbf{L}$. So,

$$\delta E_{vac} = \left[\frac{\hbar}{2} \sum_n \frac{\partial \omega_n(\mathbf{L})}{\partial \mathbf{L}} \right] \cdot \delta \mathbf{L}.$$

If the change of the vacuum energy connected with the change of the parameter $\delta\mathbf{L}$ is negative, the Casimir forces act in such direction so as to increase this change of the parameters. Thus the neutral plates in the Casimir effect attract, since the vacuum electromagnetic energy is less for smaller distances between these plates. At the same time the metallic sphere tends to increase its radius since the vacuum electromagnetic energy is less for larger sphere radius. Such behaviour is well known.

For nonstationary quantum systems, the system parameters may be time-dependent $[\mathbf{L} = \mathbf{L}(t)]$ due to the existence of external forces. Thus, the plates in the Casimir effect may be moved by external mechanical forces. The refraction index of the medium may vary with time, the geometrical dimensions of the resonator may be influenced by external mechanical forces. In all these cases the vacuum or ground state energy must change. This means that the vacuum state is continuosly reconstructed. At each time instant the state is no longer the vacuum state due to the change of the parameters. If we have a system such as photons or other quanta in a box with changing volume, it corresponds to a process with creation of quanta, for example, photons. This kind of effect in the case of time-dependent metric has been discussed as particle creation in the universe. So, due to the work against the Casimir forces one form of energy may be converted into another form. Thus, for trembling neutral plates due to mechanical external forces, the photons between the plates must be created; this means that the mechanical energy from external sources is converted into electromagnetic energy of photons. It is interesting that this effect must create the quanta of all other fields existing in nature. Thus, due to Casimir forces, we can have the generator of photons in a parametric resonator which may be called a quantum vacuum generator. As was mentioned above, the idea of quanta generation due to parametric vacuum reconstruction is not a new idea. The new element is the suggestion to realize this idea in different types of experiments. Possible experiments with a resonator, the geometrical form and dimensions of which are time-dependent, were discussed. More sensitive measurements in a resonator with stable walls have been made in [12]. A second type of analogous idea but in a different interpretation is the discussion of the semiconductor laser given in Ref. [13]. By different kinds of influence it is possible to vary with time the waveguide properties in a semiconductor that will produce the same analogue of nonstationary Casimir effect.

It is possible to discuss a third reduction of nonstationary Casimir effect using a Josephson junction. If there is no external voltage in in Josephson junction, but its parameters are time-dependent, vibrations of current and voltage will be excited in it. This suggestion [14] is based on the analogy of a Josephson junction and a conventional resonant circuit (quantized resonant circuit). In a classical resonant circuit it is impossible to excite electrical vibrations without an external source of voltage. But for a quantum resonant circuit, due to Casimir nonstationary forces, it is possible to transform mechanical energy; this may be the reason for the change of the circuit parameters into electrical energy of current vibrations.

If this idea is realized, it will be a quantum vacuum generator of electrical vibrations. The current and voltage in this case play the role of conjugate quantum observables, and in a parametric Josephson junction they may be squeezed. Thus, the quantum noise for current in a Josephson junctions may be reduced for current. In this case the voltage will have larger noise. The discussion of Josephson junction as quantum vacuum generator has been presented in Ref. [14]. The squeezed and correlated states of a Josephson junction may be also excited, if its parameter changes with time. The Berry phase can be also a subject of research for the parametric Josephson junction.

In all three realizations of the nonstationary Casimir effect, the change of parameters may be organized, for example, by the influence of external sound or laser light. In the periodic case the modes which are most sensitive to the external influence, are the modes corresponding to parametric resonance conditions. The chaotic (unstable) behaviour of the systems under study may also be checked for special time-periodic kicking corresponding to the δ-kicking situation discussed in [15].

If we have the squeezed and correlated modes of light (where the wave function or density matrix is a general Gaussian function in many dimensions), the photon distribution function in this state is explicitely expressed in terms of Hermite polynomials of several variables. Due to this it has a wavy behaviour with respect to photon labels as in the one-dimensional case [16]. The possibility to use the resonator with moving walls for generating squeezed light was discussed in Ref. [17].

20.8 References

[1] M. Moshinsky, *The Harmonic Oscillator in Modern Physics: From Atoms to Quarks* (Gordon & Breach, London, 1969).

[2] I.A. Malkin, V.I. Man'ko, and D.A. Trifonov, *Phys. Rev.,* **20**, 1371 (1970).

[3] I.A. Malkin and V.I. Man'ko, *Dynamical Symmetries and Coherent States of Quantum Mechanics* (in Russian; Nauka, Moscow,1979).

[4] I.A. Malkin and V.I. Man'ko, *Phys. Lett.* **A32**, 243 (1970).

[5] K. Husimi, *Progr. Theor. Phys.* **9**, 381 (1953).

[6] H.R. Lewis, and W. Riesenfeld, *J. Math. Phys.* **10**, 312 (1968).

[7] V.V. Dodonov and V.I. Man'ko, P.N. Lebedev Institute Proceedings, Vol. 183 (1989).

[8] R. Glauber, *Phys. Rev. Lett.* **10**, 84 (1963).

[9] E. Schrödinger, *Ber. Kgl. Akad. Wiss.*, Berlin (1930) p. 296.

[10] V.V. Dodonov, E.V. Kurmyshev, and V.I. Man'ko, *Phys. Lett.* **79A**, 150 (1980).

[11] P. Caldirola, *N. Cimento.,* **18**, 393 (1941).

[12] H. Walter, *Hyderabad Conference Proceedings on Quantum Optics* (Plenum Press, New York, 1991).

[13] A. Yablonovich, *Cortina d'Ampezzo NATO Conference Proceedings.* Ed. by P. Tombesi (Plenum Press, New York, 1991).

[14] V.V. Dodonov, O.V. Man'ko, and V.I. Man'ko, *J. Soviet Laser Research, Vol. 10,* W5 (1989).

[15] G. Karner, V.I. Man'ko, and L. Streit, *On the Theory of Quantum Kicks.* Preprint Bilios 399/89 (1989).

[16] W. Schleich, and A. Wheeler, *J. Opt. Soc. Am.* **B4**, 1715 (1987).

[17] V.V. Dodonov, A.B. Klimov, and V.I. Man'ko, *Phys. Lett.* **149A**, 225 (1990).

21

Symmetry and Dynamical Lie Algebras in Classical and Quantum Mechanics

Christianne Quesne[1]

ABSTRACT A unified definition is given for the symmetry and dynamical Lie algebras of a Hamiltonian in both classical and quantum mechanics. It is stressed that the differences observed between classical symmetry and dynamical Lie algebras and quantum ones stem from the importance of the algebra unitary irreducible representations in quantum mechanics. These concepts are illustrated on some two-dimensional rotationally-invariant Hamiltonians. By using a set of canonically conjugate variables including the Hamiltonian and the angular momentum, universal symmetry and dynamical algebras are constructed in classical mechanics and shown to have in general no counterpart in quantum mechanics.

21.1 Introduction

This symposium honours Marcos Moshinsky and it is a great pleasure for me to dedicate this talk to such a great physicist, teacher, and friend. After more than twenty years of collaboration with Marcos, I may say that he is responsible for much of my interest in the applications of group theory and symmetry principles in physics, a field to which he has contributed so much. I am also very happy to introduce hereafter the results of a work that sprang from our most recent joint paper [1].

In fact, one year ago, we got interested in a very simple, though apparently unsolved problem: given a classical two-dimensional rotationally-invariant Hamiltonian H, for which the angular momentum M is therefore a constant of the motion, is it possible to construct a set of canonically conjugate variables $\{T, \Phi, H, M\}$ including H and M as generalized momenta?

As well known, the variable T, canonically conjugate to H, is the function of r, p_r, p_φ that behaves as the time variable t [2], but the variable Φ, canonically conjugate to $M = p_\varphi$, is different from the polar angle φ. For

[1]Université Libre de Bruxelles, Belgium.

an extended class of Hamiltonians, we succeeded to show that Φ may be obtained by quadratures.

We then realized that the new set of canonically conjugate variables enable us to construct in a straightforward way universal symmetry Lie algebras for the Hamiltonians considered and to contrast the existence of the latter with the corresponding situation in quantum mechanics. The relations between classical and quantum symmetry algebras, which we did already touch on before [3], were extensively studied long ago by McIntosh and his co-workers [4]. However our new approach did drastically simplify the discussion of the problem for the special type of Hamiltonians considered.

After completion of this work, I did notice that the same approach could be used to construct universal dynamical Lie algebras for classical Hamiltonians and to discuss their translation to quantum mechanics. As far as I know, such a problem has not been considered before and is therefore very interesting. It appeared rather soon however that it was first necessary to make a definite choice amongst the wide variety of existing definitions in the literature for the notions of dynamical group and spectrum-generating algebra.

Broadly, the viewpoint adopted herein will be following that as pioneered by Malkin and Man'ko [5], Lipkin [6], Dothan [7], Barut and Rączka [8], and others (see *e.g.* Ref. [9]), according to which symmetry and dynamical Lie algebras are defined in a similar way in both classical and quantum mechanics. The main purpose of my talk is to make clear how these parallel definitions can yield very different results in practice, as to be reviewed within the next Section. The concepts introduced there will then be illustrated on some classical and quantum two-dimensional rotationally-invariant Hamiltonians in Sections 3 and 4 respectively. Finally, Section 5 will contain the conclusion.

21.2 Definition and properties of symmetry and dynamical Lie algebras

I define a symmetry Lie algebra of a classical (time-independent) Hamiltonian H as the linear span of a set of functions X_λ, $\lambda = 1, \ldots, \sigma$, of the generalized coordinates and momenta with the following four properties:

(S1) Every function X_λ is a constant of the motion with no explicit time dependence, i.e.,

$$\frac{dX_\lambda}{dt} = \{X_\lambda, H\} = 0, \tag{2.1}$$

where $\{\ \}$ denotes the Poisson bracket.

(S2) The set of X_λ's close among themselves under the Poisson bracket operation to form a (real) Lie algebra, *i.e.*,

$$\{X_\lambda, X_\mu\} = -ic_{\lambda\mu}{}^{\nu}\, X_\nu, \tag{2.2}$$

where $c_{\lambda\mu}{}^{\nu}$ are the algebra structure constants.

(S3) The Casimir operators of the Lie algebra depend upon the Hamiltonian.

(S4) The algebra is maximal in the sense that any constant of the motion with no explicit time dependence belongs to its enveloping algebra.

Similarly, I define a dynamical Lie algebra of H as the linear span of a set of functions X_λ, $\lambda = 1, \ldots, r$, of the generalized coordinates and momenta, and the time, with the following four properties:

(D1) Every function X_λ is a constant of the motion with (in general) an explicit time dependence, *i.e.*,

$$\frac{dX_\lambda}{dt} = \{X_\lambda, H\} + \frac{\partial X_\lambda}{\partial t} = 0. \tag{2.3}$$

(D2) The set of X_λ's close among themselves under the Poisson bracket operation to form a (real) Lie algebra, hence they satisfy relations similar to (2.2).

(D3) The algebra contains a symmetry Lie algebra as subalgebra, *i.e.*, the first σ generators X_λ, $\lambda = 1, \ldots, \sigma$, do not have any explicit time dependence, and $c_{\lambda\mu}{}^{\nu} = 0$ for $\lambda, \mu \leq \sigma$ and $\nu > \sigma$.

(D4) The algebra is maximal in the sense that any constant of the motion, with an explicit time dependence or not, belongs to its enveloping algebra.

The purpose of conditions (S3) and (D3), whose importance was only quite recently recognized, is to eliminate the Lie algebras with essentially no dynamical content. For symmetry Lie algebras, condition (S3) was introduced by Moshinsky *et al.* [10] in studying some two-dimensional Hamiltonians, which will be considered in the next two sections. For dynamical Lie algebras, condition (D3) leads one to reject the Heisenberg-Weyl algebra as a possible candidate for the harmonic oscillator dynamical algebra.

In the transition from classical to quantum mechanics, the functions $A, B \ldots$, of the dynamical variables become some operators $\hat{A}, \hat{B}, \ldots$, and the Poisson bracket of the former is replaced by the commutator of the latter according to the rule

$$\{A, B\} \mapsto -i[\hat{A}, \hat{B}], \tag{2.4}$$

where $\hbar$ is assumed to be equal to 1. Since in the Heisenberg picture of quantum mechanics, the equation of motion for an operator $\hat{A}$ is formally identical with that of the corresponding function A in classical mechanics, it is possible to define symmetry and dynamical Lie algebras of a quantum Hamiltonian $\hat{H}$ in a similar way as those of a classical Hamiltonian H. Their generators $\hat{X}_\lambda$ are now operators acting on the Hilbert space $\mathcal{H}$ spanned by the Hamiltonian eigenstates, and satisfying the above conditions (S1)–(S4) or (D1)–(D4) with (2.1)–(2.3) respectively replaced by the equations

$$i\frac{d\hat{X}_\lambda}{dt} = [\hat{X}_\lambda, \hat{H}] = 0, \tag{2.5}$$

$$[\hat{X}_\lambda, \hat{X}_\mu] = c_{\lambda\mu}{}^\nu \hat{X}_\nu, \tag{2.6}$$

and

$$i\frac{d\hat{X}_\lambda}{dt} = [\hat{X}_\lambda, \hat{H}] + i\frac{\partial\hat{X}_\lambda}{\partial t} = 0, \tag{2.7}$$

valid on the space $\mathcal{H}$. As already noted by Barut and Rączka [8], imposing these equations abstractly, and not only on $\mathcal{H}$, would be a more narrow definition of symmetry, leading to subalgebras of the algebras considered here.

Although the definitions look almost identical in classical and quantum mechanics, there is however a big difference. It stems from the fact that in the latter field, the algebra generators are operators acting on the Hilbert space of the Hamiltonian eigenstates, thereby giving rise to unitary irreducible representations (unirreps) of the algebras on appropriate subspaces of $\mathcal{H}$. In this respect, one may quote Barut and Rączka (Ref. [8], p. 392). Speaking of group representations in physics, they write that in classical mechanics, "only a defining representation of the groups occurs. The theory of group representations in linear spaces... is really the domain of quantum physics." It now remains to understand how the appearance of the algebra unirreps in quantum mechanics imposes very stringent conditions on the algebra themselves, which are absent in classical mechanics.

In the symmetry algebra case, the equation $[\hat{X}_\lambda, \hat{H}] = 0$, now interpreted in the Schrödinger picture of quantum mechanics, implies that if $|\psi\rangle$ is some eigenstate of $\hat{H}$ corresponding to an eigenvalue E, then $\hat{X}_\lambda|\psi\rangle$ is also an eigenstate of $\hat{H}$ associated with the same eigenvalue. Conversely, an operator transforming any eigenstate of $\hat{H}$ into a degenerate eigenstate commutes with $\hat{H}$ and, owing to condition (S4), must be expressible in terms of the $\hat{X}_\lambda$'s. Hence, the set $\{\hat{X}_\lambda | \lambda = 1, \ldots, \sigma\}$ must contain ladder operators connecting all the degenerate eigenstates of $\hat{H}$. As a result, the latter form bases for the carrier spaces of some symmetry algebra unirreps.

The usual interpretation of the symmetry algebras as explaining the "accidental" degeneracies present in some quantum mechanical spectra therefore appears as a necessary consequence of their definition. As a result of such an interpretation, the symmetry algebras must have unirreps whose dimensions agree with the degeneracies of the Hamiltonian spectra. When these degeneracies are finite, the symmetry algebras are compact and, in practice, may be determined from the dimensions of their unirreps, even without constructing their generators. When the degeneracies are infinite however, it may happen that the symmetry algebras, which are then non-compact, be not uniquely determined (see *e.g.* [11]).

In the dynamical algebra case, the equation $[\hat{X}_\lambda, \hat{H}] + i\partial\hat{X}_\lambda/\partial t = 0$, interpreted in the Schrödinger picture of quantum mechanics, implies that if $|\psi(t)\rangle$ is a solution of the time-dependent Schrödinger equation, then $\hat{X}_\lambda|\psi(t)\rangle$ is also a solution of this equation; but even when $|\psi(t)\rangle$ corresponds to a definite energy eigenvalue, $\hat{X}_\lambda|\psi(t)\rangle$ is in general a linear combination of eigenstates of $\hat{H}$ with energies different from E. Conversely, if an operator transforms any solution of the time-dependent Schrödinger equation into another solution of this equation, then it satisfies (2.7) and, owing to condition (D4), must be expressible in terms of the $\hat{X}_\lambda$'s. Hence, the set $\{\hat{X}_\lambda | \lambda = 1,\ldots,r\}$ must contain ladder operators connecting all the eigenstates of $\hat{H}$, which therefore span the carrier space of a single unirrep of the dynamical algebra.

One concludes that the definition of dynamical Lie algebras given here leads in a natural way to the properties that are often used to define this concept. For instance, Wybourne (Ref. [12], p. 286) writes about the dynamical group of a quantum Hamiltonian: "Ideally, we seek a group that can yield the energy spectrum and the degeneracies of the levels, and that contains a set of operators that determine the transition probabilities between states." In this respect, it is important to realize the crucial role played by conditions (S3) and (D3) in obtaining the first part of Wybourne's definition. Finally, one must note that a dynamical algebra can often be determined from the symmetry algebra unirreps that are present in the Hamiltonian spectrum and have therefore to be contained in one of its unirreps.

21.3 The case of two-dimensional rotationally-invariant Hamiltonians in classical mechanics

21.3.1 Variables Canonically Conjugate to the Hamiltonian and the Angular Momentum

Let us consider a four-dimensional phase space, where the variables are the polar coordinates r, φ and their conjugate momenta p_r, p_φ, and the Poisson bracket is

$$\{f, g\} = \left(\frac{\partial f}{\partial r}\frac{\partial g}{\partial p_r} - \frac{\partial f}{\partial p_r}\frac{\partial g}{\partial r}\right) + \left(\frac{\partial f}{\partial \varphi}\frac{\partial g}{\partial p_\varphi} - \frac{\partial f}{\partial p_\varphi}\frac{\partial g}{\partial \varphi}\right). \tag{3.1}$$

Rotationally-invariant Hamiltonians H do not depend on the angle φ and satisfy the relation

$$\{H, p_\varphi\} = 0, \tag{3.2}$$

proving that the angular momentum

$$M = p_\varphi \tag{3.3}$$

is a constant of the motion. Although it will be shown on some examples in Subsection 3.4 that the theory developed below may be applied to more general Hamiltonians, we shall restrict ourselves here to Hamiltonians of the type

$$H = \tfrac{1}{2}p_r^2 + W(r, p_\varphi), \tag{3.4}$$

where W is an arbitrary function of the variables indicated.

H and M may be taken as new generalized momenta. With the corresponding generalized coordinates T, Φ, they form a new set of canonically conjugate variables, in terms of which the Poisson bracket can be written as

$$\{f, g\} = \left(\frac{\partial f}{\partial T}\frac{\partial g}{\partial H} - \frac{\partial f}{\partial H}\frac{\partial g}{\partial T}\right) + \left(\frac{\partial f}{\partial \Phi}\frac{\partial g}{\partial M} - \frac{\partial f}{\partial M}\frac{\partial g}{\partial \Phi}\right). \tag{3.5}$$

The coordinate T, canonically conjugate to H, must satisfy the equations

$$\{T, H\} = 1, \tag{3.6}$$
$$\{T, M\} = 0. \tag{3.7}$$

Equation (3.7) is satisfied by any function $T(r, p_r, p_\varphi)$, while Eq. (3.6) imposes that this function fulfils the condition

$$\frac{dT}{dt} = \{T, H\} = 1. \tag{3.8}$$

In other words, T is the function of r, p_r, p_φ that behaves in the same way as the time variable t [2]. It can be found from the Hamilton equation of motion $\dot{r} = \partial H/\partial p_r = p_r$, and is given by

$$T = \int_{r_0}^{r} \frac{1}{\sqrt{2[H - W(r', p_\varphi)]}} dr'. \tag{3.9}$$

Here r_0 is an arbitrary parameter, and the integral is carried out considering H and p_φ as constants, although later H will be replaced by its value (3.4).

The coordinate Φ, completing the set of canonically conjugate variables $\{T, \Phi, H, M\}$, can be found [1] from the remaining conditions

$$\{\Phi, H\} = 0, \tag{3.10}$$
$$\{\Phi, T\} = 0, \tag{3.11}$$
$$\{\Phi, M\} = 1. \tag{3.12}$$

Condition (3.12) is automatically satisfied if we write Φ as

$$\Phi = \varphi + u(r, p_r, p_\varphi), \tag{3.13}$$

while conditions (3.10) and (3.11) lead to the two equations

$$\frac{\partial u}{\partial r} = \frac{\partial H}{\partial r}\frac{\partial T}{\partial p_\varphi} - \frac{\partial T}{\partial r}\frac{\partial H}{\partial p_\varphi} \equiv F(r, p_r, p_\varphi), \tag{3.14}$$

$$\frac{\partial u}{\partial p_r} = \frac{\partial H}{\partial p_r}\frac{\partial T}{\partial p_\varphi} - \frac{\partial T}{\partial p_r}\frac{\partial H}{\partial p_\varphi} \equiv G(r, p_r, p_\varphi). \tag{3.15}$$

From (3.6), it follows that $\partial F/\partial p_r = \partial G/\partial r$, so that (3.14), (3.15) is an integrable system and u is given by [13]

$$u(r, p_r, p_\varphi) = \int_{r_0}^{r} F(r', p_r, p_\varphi)dr' + \int_{p_{r0}}^{p_r} G(r_0, p_r', p_\varphi)dp_r', \tag{3.16}$$

where r_0, p_{r0} are arbitrary parameters [note that r_0, in (3.16) may be chosen independently of r_0, in (3.9)].

In the following two Subsections, it will be shown that the new set of canonically conjugate variables $\{T, \Phi, H, M\}$ drastically simplifies the construction of universal symmetry and dynamical Lie algebras for the Hamiltonians of type (3.4).

21.3.2 Universal Symmetry Lie Algebras

According to Section 2, the generators of a symmetry algebra are some functions $X_\lambda(r, \varphi, p_r, p_\varphi)$ satisfying Eqs. (2.1) and (2.2), as well as conditions (S3) and (S4). For the Hamiltonians (3.4), one may replace the set of canonically conjugate variables $\{r, \varphi, p_r, p_\varphi\}$ by $\{T, \Phi, H, M\}$ and consider functions $X_\lambda(T, \Phi, H, M)$. From (3.2), (3.6), and (3.10), it is obvious that any function of H, M, and Φ fulfils (2.1) and is therefore a constant of the motion with no explicit time dependence. If one can find three independent functions of this type that close under the Poisson bracket relations, they form a maximal Lie algebra, which will be a symmetry Lie algebra provided condition (S3) is satisfied. As is well known, such an algebra can only be of the type $su(2) \simeq so(3), su(1,1) \simeq so(2,1)$, or $e(2)$ [14].

Taking the weight generator of the form $aM + bH$, where a, b are arbitrary constants, one easily checks from (3.5) that the generators $J_0 = (J_0)^\star$, $J_\pm = (J_\mp)^\star$ of $su(2)$, $K_0 = (K_0)^\star$, $K_\pm = (K_\mp)^\star$ of $su(1,1)$, and $L_0 = (L_0)^\star$, $P_\pm = (P_\mp)^\star$ of $e(2)$ (where * stands for Hermitian conjugation) can be written as [1]

$$su(2): \quad J_0 = aM + bH, \quad J_\pm = \sqrt{C(H) - J_0^2}\exp(\pm i\Phi/a),$$
$$C(H) = J_+J_- + J_0^2, \tag{3.17}$$

$$\begin{aligned} su(1,1): \quad K_0 &= aM + bH, \quad K_\pm = \sqrt{K_0^2 - C(H)}\exp(\pm i\Phi/a), \\ C(H) &= K_0^2 - K_+K_-, \end{aligned} \tag{3.18}$$

$$\begin{aligned} e(2): \quad L_0 &= aM + bH, \quad P_\pm = \sqrt{C(H)}\exp(\pm i\Phi/a), \\ C(H) &= P_+P_-. \end{aligned} \tag{3.19}$$

In other words, they satisfy the Poisson bracket relations

$$\{J_0, J_\pm\} = \mp i J_\pm, \qquad \{J_+, J_-\} = -2iJ_0, \tag{3.20}$$

$$\{K_0, K_\pm\} = \mp i K_\pm, \qquad \{K_+, K_-\} = 2iK_0, \tag{3.21}$$

$$\{L_0, P_\pm\} = \mp i P_\pm, \qquad \{P_+, P_-\} = 0, \tag{3.22}$$

respectively. In (3.17), (3.18), and (3.19), $C(H)$ denotes the corresponding Casimir operator, which, as it satisfies the relations $\{C, X_\lambda\} = 0$, must be a function of H, which one can choose arbitrarily with the proviso that it does not reduce to an H-independent constant.

By using (3.3), (3.4), (3.13), and (3.16), the generators (3.17), (3.18), and (3.19) can be expressed in terms of r, φ, p_r, and p_φ. So there exists an infinite number of the three types of algebras that are symmetry Lie algebras of the classical Hamiltonians (3.4). Therefore, as already pointed out in Ref. [1], all classical Hamiltonians of type (3.4) have universal symmetry Lie algebras that go well beyond the geometrical symmetry $so(2)$ associated with the constant of the motion p_φ. In Section 4, this result will be contrasted with what happens for the corresponding quantum Hamiltonians.

21.3.3 Universal Dynamical Lie Algebras

Proceeding in the same way for dynamical algebras, one has to look for some functions $X_\lambda(T, \Phi, H, M, t)$ satisfying Eqs. (2.2) and (2.3), as well as conditions (D3) and (D4). By using (3.2), (3.6), and (3.10), equation (2.3) can be rewritten as

$$\frac{\partial X_\lambda}{\partial T} + \frac{\partial X_\lambda}{\partial t} = 0, \tag{3.23}$$

showing that any function $X_\lambda(T - t, \Phi, H, M)$ is a time-dependent constant of the motion.

One can now easily construct universal dynamical Lie algebras by adding to the symmetry algebra generators (3.17), (3.18), (3.19) some functions explicitly depending on the variable $T - t$. In the present Subsection, examples will be given for such universal dynamical Lie algebras containing either the $su(2)$ or the $su(1,1)$ symmetry Lie algebra, defined in (3.17) and (3.18) respectively.

For such a purpose, one notes that $su(2) \simeq so(3)$ and $su(1,1) \simeq so(2,1) \simeq sp(2,\mathbb{R})$ can be both embedded into a larger $sp(4,\mathbb{R}) \simeq so(3,2)$ algebra, and that the latter can be realized in terms of two coordinates ζ_1, ζ_2, and their conjugate momenta π_1, π_2, satisfying the Poisson bracket relations

$$\{\zeta_i, \zeta_j\} = \{\pi_i, \pi_j\} = 0 \,, \quad \{\zeta_i, \pi_j\} = \delta_{ij}, \quad i,j = 1,2. \tag{3.24}$$

The $sp(4,\mathbb{R})$ generators J_0, $J_\pm$, $\mathcal{N}$, $B_q^\star$, B_q, $q = 1, 0, -1$, or the corresponding $so(3,2)$ generators $L_{AB} = -L_{BA} = (L_{AB})^\star, A, B = 1, \dots, 5$, can indeed be written as

$$\begin{aligned}
J_0 &= L_{12} = \pi_1, \quad J_\pm = L_{31} \pm iL_{32} = \sqrt{\pi_2^2 - \pi_1^2}\exp(\pm i\zeta_1), \\
\mathcal{N} &= L_{45} = \pi_2, \quad B_0^\star = L_{34} - iL_{35} = \sqrt{\pi_2^2 - \pi_1^2}\exp(i\zeta_2), \\
B_0 &= L_{34} + iL_{35} = \sqrt{\pi_2^2 - \pi_1^2}\exp(-i\zeta_2), \\
B_{\pm 1}^\star &= \frac{1}{\sqrt{2}}[L_{15} \mp L_{24} + i(L_{14} \pm L_{25})] \\
&= \frac{1}{\sqrt{2}}(\pi_2 \pm \pi_1)\exp[i(\zeta_2 \pm \zeta_1)], \\
B_{\pm 1} &= \frac{1}{\sqrt{2}}[-L_{15} \mp L_{24} + i(L_{14} \mp L_{25})] \\
&= -\frac{1}{\sqrt{2}}(\pi_2 \mp \pi_1)\exp[-i(\zeta_2 \mp \zeta_1)].
\end{aligned} \tag{3.25}$$

Here $J_0 = (J_0)^\star, J_\pm = (J_\mp)^\star, \mathcal{N} = \mathcal{N}^\star$ are the generators of the $su(2) \oplus u(1) \simeq so(3) \oplus so(2)$ subalgebra, and $B_q^\star, Bq = (-1)^q (B_{-q}^\star)^\star$ are raising and lowering generators respectively.

From (3.24), it is straightforward to show that the functions (3.25) satisfy the defining $sp(4,\mathbb{R})$ Poisson bracket relations [15]

$$\begin{aligned}
\{\mathcal{N}, J_q\} &= 0, \quad \{\mathcal{N}, B_q^\star\} = -iB_q^\star, \qquad \{\mathcal{N}, B_q\} = iB_q, \\
\{J_q, J_{q'}\} &= (-1)^{q''+1} i\varepsilon_{qq'q''} J_{-q''}, \quad \{J_q, B_{q'}^\star\} = (-1)^{q''+1} i\varepsilon_{qq'q''} B_{-q''}^\star \\
\{J_q, B_{q'}\} &= (-1)^{q''+1} i\varepsilon_{qq'q''} B_{-q''}, \qquad \{B_q^\star, B_{q'}^\star\} = \{B_q, B_{q'}\} = 0, \\
\{B_q, B_{q'}^\star\} &= -2i(-1)^q \delta_{q,-q'}\mathcal{N} + 2i(-1)^{q''}\varepsilon_{qq'q''} J_{-q''},
\end{aligned} \tag{3.26}$$

or the $so(3,2)$ ones [12]

$$\{L_{AB}, L_{CD}\} = g_{AC}L_{BD} - g_{AD}L_{BC} - g_{BC}L_{AD} + g_{BD}L_{AC}, \tag{3.27}$$

where $J_{\pm 1} = \mp J_\pm/\sqrt{2}, \varepsilon_{qq'q''}$ is the antisymmetric tensor (in the indices 1, 0, -1), and g_{AB} is the metric tensor diag$(1,1,1,-1,-1)$. For the $so(3,2)$ generators, instead of L_{AB}, one may also use the alternative notation $L_3 = L_{12}$, $A_i = L_{3i}$, $N_\alpha = L_{\alpha 4}$, $K_\alpha = L_{\alpha 5}$, $\mathcal{N} = L_{45}$, $i = 1, 2$, $\alpha = 1, 2, 3$, thereby obtaining Poisson brackets similar to those given for $so(4,2)$ by Moshinsky and Seligman [16].

If in (3.25) one chooses

$$\pi_1 = aM + bH, \quad \pi_2 = \sqrt{C(H)}, \quad \zeta_1 = \Phi/a, \tag{3.28}$$

then J_0 and $J_\pm$ coincide with the $su(2)$ generators $J_0, J_\pm$, defined in (3.17). No problem does arise in (3.28) when taking the square root of $C(H)$ since by definition the latter is a non-negative function of H. On the other hand, if one chooses

$$\pi_1 = \sqrt{C(H)}, \quad \pi_2 = aM + bH, \quad \zeta_2 = \Phi/a, \tag{3.29}$$

then $\mathcal{N}$, $B_0^\star$, and B_0 reproduce the $su(1,1)$ generators K_0, $K_\pm$, defined in (3.18). This time however, one has to impose that $C(H)$ is a non-negative function of H since this condition is not automatically satisfied anymore. To complete the canonical transformation from $\{T, \Phi, H, M\}$ to $\{\zeta_1, \zeta_2, \pi_1, \pi_2\}$, it remains to find the explicit form of ζ_2 or ζ_1 in the case of (3.28) or (3.29) respectively. From (3.5) and (3.24), one easily obtains

$$\zeta_2 = \gamma(H)(T - b\Phi/a) \tag{3.30}$$

or

$$\zeta_1 = \gamma(H)(T - b\Phi/a), \tag{3.31}$$

where

$$[\gamma(H)]^{-1} = \frac{d}{dH}\sqrt{C(H)}. \tag{3.32}$$

As above explained, by making the substitution $T \to T - t$ in (3.30), (3.31) the functions (3.25) are converted into time-dependent constants of the motion, while their Poisson brackets remain unchanged and given by (3.26) or (3.27). Finally, by using (3.3), (3.4), (3.9), (3.13), and (3.16), they can be expressed as functions of $r, \varphi, p_r, p_\varphi$, and t. Hence there exists an infinite number of two different types of $sp(4,\mathbb{R}) \simeq so(3,2)$ algebras that are dynamical Lie algebras of the classical Hamiltonians (3.4). So all classical Hamiltonians of type (3.4) have universal dynamical Lie algebras, in contrast to what happens for the corresponding quantum Hamiltonians, as will be shown in Section 4.

Before proceeding to the quantum case, in the next Subsection the theory developed in the present one will be illustrated with some examples.

21.3.4 Some Examples

The harmonic oscillator Hamiltonian

The first example is the two-dimensional harmonic oscillator, whose Hamiltonian can be written in the form (3.4) with

$$W(r, p_\varphi) = \tfrac{1}{2}(r^{-2}p_\varphi^2 + r^2). \tag{3.33}$$

From (3.9), the variable T is given by

$$T - T_0 = \tfrac{1}{2} \arcsin \frac{r^2 - H}{\sqrt{H^2 - p_\varphi^2}}, \tag{3.34}$$

where T_0 is an arbitrary constant.

Although the variable Φ could be determined by performing the integrations in (3.16), it is easier in the present case to resort to an alternative method based on the classical counterparts of the creation and annihilation operators. The latter are defined by

$$\eta_\pm = \frac{1}{\sqrt{2}}(x_\pm - ip_\pm), \quad \xi^\pm = (\eta_\pm)^\star = \frac{1}{\sqrt{2}}(x_\mp + ip_\mp), \tag{3.35}$$

where

$$\begin{aligned} x_\pm &= \frac{1}{\sqrt{2}}(x_1 \pm ix_2) = \frac{1}{\sqrt{2}} r e^{\pm i\varphi}, \\ p_\pm &= \frac{1}{\sqrt{2}}(p_1 \pm ip_2) = \frac{1}{\sqrt{2}}(p_r \pm ir^{-1}p_\varphi)e^{\pm i\varphi}, \end{aligned} \tag{3.36}$$

and x_i, p_i, $i = 1, 2$, are the cartesian coordinates in the four-dimensional phase space. In terms of $\eta_\pm, \xi^\pm$, the variables H, M, and $S = 4\sin^2 2(T - T_0) - 2$ can be rewritten as

$$H = N = \eta_+\xi^+ + \eta_-\xi^-, \tag{3.37}$$

$$M = \eta_+\xi^+ - \eta_-\xi^-, \tag{3.38}$$

$$S = \frac{\eta_+\eta_-}{\xi^+\xi^-} + \frac{\xi^+\xi^-}{\eta_+\eta_-}. \tag{3.39}$$

The three equations to be satisfied by $\Phi(\eta_+, \eta_-, \xi^+, \xi^-)$, and equivalent to (3.10), (3.11), (3.12), are

$$\{\Phi, H\} = 0, \tag{3.40}$$

$$\{\Phi, S\} = 0, \tag{3.41}$$

$$\{\Phi, M\} = 1, \tag{3.42}$$

where the Poisson bracket is now

$$\{f, g\} = i\left[\left(\frac{\partial f}{\partial \eta_+}\frac{\partial g}{\partial \xi^+} - \frac{\partial f}{\partial \xi^+}\frac{\partial g}{\partial \eta_+}\right) + \left(\frac{\partial f}{\partial \eta_-}\frac{\partial g}{\partial \xi^-} - \frac{\partial f}{\partial \xi^-}\frac{\partial g}{\partial \eta_-}\right)\right]. \tag{3.43}$$

By taking (3.37), (3.38), (3.39) into account, equations (3.40), (3.41), (3.42) become

$$\eta_+\frac{\partial \Phi}{\partial \eta_+} - \xi^+\frac{\partial \Phi}{\partial \xi^+} + \eta_-\frac{\partial \Phi}{\partial \eta_-} - \xi^-\frac{\partial \Phi}{\partial \xi^-} = 0, \tag{3.44}$$

$$\frac{1}{\xi^+}\frac{\partial \Phi}{\partial \eta_+} + \frac{1}{\eta_+}\frac{\partial \Phi}{\partial \xi^+} + \frac{1}{\xi^-}\frac{\partial \Phi}{\partial \eta_-} + \frac{1}{\eta_-}\frac{\partial \Phi}{\partial \xi^-} = 0, \tag{3.45}$$

$$\eta_+\frac{\partial \Phi}{\partial \eta_+} - \xi^+\frac{\partial \Phi}{\partial \xi^+} - \eta_-\frac{\partial \Phi}{\partial \eta_-} + \xi^-\frac{\partial \Phi}{\partial \xi^-} = -i. \tag{3.46}$$

A solution to (3.44), (3.45), and (3.46) can be found by assuming $\Phi = i(\gamma_1 \ln \eta_+ + \gamma_2 \ln \xi^+ + \gamma_3 \ln \eta_- + \gamma_4 \ln \xi^-)$, where $\gamma_1, \gamma_2, \gamma_3, \gamma_4$ are some constants. These equations are indeed converted into a system of four linear equations in the four unknowns $\gamma_1, \gamma_2, \gamma_3, \gamma_4$, whose solution is given by $\gamma_1 = -\gamma_2 = -\gamma_3 = \gamma_4 = -\frac{1}{4}$. Hence a solution for Φ is

$$\Phi = \frac{1}{4} i \ln \frac{\eta_- \xi^+}{\eta_+ \xi^-} = \varphi + \frac{1}{2} \arctan \frac{2r^{-1} p_r p_\varphi}{p_r^2 - r^{-2} p_\varphi^2 + r^2}, \qquad (3.47)$$

where, in the last step, the polar coordinates and their conjugate momenta were reintroduced.

To be able to explicitly write the symmetry and dynamical Lie algebra generators (3.17), (3.18), (3.19) and (3.25), it remains to give the functions $\exp(iT)$ and $\exp(i\Phi)$. The latter take their simplest form in terms of η_+, η_-, ξ^+, and ξ^-. From (3.39) and (3.47), one easily finds

$$e^{iT} = \left(\frac{\eta_+ \eta_-}{\xi^+ \xi^-}\right)^{1/4}, \quad e^{i\Phi} = \left(\frac{\eta_+ \xi^-}{\eta_- \xi^+}\right)^{1/4}, \qquad (3.48)$$

where for convenience's sake T_0 has been chosen as $3\pi/4$.

Hamiltonians that are simple functions of N and M

The next two Hamiltonians of type (3.4) are simple functions of N and M, namely $H = N + M$, and $H = N + |M|$. The corresponding functions $W(r, p_\varphi)$ are given by

$$\begin{aligned} W(r, p_\varphi) &= \tfrac{1}{2}(r^{-2} p_\varphi^2 + r^2) + p_\varphi, \\ W(r, p_\varphi) &= \tfrac{1}{2}(r^{-2} p_\varphi^2 + r^2) + |p_\varphi|, \end{aligned} \qquad (3.49)$$

respectively. The symmetry Lie algebras of these Hamiltonians were recently derived in the quantum picture, and their classical limit obtained by Moshinsky *et al.* [10].

It is obvious that the corresponding variables T are still given as functions of $r, \varphi, p_r, p_\varphi$ by (3.34). On the other hand, the variables Φ can be easily found by the same procedure as that used in the harmonic oscillator case. They are listed in table 1 together with T, while the functions $\exp(iT)$ and $\exp(i\Phi)$ are listed in Table II.

Anharmonic oscillator Hamiltonians

As mentioned in Subsection 3.1, the theory developed there may be applied to more general Hamiltonians than the Hamiltonians (3.4). As an example, consider anharmonic oscillator Hamiltonians that are k^{th}-degree polynomial functions of N,

$$H = P_k(N). \qquad (3.50)$$

TABLE I. The variables T and Φ (up to some additive constants) in terms of r, φ, p_r, p_φ for some classical Hamiltonians.

H	T	Φ
N	$\frac{1}{2}\arcsin\frac{r^2-N}{\sqrt{N^2-p_\varphi^2}}$	$\varphi+\frac{1}{2}\arctan\frac{2r^{-1}p_r p_\varphi}{p_r^2-r^{-2}p_\varphi^2+r^2}$
$N+M$	$\frac{1}{2}\arcsin\frac{r^2-N}{\sqrt{N^2-p_\varphi^2}}$	$\varphi+\arctan\frac{rp_r}{r^2-p_\varphi}$
$N+\lvert M\rvert$	$\frac{1}{2}\arcsin\frac{r^2-N}{\sqrt{N^2-p_\varphi^2}}$	$\varphi+\arctan\frac{rp_r p_\varphi}{\lvert p_\varphi\rvert(r^2-\lvert p_\varphi\rvert)}$
$P_k(N)$	$\frac{1}{2}\left(\frac{dP_k}{dN}\right)^{-1}\arcsin\frac{r^2-N}{\sqrt{N^2-p_\varphi^2}}$	$\varphi+\frac{1}{2}\arctan\frac{2r^{-1}p_r p_\varphi}{p_r^2-r^{-2}p_\varphi^2+r^2}$
$\frac{1}{2}r(p^2+1)$	$\arcsin\frac{r-H}{\sqrt{H^2-p_\varphi^2}}$	$\varphi+\arctan\frac{2r^{-1}p_r p_\varphi}{p_r^2-r^{-2}p_\varphi^2+1}$

Such Hamiltonians with $k = 2$ were considered by Milburn and by Yurke and Stoler [17], who showed that non-linearity is responsible for some interesting departures between classical and quantum dynamics.

As shown in Table I, a solution to (3.6), (3.7) can be easily found by multiplying the variable T for the harmonic oscillator by an appropriate function of N, namely $(dP_k/dN)^{-1}$. On the other hand, the function Φ encountered for the harmonic oscillator remains a solution to (3.10), (3.11), (3.12) for the anharmonic ones.

The pseudo-Coulomb Hamiltonian

Another example of Hamiltonian that cannot be written in the form (3.4) is the so-called pseudo-Coulomb Hamiltonian, obtained from the Coulomb one by replacing the variable r/n by r, and by interchanging the roles of the potential energy and the energy eigenvalue. In two dimensions, it is defined by

$$H = \tfrac{1}{2}r(p_r^2 + r^{-2}p_\varphi^2 + 1). \tag{3.51}$$

In this case, the variables T and Φ can be obtained by using the canonical transformation relating the oscillator and pseudo-Coulomb problems [18]

$$\bar{r} = \sqrt{2r}, \quad \bar{\varphi} = \tfrac{1}{2}\varphi, \quad \bar{p}_r = \sqrt{2r}\,p_r, \quad \bar{p}_\varphi = 2p_\varphi, \tag{3.52}$$

where the variables with bars correspond to the former. The canonical

TABLE II. The functions $\exp(iT)$ and $\exp(i\Phi)$ in terms of $\eta_+, \eta_-, \xi^+, \xi^-$ or x_+, x_-, p_+, p_- for some classical Hamiltonians.

H	$\exp(iT)$	$\exp(i\Phi)$
N	$\left(\frac{\eta_+\eta_-}{\xi^+\xi^-}\right)^{1/4}$	$\left(\frac{\eta_+\xi^-}{\eta_-\xi^+}\right)^{1/4}$
$N+M$	$\left(\frac{\eta_+\eta_-}{\xi^+\xi^-}\right)^{1/4}$	$\left(\frac{\xi^-}{\eta_-}\right)^{1/2}$
$N+\|M\|$	$\left(\frac{\eta_+\eta_-}{\xi^+\xi^-}\right)^{1/4}$	$\frac{(M+\|M\|)\xi^- - (M-\|M\|)\eta_+}{\|M\|\sqrt{2(N-\|M\|)}}$
$P_k(N)$	$\left(\frac{\eta_+\eta_-}{\xi^+\xi^-}\right)^{1/[4P_k'(N)]}$	$\left(\frac{\eta_+\xi^-}{\eta_-\xi^+}\right)^{1/4}$
$\frac{1}{2}r(p^2+1)$	$\frac{\mathbf{r}\cdot\mathbf{p}+i(r-H)}{\sqrt{H^2-M^2}}$	$\frac{\sqrt{2}[-\frac{1}{2}x_\pm(p^2-1)+p_\pm(\mathbf{r}\cdot\mathbf{p})]}{\sqrt{H^2-M^2}}$

transformation (3.52) converts $\frac{1}{2}\bar{H} = \frac{1}{4}(\bar{p}_r^2 + \bar{r}^{-2}\bar{p}_\varphi^2 + \bar{r}^2)$ into the Hamiltonian (3.51). Hence the set of variables $\{T, \Phi, H, M\}$ for the pseudo-Coulomb problem can be found by performing the transformation (3.52) on the corresponding re-scaled set $\{2\bar{T}, 2\bar{\Phi}, \frac{1}{2}\bar{H}, \frac{1}{2}\bar{M}\}$ for the oscillator. The results for T and Φ are listed in Table I, and the corresponding results for $\exp(iT)$ and $\exp(i\Phi)$ in Table II (assuming this time $T_0 = 0$).

21.4 The case of two-dimensional rotationally-invariant Hamiltonians in quantum mechanics

21.4.1 The Transition from Classical to Quantum Mechanics

When one tries to translate to quantum mechanics the theory developed in the previous Section, there arises a problem with the quantum counterpart of the set of canonically conjugate variables $\{T, \Phi, H, M\}$. It has indeed been known for a long time that the quantization rule (2.4) cannot be applied without caution when dealing with canonically conjugate classical observables, one of which has a discrete spectrum in quantum mechanics, such as H or M. This is the so-called phase and angle quantization problem (for recent reviews see *e.g.* [19]).

In classical mechanics, some problems may arise too for angular-type variables, such as T and Φ defined in (3.34) and (3.47) respectively. If one wants to consider single-valued inverse trigonometric functions in a specified 2π interval, then T and Φ are discontinuous functions and do not satisfy the Poisson brackets (3.6) and (3.12) with H and M respectively. It is however very easy to cure this problem by considering multivalued but continuous functions, as was implicitly done in the previous Section.

In quantum mechanics, the problem is more serious because there do not exist any Hermitian operators $\hat{T}$, $\hat{\Phi}$, satisfying the commutation relations $[\hat{T}, \hat{H}] = [\hat{\Phi}, \hat{M}] = i$. It has been remarked however by Lévy-Leblond [20] that Hermitian operators define an unduly restricted framework for the theoretical description of quantum properties and that some non-Hermitian operators may be acceptable provided they preserve the probabilistic interpretation of the Hilbert space formalism.

In the present case, it may be possible to define in terms of the original canonically conjugate operators $\hat{x}_1, \hat{x}_2, \hat{p}_1, \hat{p}_2$, some operators $\widehat{\exp(i\alpha T)}$ and $\widehat{\exp(i\beta\Phi)}$ (where α and β are some constants to be defined later on) that have simple commutation relations with $\hat{H}$ and $\hat{M}$,

$$[\hat{H}, \widehat{\exp(i\alpha T)}] = \alpha \widehat{\exp(i\alpha T)}, \tag{4.1}$$

$$[\hat{M}, \widehat{\exp(i\beta\Phi)}] = \beta \widehat{\exp(i\beta\Phi)}, \tag{4.2}$$

as would result from the existence of Hermitian operators $\hat{T}, \hat{\Phi}$, canonically conjugate to $\hat{H}$, $\hat{M}$. However, the operators $\widehat{\exp(i\alpha T)}$ and $\widehat{\exp(i\beta\Phi)}$ cannot be unitary operators, but only partial isometries, and hence non-normal operators, *i.e.*,

$$[\widehat{\exp(i\alpha T)}, \widehat{\exp(-i\alpha T)}] \neq 0, \tag{4.3}$$

$$[\widehat{\exp(i\beta\Phi)}, \widehat{\exp(-i\beta\Phi)}] \neq 0, \tag{4.4}$$

where $\widehat{\exp(-i\alpha T)}$ and $\widehat{\exp(-i\beta\Phi)}$ denote the adjoints of $\widehat{\exp(i\alpha T)}$ and $\widehat{\exp(i\beta\Phi)}$ respectively. Similarly, one cannot be sure that the operators $\widehat{\exp(\pm i\alpha T)}$ and $\widehat{\exp(\pm i\beta\Phi)}$ commute with one another; so one must expect that

$$[\widehat{\exp(\pm i\alpha T)}, \widehat{\exp(\pm i\beta\Phi)}] \neq 0, \quad [\widehat{\exp(\pm i\alpha T)}, \widehat{\exp(\mp i\beta\Phi)}] \neq 0, \tag{4.5}$$

in contrast with the classical equation (3.11).

In going to quantum mechanics, one may therefore try to replace the set of canonically conjugate variables $\{T, \Phi, H, M\}$ by a set of operators $\{\widehat{\exp(\pm i\alpha T)}, \widehat{\exp(\pm i\beta\Phi)}, \hat{H}, \hat{M}\}$, satisfying (4.1), (4.2), as well as the com-

TABLE III. Symmetry Lie algebra generators for some quantum Hamiltonians.

Case 1. $su(2) \simeq so(3)$ symmetry Lie algebra.

$\hat{H}$	$\hat{J}_0$	$\hat{J}_+ = (\hat{J}_-)^\dagger$	$\hat{\mathbf{J}}^2 = C(\hat{H})$
$\hat{N}$	$\frac{1}{2}\hat{M}$	$\hat{\eta}_+\hat{\xi}^-$	$\frac{1}{4}\hat{H}(\hat{H}+2)$
$\hat{N}+\lvert\hat{M}\rvert$	$\hat{M}$	$-2^{-3/2}[(\lvert\hat{M}\rvert+1)^{-1}\sqrt{\hat{H}+2\lvert\hat{M}\rvert+2}\,\hat{\eta}_+(\hat{M}-\lvert\hat{M}\rvert)$ $-(\hat{M}+\lvert\hat{M}\rvert)\hat{\xi}^-\sqrt{\hat{H}+2\lvert\hat{M}\rvert+2}\,(\lvert\hat{M}\rvert+1)^{-1}]$	$\frac{1}{4}\hat{H}(\hat{H}+2)$
$P_k(\hat{N})$	$\frac{1}{2}\hat{M}$	$\hat{\eta}_+\hat{\xi}^-$	$\frac{1}{4}\hat{N}(\hat{H})[\hat{N}(\hat{H})+2]$
$\frac{1}{2}\hat{r}(\hat{p}^2+1)$	$\hat{M}$	$\sqrt{2}[-\frac{1}{2}\hat{x}_+(\hat{p}^2-1)+\hat{p}_+(\hat{\mathbf{r}}\cdot\hat{\mathbf{p}}+\frac{1}{2}i)]$	$\hat{H}^2-\frac{1}{4}$

Case 2. $su(1,1) \simeq so(2,1)$ symmetry Lie algebra.

$\hat{H}$	$\hat{K}_0$	$\hat{K}_+ = (\hat{K}_-)^\dagger$	$\hat{\mathbf{K}}^2 = C(\hat{H})$
$\hat{N}+\hat{M}$	$\hat{H}-\hat{M}+1$	$(\frac{3}{2}\hat{H}-\hat{M}+1)^{1/2}\hat{\eta}_-$	$\frac{1}{4}\hat{H}(\hat{H}+2)$

mutation relations

$$[\widehat{\exp(\pm i\alpha T)}, \hat{M}] = 0, \tag{4.6}$$

$$[\widehat{\exp(\pm i\beta\Phi)}, \hat{H}] = 0, \tag{4.7}$$

$$[\hat{H}, \hat{M}] = 0, \tag{4.8}$$

but suffering important departures from classical mechanics, as expressed in (4.3), (4.4), and (4.5). As will be shown in the next two Subsections, the possibility of finding such Hilbert space well-defined operators is actually connected with the existence of a symmetry or dynamical Lie algebra in quantum mechanics.

21.4.2 Symmetry Lie Algebras

The quantum Hamiltonians corresponding to the classical Hamiltonians considered in Subsection 3.4 all have a spectrum with accidental degeneracies and their symmetry Lie algebra is either $su(2) \simeq so(3)$ or $su(1,1) \simeq so(2,1)$, according to whether those degeneracies are finite or infinite [10,16]. This property reflects the finite (resp. infinite) dimensionality of the compact (resp. non-compact) Lie algebra unirreps.

The corresponding algebra generators $\hat{J}_0$, $\hat{J}_\pm$ or $\hat{K}_0, \hat{K}_\pm$ are listed in Table III and satisfy commutation relations obtained from (3.20) and (3.21) by performing the substitution (2.4). The degenerate eigenstates $\{|j\mu\rangle | \mu = -j, -j+1, \ldots, j\}$ (resp. $\{|k\mu\rangle | \mu = k, k+1, \ldots\}$) carry a $(2j+1)$-dimensional $su(2)$ unirrep characterized by j (resp. a positive discrete series $su(1,1)$ unirrep specified by k), which is a function of the energy eigenvalue since $C(E) = j(j+1)$ (resp. $C(E) = k(k-1)$). One indeed has

$$\hat{J}_0|j\mu\rangle = \mu|j\mu\rangle, \quad \hat{J}_\pm|j\mu\rangle = \sqrt{(j \mp \mu)(j \pm \mu + 1)}\,|j\mu \pm 1\rangle, \tag{4.9}$$

or

$$\hat{K}_0|k\mu\rangle = \mu|k\mu\rangle, \quad \hat{K}_\pm|k\mu\rangle = \sqrt{(\mu \pm k)(\mu \mp k \pm 1)}\,|k\mu \pm 1\rangle. \tag{4.10}$$

The operators $\hat{J}_0$, $\hat{J}_\pm$ and $\hat{K}_0$, $\hat{K}_\pm$ can be written in a form similar to their classical counterpart (3.17) and (3.18). This is the so-called polar decomposition of $su(2)$ [21] and $su(1,1)$ [22], namely

$$\begin{aligned}
\hat{J}_0 &= a\hat{M} + b\hat{H} + c, \\
\hat{J}_+ &= (\hat{J}_-)^\dagger = \sqrt{C(\hat{H}) - \hat{J}_0(\hat{J}_0 - 1)}\,\widehat{\exp(i\Phi/a)}, \\
C(\hat{H}) &= \hat{\mathbf{J}}^2 = \hat{J}_+\hat{J}_- + \hat{J}_0(\hat{J}_0 - 1),
\end{aligned} \tag{4.11}$$

and

$$\begin{aligned}
\hat{K}_0 &= a\hat{M} + b\hat{H} + c, \\
\hat{K}_+ &= (\hat{K}_-)^\dagger = \sqrt{\hat{K}_0(\hat{K}_0 - 1) - C(\hat{H})}\,\widehat{\exp(i\Phi/a)}, \\
C(\hat{H}) &= \hat{\mathbf{K}}^2 = \hat{K}_0(\hat{K}_0 - 1) - \hat{K}_+\hat{K}_-,
\end{aligned} \tag{4.12}$$

respectively. From (4.11) and (4.12), one obtains

$$\widehat{\exp(i\Phi/a)} = \frac{1}{\sqrt{\hat{J}_+\hat{J}_-}}\hat{J}_+ = \sum_j \sum_{\mu=-j}^{j-1} |j, \mu + 1\rangle\langle j\mu|, \tag{4.13}$$

and

$$\widehat{\exp(i\Phi/a)} = \frac{1}{\sqrt{\hat{K}_+\hat{K}_-}}\hat{K}_+ = \sum_k \sum_{\mu=k}^{\infty} |k, \mu + 1\rangle\langle k\mu|, \tag{4.14}$$

where $\mu = am + bE + c$ and the summations over j and k run over those values present in the spectrum, and hence in the corresponding Hilbert space $\mathcal{H}$.

It is straightforward to check that the operator $\widehat{\exp(i\Phi/a)}$ and its adjoint $\widehat{\exp(-i\Phi/a)}$ satisfy (4.2) and (4.7), and that, in the cases of (4.13) and

TABLE IV. The operators $\widehat{\exp(i\Phi/a)}$ for some quantum Hamiltonians.

$\hat{H}$	a	$\widehat{\exp(i\Phi/a)}$
$\hat{N}$	$\frac{1}{2}$	$[\hat{N}_+(\hat{N}_- + 1)]^{-1/2}\hat{\eta}_+\hat{\xi}_-$ $= \sum_{n_+=0}^{\infty}\sum_{n_-=1}^{\infty} \|n_+ + 1, n_- - 1\rangle\langle n_+ n_-\|$
$\hat{N} + \hat{M}$	-1	$(\hat{N}_-)^{-1/2}\hat{\eta}_- = \sum_{n_+=0}^{\infty}\sum_{n_-=0}^{\infty} \|n_+, n_- + 1\rangle\langle n_+ n_-\|$
$\hat{N} + \|\hat{M}\|$	1	$-\frac{1}{\sqrt{2}}[(\|\hat{M}\| + 1)^{-1}(\hat{N} - \|\hat{M}\|)^{-1/2}\hat{\eta}_+(\hat{M} - \|\hat{M}\|)$ $-(\hat{M} + \|\hat{M}\|)\hat{\xi}^-(\hat{N} - \|\hat{M}\|)^{-1/2}(\|\hat{M}\| + 1)^{-1}]$ $= \sum_{n_-=1}^{\infty}\left[\sum_{n_+=0}^{n_- -1} \|n_+ + 1, n_-\rangle\langle n_+ n_-\| + \sum_{n_+=n_-}^{\infty} \|n_+, n_- - 1\rangle\langle n_+ n_-\|\right]$
$P_k(\hat{N})$	$\frac{1}{2}$	$[\hat{N}_+(\hat{N}_- + 1)]^{-1/2}\hat{\eta}_+\hat{\xi}^-$ $= \sum_{n_+=0}^{\infty}\sum_{n_-=1}^{\infty} \|n_+ + 1, n_- - 1\rangle\langle n_+ n_-\|$
$\frac{1}{2}\hat{r}(\hat{p}^2 + 1)$	1	$[(\hat{H} - \hat{M} + \frac{1}{2})(\hat{H} + \hat{M} - \frac{1}{2})]^{-1/2}$ $\times \sqrt{2}\,[-\frac{1}{2}\hat{x}_+(\hat{p}^2 - 1) + \hat{p}_+(\hat{\mathbf{r}}\cdot\hat{\mathbf{p}} + \frac{1}{2}i)]$

(4.14), one has

$$\widehat{\exp(i\Phi/a)}\ \widehat{\exp(-i\Phi/a)} = I - \sum_j |j,-j\rangle\langle j,-j|,$$

$$\widehat{\exp(-i\Phi/a)}\ \widehat{\exp(i\Phi/a)} = I - \sum_j |jj\rangle\langle jj|, \qquad (4.15)$$

and

$$\widehat{\exp(i\Phi/a)}\ \widehat{\exp(-i\Phi/a)} = I - \sum_k |kk\rangle\langle kk|,$$

$$\widehat{\exp(-i\Phi/a)}\ \widehat{\exp(i\Phi/a)} = I, \qquad (4.16)$$

respectively, where I denotes the unit operator in $\mathcal{H}$. Hence, for the commutator of $\widehat{\exp(i\Phi/a)}$ and $\widehat{\exp(-i\Phi/a)}$, one finds a result in accordance with (4.4).

One might then fear that since $\widehat{\exp(i\Phi/a)}$ is not a unitary operator, $\hat{J}_0, \hat{J}_\pm$ (resp. $\hat{K}_0, \hat{K}_\pm$), defined in (4.11) (resp. (4.12)), do not satisfy $su(2)$ (resp. $su(1,1)$) commutation relations. This does not occur however because the extra projectors appearing on the right-hand side of (4.15), (4.16) are annihilated by root square operators such as $\sqrt{C(\hat{H}) - \hat{J}_0(\hat{J}_0 - 1)}$. All this can be checked in detail for the $\hat{N}$ and $\hat{M}$ functionally-dependent Hamiltonians, for which the explicit form of $\widehat{\exp(i\Phi/a)}$ is given in Table IV in terms of the kets

$$|n_+n_-\rangle = \frac{(\hat{\eta}_+)^{n_+}(\hat{\eta}_-)^{n_-}}{\sqrt{n_+!\, n_-!}}|0\rangle, \tag{4.17}$$

and of the corresponding bras.

The classical limit of the operator $\widehat{\exp(i\Phi/a)}$ can be obtained by disregarding all numbers, such as 1 or $\frac{1}{2}$, as compared with the expectation values of $\hat{H}$ and $\hat{M}$, since here $\hbar = 1$. The results agree with those for the function $\exp(i\Phi/a)$ deduced from Table II. In the same way, the quantum symmetry Lie algebra generators $\hat{J}_0, \hat{J}_\pm$, or $\hat{K}_0, \hat{K}_\pm$ go to their classical counterpart (3.17) or (3.18) for specific values of the constants a, b, and of the function $C(H)$, which were left arbitrary in the classical case. So, for the examples considered here, the three infinite sets of classical symmetry Lie algebras have shrunk to a single quantum symmetry Lie algebra as a consequence of the extra conditions imposed on Hilbert space operators.

21.4.3 Dynamical Lie Algebras

For the quantum Hamiltonians corresponding to the classical Hamiltonians considered in Subsection 3.4, except $\hat{H} = \hat{N} + \hat{M}$ and $\hat{H} = P_k(\hat{N})$, a dynamical Lie algebra is known: $so(3,2)$ for $\hat{H} = \frac{1}{2}\hat{r}(\hat{p}^2 + 1)$ [23], and $\hat{H} = \hat{N} + |\hat{M}|$ [24]; $wsp(4,\mathbb{R}) = w(2) \oplus sp(4,\mathbb{R})$ for $\hat{H} = \hat{N}$ [12, 25]. Here $\oplus$ denotes a semi-direct sum and $w(2)$ the Heisenberg-Weyl algebra spanned by

$$\hat{A}^\dagger_\pm = \hat{\eta}_\pm \exp(-it), \quad \hat{A}_\pm = \pm(\hat{A}^\dagger_\mp)^\dagger = \pm\hat{\xi}^\mp \exp(it), \tag{4.18}$$

and the unit operator. In each case, $sp(4,\mathbb{R}) \simeq so(3,2)$ contains as a subalgebra the symmetry Lie algebra $su(2) \simeq so(3)$, whose generators are given in Table III. The remaining $sp(4,\mathbb{R})$ generators are listed in Table V and satisfy commutation relations obtained from (3.26) by performing the substitution (2.4). In the $wsp(4,\mathbb{R})$ case, one has the additional commutation

TABLE V. Dynamical Lie algebra generators for some quantum Hamiltonians. A basis of the Lie algebra also includes the operators $\hat{J}_0$, $\hat{J}_+$, and $\hat{J}_-$, defined in Table III, and, in the case of $\hat{H} = \hat{N}$, the $w(2)$ generators defined in (4.18).

$\hat{H}$	$\hat{\mathcal{N}}$	$\hat{B}_0^\dagger = (\hat{B}_0)^\dagger$	$\hat{B}_{\pm 1}^\dagger = -(\hat{B}_{\mp 1})^\dagger$
$\hat{N}$	$\frac{1}{2}(\hat{H}+1)$	$\hat{\eta}_+\hat{\eta}_- \exp(-2it)$.	$\frac{1}{\sqrt{2}}\hat{\eta}_\pm^2 \exp(-2it)$
$\hat{N}+\lvert\hat{M}\rvert$	$\frac{1}{2}(\hat{H}+1)$	$[(\hat{H}+2\lvert\hat{M}\rvert)/\hat{H}]^{1/2} \times\hat{\eta}_+\hat{\eta}_- \exp(-2it)$.	$\pm\frac{1}{4}\hat{H}^{-1/2}\{[(\hat{H}+2\lvert\hat{M}\rvert)(\hat{H}+2\lvert\hat{M}\rvert-2)]^{1/2} \times(\hat{M}\pm\lvert\hat{M}\rvert)\hat{\eta}_\pm(\lvert\hat{M}\rvert+1)^{-1} - 2(\lvert\hat{M}\rvert+1)^{-1} \times\hat{\eta}_\pm^2\hat{\eta}_\mp(\hat{M}\mp\lvert\hat{M}\rvert)\}\exp(-2it)$
$\frac{1}{2}\hat{r}(\hat{p}^2+1)$	$\hat{H}$	$[\hat{\mathbf{r}}\cdot\hat{\mathbf{p}} - \frac{1}{2}i - \frac{1}{2}i\hat{r}(\hat{p}^2-1)]\exp(-it)$.	$[\frac{1}{2}i\hat{x}_\pm(\hat{p}^2+1) - i\hat{p}_\pm(\hat{\mathbf{r}}\cdot\hat{\mathbf{p}}+\frac{1}{2}i) + \hat{r}\hat{p}_\pm]\exp(-it)$

relations

$$\begin{aligned}
[\hat{A}_\mp, \hat{A}_\pm^\dagger] &= \mp I, & [\hat{\mathcal{N}}, \hat{A}_\pm^\dagger] &= \tfrac{1}{2}\hat{A}_\pm^\dagger, \\
[\hat{J}_0, \hat{A}_\pm^\dagger] &= \pm\tfrac{1}{2}\hat{A}_\pm^\dagger, & [\hat{J}_+, \hat{A}_-^\dagger] &= \hat{A}_+^\dagger, & [\hat{J}_-, \hat{A}_+^\dagger] = \hat{A}_-^\dagger, \\
[\hat{A}_\pm, \hat{B}_0^\dagger] &= \pm\hat{A}_\pm^\dagger, & [\hat{A}_\pm, \hat{B}_{\mp 1}^\dagger] &= \pm\sqrt{2}\hat{A}_\mp^\dagger,
\end{aligned} \tag{4.19}$$

as well as those which can be obtained from them by Hermitian conjugation.

For $\hat{H} = \hat{N} + |\hat{M}|$ or $\hat{H} = \frac{1}{2}\hat{r}(\hat{p}^2+1)$, the time-dependent eigenstates

$$|\psi_{j\mu}(t)\rangle = |j\mu\rangle \exp(-iE_j t), \tag{4.20}$$

where $E_j = 2j$ or $j + \frac{1}{2}$ respectively, correspond to all the integer j values ($j = 0, 1, 2, \ldots$) and span a single unirrep of $sp(4,\mathbb{R}) \simeq so(3,2)$. The latter can be characterized by $\langle\frac{1}{2}\frac{1}{2}\rangle$, denoting the eigenvalues of $\hat{\mathcal{N}} - \hat{J}_0$ and $\hat{\mathcal{N}} + \hat{J}_0$ corresponding to the unirrep lowest-weight state $|\psi_{00}(t)\rangle$. The action of the $sp(4,\mathbb{R})$ generators on the eigenstates (4.20) is given by the relations

$$\begin{aligned}
\hat{\mathcal{N}}|\psi_{j\mu}(t)\rangle &= (j+\tfrac{1}{2})|\psi_{j\mu}(t)\rangle, \\
\hat{B}_0^\dagger|\psi_{j\mu}(t)\rangle &= \sqrt{(j-\mu+1)(j+\mu+1)}\,|\psi_{j+1,\mu}(t)\rangle, \\
\hat{B}_{\pm 1}^\dagger|\psi_{j\mu}(t)\rangle &= \sqrt{\tfrac{1}{2}(j\pm\mu+1)(j\pm\mu+2)}\,|\psi_{j+1,\mu\pm 1}(t)\rangle,
\end{aligned} \tag{4.21}$$

together with the equations that can be obtained from them by Hermitian conjugation and with those resulting from (4.9).

For $\hat{H} = \hat{N}$, the time-dependent eigenstates (4.20), where $E_j = 2j$, correspond to all the integer and half-integer j values ($j = 0, \frac{1}{2}, 1, \frac{3}{2}, \ldots$)

TABLE VI. The operators $\widehat{\exp(i\alpha T)}$ for some quantum Hamiltonians.

$\hat{H}$	α	$\widehat{\exp(i\alpha T)}$
$\hat{N}$	2	$(\hat{N}_+\hat{N}_-)^{-1/2}\hat{\eta}_+\hat{\eta}_- = \sum_{n_+=0}^{\infty}\sum_{n_-=0}^{\infty} \|n_+ + 1, n_- + 1\rangle\langle n_+ n_-\|$
$\hat{N} + \|\hat{M}\|$	2	$(\hat{N}_+\hat{N}_-)^{-1/2}\hat{\eta}_+\hat{\eta}_- = \sum_{n_+=0}^{\infty}\sum_{n_-=0}^{\infty} \|n_+ + 1, n_- + 1\rangle\langle n_+ n_-\|$
$\frac{1}{2}\hat{r}(\hat{p}^2+1)$	1	$\frac{1}{\sqrt{(\hat{H}-\hat{M}-\frac{1}{2})(\hat{H}+\hat{M}-\frac{1}{2})}}[\hat{\mathbf{r}}\cdot\hat{\mathbf{p}} - \frac{1}{2}i - \frac{1}{2}i\hat{r}(\hat{p}^2-1)]$

and span two unirreps of $sp(4,\mathbb{R})$, characterized by $\langle\frac{1}{2}\frac{1}{2}\rangle$ and $\langle\frac{3}{2}\frac{1}{2}\rangle$, and whose lowest-weight states are $|\psi_{00}(t)\rangle$ and $|\psi_{1/2,-1/2}(t)\rangle$ respectively. They form however a single unirrep $\langle\frac{1}{2}\frac{1}{2}\rangle$ of the larger algebra $wsp(4,\mathbb{R})$, because

$$\hat{A}^\dagger_\pm|\psi_{j\mu}(t)\rangle = \sqrt{j\pm\mu+1}\,|\psi_{j+1/2,\mu\pm1/2}(t)\rangle. \tag{4.22}$$

For the three Hamiltonians $\hat{H} = \hat{N}, \hat{N}+|\hat{M}|$, and $\frac{1}{2}\hat{r}(\hat{p}^2+1)$, the $sp(4,\mathbb{R})$ generators $\hat{\mathcal{N}}$, $\hat{B}^\dagger_q$, and $\hat{B}_q, q = 1,0,-1$, can be written in a form similar to their classical counterpart given in (3.25), (3.28), (3.30), and (3.32), namely

$$\hat{\mathcal{N}} = \sqrt{C(\hat{H}) + \tfrac{1}{4}}, \tag{4.23}$$

$$\hat{B}^\dagger_0 = (\hat{B}_0)^\dagger = \sqrt{(\hat{\mathcal{N}}-\tfrac{1}{2})^{1/2} - \hat{J}_0^2}\,\widehat{\exp(i\alpha T)}\exp(-i\alpha t), \tag{4.24}$$

$$\hat{B}^\dagger_{\pm1} = -(\hat{B}_{\mp1})^\dagger = \sqrt{\tfrac{1}{2}(\hat{\mathcal{N}}\pm\hat{J}_0-\tfrac{1}{2})(\hat{\mathcal{N}}\pm\hat{J}_0-\tfrac{3}{2})} \times \widehat{\exp(\pm i\Phi/a)}\widehat{\exp(i\alpha T)}\exp(-i\alpha t). \tag{4.25}$$

Here $\hat{J}_0, C(\hat{H})$, and $\hat{\mathcal{N}}$ are the operators listed in Tables III and V, and α denotes the constant value of

$$\gamma(\hat{H}) = \left[\frac{d}{d\hat{H}}\hat{\mathcal{N}}\right]^{-1} = \left\{\frac{d}{d\hat{H}}\sqrt{C(\hat{H})+\tfrac{1}{4}}\right\}^{-1}, \tag{4.26}$$

equal to 2 for $\hat{H} = \hat{N}$ and $\hat{N}+|\hat{M}|$, and to 1 for $\hat{H} = \frac{1}{2}\hat{r}(\hat{p}^2+1)$. As a matter of fact, equation (4.24) is used to define the operator $\widehat{\exp(i\alpha T)}$ as

$$\widehat{\exp(i\alpha T)} = [(\hat{\mathcal{N}}-\tfrac{1}{2})^2 - \hat{J}_0^2]^{-1/2}\hat{B}^\dagger_0\exp(i\alpha t) = \sum_j\sum_{\mu=-j}^{j}|j+1,\mu\rangle\langle j\mu|, \tag{4.27}$$

giving the results listed in Table VI. In (4.27), the sum over j runs over those values present in the spectrum, *i.e.*, $j = 0, 1, 2, \ldots$ for $\hat{H} = \hat{N} + |\hat{M}|$ and $\hat{H} = \frac{1}{2}\hat{r}(\hat{p}^2 + 1)$, and $j = 0, \frac{1}{2}, 1, \frac{3}{2}, \ldots$ for $\hat{H} = \hat{N}$. Then one can check that, when introduced into (4.25), the operators (4.13) and (4.27) give for $\hat{B}^{\dagger}_{\pm 1}$ results in accordance with (4.21) and Table 5.

It is now straightforward to verify that the operator $\widehat{\exp(i\alpha T)}$ and its adjoint $\widehat{\exp(-i\alpha T)}$ satisfy (4.1) and (4.6), and that

$$\begin{aligned}
\widehat{\exp(i\alpha T)}\,\widehat{\exp(-i\alpha T)} &= I - |00\rangle\langle 00| - \sum_{j>0} \left[|j,-j\rangle\langle j,-j| + |jj\rangle\langle jj|\right], \\
\widehat{\exp(-i\alpha T)}\,\widehat{\exp(i\alpha T)} &= I,
\end{aligned} \tag{4.28}$$

in accordance with (4.3). Similarly, one finds

$$\begin{aligned}
\widehat{\exp(\pm i\Phi/a)}\,\widehat{\exp(i\alpha T)} &= \sum_j \sum_{\mu=-j}^{j} |j+1, \mu \pm 1\rangle\langle j\mu|, \\
\widehat{\exp(i\alpha T)}\,\widehat{exp(i\Phi/a)} &= \sum_j \sum_{\mu=-j}^{j-1} |j+1, \mu + 1\rangle\langle j\mu|, \\
\widehat{\exp(i\alpha T)}\,\widehat{exp(-i\Phi/a)} &= \sum_j \sum_{\mu=-j+1}^{j} |j+1, \mu - 1\rangle\langle j\mu|,
\end{aligned} \tag{4.29}$$

in accordance with (4.5). In particular, equations (4.29) show that the order of $\widehat{\exp(\pm i\Phi/a)}$ and $\widehat{\exp(i\alpha T)}$ on the right-hand side of (4.25) is essential to get the right result for $\hat{B}^{\dagger}_{\pm 1}$.

In the classical limit, the operator $\widehat{\exp(i\alpha T)}$ goes to the function $\exp(i\alpha T)$ that can be obtained from Table II, and the quantum $sp(4, \mathbb{R})$ generators, given in (4.11), (4.23), (4.24), and (4.25), go to their classical counterpart (3.25) for specific values of the constants a, b, and of the function $C(H)$. For the Hamiltonians $\hat{H} = \hat{N} + |\hat{M}|$ and $\hat{H} = \frac{1}{2}\hat{r}(\hat{p}^2 + 1)$, we therefore arrive at a similar result as that already obtained for the symmetry Lie algebras in the previous subsection, namely one of the members of the infinite sets of classical dynamical Lie algebras remains a dynamical Lie algebra when going to quantum mechanics. For the remaining Hamiltonians however, the situation is quite different, since none of the Lie algebras that are dynamical algebras in classical mechanics remains so in quantum mechanics. For $\hat{H} = \hat{N}$, this problem can be easily coped with by extending one of the algebras to a larger one, but for the other Hamiltonians no quantum dynamical Lie algebra is actually known. The latter situation is comparable with what occurs for some three-dimensional Hamiltonians that have a classical symmetry algebra but no quantum one (see Ref. [3] and references quoted therein).

21.5 Conclusion

In this talk, I have given unified definitions of symmetry and dynamical Lie algebras in both classical and quantum mechanics. I have also stressed that the importance of the algebra unirreps is a distinctive feature of quantum mechanics, responsible for the differences observed between classical symmetry and dynamical algebras and quantum ones.

I have illustrated these general concepts on some two-dimensional rotationally-invariant Hamiltonians by proving that they admit infinite sets of universal symmetry and dynamical algebras that, in general, have no counterpart in quantum mechanics.

This shows that the search for a symmetry or a dynamical algebra for a quantum Hamiltonian is a problem that can only be solved by quantum mechanical methods. First finding such an algebra for the corresponding classical Hamiltonian and then trying to appropriately quantize it will not in general yield a satisfactory result.

21.6 References

[1] M. Moshinsky and C. Quesne, *Phys. Lett. A* **152**, 128 (1991).

[2] H. Goldstein, *Classical Mechanics* (Addison-Wesley, Reading, Massachusetts 1969).

[3] M. Moshinsky and C. Quesne, *Ann. Phys. (NY)* **148**, 462 (1983).

[4] V.A. Dulock and H.V. McIntosh, *Am. J. Phys.* **33**, 109 (1965); V.A. Dulock and H.V. McIntosh, *J. Math. Phys.* **7**, 1401 (1966); A. Cisneros and H.V. McIntosh, *J. Math. Phys.* **10**, 277 (1969); A. Cisneros and H.V. McIntosh, *J. Math. Phys.* **11**, 870 (1970).

[5] I.A. Malkin and V.I. Man'ko, *JETP Lett.* **2**, 146 (1965).

[6] H.J. Lipkin, In *Nuclear Physics*, Ed. by C. De Witt and V. Gillet (Gordon and Breach, New York 1969) p. 644.

[7] Y. Dothan, *Phys. Rev. D* **2**, 2944 (1970).

[8] A.O. Barut and R. Rączka, *Theory of Group Representations and Applications* (PWN – Polish Scientific Publishers, Warsaw 1977).

[9] O. Castaños, A. Frank, and R. López-Peña, *J. Phys. A* **23**, 5141 (1990).

[10] M. Moshinsky, C. Quesne, and G. Loyola, *Ann. Phys. (NY)* **198**, 103 (1990).

[11] M. Moshinsky and J. Patera, *J. Math. Phys.* **16**, 1866 (1975).

[12] B.G. Wybourne, *Classical Groups for Physicists* (Wiley, New York 1974).

[13] E.L. Ince, *Ordinary Differential Equations* (Dover, New York 1956).

[14] R. Gilmore, *Lie Groups, Lie Algebras, and Some of Their Applications* (Wiley, New York 1974).

[15] E. Chacón, P.O. Hess, and M. Moshinsky, *J. Math. Phys.* **28**, 2223 (1987).

[16] M. Moshinsky and T.H. Seligman, *J. Math. Phys.* **22**, 1526 (1981).

[17] G.J. Milburn, *Phys. Rev. A* **33**, 674 (1986); B. Yurke and D. Stoler, *Phys. Rev. Lett.* **57**, 13 (1986).

[18] M. Moshinsky, T.H. Seligman, and K.B. Wolf, *J. Math. Phys.* **13**, 901 (1972).

[19] D.T. Pegg and S.M. Barnett, *Phys. Rev. A* **39**, 1665 (1989); S.M. Barnett and D.T. Pegg, *Phys. Rev. A* **41**, 3427 (1990); D. Ellinas, *J. Math. Phys.* **32**, 135 (1991).

[20] J.-M. Lévy-Leblond, *Ann. Phys. (NY)* **101**, 319 (1976).

[21] J.-M. Lévy-Leblond, *Rev. Mex. Fís.* **22**, 15 (1973).

[22] C.C. Gerry, *Phys. Rev. A* **38**, 1734 (1988).

[23] A.O. Barut and H. Kleinert, *Phys. Rev.* **156**, 1541 (1967); A.O. Barut and H. Kleinert, *Phys. Rev.* **157**, 1180 (1967); K.B. Wolf, *Suppl. Nuovo Cimento* **5**, 1041 (1967); C.P. Boyer and K.B. Wolf, *Lett. Nuovo Cimento* **8**, 458 (1973).

[24] C. Quesne, unpublished.

[25] M. Moshinsky and C. Quesne, *J. Math. Phys.* **12**, 1772 (1971).

22

Canonical Transformations in Mechanics *vis-à-vis* Those in Optics

Kurt Bernardo Wolf[1]

ABSTRACT The phase spaces of 2-dim Mechanics and of 3-dim Optics are locally the same, but globally different. While momentum is an unbounded variable in classical mechanics, in geometric optics it is constrained to a disk that is the projection of the ray direction sphere, whose radius is the local index of refraction. Canonical transformations in mechanics preserve the Heisenberg-Weyl algebra. In geometric optics, in addition, they must preserve the natural momentum range. We show that the physical phenomena of optics that produce canonical transformations are actually much richer than those in mechanics: they include refracting surfaces. In fact, through the opening coma map, the range limitation may be lifted. The quadratic Hamiltonians of mechanics provide the paraxial régime of optics. The optical aberrations of the metaxial régime are the higher-order approximations. Finally, we note that 2-dim mechanics is based on the Heisenberg-Weyl group W_2 while 3-dim optics is based on the Euclidean group ISO(3).

22.1 Introduction

After my graduate and postdoc years, I rejoined working with Professor Marcos Moshinsky in 1971. He had then just finished work with Dr. Christianne Quesne on *Oscillator Systems*, that was presented at the XV Solvay Conference held in Brussels in 1970, and published in the Proceedings [1]. Concurrently, they published the two first articles on linear canonical transformations in quantum mechanics [2], [3]. On this subject I helped him with the preparation of his notes for the 1971 Latin American School of Physics held in Mexico City [4]. Thereby involved in canonical transformations, I coauthored three articles on reductions of linear transformations to quotient spaces where they become nonlinear [5], [6], [7]. Thereafter, I should say, we parted ways and means, since I would accept only strongly-defined

[1]Instituto de Investigaciones en Matemáticas Aplicadas y en Sistemas, UNAM, Cuernavaca.

operators in Hilbert spaces [8], [9], and Marcos found little virtue in some of their non-physical edges, such as non-unique spectra [10], and non-local measures [11]. A reconcilliation is found as the conclusion of this work.

22.1.1 Phase Space

In this tribute to Professor Moshinsky I will write only about canonical transformations in classical mechanics and in geometric optics, because of reasons given briefly in this and the concluding Sections: their extension to quantum mechanics and wave optics are *not* parallel. Mechanics is a classical subject that is far from trivial and whose symplectic geometry has recently received solid mathematical attention [12], [13]. Optics is even older, but in the last decade a new Lie theoretic streak has prospered; subjectively I may point to references [14] and [15] as those that influenced my own decision to pursue the subject from 1982 on.

Both mechanics and optics are subject to the Hamiltonian formalism, that describes the evolution of systems through the identification of equal numbers of position and momentum coordinates subject to a pair of evolution —Hamilton's— equations:

$$\frac{d\ \text{position}}{d\ \text{parameter}} = \frac{\partial\ \text{Hamiltonian}}{\partial\ \text{momentum}}, \qquad \frac{d\ \text{momentum}}{d\ \text{parameter}} = -\frac{\partial\ \text{Hamiltonian}}{\partial\ \text{position}}.$$

In Section 2 we show that the first is a geometric relation and the second a dynamical one. The parameter may be arc length along the trajectory, time, or distance along an optical axis; mechanics generally uses the second and optics the third choice. In doing so we introduce what we mean by a *vis à vis* relation that, I hope, will be a useful concept. It makes a mechanical system the first order approximation to an optical system —its paraxial approximation. The harmonic oscillator is thus the linear part of an elliptic-profile fiber.

The main concept we want to convey in Section 2 is that mechanical and optical phase spaces, although locally identical, are globally *different*: the *range* of ray direction (optical momentum) is compact, and hence a mathematically well-defined *wavization* of optics must be different from the better-known *quantization* of mechanics. The latter has been used with success in place of the former in (necessarily paraxial) Fourier optics [16]. In a nutshell, we may state that the quantization of classical mechanics is based on the Heisenberg-Weyl algebra, group, and ring [17]; the wavization of optics, I believe [18], is based on the corresponding Euclidean structures.

22.1.2 Nonlinearities and Aberrations

Three classes of canonical transformations are very important, because each forms a closed algebraic structure: *(a)* linear transformations, *(b)* point transformations in position, and *(c)* point transformations in momentum.

Mechanics thrives on the first two; optics is more eclectic. Optical systems that collect rays issuing in all directions from an object point and try to form an image point out of them would belong to the intersection of *(a)* and *(b)*. But, alas! that is not actually possible, as we show in Section 3. Instead, we may easily unfocus a light beam by free propagation and/or relativistic motion, transformations that are wholly in *(c)*. Most commonly, optical design will involve all three classes of transformations; their composition does not close, but spills over into the group of *all* (nonlinear) transformations of phase space.

Perhaps the best imprint that Marcos has left on my work is the search for the hidden symmetries. Indeed, it so happens that the rather messy array of optical aberrations [19] may be classified in one-to-one correspondence with the states of the three-dimensional quantum harmonic oscillator [20]. Since aberrations also exist in mechanics in the form of higher-order approximations that describe nonlinearities, this classification of aberrations may be used for the purpose of systematizing their treatment, as has been done by A.J.Dragt and coworkers [15], [21] in magnetic accelerator design.

22.1.3 The Richness of Optics

The phase spaces of mechanics and optics may be related by a rather simple comatic map [22]. Thus, all transformations applying in mechanics also do so in optics. In particular, the Fourier transformation that is the basis for much of optical processing and measurement theory in the paraxial régime may be extended to 4π Euclidean optics [23]. Also, since coset manifolds of the Euclidean group ISO(3) are subject to deformation by the Lorentz and other groups [24], a relativistic group of canonical transformations appears in a natural way in geometric optics [25]. It seems surprising that this optical effect has not been noticed before.

Optics is manifestly a richer field than mechanics for the action of canonical transformations: *lenses* are part and parcel of most real optical systems. They produce transformations of the space of rays that are canonical and have no counterpart in mechanics. Nonflat refracting surfaces produce a class of canonical transformations that we are still in the process of exploring thoroughly. They have several peculiar and distinguishing properties that are sketched in Section 4. They exhibit a unique factorization into two *root* transformations [26] that are canonical and exhibit selection rules in their aberration coefficients [27]. The concluding Section will indicate some other lines on which the author's study in optics extends, some of its loose ends, bits and pieces. From time to time, it is a welcome opportunity to present cross sections of one's work following definite directions. Doing it along the momentum vector given by Professor Marcos Moshinsky is a pleasure.

22.2 The phase space of mechanics and that of optics

We assume that the physical objects whose phenomena we want to describe, are defined by the coordinates of a *position* vector $\vec{q} \in \Re^N$, and a canonically conjugate *momentum* vector $\vec{p} \in \Re^N$. We assume there is an evolution in the space so that in a neigborhood of $(\mathbf{p}, \mathbf{q}) \in \Re^{2N}$ the arc length parameter is $s \in \Re^+$ for the trajectories $(\vec{p}(s), \vec{q}(s))$, which are on Hamiltonian surfaces $\mathcal{H}(\mathbf{p}, \mathbf{q}) = \mathcal{H}^{[s]} = \mathcal{H}^{[0]} =$ constant [28]. Finally, for the velocity-independent and isotropic optical systems that we consider here, we assume that $\vec{p}(s)$ is tangent to the curve $\vec{q}(s)$.

22.2.1 The First Hamilton Equation in Arc Length

The above assumptions imply that in the following equalities, the first two unit vectors are equal

$$\frac{d\vec{q}}{ds} = \frac{\vec{p}}{|\vec{p}|} = \frac{\partial \mathcal{H}^{[s]}}{\partial \vec{p}}, \tag{2.1}$$

where $ds = |d\vec{q}| = \sqrt{d\vec{q} \cdot d\vec{q}}$. The last member of (2.1) *proposes* an *s-in*dependent Hamiltonian function $\mathcal{H}^{[s]}(\vec{q}, \vec{p})$ (modulo additive constants) that *generates* this evolution along s. Then, partial integration on $\vec{p}$ leads to

$$\mathcal{H}^{[s]} = |\vec{p}| - n(\vec{q}), \tag{2.2}$$

where $n(\vec{q})$ is an as yet undetermined function of position alone, that we shall examine below. We see thus that the *first Hamilton equation* (2.1) is a direct consequence of the previous geometric assumptions.

22.2.2 The Second Hamilton Equation in Arc Length

The *dynamic* laws of a system are in the *second* Hamilton equation, that speaks of the departure of the trajectory from a straight line, *i.e.*, of $d\vec{p}$. From (2.2), this equation is

$$\frac{d\vec{p}}{ds} = \frac{\partial n(\vec{q})}{\partial \vec{q}} = -\frac{\partial \mathcal{H}^{[s]}}{\partial \vec{q}}, \tag{2.3}$$

and we are left to specify the physical laws by the gradient of the function n. Here, mechanics and optics depart; mechanics follows Newton's law, and optics, Snell's. In either case, the value of $\mathcal{H}^{[s]}$ on the trajectories (2.2) will be constant and congruent with zero. At each point $\vec{q}$, therefore, $\vec{p}$ will be constrained to a *sphere* of radius

$$|\vec{p}| = n(\vec{q}). \tag{2.4}$$

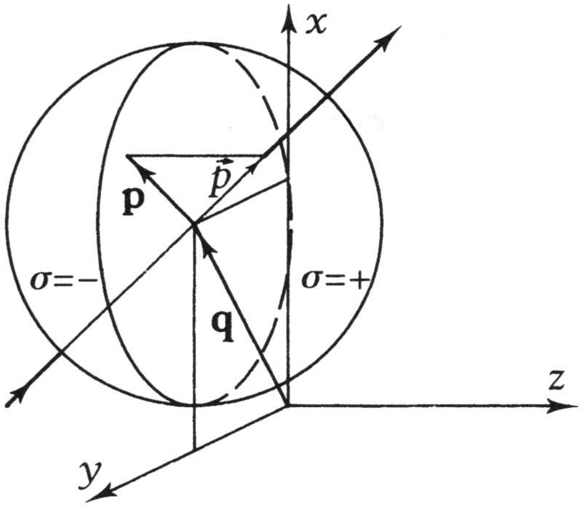

FIGURE 22.1. Phase space $(\mathbf{p}, \mathbf{q})$ describes a ray through a *position* coordinate $\mathbf{q}$ that is the intersection of the ray with a screen at z, and a *momentum* coordinate $\mathbf{p}$ that is the projection on the screen of the ray direction vector $\vec{p}$. The chart index σ distinguishes between a ray and its reflection by the screen, that have the same value of $\mathbf{p}$.

In optics, $n(\vec{q})$ is the local refractive index of the medium; it is also the radius of the *Descartes* sphere that is familiar from the French philosopher's sketch of the refraction process, Figure 22.1.

22.2.3 The Hamilton Equations of Mechanics in Time

In mechanics, this formulation of the Hamilton equations with respect to the arc length parameter is not the most familiar; Newton's law (time derivative of momentum equals minus potential gradient) can be recognized if we write, for $m > 0$ and E constants, and V a function of $\vec{q}$,

$$n(\vec{q}) = \sqrt{2m(E - V(\vec{q}))}, \qquad V(\vec{q}) = E - n(\vec{q})^2/2m. \tag{2.5}$$

Then, the dynamical equation (2.3) becomes

$$\frac{d\vec{p}}{ds} = \frac{\partial n(\vec{q})}{\partial \vec{q}} = -\frac{m}{|\vec{p}|}\frac{\partial V(\vec{q})}{\partial \vec{q}}. \tag{2.6}$$

We may incorporate the trajectory constant $m/|\vec{p}|$ into the arc length parameter s to define a *time* parameter t such that

$$\frac{d\vec{p}}{dt} = -\frac{\partial V(\vec{q})}{\partial \vec{q}}, \qquad \text{where} \quad \frac{ds}{dt} = \frac{|\vec{p}|}{m}. \tag{2.7}$$

With this change of variables we recognize V as the potential function, and the the first Hamilton equation becomes the definition of vector momentum

$\vec{p}$ as m (mass) times velocity $d\vec{q}/dt$. The change from s to t is singular when $|\vec{p}| = 0$, *i.e.*, when the particle stops at the space surfaces $V(\vec{q}) = E$; the constant E is the familiar *energy* of the trajectory,

$$\mathcal{M}^{[t]}(\vec{p},\vec{q}) = -E + \frac{1}{2m}|\vec{p}|^2 + V(\vec{q}) = -E + H^{\text{mech}}(\vec{p},\vec{q}) = 0. \qquad (2.8)$$

In the time parameter, thus, the Hamilton equations of motion have the well known form [28]

$$\frac{d\vec{q}}{dt} = \frac{\vec{p}}{m} = \frac{\partial \mathcal{M}^{[t]}}{\partial \vec{p}}, \qquad (2.9)$$

$$\frac{d\vec{p}}{dt} = -\frac{\partial V(\vec{q})}{\partial \vec{q}} = -\frac{\partial \mathcal{M}^{[t]}}{\partial \vec{q}}. \qquad (2.10)$$

22.2.4 The Hamilton Equations of Optics Along an Axis

The Hamilton equations of optics in the arclength parameter s are (2.1) and (2.3) with $|\vec{p}| = n(\vec{q}) > 0$, the refractive index which normalization and Nature restrict to $1 \leq n \lesssim 2.4$ (for diamond). We may find a corresponding set of Hamilton equations in time through the statement that the velocity of 'optical particles' is $ds/dt = c/n(\vec{q})$, and with a corresponding Hamiltonian. Practice with optical systems and minimality arguments, however, favor another parametrization: by *length* along an optical axis, traditionally the Nth axis (called z). Provided we can invert $z = q_N(s)$ to $s(z)$, this leaves us with the $N-1$ coordinates $\mathbf{q}(z) = \{q_k(z)\}_{k=1}^{N-1}$ and $\mathbf{p}(z) = \{p_k(z)\}_{k=1}^{N-1}$. The Nth momentum component, p_N, is a dependent variable (up to a sign σ) because of the restriction to the Descartes sphere (2.4), *i.e.*,

$$p_N = \sigma\sqrt{n(\vec{q})^2 - |\mathbf{p}|^2}, \qquad |\mathbf{p}|^2 = \sum_{k=1}^{N-1} p_k^2. \qquad (2.11)$$

Observing that ds, dz (and $d\mathbf{q}$) are along $\vec{p}$, p_N (and $\mathbf{p}$), it follows by similar triangles that $ds/dz = n(\mathbf{q}, z)/p_N$.

Thus, in a space of $2(N-1)$ coordinates $(\mathbf{p}(z), \mathbf{q}(z))$ we write the Hamilton equations of optics as

$$\frac{d\mathbf{q}}{dz} = \frac{\mathbf{p}}{p_N} = \frac{\partial \mathcal{G}^{[z]}}{\partial \mathbf{p}}, \qquad (2.12)$$

$$\frac{d\mathbf{p}}{dz} = \frac{n}{p_N}\frac{\partial n}{\partial \mathbf{q}} = -\frac{\partial \mathcal{G}^{[z]}}{\partial \mathbf{q}}, \qquad (2.13)$$

with the Hamiltonian function

$$\mathcal{G}^{[z]}(\mathbf{p},\sigma,\mathbf{q};z) = -p_N = -\sigma\sqrt{n(\mathbf{q},z)^2 - |\mathbf{p}|^2} \qquad (2.14)$$

$$= -\sigma n + \frac{\sigma}{2n}|\mathbf{p}|^2 + \frac{\sigma}{8n^3}|\mathbf{p}|^4 + \frac{\sigma}{16n^5}|\mathbf{p}|^6 + \cdots. \qquad (2.15)$$

Optical phase space has thus $2(N-1)$ coordinates $(\mathbf{p}, \mathbf{q})$ and a chart index $\sigma \in \{+, 0, -\}$ that distinguishes between rays along, across, and counter the optical $(+z)$ axis.[2] The accessible region[3] is the full screen $\mathbf{q} \in \Re^{N-1}$, while momentum $\mathbf{p}$ is constrained by $|\mathbf{p}| \leq |\vec{p}| = n(\vec{q})$. When we stay away from $\sigma = 0$ submanifold we may concentrate on the $\sigma = +$ chart of 'forward-z' rays. Then, we may state that the range of the optical momentum coordinates is the interior of the $(N-1)$-*sphere*, indicated $\bar{\mathcal{S}}_{N-1}$.

22.2.5 The *vis-à-vis* Relation Between Mechanics and Optics

Consider the case of *homogeneous* media, *i.e.*, where $n =$ constant. Since the second Hamilton equation is identically zero, the particle (optical or mechanical) moves with *free flight* $\vec{p} =$ constant. The first Hamilton equation (2.1) in s integrates to

$$\vec{q}(s) = \vec{q}(0) + s\vec{p}/n, \tag{2.18}$$

the Hamilton equation (2.9) in t integrates to

$$\vec{q}(t) = \vec{q}(0) + t\vec{p}/m, \tag{2.19}$$

and, lastly, the Hamilton equation (2.12) in z integrates[4] to

$$\mathbf{q}(z) = \mathbf{q}(0) + z\frac{\mathbf{p}}{\sigma\sqrt{n^2 - |\mathbf{p}|^2}} \tag{2.20}$$

$$= \mathbf{q}(0) + \sigma\frac{z}{n}\mathbf{p} + \sigma z\left(\frac{1}{2n^3}|\mathbf{p}|^2\mathbf{p} + \frac{3}{8n^5}|\mathbf{p}|^4\mathbf{p} + \frac{5}{16n^7}|\mathbf{p}|^6\mathbf{p} + \cdots\right). \tag{2.21}$$

Let us suggest the following relation between mechanics and optics. We consider an $(N-1)$-dimensional classical mechanical system evolving

---

[2]Alternatively, if we denote by θ the angle between the ray $\vec{p}$ and the z-axis $(0 \leq \theta \leq \pi)$, then

$$|\mathbf{p}| = n(\mathbf{q}, z)\sin\theta, \qquad p_N = n(\mathbf{q}, z)\cos\theta. \tag{2.16}$$

The $\sigma = +$ chart corresponds to $\theta < \frac{1}{2}\pi$ rays.

[3]The form (2.12)–(2.13) of the Hamilton equations is valid also in mechanics, provided the refractive index and the potential are related by (2.5). The corresponding mechanical Hamiltonian is

$$\mathcal{M}^{[z]}(\mathbf{p}, \sigma, \mathbf{q}; z) = -p_N = -\sigma\sqrt{2m(E - V(\mathbf{q}, z)) - |\mathbf{p}|^2}. \tag{2.17}$$

[4]The second summand in (2.20) is $z\tan\theta$ in terms of the ray angle (2.16), as is geometrically obvious.

in t governed by the Hamilton equations (2.9)–(2.10) [with the Hamiltonian (2.8) and free flight (2.19)]; its canonical coordinates are $(\mathbf{p},\mathbf{q}) \in \Re^{2(N-1)}$, the region of phase space accessible to the particle moving on submanifolds of constant energy E unbounded from above. On the other hand, we consider a geometric optical N-dimensional system described on $(N-1)$-dimensional screens moving along z through the Hamilton equations (2.12), (2.13) [Hamiltonian (2.14)–(2.15) and free flight (2.20)–(2.21)].

Although mechanical and optical phase spaces are *globally* quite different, *locally*, near to the optical axis and center $(\mathbf{p},\mathbf{q}) = (\mathbf{0},\mathbf{0})$, they are indistinguishable. We say that a mechanical system with potential $V(\mathbf{q},t)$ is *vis à vis* an optical system with refractive index $n(\mathbf{q},z)$ when they are related by equation (2.5). A mechanical system *vis à vis* an optical system is its *paraxial* approximation (for $|\mathbf{q}| \ll 1, |\mathbf{p}| \ll 1$). For homogeneous media (n and V constant), the mechanical Hamiltonian (2.8) is the up-to-quadratic part of the optical one (2.15), and free motion (2.19) is the linear part of free propagation (2.21). In this context, the mechanical model yields the results of paraxial optics ($\theta \ll \frac{1}{2}\pi$) and ultimately sustains the quantization of the former [1]–[4] to describe the wave version of the latter [29], [30].

22.2.6 The Harmonic Oscillator *vis-à-vis* the Elliptic-Profile Fiber

A widely used application is the one-dimensional harmonic oscillator* in the paraxial model of fiber optics. Let us use capitals to denote the coordinates of the mechanical harmonic oscillator of mass m and Hooke's constant k, so $V_{\text{ho}}(Q) = \frac{1}{2}kQ^2$. Then, the equations of motion are

$$\frac{dQ}{dt} = \frac{P}{m}, \qquad \frac{dP}{dt} = -kQ, \tag{2.22}$$

and their solution is thoroughly known:

$$Q = Q_0 \cos\omega t + P_0 \frac{1}{\sqrt{km}} \sin\omega t, \tag{2.23}$$

$$P = -Q_0\sqrt{km}\sin\omega t + P_0 \cos\omega t, \qquad \omega = \sqrt{\frac{k}{m}}. \tag{2.24}$$

It is a canonical transformation of the P–Q plane foliated by concentric, equal-energy ellipses $E = \frac{1}{2}(P^2/m + kQ^2) \geq 0$, such that points slide around the origin by the *same* angle ωt, as shown (for the case of a circle when $mk = 1$) in Figure 22.2 *(a)*.

*This habit has been ingrained by Professor Marcos Moshinsky. Many times when I turned to him to discuss some big but half-baked model, Marcos would propose ... *let's see how it works with the one-dimensional harmonic oscillator...*

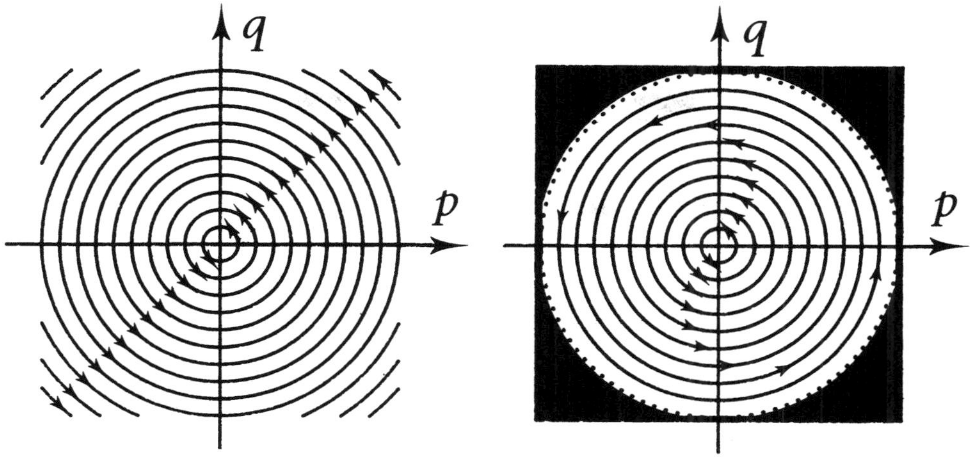

FIGURE 22.2. *(a)* Phase space time evolution for the one-dimensional harmonic oscillator is a rigid rotation of the full plane. *(b)* The evolution of optical phase space along the axis of an elliptic index-profile fiber is a (*non*-rigid) rotation of the interior of the circle; the region outside has no physical meaning.

What is the corresponding optical system *vis à vis* the harmonic oscillator? The relation (2.5) directs us to a two-dimensional optical medium of refractive index $n(q, z)$ that is independent of z, and of the form [31]

$$n_{\text{ho}}(q, z) = +\sqrt{n_o^2 - \kappa^2 q^2}, \qquad n_0^2 = 2mE, \quad \kappa^2 = mk. \tag{2.25}$$

The Hamilton equations of optics in the one-dimensional screen coordinates (p, q) are (2.12)–(2.13), *viz.*

$$\frac{dq}{dz} = \frac{p}{h}, \qquad \frac{dp}{dz} = -\frac{\kappa^2 q}{h}, \qquad h = \sigma\sqrt{n_o^2 - (p^2 + \kappa^2 q^2)}. \tag{2.26}$$

Their solution is

$$q(z) = q_0 \cos vz + p_0 \frac{1}{\kappa} \sin vz, \tag{2.27}$$

$$p(z) = -q_0 \kappa \sin vz + p_0 \cos vz, \qquad v = \frac{\kappa}{\sqrt{n_0^2 - (p_0^2 + \kappa^2 q_0^2)}}. \tag{2.28}$$

Here the transformation of phase-space points with z is also canonical and along ellipses $0 \leq e = p^2 + \kappa^2 q^2 = \text{constant}$, but with a *dispersion* of angles around the origin, as shown in Figure 22.2 *(b)* for circles ($\kappa = 1$). At z, the angle is $vz = \kappa z / \sqrt{n_o^2 - e}$; this increases with e up to a critical $e_c = n_o^2$ where it becomes infinite, and is physically meaningless beyond. In this manner, the *optical* phase space of the fiber (2.25) is the interior of the ellipse representing rays that travel and stay within the fiber.

The one-dimensional harmonic oscillator is the mechanical system *vis à vis* the 2-dim optical world with the refractive index (2.25). The profile in the n–q plane is a half-ellipse, and the fiber is called an *elliptic-profile* fiber. The statement that the mechanical system is the paraxial approximation of the optical system means that we can write the Hamiltonian of the latter in a series where the first (non-constant) term is the Hamiltonian of the former. From (2.26) for $h(p,q)$ and from (2.8) for $H^{\rm mech}(p,q)$,

$$\begin{aligned} h(p,\sigma,q) &= -\sigma\sqrt{n_0^2-(p^2+\kappa^2q^2)} = -\sigma\sqrt{n_o^2-2mH^{\rm mech}(p,q)} \\ &= -n_0+\frac{m}{n_0}H^{\rm mech}+\frac{m^2}{2n_o^3}(H^{\rm mech})^2+\frac{m^3}{2n_o^5}(H^{\rm mech})^3+\cdots \quad (2.29) \end{aligned}$$

The harmonic oscillator serves to model light in a fiber in the paraxial régime with a refractive index given by a *parabola* $n_{\rm ho}(q,z) = n_o - (mk/2n_o)q^2$; this is the *parabolic-profile* fiber that models, to first approximation, actual fibers whose dense core keeps light near to the optical axis.

22.3 Transformations in mechanical *vis-à-vis* optical phase space

In last Section we contrasted and related classical mechanical and geometric optical systems and phase space ranges. To wit: the phase space of mechanics is $(\mathbf{p},\mathbf{q}) \in \Re^{2(N-1)}$ and in optics it is $(\mathbf{p},\mathbf{q})_\sigma \in \bar{S}_{N-1}\times\Re^{N-1}$, $\sigma=\pm$. The former is the 'paraxial approximation' of the latter. In what follows we shall consider the 'physical' case of optics in two-dimensional screens and two-dimensional classical mechanics, and concentrate on $\sigma = +$ 'forward' rays; we may use the same letters $(\mathbf{p},\mathbf{q})$ for both spaces since they are distinguished only by their global connectivity.

Canonical transformations are defined by analytic functions $\mathbf{p}'(\mathbf{p},\mathbf{q})$, $\mathbf{q}'(\mathbf{p},\mathbf{q})$, that are a locally canonical set of coordinates of a symplectic-metric space, just as $(\mathbf{p},\mathbf{q})$. This property may be introduced through the well-known preservation of the Poisson bracket [28]

$$\{f,g\} = \frac{\partial f}{\partial \mathbf{q}}\cdot\frac{\partial g}{\partial \mathbf{p}} - \frac{\partial f}{\partial \mathbf{p}}\cdot\frac{\partial g}{\partial \mathbf{q}}. \quad (3.1)$$

So that

$$\{p_i',q_j'\} = -\delta_{i,j}, \quad \{p_i',p_j'\} = 0, \quad \{q_i',q_j'\} = 0,$$

for $i,j=1,2$.

22.3.1 Linear Canonical Transformations

The simplest transformation subgroup is that of inhomogeneous linear transformations,

$$\begin{pmatrix} \mathbf{p} \\ \mathbf{q} \\ 1 \end{pmatrix} \mapsto \begin{pmatrix} \mathbf{p}' \\ \mathbf{q}' \\ 1 \end{pmatrix} = \begin{pmatrix} \mathbf{A} & \mathbf{B} & \mathbf{e} \\ \mathbf{C} & \mathbf{D} & \mathbf{f} \\ 0 & 0 & 1 \end{pmatrix}^{-1} \begin{pmatrix} \mathbf{p} \\ \mathbf{q} \\ 1 \end{pmatrix}$$
$$= \begin{pmatrix} \begin{pmatrix} \mathbf{D}^\top & -\mathbf{B}^\top \\ -\mathbf{C}^\top & \mathbf{A}^\top \end{pmatrix} \left[\begin{pmatrix} \mathbf{p} \\ \mathbf{q} \end{pmatrix} - \begin{pmatrix} \mathbf{e} \\ \mathbf{f} \end{pmatrix} \right] \\ 1 \end{pmatrix}, \quad (3.2)$$

where $\mathbf{A}, \ldots, \mathbf{D}$ are 2×2 submatrices of a 4×4 real symplectic matrix, $\mathbf{e} \in \Re^2$ a momentum translation two-vector and $\mathbf{f} \in \Re^2$ a position translation two-vector. These linear transformations form a finite-parameter group that is the semidirect product between the real symplectic group $\mathrm{Sp}(4, \Re)$ of linear transformations, and of the group of translations of phase space T_4. A distinguished member of $\mathrm{Sp}(4, \Re)$ is the *Fourier transformation*, a $\frac{1}{2}\pi$ rotation between positions and momenta:

$$\mathcal{F} : \begin{pmatrix} \mathbf{p} \\ \mathbf{q} \end{pmatrix} \mapsto \begin{pmatrix} \mathbf{p}^{\mathrm{F}} \\ \mathbf{q}^{\mathrm{F}} \end{pmatrix} = \begin{pmatrix} -\mathbf{q} \\ \mathbf{p} \end{pmatrix} = \begin{pmatrix} 0 & -1 \\ 1 & 0 \end{pmatrix} \begin{pmatrix} \mathbf{p} \\ \mathbf{q} \end{pmatrix}. \quad (3.3)$$

The group of transformations (3.2) is well known to be the homomorphism group of the basic Heisenberg-Weyl algebra of generators p_i, q_j, and 1. The latter is noncompact and the ranges of the observables $\mathbf{p}$ or $\mathbf{q}$ are infinite. In optics, the position translation $\mathbf{f}$ on the $\Re^2$ manifold of the screen encounters no problem. However, the restriction $|\mathbf{p}| \leq n$ is *not* invariant under translations in momentum e; in optics, we may **not** legitimately translate momenta; consequently, the translation generators q_i will *not* have legitimate self-adjoint wave-optical operators; there are *no* Dirac δ's in position for wave optics. The allowed linear transformations in global optics are the proper subgroup

$$\mathbf{p} \mapsto \mathbf{p}' = \mathbf{R}_2 \mathbf{p}, \qquad \mathbf{R}_2 \in \mathrm{SO}(2), \quad (3.4)$$
$$\mathbf{q} \mapsto \mathbf{q}' = \mathbf{R}_2^{-1} \mathbf{q} - \mathbf{C}^\top \mathbf{p} + \mathbf{f}. \quad (3.5)$$

22.3.2 Phase and Point Transformations

Two well-known classes of functions of phase space that in quantum mechanics have scheme-independent quantization are: *(a)* functions of position variables *only*, $F(\mathbf{q})$, and *(b)* functions that are *linear* in one of the variables, $G(\mathbf{p}, \mathbf{q}) = \mathbf{p} \cdot \mathbf{G}(\mathbf{q})$, with $\mathbf{G} = (G_1, G_2)$ a two-vector of functions of q_i. Similar statements, exchanging $\mathbf{p}$ and $\mathbf{q}$ are also valid.

For a smooth function $K(\mathbf{p}, \mathbf{q})$ we build its Lie operator $\{K, \circ\}$ and therefrom its Lie transformation [32] $\mathcal{K}_\alpha = \exp \alpha\{K, \circ\} = 1 + \alpha\{K, \circ\} +$

$\frac{1}{2!}\alpha^2\{\{K,\circ\},\circ\}+\cdots$ on phase space. The Lie transformation generates a flow of phase space parametrized by α along lines $K(\mathbf{p},\mathbf{q})=$ constant and is, by construction, canonical. When the generating function is of position only, its action is

$$\exp\alpha\{F(\mathbf{q}),\circ\}:\begin{pmatrix}\mathbf{p}\\ \mathbf{q}\end{pmatrix}=\begin{pmatrix}\mathbf{p}+\alpha\frac{\partial F(\mathbf{q})}{\partial\mathbf{q}}\\ \mathbf{q}\end{pmatrix}. \tag{3.6}$$

Since in quantum mechanics this will put a phase $\exp i\alpha F(\mathbf{q})$ to wavefunctions, we may call these *phase* transformations of (classical) phase space.

For linear functions $\mathbf{p}\cdot\mathbf{G}(\mathbf{q})$ we can see from general homogeneity arguments that the Lie transformation generated is of the form

$$\exp\alpha\{\mathbf{p}\cdot\mathbf{G}(\mathbf{q}),\circ\}:\begin{pmatrix}\mathbf{p}\\ \mathbf{q}\end{pmatrix}=\begin{pmatrix}\mathbf{p}'(\mathbf{p},\mathbf{q};\alpha)\\ \mathbf{q}'(\mathbf{q};\alpha)\end{pmatrix}=\begin{pmatrix}\mathbf{p}\boldsymbol{\Psi}(\mathbf{q};\alpha)\\ \boldsymbol{\Phi}(\mathbf{q};\alpha)\end{pmatrix}, \tag{3.7}$$

where $\boldsymbol{\Phi}$ is a two-vector of functions of $\mathbf{q}$ and α, and $\boldsymbol{\Psi}$ a 2×2 matrix of such functions. Linear functions are thus the generators of the *point* transformations of mechanics [17], [33], whose action one finds (in principle) solving $\boldsymbol{\Phi}$ from the coupled partial differential equations

$$\begin{aligned}\frac{\partial\boldsymbol{\Phi}(\mathbf{q};\alpha)}{\partial\alpha} &= \{\mathbf{p}\cdot\mathbf{G},\circ\}\exp\alpha\{\mathbf{p}\cdot\mathbf{G},\circ\}\mathbf{q}\\ &= \{\mathbf{p}\cdot\mathbf{G},\boldsymbol{\Phi}\}=-\left[\mathbf{G}\cdot\frac{\partial}{\partial\mathbf{q}}\right]\boldsymbol{\Phi}(\mathbf{q};\alpha), \qquad (3.8)\\ \frac{\partial\boldsymbol{\Psi}(\mathbf{q};\alpha)}{\partial\alpha} &= \left[\boldsymbol{\Psi}\frac{\partial}{\partial\mathbf{q}}\right]\mathbf{p}\cdot\mathbf{G}-\left[\mathbf{G}\cdot\frac{\partial}{\partial\mathbf{q}}\right]\boldsymbol{\Psi}\mathbf{p}. \qquad (3.9)\end{aligned}$$

A change of variables $(q_1,q_2)\mapsto(\rho,\varphi)$ may transform (3.8) into a simple translation by α in the new ρ coordinate. Given the vector $\mathbf{G}$ and $G^2=G_1^2+G_2^2$, we must find by quadrature $d\rho=\mathbf{G}\cdot d\mathbf{q}/G^2$ and $d\varphi=\mathbf{G}\times d\mathbf{q}/G^2$; finally, and also by quadrature we must find the matrix function $\boldsymbol{\Psi}$ for the canonically conjugate variable.

In one dimension the situation is much simpler since only $d\rho=dq/G$ needs solution. This leads to translations of the scaled variable $\rho(q'(q;\alpha))=\rho(q)-\alpha$. The full solution to (3.7) is then

$$\begin{aligned}\Phi(q;\alpha) &= \rho^{-1}(\rho(q)-\alpha),\quad \rho(q)=\int^q\frac{dx}{G(x)}, \qquad (3.10)\\ \Psi(q;\alpha) &= 1\Big/\frac{\partial\Phi(q,\alpha)}{\partial q}, \qquad (3.11)\end{aligned}$$

provided $x=\rho(q)$ can be inverted to $q=\rho^{-1}(x)$ in some range. The flow of phase space will be along constant values of the generating function $F=p\,G(q)$, or lines $p=1/G(q)$. We do expect that points in phase space representing mechanical particles in motion will map on other mechanical particles, perhaps in a different potential, but corresponding to valid points in the manifold. In particular the *position* of the latter will be a function *only* of the position of the former.

In optics, the transformation (3.7) characterized by $\mathbf{q} \mapsto \mathbf{\Phi}(\mathbf{q})$ of ray positions on the screen is generically called aberration of *distortion*. Since it is a canonical transformation, the ray direction (momentum) must also change. It does so by a $\mathbf{q}$-dependent scale transformation $\mathbf{p} \mapsto \mathbf{p}\mathbf{\Psi}$; inevitably thus, some physical rays $|\mathbf{p}| < n$ will exchange places with unphysical ones $|\mathbf{p}| > n$. The point transformations of mechanics are *vis à vis* distortions in optics, but they are *globally* **not** allowed in optics. Regarding the phase transformations (3.6), these do leave the position of the rays $\mathbf{q}$ on the screen unaffected, but change their direction by $\mathbf{p} \mapsto \mathbf{p} + \mathbf{F}(\mathbf{q})$. This transformation is of little import to the design of lenses and has apparently received no name and little mention among the aberrations. Since it has $\mathbf{p}$-unfocusing properties we have playfully called it *pocus* [34]. It is also **not** globally allowed in 4π optics.

22.3.3 Spherical Aberration

There are two families of *proper* canonical transformations that map optical phase space on itself: spherical aberration and coma. These are phase and point transformations of the *momentum* variable $\mathbf{p} \in \bar{S}_2$ (ray direction on the 4π sphere).

The Fourier conjugate to the phase transformation (3.6) is

$$\exp \alpha\{F(\mathbf{p}), \circ\} : \begin{pmatrix} \mathbf{p} \\ \mathbf{q} \end{pmatrix} = \begin{pmatrix} \mathbf{p} \\ \mathbf{q} - \alpha \frac{\partial F(\mathbf{p})}{\partial \mathbf{q}} \end{pmatrix}, \tag{3.12}$$

and it evidently preserves the optical phase space range. In particular, free flight in a homogeneous optical medium, generated by $\mathcal{G}^{[z]} = -\sigma\sqrt{n^2 - |\mathbf{p}|^2}$ [Eqs. (2.12)–(2.15) when n = constant], does not change ray directions ($\mathbf{p}' = \mathbf{p}$), but only *unfocuses* images, as shown in (2.20)–(2.21). Each term in the $(|\mathbf{p}|^2)^k$ expansion of $\mathcal{G}^{[z]}$ corresponds to a term in the $(|\mathbf{p}|^2)^{k-1}\mathbf{p}$ expansion of $\mathbf{q}(z)$ (except for the constant in the Hamiltonian). In the latter, the linear term is 'paraxial free flight' while the higher-order terms $\simeq (|\mathbf{p}|^2)^{k-1}\mathbf{p}$, $k = 2, 3, \ldots$ are called spherical aberrations of order $A = 2k - 1 = 3, 5, \ldots$. Simple Euclidean free flight, therefore, contains a series of spherical aberrations (of the special kind generated by functions of $|\mathbf{p}|^2$) of increasing order, with a fixed set of coefficients.

A *basis* for all generators of spherical aberrations is given by the linearly independent set of functions $F(p_x, p_y)$, organized by degree of homogeneity. We call *rank* to the integer or half-integer k such that

$$F_\lambda^k(\mathbf{p}) = p_x^{k+\lambda} p_y^{k-\lambda}, \quad \lambda = -k, 1-k, \ldots, k. \tag{3.13}$$

We thus have a *multiplet* of $2k + 1$ spherical aberrations, that constitute a finite-dimensional irrep of $\mathrm{Sp}(2, \Re) = \mathrm{Sl}(2, \Re)$ that naturally reduces by *helicity* [35] under SO(2) plane rotations of the screen around the optical axis. For $k = \frac{1}{2}$, the two functions $F^{1/2}_{1/2}$ and $F^{1/2}_{-1/2}$ generate screen translations; for $k = 1$ they generate three linear transformations of the kind (3.4)–(3.5). For $k = \frac{3}{2}, 2, \ldots$ they properly generate ***aberrations*** of order $A = 2k - 1 = 2, 3, \ldots$. For aberration order three ($k = 2$) there is a quintuplet of spherical aberrations, of which only the linear combination $|\mathbf{p}|^4 = F^2_2 + 2F^2_0 + F^2_{-2}$ appears in Euclidean free flight.

22.3.4 Coma

Coma is a fascinating aberration in optics. It derives its name from the shape of the tail of a comet, Latin *coma*, and is especially deleterious for optical imaging. It appears on the screen when we *distort* the ray directions on their sphere: $\vec{p}(\theta,\phi) \mapsto \vec{p}(\theta'(\theta,\phi), \phi'(\theta,\phi))$. The sphere can be rotated rigidly by an SO(3) group of *Euclidean* transformations, or deformed by a relativistic boost, as we shall see below. It may also be subject to more complicated analytic maps (or *flows*) that leave the sphere manifold invariant.

If we search for a *basis* of comatic transformations, however, we recognize that, as with distortion, there does not seem to be a good characterization of those two-vector functions $G(\mathbf{p},\mathbf{q}) = \mathbf{q}\cdot\mathbf{F}(\mathbf{p})$ that comply with the momentum restriction of optics and those that do not. If we ***disregard*** this restriction, as in the case of $\Re^4$-phase space mechanics, there *is* a simple classification: along the lines of (3.13), by the degrees of homogeneity of the four phase space variables $(p_x, p_y; q_x, q_y)$. Rank k is one-half the sum of powers (where the power of q_i is unity):

$$G^k_\lambda(\mathbf{p},\mathbf{q}) = p_x^{k-\frac{1}{2}+\lambda}\, p_y^{k-\frac{1}{2}-\lambda} \times (q_x \text{ or } q_y), \quad \lambda = -k+\tfrac{1}{2}, -k+\tfrac{3}{2}, \ldots, k-\tfrac{1}{2}. \tag{3.14}$$

Since there are now $2k$ possible p_i-factors, there are $4k$ distinct comas in aberration order $A = 2k - 1$. In third aberration order there are thus eight asymmetric comas; these further reduce into an $\mathrm{Sp}(2,\Re)$ invariant *triplet* characterized by the invariant common factor $\mathbf{p}\times\mathbf{q} = p_x q_y - p_y q_x$ (times $p_x^{1+m} p_y^{1-m}$), and a *quintuplet* that is its orthogonal complement in the set (3.14). So $|\mathbf{p}|^2\mathbf{p}\cdot\mathbf{q}$ and $|\mathbf{p}|^2\mathbf{p}\times\mathbf{q}$ are the screen-rotation invariants in each multiplet. The former is also invariant under reflections across planes that contain the optical axis; the latter is not, and is therefore absent from axis-symmetric optical systems.

For integer rank k, we call *circular coma* to the aberration generated by the monomials ${}^k\chi^k_{k-1} = |\mathbf{p}|^{2(k-1)}\mathbf{p}\cdot\mathbf{q}$. For it (and for rank-$k$ distortion by Fourier conjugation) we offer here the following integrated Lie transformations of phase space:

$$\begin{aligned} \exp\alpha\{|\mathbf{p}|^{2(k-1)}\mathbf{p}\cdot\mathbf{q}, \circ\} \quad &: \quad \mathbf{p} \mapsto \mathbf{p}'(\mathbf{p};\alpha) \\ &= \quad [1 - 2\alpha(k-1)|\mathbf{p}|^{2(k-1)}]^{-1/2(k-1)}\mathbf{p}, \end{aligned} \tag{3.15}$$

$$\begin{aligned} \exp \alpha\{|\mathbf{p}|^{2(k-1)}\mathbf{p}\cdot\mathbf{q}, \circ\} \quad : \quad & \mathbf{q} \mapsto \mathbf{q}'(\mathbf{p},\mathbf{q};\alpha) \\ = \quad & [1-2\alpha(k-1)|\mathbf{p}|^{2(k-1)}]^{1/2(k-1)} \\ & \times[\mathbf{q}-2\alpha(k-1)|\mathbf{p}|^{2(k-2)}\mathbf{p}\cdot\mathbf{q}\,\mathbf{p}]. \end{aligned} \tag{3.16}$$

The most popular coma is third order. Formulas simplify somewhat and we may look at the expansion to seventh aberration order:

$$\mathbf{p}' = \exp(\alpha\,|\mathbf{p}|^2\mathbf{p}\cdot\mathbf{q})\,\mathbf{p} = \mathbf{p}/\sqrt{1-2\alpha p^2} \tag{3.17}$$

$$= \mathbf{p} + \alpha p^2\mathbf{p} + \tfrac{3}{2}\alpha^2(p^2)^2\mathbf{p} + \tfrac{5}{2}\alpha^3(p^2)^3\mathbf{p} + \cdots, \tag{3.18}$$

$$\begin{aligned} \mathbf{q}' &= \exp\alpha\{|\mathbf{p}|^2\mathbf{p}\cdot\mathbf{q},\circ\}\,\mathbf{q} \\ &= \sqrt{1-2\alpha p^2}\,\mathbf{q} - 2\alpha\sqrt{1-2\alpha p^2}\,\mathbf{p}\cdot\mathbf{q}\,\mathbf{p} \end{aligned} \tag{3.19}$$

$$\begin{aligned} &= \mathbf{q} - \alpha(p^2\mathbf{q} - 2\mathbf{p}\cdot\mathbf{q}\,\mathbf{p}) \\ &\quad -\alpha^2p^2(\tfrac{1}{2}p^2\mathbf{q} - 2\mathbf{p}\cdot\mathbf{q}\,\mathbf{p}) - \alpha^3(p^2)^2(\tfrac{1}{2}p^2\mathbf{q} - \mathbf{p}\cdot\mathbf{q}\,\mathbf{p}) + \cdots \end{aligned} \tag{3.20}$$

22.3.5 Phase Space Deformations —Spot Diagrams

In a two-dimensional world, screens are one-dimensional. Phase space is then two-dimensional and its deformations under canonical maps may be easily drawn. Thus, Figure 22.3 shows the map of two-dimensional phase space under third order coma, $\exp\alpha\{p^3q,\circ\}$ for some fixed α, *within third aberration order*. The latter means that we keep up-to-third homogeneous degree in the above aberration expansions: $p' = p + \alpha p^3$, $q' = q + \alpha p^2 q$. Now, vertical lines $p = c_o$ (c's denote constants here) represent a beam of parallel rays with angle $\theta = \arcsin(c_o/n)$ in any range we may impose on q; these shift to other vertical lines lines $p' = c_1 = c_o + \alpha c_o^3$, $q' = c_2 q = q(1+\alpha c_o^2)$. Horizontal lines $q = c_3$ in some range of p represent a pencil of rays intersecting the screen at q with a corresponding range of angles. These deform to parabolas $q' = c_3(1+\alpha p^2)$, as shown in the second half of Figure 22.3.

Through coma, a sharp image forming on a 1-dim screen will have a caustic point (the tip of the parabola) and extend to one side of it (up for $\alpha > 0$). Furthermore, each point q_* of the aberrated image will come from *two* rays of opposite angle, $\pm p = \sqrt{(q_*/c_3 - 1)/\alpha}$. This bivaluation is the most striking characteristic of coma. In 1-dim screens, an image point is formed when a pencil of rays intersect at it. This pencil is represented in Figure 3*(b)* by a horizontal segment, *i.e.* points with the same image ordinate q and a range of momenta $p = n\sin\theta$. When phase space is deformed this horizontal line becomes in general a segment of a curve with a vertical range of q''s and horizontal domain of p''s, *i.e.*, we *project* the curve on the vertical screen and we have the aberrated image *spot*. If we define a geometric-optical density distribution function of rays $W(p,q)$, such as a rectangular or Gaussian

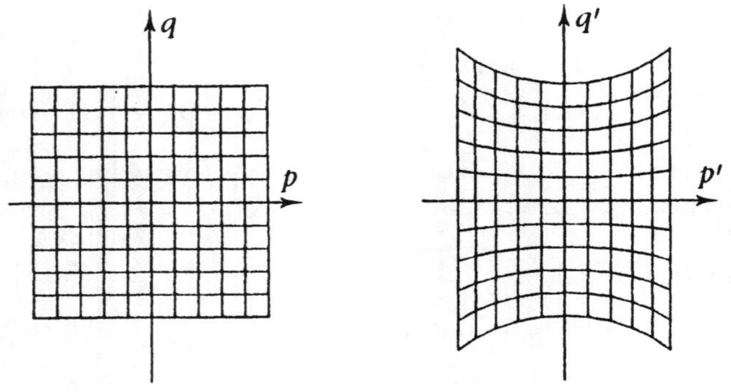

FIGURE 22.3. *(a)* Optical phase space in one-dimensional screens: vertical lines of abscissa $p = n \sin\theta$ represent beams of parallel rays with a fixed angle θ to the optical axis; horizontal lines with ordinate q, a pencil of rays through that point on the screen. The vertical axis may be pictured as the screen on which we may project all ray directions to form an optical image. *(b)* After a pure third-order coma transformation, rays away from the optical axis ($p \neq 0$) change their angles and screen positions. On the screen we see the comatic image by the projection of the point-transformed ray angles. The curves are parabolas. On this space lives the optical Wigner distribution $W(\mathbf{p}, \mathbf{q})$ representing ray density in a beam.

function, but truncated to the optical range of angles, then the projection of this linear density on the vertical q-axis, $D(q) = \int dp\, W(p'(p,q), q'(p,q))$ will be the luminous density of the spot on the 1-dim screen.

Since actual optical systems are 2-dim and phase space is four-dimensional, the *spot* diagram is very useful. The pencil of rays that initially intersect at the screen to form the perfect point object q, can be marked by the polar coordinates of the ray direction sphere, where the $+z$ (North) pole is along the optical axis perpendicular to the screen (opticists always picture this axis extending to the *right*). After a comatic transformation the pencil of rays no longer falls on a point but spreads to the spot as shown by (6.13). To third aberration order this is $\mathbf{q}'(\mathbf{p}) = [1 - \alpha p^2 + 2\mathbf{p}\mathbf{p}\cdot]\mathbf{c}_3$. The meridians of the ray direction sphere map 2:1 on straight lines that reach the point $\mathbf{c}_3$ spreading 30° to either side. Each $\theta =$ constant (geographical parallel) maps 2 : 1 onto a circle on the screen whose radius and center depend on θ in such a way that they fall into this 60° cone. In Figures 4 we show the traditional *spot diagram* of optics for third-, fifth-, and seventh-order coma. The half-angles of the comets are 30°, 41.8104°, 48.5904°, ..., $\arctan[(k-1)/\sqrt{2k-1}]$, as can be shown with little effort [36].

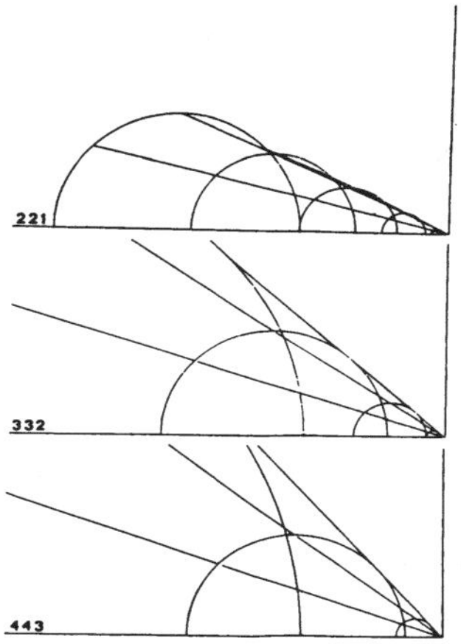

FIGURE 22.4. The traditional *spot diagram* of optics for third-, fifth-, and seventh-order circular coma. The comet shapes are 2 : 1 maps from a polar-coordinated cap of the incident ray direction sphere. The half-angles of the comets are $\arctan[(k-1)/\sqrt{2k-1}]$ for aberration order $A = 2k-1$, $k = 2,3,4$. The labels are (k, j, m) characterizing the symplectic classification of aberrations in Eq. (3.27), ahead.

22.3.6 Aberration Expansions

There is a theorem due to Dragt and Finn [37] that states that if $\mathcal{M}$ is any reasonably smooth canonical transformation sending the ray $\mathbf{p} = 0$, $\mathbf{q} = 0$ along the optical axis, onto itself, then the following *factored product* approximating sequence exists

$$\mathcal{M} = \cdots \exp\{f_4, \circ\} \exp\{f_3, \circ\} \exp\{f_2, \circ\}, \tag{3.21}$$

where each function $f_k(\mathbf{p}, \mathbf{q})$ is a homogeneous polynomial of degree k in the components of $(\mathbf{p}, \mathbf{q})$. The partial converse of this statement is that the map $\mathcal{M}$ is canonical for any set of polynomials, and if the product is truncated at any stage, the map is still canonical. Truncation of the *representation* space of *functions* $f(\mathbf{p}, \mathbf{q})$ to polynomials, and reduction of the Poisson Lie brackets *modulo* polynomials of a highest degree (aberration order) realizes the finite groups of Ath order aberrations. Its composition functions (of 35 parameters for seventh-order aberrations) we have taken the task to determine through symbolic computation. They appear written out in full in reference [35].

Thus, by considering the realization of canonical transformations on phase space *modulo* a highest aberration order, we define canonicity to hold to that order. Thus we reduce the problem of the classification of aberrations to that of the classification of polynomials of degree $2k+1$ in $(\mathbf{p}, \mathbf{q})$. These are nothing more than irreducible representation spaces for $Sp(4,\Re)$ and their classification will follow suit [35]. Since axially symmetric optical systems are of central importance, it makes sense to expand asymmetric aberrations around symmetric ones, labelling them by $Sp(4,\Re) \supset Sp(2,\Re)$. For the latter ones, a corollary of the Dragt-Finn theorem directs us to consider functions only of p^2, $\mathbf{p}\cdot\mathbf{q}$, and q^2 with the restriction that $(\mathbf{p}\times\mathbf{q})^2 = p^2q^2 - (\mathbf{p}\cdot\mathbf{q})^2$ is an invariant. We are thus directed by complex canonical transformations to introduce the coordinates

$$\begin{array}{rclrcl} \xi_+ &=& -\frac{1}{\sqrt{2}}(\xi_1 + i\xi_2), & \xi_1 &=& \frac{1}{2}(p^2 - q^2), \\ \xi_0 &=& \xi_3, & \xi_2 &=& \frac{i}{2}(p^2 + q^2), \\ \xi_- &=& \frac{1}{\sqrt{2}}(\xi_1 - i\xi_2), & \xi_3 &=& \mathbf{p}\cdot\mathbf{q}, \end{array} \tag{3.22}$$

and consider functions on the $\vec{\xi}$-sphere. Indeed, the norm of the three-vector $\vec{\xi}$ is the Casimir invariant of $Sp(2,\Re)$ (the *Petzval* of optics):

$$\xi^2 = \xi_1^2 + \xi_2^2 + \xi_3^2 = \xi_0^2 - 2\xi_+\xi_- = (\mathbf{p}\cdot\mathbf{q})^2 - p^2q^2 = -(\mathbf{q}\times\mathbf{p})^2. \tag{3.23}$$

The functions $f_k(\vec{\xi})$ may be written in a *monomial* or in a *symplectic* basis, as in the case of the states of the three-dimensional harmonic oscillator,

$$f_k(\vec{\xi}) = \sum_{k_+ + k_0 + k_- = k} A^{\text{mon}}_{k_+,k_0,k_-} M_{k_+,k_0,k_-} \tag{3.24}$$

$$= \sum_{j=k(-2)}^{1 \text{ or } 0} \sum_{m=-j}^{j} A^{\text{sym}}_{kjm}\, {}^k\mathcal{X}^j_m(\vec{\xi}). \tag{3.25}$$

We thus have the aberration coefficients $A^{\text{mon}}_{k_+,k_0,k_-}$ as classified along orthogonal axes with the *monomial* functions

$$M_{k_+,k_0,k_-} = 2^{(k_+ + k_-)/2}\, \xi_+^{k_+} \xi_0^{k_0} \xi_-^{k_-}. \tag{3.26}$$

The *symplectic* aberration coefficients are A^{sym}_{kjm}, and the *symplectic* functions are essentially the solid spherical harmonics,

$${}^k\mathcal{X}^j_m(\vec{\xi}) = (\xi^2)^{(k-j)/2} \frac{(j+m)!\,(j-m)!}{2^{m/2}(2j-1)!!} \sum_n \frac{1}{2^n} \frac{\xi_+^{m+n}}{(m+n)!} \frac{\xi_0^{j-m-2n}}{(j-m-2n)!} \frac{\xi_-^n}{n!}. \tag{3.27}$$

The coefficients of ${}^k\mathcal{X}^j_m(\vec{\xi})$ were chosen to contain no square roots because these are inconvenient for symbolic computation. We have proposed the *Seidel-Lie* classification of optical aberrations in references [20], [34]; their relation to the traditional Seidel names [19], [38] is shown in the Table. Thus we classify also all *axially-symmetric* canonical maps of four-dimensional phase space in 2-dim mechanics. These are systems that depend only on the relative orientation of position and momentum through $\mathbf{p}\cdot\mathbf{q}$. We also

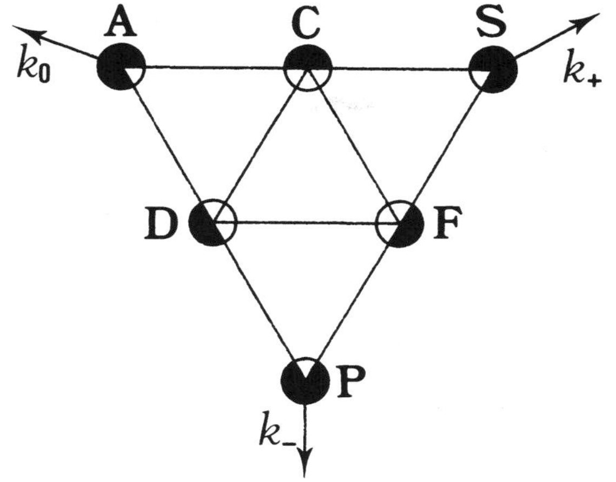

FIGURE 22.5. The triangular multiplets of optical aberrations are in one-to-one correspondence with the three-dimensional harmonic oscillator states. We identify them by spherical aberration **S**, coma **C**, astigmatism **A**, curvature of field **F**, distorsion **D**, and pocus **P**.

have an attractive diagram for aberrations as triangular multiplets. This is shown for the third-order aberrations in Figure 22.5. In particular, *pocus* occupies the corner of the Ω_- particle in the SU(3) model of elementary particles (not in the decuplet, though, but in the sextet). A major standing problem is to evaluate the Poisson bracket between two phase-space symplectic harmonics in general, *i.e.*, to find the X-coefficients in the *single* sum

$$\{{}^k\chi^j_m, {}^{k'}\chi^{j'}_{m'}\} = \sum_{\substack{j''=|m+m'| \\ k+k'+j''\ \text{odd}}}^{j+j'-1} X^{j,j',j''}_{m,m'}\ {}^{k+k'-1}\chi^{j''}_{m+m'}. \tag{3.28}$$

The form of the sum contains selection rules on the ranks k and Seidel indices m, with a single sum over the j-index in the ranges that are indicated. The coefficients $X^{j,j',j''}_{m,m'}$ include SO(3) Wigner coefficients in the only closed formula known to the author [20]. These coefficients are the structure constants of the aberration algebra of rank k. These have been found numerically and are very useful because they increase the efficiency in the computation of Poisson brackets between polynomials. To seventh aberration order, they cut computation time in half [39].

TABLE I. The Lie-Seidel index classification of aberrations.

Seidel index	Degeneracy	classical name
$m = k$	$j = k$	spherical aberration,
$m = k - 1$	$j = k$	circular coma,
$m = k - 2$	$j = k,\ k - 2$	oblique spherical aberration,
$m = k - 3$	$j = k,\ k - 2$	(nameless)
...	...	...
$m = 0$	$j = k,\ k - 2,\ k - 4 \ldots, 1$ or 0	(nameless)
...	...	...
$m = 3 - k$	$j = k,\ k - 2$	elliptical coma,
$m = 2 - k$	$j = k,\ k - 2$	curvature of field / astigmatism,
$m = 1 - k$	$j = k$	distortion,
$m = -k$	$j = k$	pocus.

22.4 The canonical transformations that are specific to optics

A purpose of this contribution to the Symposium is to convince the quantum-mechanically-bred group-theorists, that optics offers a scope at least equally rich for canonical transformations. Certainly, the physically existing ones in optics are more than those in mechanics. This statement may seem rash in view of the last Section, where we dealt with linear transformations and aberration expansions that could well apply in mechanics, but were difficult to fit into the range of phase space peculiar to optics. For most actual 2-dim classical mechanical systems that are point-particles in potential fields, the physically meaningful canonical transformation is the time evolution. (This is extremely important for quantization.) Time-independent potentials correspond *vis à vis* to optical fiber systems. Slow change in the potential is usually meant when time dependence is present. But time-dependent potentials with characteristic intervals that are of the order of the orbit period are rarely used or solved exactly.

By contrast, optics —particularly that of imaging systems— is built on elements that are highly and crucially z-dependent, such as the refracting surfaces of lenses. These transformations have very awkward *vis à vis* mechanical counterparts. In this Section, we shall first show that true optical canonical transformations (*i.e.*, those respecting the ray direction sphere) are quite plentiful. Then we shall sketch the refracting and reflecting surface transformations as canonical maps.

22.4.1 THE OPENING COMA MAP

In reference [40], Vladimir I. Man'ko and the author explored the properties of the third-order coma map (3.17)–(3.20) for $\alpha = 1/2n^2$. We denoted as the *opening coma* map by

$$\mathcal{C} = \exp\left(- 1/2n^2\right)\{|\mathbf{p}|^2\mathbf{p}\cdot\mathbf{q}, \circ\}. \tag{4.1}$$

Using capitals instead of primes for the image space, its action is

$$\mathbf{p} \mapsto \mathbf{P} = \mathcal{C}\,\mathbf{p} = \frac{\mathbf{p}}{\sqrt{1 - p^2/n^2}}, \tag{4.2}$$

$$\mathbf{q} \mapsto \mathbf{Q} = \mathcal{C}\,\mathbf{q} = \sqrt{1 - p^2/n^2}\,(\mathbf{q} - \tfrac{1}{n^2}\mathbf{p}\cdot\mathbf{q}\,\mathbf{p})\,. \tag{4.3}$$

This transformation *opens* the optical momentum range $|\mathbf{p}| < n$ to the full mechanical momentum plane $\Re^2$. The inverse transformation $\mathcal{C}^{-1}$ is obtained from (3.17)–(3.20) for $\alpha = -1/2n^2$ and *closes* the mechanical momentum plane to the optical one.

This rather simple map intertwines all continuous canonical transformations of mechanical phase space, $\exp\{F,\circ\}(\mathbf{P},\mathbf{Q}) \mapsto (\mathbf{P}',\mathbf{Q}')$ onto proper (range-preserving), canonical transformations of optics $\exp\{f,\circ\}(\mathbf{p},\mathbf{q}) \mapsto (\mathbf{p}',\mathbf{q}')$. This is through

$$\exp\{f,\circ\} = \mathcal{C}^{-1}\exp\{F,\circ\}\,\mathcal{C} = \exp\{\mathcal{C}^{-1}\,F,\circ\}. \tag{4.4}$$

Thus, we find a realization of the Heisenberg-Weyl nilpotent Lie algebra that preserves optical phase space:

$$P_i = \frac{p_i}{\sqrt{1 - p^2/n^2}}, \quad Q_j = \sqrt{1 - p^2/n^2}\left(q_j - \frac{\mathbf{p}\cdot\mathbf{q}\,p_j}{n^2}\right), \quad 1, \tag{4.5}$$

for $i, j = x, y$.

For axis-symmetric optical and mechanical systems, the sp$(2,\Re)$ algebra generated by the quadratic Hamiltonians of mechanics is similarly generated in optics by

$$P^2 = p^2/(1 - p^2/n^2), \tag{4.6}$$

$$\mathbf{P}\cdot\mathbf{Q} = (1 - p^2/n^2)\,\mathbf{p}\cdot\mathbf{q}, \tag{4.7}$$

$$Q^2 = (1 - p^2/n^2)\Big(q^2 - 2(\mathbf{p}\cdot\mathbf{q})^2/n^2 + p^2(\mathbf{p}\cdot\mathbf{q})^2/n^4\Big). \tag{4.8}$$

These are *bona fide* transformations in optics $(\mathbf{p},\mathbf{q})$ that in the *paraxial* régime (in first order approximation and near to the optical axis and center) are linear and match with their *vis à vis* mechanical transformations (3.2).

22.4.2 THE OPTICAL FOURIER TRANSFORM

The Fourier transform is the basis for a whole field called *Fourier optics*. It applies from the construction of phase-contrast microscopes[4] to signal

[4]This apparatus earned the first Nobel Prize awarded in optics.

analysis [41], [42] and wave optics in the paraxial regime [14], [30]. An optical Fourier transformer will produce an *image* of the incoming ray *directions* [16]. Now, the classical Fourier transform $\mathcal{F}$ in (3.3) is *not* a proper optical transformation since the range of $|\mathbf{p}|$ is not invariant. We have thus proposed an *optical Fourier transform* $\mathcal{F}_\circ$ defined with the aid of the opening coma transformation (4.1) [23]:

$$\mathcal{F}_\circ = \mathcal{C}\,\mathcal{F}\,\mathcal{C}^{-1}. \tag{4.9}$$

Its action on phase space will evidently conserve the optical restriction $|\mathbf{p}| \leq n$, and is

$$\mathcal{F}_\circ : \mathbf{p} \quad \mapsto \quad \mathbf{p}^{\mathrm{F}} = \frac{-\sqrt{1-p^2}}{\sqrt{1+(1-p^2)(\mathbf{q}-\mathbf{p}\cdot\mathbf{q}\,\mathbf{p})^2}}(\mathbf{q}-\mathbf{p}\cdot\mathbf{q}\,\mathbf{p}), \tag{4.10}$$

and correspondingly,

$$\mathcal{F}_\circ : \mathbf{q} \quad \mapsto \quad \mathbf{q}^{\mathrm{F}} = \frac{\sqrt{1+(1-p^2)(\mathbf{q}-\mathbf{p}\cdot\mathbf{q}\,\mathbf{p})^2}}{\sqrt{1-p^2}} \times [\mathbf{p}+\mathbf{p}\cdot\mathbf{q}\,(1-p^2)^2\,(\mathbf{q}-\mathbf{p}\cdot\mathbf{q}\,\mathbf{p})]. \tag{4.11}$$

In the paraxial region this is the classical Fourier transform $\mathbf{p}^{\mathrm{F}} \approx -\mathbf{q}$, $\mathbf{q}^{\mathrm{F}} \approx \mathbf{p}$. In Figure 22.6 we show the deformation of phase space under this transformation in one dimension of position and one of momentum. We note to the reader the trivaluation phenomenon that occurs in the metaxial region. In two-dimensional screens, the spot diagrams exhibit a fourfold loop of circles. The aberrations of a wide-angle Fourier transformer were analyzed in reference [23].

22.4.3 Rotations and Boosts: the Relativistic Coma Aberration

Quite trivially, the sphere of ray directions in a homogeneous, isotropic medium may be rotated. These are the SO(3) Lie transformations generated by the three functions

$$r_x = q_y p_z, \qquad r_y = -q_x p_z, \qquad r_z = q_x p_y - q_y p_x = \mathbf{q} \times \mathbf{p}, \tag{4.12}$$

where $p_z = \sigma\sqrt{n^2 - |\mathbf{p}|^2}$ They close into the Lie algebra so(3) under Poisson brackets. These transformations of phase space are easy to visualize but somewhat long to write (so we omit them here). Geometric rotations are purely *comatic* because they are purely point transformations of optical momentum space (and involve both signs of σ).

Nontrivially, however, the sphere of ray directions may be distorted by Lorentz transformations stemming from relativistic motion. Astronomers know this as *stellar aberration*, that is, the change of direction to the fixed stars due to the earth's motion in its orbit. We have investigated this phenomenon *globally* (*i.e.*, for the 4π range of angles) both in geometric and

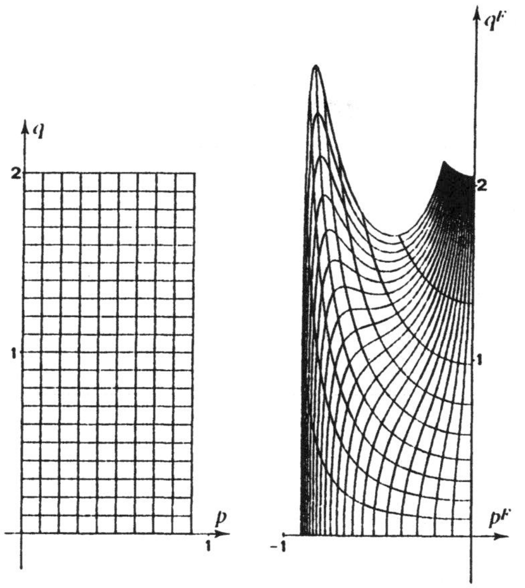

FIGURE 22.6. Whereas the classic-mechanical Fourier transform rigidly rotates phase space by $+\pi/2$ (from the p to the q axis), the *optical* Fourier transform here respects the optical momentum range $|p| \leq n$. Shown is the image of a first-quadrant rectangular patch of phase space. Note the caustics that develop on the screen when we form an image of a beam far from the optical axis.

wave optics, using group deformation techniques [24]. Quite evidently, when a boy on a relativistic motorcycle sees his sphere of ray directions contract towards the axis of motion, the canonically conjugate position variable that forms image on his retinas must expand from that direction. Otherwise, the phase space transformation would not be canonical.

If an imaging device (with a screen) travels at a relativistic velocity v in the z direction, and we let $\alpha = \operatorname{arctanh} v/c$ so that lightlike four-vectors (l_x, l_y, l_z, l_o) transform as usual ($l_z \mapsto l'_z = l_z \cosh\alpha + l_0 \sinh\alpha$, $l_0 \mapsto l'_0 = l_z \sinh\alpha + l_0 \cosh\alpha$), then the optical phase space will transform through [25]

$$\exp\alpha\{b_z, \circ\}\,\mathbf{p} = \frac{\mathbf{p}}{\cosh\alpha + p_z/n\,\sinh\alpha}, \tag{4.13}$$

$$\exp\alpha\{b_z, \circ\}\,\mathbf{q} = (\cosh\alpha + p_z/n\,\sinh\alpha) \times \left(\mathbf{q} - \frac{\mathbf{p}\cdot\mathbf{q}\,\mathbf{p}/n}{n\sinh\alpha + p_z\cosh\alpha}\right), \tag{4.14}$$

where b_z is the third component of $\vec{b} = \vec{q} \times \vec{p}$, or

$$\mathbf{b} = n\mathbf{q} - \mathbf{p}\cdot\mathbf{q}\,\mathbf{p}/n, \qquad b_z = p \cdot q\, p_z/n. \tag{4.15}$$

The six functions $\vec{r}, \vec{b}$ under the Poisson bracket generate the Lorentz algebra so(3,1). The two Lorentz invariants are $\vec{b}^2 - \vec{r}^2 = 0$ and $\vec{r} \cdot \vec{b} = 0$.

It may seem paradoxial that geometric optics, which does not involve a time variable, should be subject to a relativistic transformation. In reference [25] we remark that we may be describing, for example, disintegration of relativistic particles with wide-angle decay products. That such construction can be made also for wave optics is to be expected.

22.4.4 The Refracting Surface Transformation

Exclusive and central to optics is the transformation of phase space due to a refracting surface $z = \zeta(\mathbf{q})$ between two homogeneous optical media of refractive indices n and n',

$$\mathcal{S}_\zeta^{n,n'} : (\mathbf{p}, \mathbf{q}) \mapsto (\mathbf{p}', \mathbf{q}'). \tag{4.16}$$

There does not seem to be any obvious mechanical counterpart to this process. Step potentials in time do not lead to good equations of motion in classical mechanics, but evolution in time of such potentials somewhat imperfectly models the transformation due to a refracting surface. In optics, the refracting surface is the Queen of Maps.

We have a simple, smooth surface represented by a function $S(x, y, z) = \zeta(x, y) - z = 0$, whose normal is $\vec{\Sigma}(x, y) = (\partial\zeta/\partial x, \partial\zeta/\partial y, -1) = (\mathbf{\Sigma}, -1)$. On S, the coordinates are indicated by $\vec{q}|_S = (\mathbf{q}|_S, q_z|_S) = (\bar{\mathbf{q}}, \zeta(\bar{\mathbf{q}}))$. The incident and refracted rays *(a)* intersect, and *(b)* comply with Snell's law. This is shown schematically in Figure 22.7. It entails

$$\vec{q}|_S = \vec{q}'|_S, \tag{4.17}$$

$$\vec{\Sigma} \times \vec{p}|_S = \vec{\Sigma} \times \vec{p}'|_S, \qquad |\vec{p}| = n, \quad |\vec{p}'| = n'. \tag{4.18}$$

When we write in coordinates $(\mathbf{p}, \mathbf{q})$ of the ray at the $z = 0$ screen, $\mathbf{q}|_S$ is free flight by $\zeta(\bar{\mathbf{q}})$ of $\mathbf{q}$, and the cross product written out in the x–y components lead to the following two *conservation* equations

$$\mathbf{q} + \zeta(\bar{\mathbf{q}})\mathbf{p}/p_z = \bar{\mathbf{q}} = \mathbf{q}' + \zeta(\bar{\mathbf{q}})\mathbf{p}'/p'_z, \tag{4.19}$$

$$\mathbf{p} + \mathbf{\Sigma}(\bar{\mathbf{q}})p_z = \bar{\mathbf{p}} = \mathbf{p}' + \mathbf{\Sigma}(\bar{\mathbf{q}})p'_z. \tag{4.20}$$

Here we define $\bar{\mathbf{p}}$ and recall the definition of $p_z = \sigma\sqrt{n^2 - |\mathbf{p}|^2}$, and similarly for the primed quantities. The map $(\mathbf{p}, \mathbf{q}) \mapsto (\mathbf{p}', \mathbf{q}')$ is the refracting surface transformation. We are not detailing here the domains and ranges, nor the fate of the ray sign σ [43].

The simplicity of Eqs. (4.19)–(4.20) is somewhat deceptive. They have the structure of *implicit* equations, $R_n(\mathbf{p}, \mathbf{q}) = R_{n'}(\mathbf{p}', \mathbf{q}')$. They exhibit a peculiar *factorization* property that will be given below. Equation (4.19) must be solved for $\bar{\mathbf{q}}$, the point of impact, and replaced into (4.20), that is

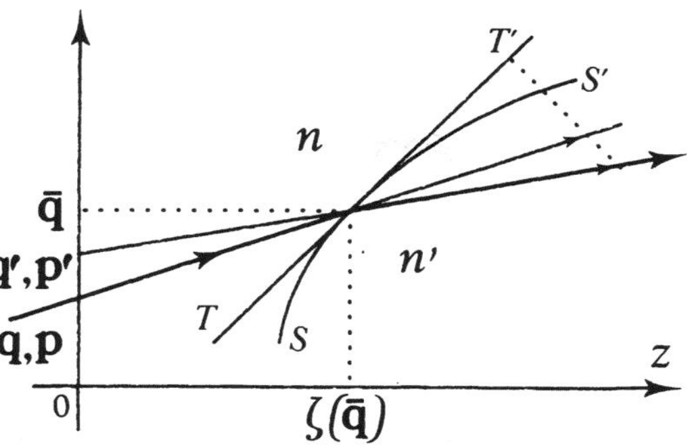

FIGURE 22.7. The process of refraction between media n and n' separated by a surface S–S' is a map between unprimed and primed phase spaces, that conserves the point of impact $(\mathbf{q}, \zeta(\bar{\mathbf{q}}))$ on the surface. and the momentum component along its tangent T–T' at the point. It is a canonical map that can be factorized into two canonical *roots*. This corresponds to a 'forth and back' propagation to the reference screen $z = 0$ in the diagram.

then explicit. The equations may be solved explicitly only for rather simple surfaces (inclined planes, revolution conics, and —in principle— for up-to-quartic Cartesian ovals). The equations are extremely useful, moreover, to perform aberration expansions and find aberration coefficients. Aberration order 3 may be managed by hand, but orders 5 and 7 definitely require machine symbolic computation [44]. Their solution and concatenation with other optical element occupies a major part of our endeavours to apply group theory to optics.

The refracting surface transformation $\mathcal{S}_\zeta^{n,n'}$ is unity when $n = n'$. When $n' = n + dn$, we have

$$\mathcal{S}_\zeta^{n,n+dn} = \mathcal{I} + dn\{T_\zeta^n, \circ\}, \tag{4.21}$$

where the Lie generator of the surface transformation is a function of the *point of impact*

$$T_\zeta^n = -\frac{n\zeta(\bar{\mathbf{q}})}{p_z} = -\zeta(\bar{\mathbf{q}}) \sec\theta. \tag{4.22}$$

22.4.5 The Root Transformation

The structure of equations (4.19)–(4.20), and a similar set for *reflection*, shows that the refracting and reflecting surface transformations, $\mathcal{S}_\zeta^{n,n'}$ and $\mathcal{M}_\zeta^n$ are the product of two *root* transformations,

$$\mathcal{S}_\zeta^{n,n'} = \mathcal{R}_\zeta^n (\mathcal{R}_\zeta^{n'})^{-1}, \tag{4.23}$$

$$\mathcal{M}_\zeta^n = \mathcal{R}_\zeta^n (\mathcal{R}_\zeta^{-n})^{-1}, \tag{4.24}$$

The *root* transformation $\mathcal{R}^n_\zeta$ was introduced in reference [26]. It maps $(\mathbf{p}, \mathbf{q})$ onto $(\bar{\mathbf{p}}, \bar{\mathbf{q}})$ through

$$\mathcal{R}^n_\zeta : \mathbf{q} \mapsto \bar{\mathbf{q}} \quad = \quad \mathbf{q} + \zeta(\bar{\mathbf{q}}) \frac{\mathbf{p}}{\sigma\sqrt{n^2 - |\mathbf{p}|^2}}, \tag{4.25}$$

$$\mathcal{R}^n_\zeta : \mathbf{p} \mapsto \bar{\mathbf{p}} \quad = \quad \mathbf{p} + \Sigma(\bar{\mathbf{q}})\sigma\sqrt{n^2 - |\mathbf{p}|^2}. \tag{4.26}$$

For reflection σ' changes sign and by interpetation of notation we indicate by $-n$ the form of (4.25)–(4.26) when the root function is taken with its minus sign.

The root transformation is *canonical*, as may be rather easily shown by use of the point-angle Hamilton characteristic function [19] ($-F_3$ in Goldstein's notation [28]). This is a function of the 'old' p_i and the 'new' $\bar{q}_j$, such that

$$q_k = \frac{\partial R(\bar{q}, p)}{\partial p_k}, \qquad \bar{p}_k = \frac{\partial R(\bar{q}, p)}{\partial \bar{q}_k}, \tag{4.27}$$

is the root transformation (4.25)–(4.26). Recall that the identity transformation has the characteristic function $I(\bar{q}, p) = \sum_{k=x,y,z} \bar{q}_k p_k = \vec{\bar{q}} \cdot \vec{p}$. If we restrict the system to optical rays that have direction vectors on the Descartes sphere [so $p_z = \sigma\sqrt{n^2 - p_x^2 - p_y^2}$], and cut the new position space by the refracting surface [$\bar{q}_z = \zeta(\bar{q}_x, \bar{q}_y)$], the projection that remains, we can define the root transformation (4.25)–(4.26) through the characteristic function[5]

$$R_{\zeta;n}(\bar{\mathbf{q}}, \mathbf{p}, \sigma) = \bar{\mathbf{q}} \cdot \mathbf{p} + \zeta(\bar{\mathbf{q}})\sigma\sqrt{n^2 - |\mathbf{p}|^2}. \tag{4.28}$$

This proves that the root transformation is canonical, and hence so is the refracting surface. The root transformation does *not* preserve optical phase space, and introduces a complex parametric extension of it to complex values. The composition to a surface transformation through (4.23)–(4.24) still lets optical rays 'escape' into the complex plane. Refraction *and reflection* must be considered to fulfill the physics of the refracting surfaces that do not create nor extinguish rays; the two processes together must allow for total internal reflection and multiple refraction (see the possibilities of this in reference [45]).

[5]It was Joaquín Delgado (student of the Departamento de Matemáticas, Universidad Autónoma Metropolitana) who pointed out to me this very interesting and compact result.

22.5 Outlook: canonical transformations in wave optics

In the Heisenberg-Weyl algebra w, group W, and ring [17], we find models for classical and quantum mechanics. We call the generators p, q, and h ($[q,p] = h$ and h the central element), and their corresponding generated subgroups W_P, W_Q, and W_H. Then, classical mechanics realizes its elementary objects (points of definite position and momentum but no phase) as elements of the coset space $W_H\backslash W$. Quantum mechanics handles its elementary objects (of definite position and phase, and containing all momenta) as points in the coset space $W_P\backslash W$ (or as objects of definite momentum and phase, but extended over all positions, $W_Q\backslash W$). We may now sketch our corresponding construction for geometric and wave optics.

22.5.1 The Euclidean Models of Optics

What are the elementary objects of geometric optics? In reference [18] we proposed the three-dimensional, six-parameter Euclidean group $\mathrm{ISO}(3) = T_3 \triangleleft \mathrm{SO}(3)$ and the coset spaces $H^{\mathrm{sym}}\backslash\mathrm{ISO}(3)$, where H^{sym} is the symmetry group of the elementary object. In ordinary geometric optics, rays are idealized by straight lines whose symmetry group is $T_1 \times \mathrm{SO}(2)$: translations along the ray in direct product with rotations around it. The coset space is thus four-dimensional and parametrized by $(\mathbf{p}, \mathbf{q})$; the Haar measure of the Euclidean group subduces the symplectic measure $d\mathbf{p}\, d\mathbf{q}$ on the coset space, that is conserved under Euclidean transformations. Geometric optics with polarization [46] breaks the last symmetry and has only $H^{\mathrm{sym}} = T_1$; the Euclidean polarization parameter is the third Euler angle.

The elementary objects for monochromatic wave optics [18] are planes with the two-dimensional Euclidean symmetry $H^{\mathrm{sym}} = \mathrm{ISO}(2)$, reduced by Fourier analysis along the ray to a single wavelength component. The nonzero Casimir invariant then becomes the Helmholtz equation for that wavelength. From the Hilbert space of square-integrable functions on the group we subduce a space of square-integrable functions on the direction sphere $\mathcal{L}^2(\mathrm{S}_2)$; plane waves are generalized basis elements represented by Dirac δ's. In the screen coordinates we obtain the Hilbert space of oscillatory solutions $F(\mathbf{q})$ to the Helmholtz equation and their normal derivatives $F'(\mathbf{q})$ at the screen. This Hilbert space has a unique [22], [47] Euclidean-invariant measure that is *nonlocal*. Its nonlocality is described by spherical Bessel functions $j_0(x)$ and $j_1(x)/x$,

$$(\mathbf{F}_1, \mathbf{F}_2)_{\mathcal{H}_k} = \int_{\Re^2} d^2\mathbf{q} \int_{\Re^2} d^2\mathbf{q}'\ \mathbf{F}_1(\mathbf{q})^\dagger \mathbf{H}_k(|\mathbf{q}-\mathbf{q}'|)\mathbf{F}_2(\mathbf{q}'), \tag{5.1}$$

$$\mathbf{F}_1(\mathbf{q})^\dagger = (F_1(\mathbf{q})^* F_1'(\mathbf{q})^*), \tag{5.2}$$

where $\mathbf{H}$ is the *measure matrix*

$$\mathbf{H}_k(|\mathbf{q}-\mathbf{q}'|) = \frac{1}{4\pi}\begin{pmatrix} k^2\dfrac{j_1(k|\mathbf{q}-\mathbf{q}'|)}{k|\mathbf{q}-\mathbf{q}'|} & 0 \\ 0 & j_0(k|\mathbf{q}-\mathbf{q}'|)\end{pmatrix}. \tag{5.3}$$

22.5.2 Strongly-Defined Operators and Nonlocality in Optics

The Euclidean models of optics rest on the generators $\{p_x, p_y, p_z; r_x, r_y, r_z\}$ of space translations and rotations. In geometric optics we use Poisson as the Lie bracket and the realization (4.12). A 'pure position' observable (q_x, q_y) is *not* among the *strongly* defined operators (polynomial or analytic functions convergent uniformly everywhere). Although we may write $q_x = -r_y/p_z$ and $q_y = r_x/p_z$, we see that *zero* is in the spectrum of p_z; we should *not* divide by it, therefore. In geometric optics, without a Hilbert space, we may relax this restriction if we take good account of the chart problem at $\sigma = 0$.

There is a good physical reason for *not* searching for a position operator in monochromatic wave optics: Dirac δ's are *impossible* to form on a screen with plane-wave superpositions. The minimum-spread functions that *do* occur in wave optics (such as sinc, Bessel, Gaussian) show that there is a q_x–q_y complementarity that must be respected in the mathematics. Ongoing endeavours associated with this paradigm include the solutions to some rather important problems such the optimum sampling theorem for point sensors on screens. Rectangular arrays are easy to fabricate, but are not optimal.

22.5.3 Refraction in Groups

For a transverse geometric polarization vector $\vec{u}$ whose tangential component to the surface is conserved, the conservation laws of refraction (4.17)–(4.18) may be extended by $\vec{\Sigma} \times \vec{u}|_S = \vec{\Sigma} \times \vec{u}'|_S$. This is 'Euclidean' polarization —not yet 'Maxwell' polarization, which must contain reflection together with refraction (including total internal reflection, but no reflected-refracted ray [18], [46]). This extends the refraction map to the five-parameter coset space of polarization optics. Lastly, the coset parameter s that measures length along the ray from the screen can be seen from Figure 22.7 to jump from $s = \zeta(\bar{\mathbf{q}})\sec\theta$ to $s' = \zeta(\bar{\mathbf{q}})\sec\theta'$. In this sense we can say that we have a map of the six-dimensional manifold of the Euclidean group due to the presence of a refracting surface. Ditto for reflection. These maps are not group homomorphisms, and are not onto. Similar maps should also exist for other inhomogeneous classical groups over other fields.

Defining refraction for the full group manifold is important because if we have calculated the action of an optical system on geometric optical rays, we can deduce its action on the full Euclidean group, and from there predict its action on the coset spaces that model polarized or wave optics. In particular, the wavization of the aberration expansions will be interesting, because if we follow the usual quantization recipes we are inevitably faced with ordering problems [such as curvature of field and astigmatism, generated by p^2q^2 and $(\mathbf{p}\cdot\mathbf{q})^2$ respectively]. There should be no 'wavization-scheme' ambiguities in optics at all. We can measure relative phases in optical fields through interference experiments and settle any such problems.

Acknowledgements: In my approach to the subject of optics I am very much guided by the regard Professor Moshinsky taught me for symmetry, manifest, hidden, and broken [48]. In the end, though, the group theoretical formalism must yield computable numbers to compare with data and, if possible, both illuminate and abbreviate the process. My aim is to carry on with his work in this new field fittingly.

On that way, there have been several colleagues whom I thank for bringing me to other subjects near to canonical transformations: Thomas H. Seligman, Charles P. Boyer, Tetsundo Sekiguchi, Vladimir I. Man'ko, Natig Atakishiyev, Wolfgang Lassner, Octavio Castaños, François Leyvraz, Alejandro Frank, Stanly Steinberg, Alex J. Dragt and Etienne Forest, by order of association. Students I involved in this research were Miguel Navarro Saad, Enrique López Moreno, Joaquín Delgado, and Guillermo Krötzsch; the last I separately thank for computation and technical assistance in the preparation of this article.

22.6 References

[1] M.Moshinsky and C. Quesne, Oscillator Systems. In *Proceedings of the XV Solvay Conference in Physics* (Brussels, 1970), edited by E. Prigogine (Gordon & Breach, New York, 1974).

[2] C. Quesne and M. Moshinsky, Linear canonical transformations and their unitary representations, *J. Math. Phys.* **12**, 1772–1780 (1971).

[3] M. Moshinsky and C. Quesne, Canonical transformations and matrix elements, *J. Math. Phys.* **12**, 1780–1783 (1971).

[4] M. Moshinsky, *Canonical Transformations and Quantum Mechanics*, Lectures of the 1971 Latin American School of Physics, Mexico DF.

[5] M. Moshinsky, T.H. Seligman, and K.B. Wolf, Canonical transformations and the radial oscillator and Coulomb problems. *J. Math. Phys.* **13**, 901–907 (1972).

[6] J.D. Louck, M. Moshinsky, and K.B. Wolf, Canonical transformations and accidental degeneracy. I. The anisotropic harmonic oscillator, *J. Math. Phys.* **14**, 692–695 (1973).

[7] J.D. Louck, M. Moshinsky, and K.B. Wolf, Canonical transformations and accidental degeneracy. II. The isotropic oscillator in a sector, *J. Math. Phys.* **14**, 696–700 (1973).

[8] P.A. Mello and M. Moshinsky, Nonlinear canonical transformations and their representation in quantum mechanics. *J. Math. Phys.* **16**, 2017–2022 (1975).

[9] M. Moshinsky and T.H. Seligman, Canonical transformations to action and angle variables and their representation in quantum mechanics, *Ann. Phys.* **114**, 243–258 (1978).

[10] K.B. Wolf, Equally-spaced energy spectra: the harmonic oscillator with a centrifugal barrier or with a centripetal well. *Kinam* **3**, 323–346 (1981).

[11] K.B. Wolf, Canonical transformations to phase variables in quantum oscillator systems. A group theoretic solution. *Kinam* **4**, 293-332 (1982).

[12] J.M. Souriau, *Structure des Systèmes Dynamiques* (Dunod, Paris, 1970).

[13] V. Guillemin and S. Sternberg, *Symplectic Techniques in Physics* (Cambridge U. Press, Cambridge, 1984).

[14] M. Nazarathy and J. Shamir, Fourier optics described by operator algebra, *J. Opt. Soc. Am.* **70**, 150–158 (1980); *ibid.*, First order optics —a canonical operator representation. I. Lossless systems. *J. Opt. Soc. Am.* **72**, 356–364 (1982).

[15] A.J. Dragt, Lie algebraic theory of geometric optics and optical aberrations, *J. Opt. Soc. Am.* **72**, 372–379 (1982); *ibid.*, *Lectures on Nonlinear Orbit Dynamics*, AIP Conference Proceedings N° 87 (American Institute of Physics, New York, 1982).

[16] J.W.Goodman, *Introduction to Fourier Optics*, (Mc Graw-Hill, New York, 1968).

[17] K.B. Wolf, The Heisenberg-Weyl ring in quantum mechanics. In: *Group Theory and its Applications*, Vol. 3, edited by E.M.Loebl (Academic Press, New York, 1975), pp. 189–247.

[18] K.B. Wolf, Elements of Euclidean optics. In: *Lie Methods in Optics, second workshop*, Lecture Notes in Physics, Vol. 352 (Springer Verlag, Heidelberg, 1989), pp. 116–162.

[19] H. Buchdahl, *Optical Aberration Coefficients* (Dover, New York, 1968); *ibid.*, *An Introduction to Hamiltonian Optics* (Cambridge University Press, 1970).

[20] A.J. Dragt, E. Forest, and K.B. Wolf, Foundations of a Lie algebraic theory of geometrical optics. In: *Lie Methods in Optics*, Springer Lecture Notes in Physics, Vol. 250, edited by J.Sánchez Mondragón and K.B. Wolf(Springer Verlag, Heidelberg, 1986); K.B. Wolf, Symmetry-adapted classification of aberrations, *J. Opt. Soc. Am. A* **5**, 1226–1232 (1988).

[21] E. Forest and M. Berz, Canonical integration and analysis of periodic maps using non-standard analysis and Lie methods; L.M.Healy and A.J.Dragt, Concatenation of Lie algebraic maps. In: *Lie Methods in Optics, second workshop, op.cit.*

[22] V.I. Man'ko and K.B. Wolf, The map between Heisenberg-Weyl and Euclidean optics is comatic. In: *Lie Methods in Optics, second workshop, op.cit.*

[23] K.B. Wolf, The Fourier transform in metaxial geometric optics. *J. Opt. Soc. Am.* **8**, 1399–1405 (1991).

[24] C.P. Boyer and K.B. Wolf, Deformations of inhomogeneous classical Lie algebras to the algebras of the linear groups, *J. Math. Phys.* **14**, 1853–1859 (1973); *ibid.* The algebra and group deformations $I^m[SO(n) \otimes SO(m)] \Longrightarrow SO(n,m)$, $I^m[U(n) \otimes U(m)] \Longrightarrow U(n,m)$, and $I^m[Sp(n) \otimes Sp(m)] \Longrightarrow Sp(n,m)$, for $1 \leq m \leq n$. *J. Math. Phys.* **15**, 2096–2100 (1974).

[25] N.M. Atakishiyev, W. Lassner, and K.B. Wolf, The relativistic coma aberration. I. Geometrical optics, *J. Math. Phys.* **30**, 2457–2462 (1989); *ibid.*, The relativistic coma aberration. II. Helmholtz wave optics *J. Math. Phys.* **30**, 2463–2468 (1989).

[26] M. Navarro-Saad and K.B. Wolf, Factorization of the phase-space transforamtion produced by an arbitrary refracting surface, *J. Opt. Soc. Am. A* **3**, 340–346 (1986).

[27] K.B. Wolf, Symmetry in Lie optics, *Ann. Phys.* **172**, 1–25 (1986).

[28] H. Goldstein, *Classical Mechanics*, 2nd edition (Addison Wesley, Reading, Mass., 1980).

[29] H. Bacry and M. Cadilhac, Metaplectic group and Fourier optics, *Phys. Rev. A* **23**, 2533–2536 (1981).

[30] O. Castaños, E. López-Moreno, and K.B. Wolf, Canonical transforms for paraxial wave optics. In: *Lie Methods in Optics, op. cit.*

[31] T. Sekiguchi and K.B. Wolf, The Hamiltonian formulation of optics, *Am. J. Phys.* **55**, 830–835 (1987).

[32] S.Steinberg, Lie series, Lie transformations, and their applications. In *Lie Methods in Optics*, *op.cit.*, pp. 45–103.

[33] K.B. Wolf, Point transformation in quantum mechanics. *Rev. Mex. Fís.* **22**, 45–74 (1973).

[34] M. Navarro-Saad and K.B. Wolf, The group–theoretical treatment of aberrating systems. I. Aligned lens systems in third aberration order *J. Math. Phys.* **27**, 1449–1457 (1986).

[35] K.B. Wolf, The group–theoretical treatment of aberrating systems. III. The classification of asymmetric aberrations. *J. Math. Phys.* **28**, 2498–2507 (1987).

[36] K.B. Wolf, Nonlinearity in aberration optics. In: *Symmetries and Nonlinear Phenomena*, Proceedings of the International School in Applied Mathematics (Centro Internacional de Física, Paipa, Colombia). Edited by D.Levi and P.Winternitz (World Scientific, Singapur, 1988), pp. 376–429.

[37] A.J. Dragt and J. Finn, Lie series and invariant functions for analytic symplectic maps. *J. Math. Phys.* **17**, 2215–2227 (1976).

[38] L. Seidel, Zur Dioptrik. *Astronomische Nachrichten* N^o **871**, 105–120 (1853).

[39] K.B. Wolf and G. Krötzsch, Group-classified polynomials of phase space in higher-order aberration expansions, Comunicaciones Técnicas IIMAS preprint, Serie Investigación # 563. *Journal of Symbolic Computation* , in press (1991).

[40] V.I. Man'ko and K.B. Wolf, The map between Heisenberg-Weyl and Euclidean optics is comatic. In: *Lie Methods in Optics, second workshop*, Lecture Notes in Physics, Vol. 352 (Springer Verlag, Heidelberg, 1989), pp. 163–197.

[41] H. Raszillier and W. Schempp, Fourier optics from the perspective of the Heisenberg group. In *Lie Methods in Optics*, *op.cit.*, pp. 29–44.; *ibid.* Holographic image processing, coherent optical computing, and neural computer architecture for pattern recognition. In: *Lie Methods in Optics, second workshop*, Lecture Notes in Physics, Vol. 352 (Springer Verlag, Heidelberg, 1989), pp. 19–46.

[42] R. Lenz, *Group Theoretical Methods in Image Processing*, Lecture Notes in Computer Science, Vol. 413 (Springer Verlag, Heidelberg, 1987).

[43] G. Krötzsch and K.B. Wolf, La transformación raíz de superficies refractantes y espejos, *Rev. Mex. Fís.* **37**, 540–554 (1991).

[44] M. Navarro-Saad and K.B. Wolf, Applications of a factorization theorem for ninth-order aberration optics, *Journal of Symbolic Computation* **1**, 235–239 (1985).

[45] H.M.S. Nussenzweig, Complex angular momentum theory of the rainbow and the glory. *J. Opt. Soc. Am.* **69**, 1069–1098 (1979).

[46] K.B. Wolf, The Euclidean root of Snell's law. I. Geometric polarization optics. Reportes de Investigación IIMAS preprint **1**, N° 1 (May 1991), to appear in *J. Math. Phys.* (1992).

[47] S. Steinberg and K.B. Wolf, Invariant inner products on spaces of solutions of the Klein-Gordon and Helmholtz equations, *J. Math. Phys.* **22**, 1660–1663 (1981).

[48] K.B. Wolf, *Estudios Teóricos sobre la Estructura Nuclear del Flúor 19*, Thesis, Facultad de Ciencias, Universidad Nacional Autónoma de México (1965). Directed by Marcos Moshinsky.

23

Science and Technology in Latin America

Round Table

ABSTRACT The various processes by which Science and Technology grew in the North have not been reproduced elsewhere. The Latin American region in particular, has been trying hard to catch up. Plans and schemes to do so have not been lacking; yet the aim is quite elusive. The problems, means, and results thus far are as varied as the mosaic of countries that compose the region. Discussions on this theme have taken place in many forums, among them the Third World Academy of Sciences conferences. The Cocoyoc meeting reserved a prominent time niche for a critical comparison by actual policy makers or leaders from Brazil, Colombia, and Mexico. The Round Table was chaired by Dr.[1] JORGE FLORES, Director of the Centro Universitario de Comunicación de la Ciencia, Universidad Nacional Autónoma de México; the participants, whose contributions we reproduce below[2] were Drs. GERHARD JACOB, President of the Brazilian Conselho Nacional de Desenvolvimento Científico e Tecnológico, MIGUEL JOSÉ YACAMÁN, Director for Scientific Research of the Consejo Nacional de Ciencia y Tecnología, México, GALILEO VIOLINI, Founding Director of the Centro Internacional de Física, Colombia, and MARCOS MOSHINSKY —the only *civilian* in the panel— in whose honor we add excerpts from a letter by Professor Abdus Salam.

SCIENCE AND TECHNOLOGY DEVELOPMENT IN BRAZIL
by *Gerhard Jacob*

The role of basic Science in Latin America, and in developing countries in general, is of absolute need; it is not a superflous function. The arguments for this position are the following: Science is necessary to establish the appropriate **atmosphere** and environment that is essential to keep **qualified people** in the country, thus guaranteeing **modern** and updated research and teaching, to educate promising students that will keep up with modern

---

[1] In Latin America the academic title of *Profesor*, indicating tenure or titularity as in the US or Europe, does not exist. *Doctor* is the proper and only title for the bearer of a Ph.D. degree.

[2] The original manuscripts were adapted by the editors to a common format for inclusion in this proceedings volume.

technological developments and immediately assimilate its breakthroughs.

The milestones in scientific development in Brazil have been: in 1934, the creation of the University of São Paulo; in 1951, the creation of the National Research Council (CNPq), funding basic and applied research, and the Coordination for Improvement of University Teachers (CAPES), funding the education of graduate students. 1969 saw the creation of the National Fund for Science and Technology (FNDCT), administered by the Financing Agency for Studies and Projects (FINEP), the main funding agency for basic and applied, institutional scientific resarch, "big science" and industrial laboratories. At the same time, strict rules for graduate studies were established. Finally, in 1985 the Ministry of Science and Technology was created.

The organization of the state funding agencies in Brazil is the following: the Secretary of Science and Technology perpasses all Ministries and is responsible for the coordination of its policies and priorities. His advisory and policy making boards are the National Council for Science and Technology and the Council for Policy in Informatics. The executive arms are the FINEP and the CNPq. The human resources are fostered by the activities of the Ministry of Education through the universities and technical institutes, that provide the policies for higher and other levels of education. The executive arms for the latter are Coordination for Improvement of University Teachers and the Institute for Studies and Research in Education.

In Brazil there is good basic research by international standards, yet much too few scientists and even less applied research. Industrial research is practically nonexistent; the reasons are that multinational industries do their research "at home" while national industries find it more convenient to import technology, because of lack of qualified people to carry out industrial research. Although technological research centers are few and the fostering of applied research in universities negligible, there are two notable exceptions: informatics, an area in which national industries were practically born in universities and research in agronomy and veterinary, that resulted from extramural activities of universities.

The challenges for Brazil's future may be summarized in the following figures: only 0.7% of the gross national product is spent on Science and Technology, and of that 94% is governmental. The proportion of scientists by population is 400 per million and the average age of Ph.D. students when they receive their degrees is 40 years. The infrastructure of research centers and most industries is obsolete, needing modernization in science and competitiveness in industries. The latter have to be convinced that resarch is investment and that know-how is international merchandise. Social contrasts in Brazil can be overcome only through overall development; all these challenges are valid for Latin America as a whole.

The Brazilian response to these challenges have been a clear policy for scientific, technological and industrial developments, with scientific priorities in relevant areas meriting additional allocation of funds. There should be an increase of 15% per year in the percentage for research and development of the gross national product, coming largely from private funds. This means a five year doubling span. Special funds are destined to increase the development of human resources. Funds for modernization and competitiveness are allocated 1/3 for basic science and 2/3 in technology. Fiscal incentives will be given to industries to invest in research with the opening of the Brazilian market to international competition. A prime example of this is the field of informatics. Stronger collaboration between the universities and the productive sector is fostered and the fundamental role of international cooperation is realized.

The recent changes in Brazil have been economical, political, and administrative. There have been deep changes in scientific, technological, and industrial policies. The problems and developments in the universities during the last five decades and the consequent response of the scientific community are clearly stated, and the private productive sector has understood its responsibilities. Brazil can be considered as an example for Latin America as far as the universities, Science and Technology are concerned. SOME

REFLECTIONS ON SCIENCE IN LATIN AMERICA
by *Galileo Violini*

It is a great honor for me to join this tribute to one of the founding fathers of physics in this subcontinent. The creation of a scientific structure in Latin America, and in particular in Mexico, has been one one of the most important components of the first fifty years of activity of Marcos Moshinsky. As a minor testimony of this at a continental scale, I would like to recall my first conversation with Marcos, some ten years ago, on occasion of a visit to Colombia, when my proposal of creating the Centro Internacional de Física was at its origin.

I shall limit myself to discuss Science in Latin America; this by no means indicates that I consider Technology a minor issue for our continent. However, in this audience the statement that technology is impossible without a strong basis in Science [1] does not require special justification or comment. In particular, I will discuss some problems referring to the establishment of this necessary and strong basis that in our continent is relatively young [2]. Nevertheless, the substantial efforts in training people during the last decade have created an increasing capacity of research. As a result of this increase, the scientific policy problems have become very different from what they appeared to be two or three decades ago. In support of this statement I present a few examples.

The first example is the antithesis between theoretical and experimental physics. The importance of developing experimental physics was recognized long ago [3], but this was to a large extent an isolated position. Today, even if one still faces difficulties, nobody questions the right of existence of experimental physics in our continent. Of course, some prejudices remain, as that of the "high" cost of laboratories, but it is nowdays relatively easy to argue that these costs are often higher than theoretical research only as a consequence of the unfortunate underfinancing of the latter.

A second example of these changes concerns the type of physics which can be done. In the past, strong feelings against hard science, and in particular high energy physics, were very common. Their basis was essentially the cost of this field of research and supposed absence of social benefits that characterize other areas of physics. This has changed. The right of existence of experimental high-energy groups is no longer questioned. Many groups exist now in Brazil, Argentina, the Andean Region, and of course Mexico. For this country I would like to recall the role played by our colleague Clicerio Avilez, who prematurely passed away in May 1991. This change reflects a clearer understanding of the structure of costs of experimental physics [4], a less prejudiced analysis of the connection between investments in research and the actual useful results, and the recognition that the interactions between scientific research and production are complex. They cannot be solved simply by developing certain areas such as solid state physics which, under certain conditions in the most advanced countries of the First World, have shown to be of the utmost importance for industry.

The third example of this change is the increasing consciousness about the brain drain problem [5], [6]. This increased consciousness is the result of the increased university enrolment and larger production of young scientists. The cost of this brain drain on our economies [7] is now being understood to be comparable to the external debt [6]. This understanding provides a stronger basis for the political decisions necessary to attack its roots, such as scientific isolation, imbalance of resarch and teaching in universities, low salaries, and limited funding for research.

These examples show that the problems in science policy have drastically changed in the last decades. However, this does not imply a significant change of the actions necessary for scientific development of our region though, of course, some of their characteristics may require an adaptation to the new situation. In my opinion, Latin American scientific development requires four actions:

- creation of infrastructure
- strengthening of regional and international collaboration
- increase of financing
- establishment of links with the political and economic sectors of society.

As I said, none of these actions is new. However, under the referred changes, certainly new aspects must be stressed.

Concerning the creation of infrastructure, the increase in university enrolment (supported by broad UN policies such as that of UNESCO [8]) makes it inconceivable to insure it adequately for all in the subcontinent: there are now more than 500 Universities [9]. Of course, only a fraction of them are of such a level as to justify investment for research. But even so, one can foresee that universities with qualified personnel will not be able to offer adequate facilities for research. Thus, alternatives must be found to make research possible, to allow good people to do it, and not depriving the universities of the potential these people represent. Of course, I have in mind the establishment of Regional Centers, such as the one to which I am linked [10]. But, at the same time, I think that this may be an excellent area for international collaboration, in view of its linkage with the brain drain problem [11].

Indeed, the action of collaboration requires a new point of view. Here I would like to stress mainly the political aspects of collaboration in our region between universities and research centers. Occasionally I have felt that some people think that the limited availability of funds favors competition instead, for what concerns their applications to international funding agencies. I totally disagree with this view. The aim is to exert pressure on these agencies to force them to take the appropriate policies in favor of the scientific and technological development of our region. We all know how in the present situation these policies are at least inadequate. The collaboration I suggest takes advantage of the "financial elasticity" of the system and should lead to the necessary changes of policies. The objects of such pressure should be our governments, the Organization of American States, the International Development Bank, and similar agencies, who certainly would pay more attention to the pressure of a federation of Centers and Universities than to individual actions, since a federation would obviously reflect a broad regional need. It is totally obvious to this audience of scientists that this political collaboration must be accompanied by the scientific one.

The problem of the financing of Science, and Physics in particular, also requires a new approach. Much more than the money needed, the data on the brain drain cost should worry our Ministers of Treasure. It is a fact that the level of financing of Science in our region is terrifically low. Unfortunately, this fact is somewhat hidden by the common reference to the percentage of the gross national product, instead of its absolute value.

In my opinion, it is not only the level of financing that must worry the scientific community; other problems that also require attention concern the role of the State in the scientific enterprise. It is dangerous to create too many illusions on the participation of industry, often too weak and not especially convinced. A country like Brazil, with 0.7% of its gross national product invested in research and a reasonably developed industry, bases 94% of its research funding on the State. To bring ideas as the economical opening to their extreme consequences in the self-financing of research will have, in my opinion, dramatic final results. The functioning of scientific research is a duty of the State, and cannot be delegated to the economical sector of a country. I shall return to this point below, but here I only want to observe how this problem is strictly related to that of the establishment of "prioritary areas" of research, a subject on which both Drs. Jacob [12] and José-Yacamán [13] have presented their comments.

From what has been presented here, a number of possible lines of approach and subjects of discussion with the political sectors are sketched. Concerning the economical sector, scientists must make an effort to understand its needs but, at the same time, the reasons for this collaboration must reach the industrialists. They should certainly find interesting the current elaboration of ideas about the research-industry relation in our continent. For example: the development of incubators of high-technology industries [14] cannot be disregarded in the near future. However, the participation of industry in the funding of research should take place by mechanisms that do not replace the role of the State. A very interesting scheme is that mentioned by Dr. Jacob [12]: that a substantial fraction (2/3) of the university budget be a loan fund provided by industry.

It would be naïve to believe that one can promote the connection between the academic and productive sectors forgetting the interrelation with the political sector. Indeed, only politicians may create the economic and legal conditions required to stimulate the industrialists to include Science in their programs, both at national and regional scales. The goals outlined are not out of reach. All the meetings of our Presidents and Ministers stress the importance of developing Science and Technology. It is a duty of all of us to make these statements materialize, and not to remain simply on paper. And let me add that they must materialize within a short time [8], [15]. For that, we cannot wait until the MM240 Symposium, when I wish Marcos Moshinsky will be witness of our successes.

References

[1] A. Salam, *Science, Technology and Science Education in the Development of the South* (Trieste, 1990).

[2] G. Violini, Status and problems of physics in Latin America. In *Status and Problems of Science in Latin America and the Caribbean*, M.L. Guardiola, J.L. Villaveces, and G. Violini, Eds. (Bogotá, 1990), p. 209.

[3] J. Leite Lopes, private communication

[4] B. Gómez, J.P. Negret, S. Torres, and G. Violini, High energy physics in developing countries, problems and perspectives. CIF preprint (Bogotá, 1991).

[5] G. Violini, Some considerations on brain drain, the Colombian case, to be published (Trieste, 1991).

[6] G. Violini, Brain drain and its implications for the policies of higher educiation, to be published (Caracas, 1991).

[7] A. Papic, Brain drain from the South, to be published (Trieste, 1991).

[8] F. Mayor, to be published (Caracas 1991).

[9] UNESCO–CRESALC, Visión cuantitativa de la educación superior en América Latina y el Caribe. (Caracas, 1991).

[10] A. Pantoja, E. Posada, and G. Violini, The Centro Internacional de Física. In *Status and Problems of Science in Latin America and the Caribbean*, M.L. Guardiola, J.L. Villaveces, and G. Violini, Eds. (Bogotá 1990), p. 438.

[11] C. Martelli, to be published (Trieste, 1991).

[12] G. Jacob, contribution to this Round Table.

[13] M. José-Yacamán, contribution to this Round Table.

[14] E. Posada *et al.*, Incubadora de empresas de base tecnológica en la Ciudad de Bogotá. CIF report (Bogotá, 1991), and references quoted therein, see Annex 1.

[15] M.L. Guardiola, J.L. Villaveces, and G. Violini, Some thoughts about Third World policies on science and technology. In *Status and Problems of Science in Latin America and the Caribbean*, M.L. Guardiola, J.L. Villaveces, and G. Violini, Eds. (Bogotá, 1990), p. 3.

THE NEW SCIENCE POLICY IN MEXICO

by *Miguel José-Yacamán*

In the last few years the Mexican government has launched a great effort to modernize the country, that is, to transform its structure from a Third World model to a First World one. This implies that Science and Technology should also be brought up to date. In this aspect, CONACYT (Consejo Nacional de Ciencia y Tecnología), the Mexican state organization that oversees Science and Technology, is being restructured along the following lines:

- Science in Latin America should be of the same quality as science everywhere. Therefore, the peer review should be at the same level of international standards.
- The only science which should be supported is that with academic excellence at international levels.
- There should be support for the interaction between the academic community and industry. However, the kind of research that is not

connected to industry, but is of potential application, should also be supported, but on the same basis as that of basic research.

We believe that at this time the most important issue is to stop the brain drain, and to promote the formation of new researchers. That implies offering better working conditions for our scientists and reinforce the national graduate studies programs. At this particular time it is most important to support basic research of high quality, regardless of the topics; all science is important and there should be no "prioriy areas". In this sense we must avoid the distortions that are introduced into the scientific community when it is stated by government agencies that some topics of science will enjoy more support than others.

Mexico is a country that has now the project of becoming part of the developed world. In this sense, Mexican science should move in the same direction. This means to leave out the concepts related to the way science is done in poor countries. Concepts such as "national" science or "local interest" problems should be taken out of our dictionary.

Marcos Moshinsky and the future of science in Mexico
(excerpts from a letter by Abdus Salam)

...I had known Professor Moshinsky even before I met him. This was because Professor Manuel Sandoval Vallarta was the head of the Scientific Council of ICTP. His conversation was always about Professor Moshinksy and his great deserts in physics, particularly his work on symmetries in nuclear physics. So when I did meet him it was no suprise to me to find him so erudite and so full of learning. But it did come as a surprise to me to find him so fully dedicated to Mexico and its science, and carrying no real rancour against the government which had beggared his and his colleagues' careers by its insensitivity in not having given enough funds for research.

When I visited Mexico in 1980, Marcos Moshinsky took me to his beautiful home in the outskirts of Mexico City. This was the home of a man of learning as well as that of a man of taste, who liked the good things of life... Subsequently I met him here at Trieste as well as at the Pontificial Academy of Sciences in Rome. This is the most prestigious Academy of Sciences which recruits members without consideration of their religion of their choice, and thus it is the most international of all Academies I have the honor to belong to...

In his 50 years of research, Professor Moshinsky has published some 200 papers in prestigious journals and four books, and has consistently been the most cited scientist in Mexico. In addition to his world-class contributions to physics, he has contributed most to the advancement of science in Mexico, particularly through the training of several generations of young scientists. In fact, some 40% of the staff in the Institute of Physics, UNAM received their degrees under his supervision.

There is no question but that Mexico, in spite of its North American location, has not made similar progress in sciences as the other Northern countries. This is because it spends no more than 0.6% of its gross national product (GNP) on sciences and technologies. Mexico spends relatively little on its defence (0.6%); it spends more on health education and health matters (1.7%) as compared to the developing countries (1.6%). It spends 2.8% only, on education, which is very small considering that its GNP is of the order of US$2,678 per capita. Mexico is not a small country, being the eighth country in population size in the whole world. Thus it comes as a disappointment to see that the total numbers of its scientists and engineers amounted to no more than 70 per million during 1970, as compared to the developed countries' average of 2,900 (no figures are available for 1988 or even for 1980 for Mexico, at least the UNESCO book shows none.) Thus there is a need for considerably more expenses on Science and Technology —at least 1% of the GNP according to the UNESCO formula. Also, there is need for more scientists and engineers than 70 per million: this numbers should be increased to at least 700 per million soon.

An appeal has been made for a fund for the construction of a building to be sited at the grounds of the National University, UNAM, at the campus of the Morelos State University (UAEM) in Cuernavaca. Leon Lederman, who I understand will be with you, has recently made a wonderful plea for more Science and Technology at the national level in the United States. I am sure he will take up cudgels on your behalf, and his fertile brain will come up with some new scheme for integrating the North American community of scientists with the Mexican community. In fact, it does not seem to be impossible that Mexico should have a bright future in view of the great beauty of the country, its monuments, and the desire of US scientists to travel to such places. I am sure you will all be thinking on this. Science is the one profession which does not lead to any patronizing; it is a universal profession and, as such, sciences should be developed within Mexico. I wish we could do more..., but Trieste is a whole ocean away from Mexico.

It is thus appropriate that we should all salute Marcos Moshinsky, and hope that his contributions to the subject will go on for many more years.

AUTHOR INDEX
FOR THIS VOLUME

SPRINGER NATURE

GPSR Compliance

The European Union's (EU) General Product Safety Regulation (GPSR) is a set of rules that requires consumer products to be safe and our obligations to ensure this.

If you have any concerns about our products, you can contact us on ProductSafety@springernature.com

In case Publisher is established outside the EU, the EU authorized representative is:

Springer Nature Customer Service Center GmbH
Europaplatz 3
69115 Heidelberg, Germany

FSC
www.fsc.org
MIX
Papier | Fördert gute Waldnutzung
FSC® C083411

Kolibri 360 GmbH
Ferdinand-Jühlke-Straße 7
99095 Erfurt, Deutschland
produktsicherheit@kolibri360.de